Lecture Notes in Physics

The Lecture Notes in Physics

The series Lecture Notes in Physics (LNP), founded in 1969, reports new developments in physics research and teaching – quickly and informally, but with a high quality and the explicit aim to summarize and communicate current knowledge in an accessible way. Books published in this series are conceived as bridging material between advanced graduate textbooks and the forefront of research to serve the following purposes:

• to be a compact and modern up-to-date source of reference on a well-defined topic;

• to serve as an accessible introduction to the field to postgraduate students and nonspecialist researchers from related areas;

• to be a source of advanced teaching material for specialized seminars, courses and schools.

Both monographs and multi-author volumes will be considered for publication. Edited volumes should, however, consist of a very limited number of contributions only. Proceedings will not be considered for LNP.

Volumes published in LNP are disseminated both in print and in electronic formats, the electronic archive is available at springerlink.com. The series content is indexed, abstracted and referenced by many abstracting and information services, bibliographic networks, subscription agencies, library networks, and consortia.

Proposals should be sent to a member of the Editorial Board, or directly to the managing editor at Springer:

Dr. Christian Caron
Springer Heidelberg
Physics Editorial Department I
Tiergartenstrasse 17
69121 Heidelberg/Germany
christian.caron@springer-sbm.com

Sadrilla S. Abdullaev

Construction of Mappings for Hamiltonian Systems and Their Applications

 Springer

Author

Sadrilla S. Abdullaev
Institut für Plasmaphysik
Forschungszentrum Jülich GmbH
52425 Jülich, Germany
E-mail: s.abdullaev@fz-juelich.de

S.S. Abdullaev, *Construction of Mappings for Hamiltonian Systems and Their Applications*, Lect. Notes Phys. 691 (Springer, Berlin Heidelberg 2006), DOI 10.1007/b11554707

Library of Congress Control Number: 2005937897

ISSN 0075-8450
ISBN-10 3-540-30915-2 Springer Berlin Heidelberg New York
ISBN-13 978-3-540-30915-4 Springer Berlin Heidelberg New York

Springer is a part of Springer Science+Business Media
springer.com
© Springer-Verlag Berlin Heidelberg 2006
Printed in The Netherlands

Typesetting: by the author and TechBooks using a Springer LATEX macro package

Printed on acid-free paper   SPIN: 11554707   57/TechBooks   5 4 3 2 1 0

Preface

Mappings constitute a powerful method for studying dynamical systems. They are fundamentally based on a formulation of dynamical equations governing them as a system of first-order ordinary differential equations. According to the theorem of Cauchy, the solutions of these dynamical equations are unique and are completely determined by the initial conditions, i.e., there exists the unique transformation or *mapping* of the initial conditions into the final conditions. The surface-to-surface maps (*Poincare return map*) and stroboscopic maps introduced by Poincaré (1892–99) replace the dynamics of a continuous system by a discrete one. These maps have important advantages in a study of dynamical systems. First, they reduce dimensions of the system at least by one. They allow one to visualize the dynamics of the system at certain sections (Poincaré sections) of phase space and thereby display the global behavior of the system. Many concepts of continuous systems become more clear when they are formulated using Poincaré maps. For instance, the study of stability of periodic orbits can be simply reduced to a study of stability of fixed points of the mappings.

The Hamiltonian formulation of dynamical equations of physical systems of different nature had a deep impact on the study of dynamical systems (Hamilton, 1834; Goldstein, 1980; Arnold, 1989). A system with N degrees of freedom can be described by $2N$ ordinary differential equations of first order in the phase space of the canonical coordinates $q = (q_1, \ldots, q_N)$ and momenta $p = (p_1, \ldots, p_N)$, and are determined by a single scalar master function, known as Hamilton function H. One of the features of Hamiltonian systems is that it conserves certain invariants in phase space, which constitute phase space as a *symplectic* space.

Whenever dissipation is negligible, most fundamental models of physics and mechanics are described by Hamiltonian systems. Hamiltonian systems have been the subject of numerous studies during the last two centuries in physics, mechanics, and astronomy, in problems ranging from the dynamics of elementary particles in accelerators to the dynamics of planetary objects in a space (Poincaré, 1892–99; Lichtenberg and Lieberman, 1992; MacKay and Meiss, 1987; Arnold et al., 1988).

Standard numerical methods of integrating systems of ordinary differential equations are not ideal for the purposes of solving Hamiltonian systems because the numerical approximation introduces non-Hamiltonian perturbations that completely change long-term behavior of the solutions. For this reason, special integration tools, known as symplectic integrators, have been developed for the numerical study of Hamiltonian systems (see, for example, reviews Sanz-Serna (1992); Sanz-Serna and Calvo (1993, 1994); Feng (1994)). The methods are constructed to preserve the symplectic properties of Hamiltonian systems by arranging each integration step to be a canonical transformation. Symplectic integration methods play an important role in the study of the long-term evolution of Hamiltonian systems.

Mappings are a powerful tool for studying Hamiltonian systems (see, e.g., Lichtenberg and Lieberman, 1992; MacKay and Meiss, 1987; Chirikov, 1979; Zaslavsky, 1985; Sagdeev et al., 1988; Zaslavsky et al., 1991). These maps are inherently constructed in symplectic from, and thereby preserve properties of Hamiltonian systems. This approach is most ideal to study the long-term evolution of a system, especially in cases where the system exhibits chaotic behavior caused by exponential divergency of orbits with close initial coordinates in phase space. Symplectic maps have been successfully employed in many problems of astronomy, plasma physics, fluid dynamics, accelerator physics, and others.

In spite of the extensive use of symplectic maps for many Hamiltonian problems during the last four decades, the derivation of generic symplectic maps from given Hamiltonian equations still remains somehow elusive. There are several approaches to construct symplectic maps from the continuous formulation of systems. One approach is based on the a priori assumption that the map has a symplectic form and the generating functions associated with the map are found from the equations of motion (Lichtenberg and Lieberman, 1992). Another method to construct symplectic maps is based on the assumption that a time-periodic perturbation acting on the integrable system may be replaced by periodic delta functions, which is equivalent to adding fast oscillating terms to the Hamiltonian (Wisdom, 1982; Zaslavsky, 1985; Sagdeev et al., 1988; Zaslavsky et al., 1991). Integration of the equations of motion along delta functions gives symplectic maps with the time-step equal to the period of perturbation. In particular, this method was used by Chirikov to derive the celebrated *standard map* (Chirikov, 1979; Lichtenberg and Lieberman, 1992). However, these methods have significant shortcomings and difficulties, and they do not have a good mathematical justification. Particularly, they do not establish more general forms of the maps, estimate their accuracy, and establish relations between variables of the original system and of the mapping.

Recently in Abdullaev (1999, 2002) a mathematically rigorous method to derive symplectic maps has been developed. Based on the Hamilton–Jacobi theory and the classical perturbation theory, it allows one to construct symplectic mappings for generic Hamiltonian systems in a rigorous and consistent way. It does not encounter the difficulties of more traditional methods.

The present book is devoted to the systematic theory of symplectic mappings for Hamiltonian systems and its application to different Hamiltonian problems. The method is based on the Hamilton–Jacobi method and perturbation theory of classical mechanics. This book compresses 13 chapters. The theory of construction of Hamiltonian maps is given in the first five chapters. Application of mapping methods to study physical problems described by Hamiltonian systems are given in Chaps. 6–13.

The first chapter contains the essential elements of Hamiltonian dynamics including the different formulations of Hamiltonian equations, constant of motion, the Hamilton–Jacobi method, and the formalism of action-angle variables. In the second chapter we have presented the methods of classical perturbation theory. Time-dependent perturbation theory that constitutes the basis for the construction of symplectic mappings has been also reiterated in this chapter. The current methods to construct the symplectic maps for generic Hamiltonian problems are discussed in the third chapter. The Hamilton–Jacobi method or the method of canonical transformation to construct Hamiltonian mappings is presented in the fourth chapter. There we also discussed the different forms of symplectic maps, their accuracy, and how they compare with standard numerical symplectic integration methods. Mappings near separatrix of Hamiltonian systems are constructed in Chap. 5 using canonical transformations of the variables. The construction of mappings near separatrix is illustrated in Chap. 6 for several Hamiltonian systems. In Chap. 7 we have applied the mapping methods to analyze some non-standard issues of Hamiltonian dynamics, namely, regular and chaotic dynamics in non-twist and non-smooth Hamiltonian systems. The rescaling invariance properties of Hamiltonian systems near the hyperbolic saddle points are discussed in Chap. 8. Chaotic transport in a stochastic layer and $\log \epsilon$-periodicity (ϵ is a perturbation amplitude) in $1\frac{1}{2}$-degrees of freedom Hamiltonian systems are studied in Chap. 9. Applications of symplectic mappings to the study of magnetic field lines in magnetically confinement devices are presented in the next three chapters. Particularly, in Chap. 10 the Hamiltonian formulation of magnetic field line equations in magnetically confinement devices, namely in tokamaks. Particularly, we discuss also the mapping methods to integrate magnetic field line equations, and mapping models of field lines in toroidal system. Chapters 11 and 12 are devoted to the application of mapping methods to study the magnetic structure in special devices of magnetic confinement, namely, in ergodic and poloidal divertors. In Chap. 13 other areas of physics, namely, wave propagation problems, accelerator physics and dynamical astronomy, where mapping methods play an important role, are discussed.

The book is intended for postgraduate students and researchers, physicists, and astronomers working in the areas of Hamiltonian dynamics and chaos, and its applications to plasma physics, hydrodynamics, celestial mechanics, dynamical astronomy, and accelerator physics. It should also be useful for applied mathematicians involved in analytical and numerical studies of dynamical systems. Readers are supposed to be familiar with the methods of classical mechanics on the level of Chaps. 1–3 and 7–9 of the book *Mathematical methods of classical mechanics* (Springer-Verlag, 1989) by V.I. Arnold.

Acknowledgments. The participation in the project "Dynamic Ergodic Divertor" for the Jülich tokamak TEXTOR gave me an opportunity to work in the area of Hamiltonian dynamics and chaos and their applications to the plasma physics. For this I indebted to Professor Joachim Treusch, Professor Gerd H. Wolf, Professor Ulrich Samm and Professor Gert Eilenberger who invited me to this project and supported my activities in these areas. Professor G. Eilenberger read the manuscript and offered valuable suggestions and comments for improvements. Fruitful cooperations with Dr. Karl-Heinz Finken and Professor Karl-Heinz Spatschek were very beneficial for me. I have greatly benefited from many discussions with Professor Robert Wolf, Professor Detlev Reiter, Dr. André Rogister, Professor Radu Balescu, Professor Dominique Escande, Dr. Marcin Jakubowski, Mr. Armin Kaleck, Dr. Masahiro Kobayashi, Dr. Michael Lehnen, Dr. Albert Nicolai, Dr. Hartmut Gerhauser, Dr. Dirk Reiser, Dr. Mikhail Tokar', Dr. Bernard Unterberg, Dr. Todd Evans, Dr. Raymond Koch, Professor Niek Lopes Cardozo, Dr. Boris Weyssow. The work has been partially performed in the frame of the Sonderforschungsbereich (SFB) 591 of the Deutsche Forschungsgemeinschaft led by Professor Reinhard Schlickeiser. I am also grateful to Professor George Zaslavsky with whom I started my first steps into the area of chaos theory and cooperated with him during a long period time.

Jülich   *Sadrilla S. Abdullaev*
August, 2005

Contents

1   Basics of Hamiltonian Mechanics 1
  1.1   Hamilton Equations 1
     1.1.1   Invariants of Motion 2
     1.1.2   Hamiltonian Equations in Extended Phase Space 3
     1.1.3   Formulation of Hamiltonian Equations
            with One Coordinate as Independent Variable Instead
            of t 4
  1.2   The Hamilton–Jacobi Method 4
     1.2.1   Canonical Change of Variables 4
     1.2.2   The Hamilton–Jacobi Equation 5
     1.2.3   The Jacobi's Theorem 6
  1.3   Action-Angle Variables 7
     1.3.1   Integrable Hamiltonian Systems 7
     1.3.2   Hamiltonian in Action-Angle Variables 8
     1.3.3   Systems with One Degree of Freedom 9
     1.3.4   Many-Dimensional Systems with Separable Variables . 9
  1.4   Particles in a Wave Field 12
     1.4.1   Trapped Motion of Particles 13
     1.4.2   Untrapped Motion of Particles: $H > \omega_0^2$ 15
     1.4.3   Motion on Separatrix 16
  1.5   On Symplectic Numerical Integration of Hamiltonian Systems 17
  1.6   Bibliographic Notes 18

2   Perturbation Theory for Nearly Integrable Systems 21
  2.1   Perturbation Methods in Infinite Time Intervals 21
     2.1.1   A Fundamental Problem of Dynamics 22
     2.1.2   The Main Idea of Averaging Procedure 22
     2.1.3   Determination of the Generating Function 23
     2.1.4   von Zeipel's Method 25
  2.2   Lie Transform Methods 26
  2.3   Time-Dependent Perturbation Series 28
     2.3.1   Cauchy Problem 29
     2.3.2   Generation Functions 29
     2.3.3   Remarks 31
     2.3.4   Time-Dependent Lie Transform Method 33

2.3.5   Example Duffing Oscillator 33
2.4   Method of Successive Transformations 36
2.5   Bibliographic Notes 37

3   **Mappings for Perturbed Systems** 39
3.1   Poincaré Mappings 39
3.2   Method of a priori Assumption 41
3.3   Method of Delta Functions 44
3.4   The Standard Map 47
3.5   Exact Mappings of Hamiltonian Systems 49
3.6   Difficulties in Constructing Mappings 50

4   **Method of Canonical Transformation**
for Constructing Mappings 53
4.1   Canonical Transformation and Mapping 54
        4.1.1   Nonsymmetric Forms of Maps 55
        4.1.2   Symmetric Map 56
        4.1.3   The Generating Function of Mappings 57
4.2   Accuracy of Maps 58
        4.2.1   Particles in a Single-Frequency Wave Field 59
        4.2.2   Accuracy of the Symmetric Map 62
4.3   Mappings for Hamiltonian Systems
        with a Broad Perturbation Spectrum 64
        4.3.1   Non-Symmetric Forms of Maps 68
        4.3.2   Standard Hamiltonian and Corresponding Mappings .. 69
4.4   Mappings with Lie Generating Functions 71
4.5   Poincaré Maps at Arbitrary Sections of Phase-Space 73
4.6   Method of Successive Canonical Transformations 78
4.7   Summary 80

5   **Mappings Near Separatrix. Theory** 83
5.1   Separatrix and Mappings 83
        5.1.1   The Conventional Separatrix (Whisker) Map 85
        5.1.2   Shortcomings of Conventional Separatrix Mappings ... 87
5.2   The Hamilton–Jacobi Method
        to Construct Maps Near a Separatrix 88
        5.2.1   Mapping Along Single Saddle–Saddle Connection 90
        5.2.2   Calculation of the Generating Function 93
        5.2.3   Symmetric Mappings 94
        5.2.4   Nonsymmetric Mappings 94
        5.2.5   Properties of the Melnikov Type Integrals $K_n(h)$
                and $L_n(h)$ 96
5.3   Simplification of Mappings 97
        5.3.1   "Primary Resonant" Approximation 98
        5.3.2   Simplified Form of Mappings 98

        5.3.3   Separatrix Mapping Approximation 99
  5.4   Mapping at Arbitrary Sections of Phase Space 100
        5.4.1   Mapping to a Section Σ_c 101
        5.4.2   Mapping to Section $\Sigma_{\vartheta=\eta}$ 103
  5.5   Conclusion 104

6   **Mappings Near Separatrix. Examples** 105
  6.1   Motion in a Perturbed Double–Well Potential 105
  6.2   Mapping for the Periodically Driven Pendulum 114
        6.2.1   Behavior of Integrals $K_n(h)$ and $L_n(h)$ (5.31) 116
        6.2.2   Mapping to Sections Σ_s 118
        6.2.3   Mapping to Sections Σ_c 122
  6.3   Mapping for the Periodic–Driven Morse Oscillator 125
        6.3.1   Mapping 127
        6.3.2   The Symmetric Mapping 127
        6.3.3   A Nonsymmetric Mapping 130
        6.3.4   Comparison with a Numerical Integration 131
  6.4   The Kepler Map 132
  6.5   Comments on Separatrix Map Methods 136
  6.6   Bibliographic Notes 137

7   **The KAM Theory Chaos Nontwist and Nonsmooth Maps** 139
  7.1   Conservation of Conditionally Periodic Motions.
        The KAM Theory 139
        7.1.1   Invariant Tori for Mapping 140
        7.1.2   Destruction of Resonant Orbits: Nonlinear Resonance . 142
        7.1.3   Chaotic Layer Near a Separatrix 144
        7.1.4   Lyapunov Exponents 146
  7.2   Applicability of KAM Theory 148
        7.2.1   On the Smallness of Perturbations 148
        7.2.2   On the Smoothness of Perturbations 150
  7.3   Non-Twist Maps 150
        7.3.1   Dynamics of Systems with a Non-Monotonic Frequency 151
        7.3.2   Behavior Near Local Maxima or Minima 152
        7.3.3   Behavior Near a Bending Point 155
        7.3.4   Non-Twist Standard Maps 155
        7.3.5   Fixed Points and Transit to Chaos 159
  7.4   Non-Smooth Mappings 160
        7.4.1   Intermittence in Nontwist Systems 162
        7.4.2   Simplified Non-Smooth Mappings 168
        7.4.3   A Nontwist Map and Intermittency 169
  7.5   Suppression of Chaos in Smooth Hamiltonian Systems 173
  7.6   Bibliographic Notes 174

**8   Rescaling Invariance of Hamiltonian Systems
Near Saddle Points** 175
   8.1   Rescaling Invariance Near Saddle Points and Separatrix Maps 176
        8.1.1   Structure of Phase Space Near Saddle Points 176
        8.1.2   Universal Rescaling Invariance 177
        8.1.3   Proof of the Rescaling Invariance
               of Hamiltonian Equations 179
        8.1.4   Separatrix Mapping Approach 181
        8.1.5   Rescaling Invariance in Parameter Space 182
   8.2   Rescaling Invariance due to the Symmetry of Hamiltonians .. 183
        8.2.1   Separatrix Mapping Analysis 184
   8.3   2D-Periodic Vortical Flow 187
        8.3.1   Model 187
        8.3.2   Rescaling Invariance Property.................... 189
        8.3.3   Separatrix Maps of the System................... 191
        8.3.4   On the Validity Conditions
               of the Rescaling Invariance Property 193
   8.4   Summary.. 195

9   Chaotic Transport in Stochastic Layers 197
   9.1   Statistical Description of Chaotic Dynamical Systems 197
        9.1.1   Ergodicity and Mixing 197
        9.1.2   Kinetic Description 199
        9.1.3   Anomalous Diffusion............................. 201
   9.2   Non-Gaussian Statistics in Stochastic Layers 202
        9.2.1   Mean Residence Time............................. 202
        9.2.2   Statistics of Poincaré Recurrences 203
   9.3   Chaotic Transport in Stochastic Layers.
         Three-Wave Field Model 206
        9.3.1   Advection 206
        9.3.2   Anomalous Diffusion............................. 206
        9.3.3   Probability Density Function 208
   9.4   Chaotic Transport in 2D-periodic Vortical Flow 210
        9.4.1   Variation of Diffusion Regimes 211
        9.4.2   Superdiffusive Regime. Levý Flights 213
        9.4.3   Fixed Points of Flight Islands 215
   9.5   Conclusions... 217

10   Magnetic Field Lines in Fusion Plasmas 219
   10.1 Magnetic Field Lines as Hamiltonian System.............. 219
        10.1.1 Equilibrium Magnetic Field........................ 220
        10.1.2 Hamiltonian Field Line Equations 220
        10.1.3 Hamiltonian Formulation of Field Line Equations
               in a Toroidal System.............................. 221
        10.1.4 The Standard Magnetic Field 224

10.1.5 Equilibrium Magnetic Field with the Shafranov Shift . 225
10.2 Hamiltonian Equations in the Presence
     of Magnetic Perturbations 229
     10.2.1 Cylindrical Model of Plasmas 230
     10.2.2 Magnetic Perturbations in Toroidal Plasmas 232
     10.2.3 Asymptotics of the Transformation Matrix
            Elements $S_{mm'}(\psi)$ 233
     10.2.4 Asymptotic Behavior of $H_{mn}(\psi)$ 236
10.3 Mapping of Field Lines 237
10.4 Mappings as Models for Magnetic Field Lines 240
     10.4.1 The Standard Map and its Generalizations 240
     10.4.2 The Wobig–Mendonça Map 241
     10.4.3 The Tokamap 242
10.5 Continuous Hamiltonian System and Tokamap 244
     10.5.1 The Symmetric Tokamap 246
     10.5.2 Comparison of the Tokamap
            and the Symmetric Tokamap 247
     10.5.3 The Revtokamap and the Symmetric Revtokamap 251
10.6 Other Mapping Models of Field Lines 252
     10.6.1 Analytical Models 252
     10.6.2 Numerical Mapping Models 253
10.7 Conclusions 254

11 Mapping of Field Lines in Ergodic Divertor Tokamaks 255
11.1 Ergodic Divertor Concept 255
     11.1.1 Mappings to Study Ergodic Divertors 256
11.2 Magnetic Structure of the DED 257
     11.2.1 Set of Divertor Coils and Magnetic Perturbations..... 257
11.3 Spectrum of Magnetic Perturbations 259
11.4 Formation of the Ergodic Zone 263
11.5 Statistical Properties of Field Lines 264
     11.5.1 Global and Local Diffusion Coefficients 265
     11.5.2 Quasilinear Diffusion Coefficients 266
     11.5.3 Numerical Calculation
            of Field Line Diffusion Coefficients.................. 266
11.6 Ergodic Divertor as a Chaotic Scattering System 268
     11.6.1 Basin Boundary Structure at the Plasma Edge 269
     11.6.2 Magnetic Footprints 270
11.7 Conclusion 273

12 Mappings of Magnetic Field Lines
    in Poloidal Divertor Tokamaks 275
12.1 Field Lines in Equilibrium Plasmas Near the Separatrix 277
     12.1.1 Magnetic Perturbations 280
     12.1.2 Separatrix Map 280

12.1.3 Mappings to the Divertor Plates 284
12.2 Two-Wire Model of the Plasma 284
12.2.1 Magnetic Field Perturbations 288
12.2.2 The Structure of the Stochastic Layer 292
12.2.3 Structure of Magnetic Footprints 296
12.3 Conclusion 298

13 Miscellaneous 299
13.1 Ray Dynamics in Waveguide Media 299
13.1.1 Rays as a Hamiltonian System 299
13.1.2 Mapping Models of Ray Propagation
        in Waveguide Media 302
13.1.3 Ray Dynamics in the Waveguide Model 305
13.1.4 Other Mapping Models of Rays 308
13.2 Mapping Methods in Accelerator Physics 308
13.3 Mappings in Dynamical Astronomy 313

A   The Second Order Generating Function 317

B   Asymptotic Estimations of the Integral $K(h)$
    and $L(h)$ Near Separatrix 321
B.1 General Structure of Integrals 321
B.1.1 Unperturbed Orbits Near the Separatrix 322
B.1.2 Perturbation Hamiltonian in Normal Coordinates
       ξ, η Near the Saddle Points 324
B.1.3 Integrals Over the Powers of Orbits $\xi(t,t), \eta(t,h)$
       Near the Separatrix 325
B.1.4 Oscillatory Parts of $R(h)$ 327
B.2 Periodically–Driven Pendulum 330
B.3 The Integral $K(h)$ in the Problem of Driven Morse Oscillator 332

C   Proof of Rescaling Invariance of the Equations of Motion . 335
C.1 The Case of Linear Approximation 335
C.2 The Case of Nonlinear Approximation 338

D   Relation Between ϑ and θ 341

E   Asymptotic Estimation of the Integral $S_{mm'}$ (10.49) 345

F   Sample Program for Implementing a Mapping Procedure . 349

References 359

Index 377

1 Basics of Hamiltonian Mechanics

In this chapter we shall briefly recall the fundamental principles and methods of Hamiltonian mechanics which will be used throughout the book. This is for convenience of the reader and to fix notation. For more details, the reader might consult Arnold (1989). In particular, we shall give different formulations of Hamiltonian equations, and recall the invariants of motion. Special emphasis will be given to the Hamilton–Jacobi method and the action-angle formalism to integrate the equations of motion. These methods will be illustrated with the example of the pendulum. Finally, we shall shortly discuss modern methods of numerical symplectic integration of Hamiltonian systems.

1.1 Hamilton Equations

Consider a classical system with N degrees of freedom with q_i $(i = 1, \ldots, N)$ being the position coordinates of the particles of the system. In the classical (Newtonian) formulation the equations governing the time-evolution of the system are a set of second order ordinary differential equations for the positions q_i.

In the Hamiltonian formulation of classical mechanics the state of the system is characterized not only by its positions q_i, but also its momenta p_i, i.e., it is determined by coordinates in the so-called $2N$-dimensional *phase space* (q, p): N-coordinates $q = (q_1, \ldots, q_N)$ and $N-$ momenta $p = (p_1, \ldots, p_N)$. The time-evolution of the system is then governed by a set of $2N$ ordinary differential equations of first order in time t Hamilton (1834):

$$\frac{dq_i}{dt} = \frac{\partial H}{\partial p_i}, \qquad \frac{dp_i}{dt} = -\frac{\partial H}{\partial q_i}, \qquad (i = 1, \ldots, N) \tag{1.1}$$

known as *Hamilton equations*, and determined by only one scalar master function $H = H(q, p, t)$ known as *Hamilton's function* (or *Hamiltonian*). The positions q_i and momenta p_i are called *canonical variables* and time t is an independent variable.

The Hamilton equations (1.1) with given initial conditions $q^{(0)} = (q_1^{(0)}, \ldots, q_N^{(0)})$ and $p^{(0)} = (p_1^{(0)}, \ldots, p_N^{(0)})$ at the moment $t = 0$ have unique solution

S.S. Abdullaev: *Construction of Mappings for Hamiltonian Systems and Their Applications*, Lect. Notes Phys. **691**, 1–19 (2006)
www.springerlink.com

$$q_i(t) = q_i(t, q_1^{(0)}, \ldots, q_N^{(0)}, p_1^{(0)}, \ldots, p_N^{(0)}) \, ,$$

$$p_i(t) = p_i(t, q_1^{(0)}, \ldots, q_N^{(0)}, p_1^{(0)}, \ldots, p_N^{(0)}) \, , \qquad (1.2)$$

$(i = 1, \ldots, N)$ at any arbitrary time instant $t > 0$ or $t < 0$.

Geometrically, the trajectories (1.2) may be considered as a *flow* of a $2N$-dimensional fluid in the phase space Lanczos (1962); Guckenheimer and Holmes (1983). The velocity field $\mathbf{v}$ of this fluid flow is $\mathbf{v} = (\dot{q}_1, \ldots, \dot{q}_N, \dot{p}_1, \ldots, \dot{p}_N)$. Below we shall see that this flow preserves some invariants of motion which are important in construction of mappings.

1.1.1 Invariants of Motion

Invariants (or *integrals*) of motion are most important to study the evolution of Hamiltonian systems. A function $F = F(q, p, t)$ is called an integral of motion if it does not change its initial value during the time evolution of the system. Using the Hamiltonian equations (1.1) it can be formally written as

$$\frac{dF}{dt} = \frac{\partial F}{\partial t} + \{F, H\} = 0 \, , \qquad (1.3)$$

where the notation $\{F, \Phi\}$ stands for the Poisson bracket

$$\{F, \Phi\} = \sum_{i=1}^{N} \left(\frac{\partial F}{\partial q_i} \frac{\partial \Phi}{\partial p_i} - \frac{\partial F}{\partial p_i} \frac{\partial \Phi}{\partial q_i} \right) \, . \qquad (1.4)$$

The first integral of the system is the energy of a conservative system if the Hamiltonian H does not explicitly depend on time t, i.e., $H = H(q_1, \ldots, q_N; p_1, \ldots, p_N)$. It follows from (1.3) that $dH/dt = 0$ since $\partial H/\partial t = 0$ and $\{H, H\} \equiv 0$, and thus the energy of the conservative system is an integral of motion, $H = E = $ const.

Another invariant property of Hamiltonian motion (or flow) comes from its similarity with a "incompressible fluid", i.e., an arbitrary volume of fluid element is unchanged during the motion. The condition of incompressibility for the phase fluid,

$$\nabla \cdot \mathbf{v} = \sum_{i=1}^{N} \left(\frac{\partial \dot{q}_i}{\partial q_i} + \frac{\partial \dot{p}_i}{\partial p_i} \right) = 0 \, , \qquad (1.5)$$

is satisfied for the canonical equations (1.1) with an arbitrary Hamiltonian $H = H(q, p, t)$, for conservative, as well as for non-conservative systems. This property of the Hamiltonian flow leads to conservation of any closed volume $\Omega(t)$ of phase space, i.e.,

$$V = \int_{\Omega(t)} dq_1 \ldots dq_N dp_1 \ldots dp_N = \text{const} \, . \qquad (1.6)$$

This constitutes the "Liouville's theorem" which states that the shape of Ω may be deformed during the motion but *its volume remains unchanged.* From (1.6) it follows that the Jacobian of the transformation from initial coordinates $P_0 = (q_1^{(0)}, \ldots, q_N^{(0)}, p_1^{(0)}, \ldots, p_N^{(0)})$ to final state coordinates $P = (q_1(t), \ldots, q_N(t), p_1(t), \ldots, p_N(t))$ equals one:

$$J = \left| \frac{\partial q_1(t), \ldots, q_N(t), p_1(t), \ldots, p_N(t)}{\partial (q_1^{(0)}, \ldots, q_N^{(0)}, p_1^{(0)}, \ldots, p_N^{(0)})} \right| = 1 \; . \tag{1.7}$$

The property of Hamiltonian systems given by the Liouville's theorem (1.6) and (1.7) constitutes one of the invariants of motion. A number of invariants of motion may be as many as degree of freedom (see for details Arnold (1989)). The transformation $P_0 \rightarrow P$ is called a *volume-preserving map* (or a *symplectic map*) when it conserves the certain symplectic structure of system along with the property (1.7). Below we consider one of them.

Let C be a closed curve in the $(2N + 1)$-dimensional extended phase space (q, p, t) consisting points at the different time instants. Then from the Hamiltonian equations (1.1) it follows that the integral

$$J_C = \oint_C \left(\sum_{i=1}^{N} p_i dq_i - H dt \right) = \text{const} \; , \tag{1.8}$$

taken along the closed curve C is a constant of motion, i.e., $dI_C/dt = 0$. The integral (1.8) is known as *Poincaré-Cartan's integral invariant.*

1.1.2 Hamiltonian Equations in Extended Phase Space

In some situations it is useful to formulate the Hamiltonian equations in the extended phase-space $(t, q_1, \ldots, q_N, p_0, p_1, \ldots, p_N)$ which also includes the time t and a new canonical momentum p_0 conjugated to it in the new Hamiltonian function $\mathcal{H}$

$$\begin{aligned} \mathcal{H} &= \mathcal{H}(t, q_1, \ldots, q_N, p_0, p_1, \ldots, p_N) \\ &= p_0 + H(q_1, \ldots, q_N, p_1, \ldots, p_N, t) \; . \end{aligned} \tag{1.9}$$

Let τ be a new independent time-variable. Then from the Hamiltonian equations for the canonical variables (t, p_0) one obtains:

$$\frac{dt}{d\tau} = \frac{\partial \mathcal{H}}{\partial p_0} = 1 \; , \quad \frac{dp_0}{d\tau} = -\frac{\partial \mathcal{H}}{\partial t} = -\frac{\partial H}{\partial t} \; , \tag{1.10}$$

thus $t = \tau$ and $p_0 = -H$. The equations for the other variables $(q_1, \ldots, q_N, p_1, \ldots, p_N)$ are given by the Hamiltonian equations (1.1) with Hamiltonian $\mathcal{H}$ (1.9) instead of H. Such a formulation will be used in Sect. 5 to construct maps near a separatrix.

1.1.3 Formulation of Hamiltonian Equations
with One Coordinate as Independent Variable Instead of t

For some problems it is convenient to choose one of the coordinates as an independent time like variable. Suppose, that the coordinate q_1 obeys the following condition: $\dot{q}_1 = \partial H/\partial p_1 \neq 0$ in the some region of the phase-space (q, p) of the system. Then defining t and $p_0 = -H(q, p, t)$ as new canonical variables, coordinate and momentum, supplemented to the rest of canonical variables $(q_2, \ldots, q_N, p_2, \ldots, p_N)$, one can obtain Hamilton equations for the new canonical variables (see, e.g., Arnold (1989))

$$\frac{dt}{dq_1} = \frac{\partial K}{\partial p_t}, \quad \frac{dp_0}{dq_1} = -\frac{\partial K}{\partial t}, \tag{1.11}$$

$$\frac{dq_i}{dq_1} = \frac{\partial K}{\partial p_i}, \quad \frac{dp_i}{dq_1} = -\frac{\partial K}{\partial q_i}, \quad (i = 2, \ldots, N), \tag{1.12}$$

where

$$K = K(t, q_2, \ldots, q_N, p_0, p_2, \ldots, p_N, q_1) = -p_1. \tag{1.13}$$

is a new Hamiltonian function found by inversion of the relation $p_t = -H(q, p, t)$ with respect to p_1. Such an inversion exists since $\partial H/\partial p_1 \neq 0$.

Formulation of Hamiltonian equations with the q_1-coordinate as an independent variable is useful to construct Poincaré maps at sections of phase-space where q_1 is constant.

1.2 The Hamilton–Jacobi Method

A powerful method to integrate Hamiltonian equations, the Hamilton–Jacobi method, is based on a change of variables in the Hamilton equations (see Goldstein (1980); Arnold (1989)). The change of variables $(q, p) \rightarrow (Q, P)$ must be a *canonical*, i.e., it must preserve the invariants of motion and the Hamiltonian form of equations. The idea of the Hamilton–Jacobi method consists of finding such a canonical change of variables which reduces the Hamiltonian function to a form that the Hamiltonian equations can be easily integrated. It is important that any canonical transformation of variables is determined by a so called *generating function* satisfying the Hamilton–Jacobi partial differential equation. If we succeed to find a complete integral, i.e., the solutions of this equation depending on N independent constants of motion ($2N$ is the number of variables), then the time evolution of the system is completely determined by the generating function.

1.2.1 Canonical Change of Variables

Consider the change of variables $(q_1, \ldots, q_N, p_1, \ldots, p_N)$ to new ones $(Q_1, \ldots, Q_N, P_1, \ldots, P_N)$ given by the $2N$ functions $Q_i(q_1, \ldots, q_N, p_1, \ldots, p_N)$,

$P_i(q_1, \ldots, q_N, p_1, \ldots, p_N)$ of the $2N$ variables $(q_1, \ldots, q_N, p_1, \ldots, p_N)$. Such a transformation of variables, if canonical, must preserve the invariants of the system, in particularly, the Poincaré-Cartan integrals (1.8) and the form of canonical equations (1.1):

$$\frac{dQ_i}{dt} = \frac{\partial \mathcal{H}}{\partial P_i} , \qquad \frac{dP_i}{dt} = -\frac{\partial \mathcal{H}}{\partial Q_i} , \qquad (1.14)$$

with the new Hamiltonian $\mathcal{H} = \mathcal{H}(Q_1, \ldots, Q_N, P_1, \ldots, P_N, t)$. From the invariance of the Poincaré – Cartan integral invariant (1.8) follows that the difference

$$\sum_{i=1}^{N} p_i dq_i - H dt - \sum_{i=1}^{N} P_i dQ_i + \mathcal{H} dt = d\Phi(q, p, t) \qquad (1.15)$$

is a total derivative of some function Φ. The function Φ must depend on N of the old $q = (q_1, \ldots, q_N)$, $p = (p_1, \ldots, p_N)$ and N of the new $Q = (Q_1, \ldots, Q_N)$, $P = (P_1, \ldots, P_N)$ variables, and on time t.

There are many possible pairs of independent variables of generating functions F. In many cases one considers four combinations: (q, Q), (q, P), (p, Q), and (p, P). Below we specifically consider the second pair (q, P) as independent variables, i.e., $F = F(q, P)$ since it allows to obtain the identity transformation, $q = Q, p = P$ choosing $F = q \cdot P$. Such generating functions will be used in the next chapters for perturbation theory and to construct mappings. In this case (1.15) can be rewritten as

$$dF(q, P, t) = \frac{\partial F}{\partial t} dt + \sum_{i=1}^{N} \left(\frac{\partial F}{\partial p_i} dq_i + \frac{\partial F}{\partial P_i} dP_i \right)$$
$$= (\mathcal{H} - H) dt + \sum_{i=1}^{N} (p_i dq_i + Q_i dP_i) , \qquad (1.16)$$

where the generating function $F(q, P, t)$ is related to $\Phi(q, P, t)$, i.e., $F(q, P, t) = \sum_{i=1}^{N} P_i Q_i + \Phi(q_1, \ldots, q_N, P_1, \ldots, P_N, t)$. From (1.16) follows the relation between old and new variables:

$$p_i = \frac{\partial F}{\partial q_i} , \qquad Q_i = \frac{\partial F}{\partial P_i} , \qquad \mathcal{H} = H + \frac{\partial F}{\partial t} , \qquad (1.17)$$

where

$$F = F(q_1, \ldots, q_N, P_1, \ldots, P_N, t) .$$

1.2.2 The Hamilton–Jacobi Equation

Suppose that the canonical change of variables $(q, p) \rightarrow (Q, P)$ transforms the Hamiltonian system (1.1) to the new one such that $\mathcal{H}$ depends only on the canonical momenta $P_1, \ldots, P_N$, i.e., $\mathcal{H} = \mathcal{H}(P_1, \ldots, P_N)$. Then one can immediately integrate the Hamiltonian equations (1.14) which gives

$$Q_i = Q_{i0} + \omega_i(t - t_0), \qquad P_i = \text{const}. \tag{1.18}$$

where $\omega_i = \omega_i(P_1, \ldots, P_N) = \partial\mathcal{H}(P_1, \ldots, P_N)/\partial P_i$ ($i = 1, \ldots, n$) are the oscillation frequencies of the system in the new variables ($Q_1, \ldots, Q_N$, $P_1, \ldots, P_N$).

Using (1.17) one obtains the equation for the generating function $F = F(q_1, \ldots, q_N, P_1, \ldots, P_N, t)$ which implements such a transformation of variables:

$$H\left(q_1, \ldots, q_N, \frac{\partial F}{\partial q_1}, \ldots, \frac{\partial F}{\partial q_N}, t\right) + \frac{\partial F}{\partial t} = \mathcal{H}(P_1, \ldots, P_N). \tag{1.19}$$

It is known as the *Hamilton–Jacobi equation*. As a partial differential equation it usually has a large number of solutions. The solution $F(q_1, \ldots, q_N, P_1, \ldots, P_N, t)$ of the Hamilton–Jacobi equation (1.19) depending on the N independent constants $P_1, \ldots P_N$ is called a *complete integral* of the equation if the condition:

$$\det\left|\frac{\partial^2 F}{\partial q_i \partial P_j}\right| \neq 0.$$

is satisfied.

1.2.3 The Jacobi's Theorem

The *Jacobi's theorem* (see, e.g., Arnold (1989)) states that *if a solution* $F(q, P, t)$ *is a complete integral of the Hamilton–Jacobi equation, then solutions of the canonical equations (1.1) may be presented as*

$$p_i = \frac{\partial F(q_1, \ldots, q_N, P_1, \ldots, P_N, t)}{\partial q_i},$$

$$Q_i = \frac{\partial F(q_1, \ldots, q_N, P_1, \ldots, P_N, t)}{\partial P_i}. \tag{1.20}$$

It means that the time-evolution of the system is determined completely by the generating function $F(q, P, t)$. From the implicit algebraic equations (1.20) one can construct the *mapping* of the initial conditions ($q_i^{(0)}, p_i^{(0)}$) ($i = 1, \ldots, n$) of the system at the time $t = t_0$ to the final state ($q_i(t), p_i(t)$) at an arbitrary time instant t: the equation (1.20) at time $t = t_0$ determines the canonical transformation from the initial conditions ($q_i^{(0)}, p_i^{(0)}$) ($i = 1, \ldots, n$) to the new variables ($Q_i^{(0)}, P_i^{(0)}$) ($i = 1, \ldots, n$). Since the time-evolution of these variables is known by (1.18), then the original variables ($q(t), p(t)$) at any time $t > 0$ may be found by the backward canonical transformation ($Q(t), P(t)$) $\rightarrow$ ($q(t), p(t)$) with (1.20). We shall use this idea of Jacobi's theorem in Chap. 4 to construct mappings for generic Hamiltonian systems.

1.3 Action-Angle Variables

In this section we shortly recall the action-angle variables for the integrable Hamiltonian systems. The formulation of Hamiltonian equations in action-angle variable is most convenient to study Hamiltonian systems in the presence of perturbations and to construct symplectic maps. This formalism will be used throughout the text.

1.3.1 Integrable Hamiltonian Systems

First we recall the notion of integrability of Hamiltonian systems. The integrability problem was solved by the Liouville theorem (the modern formulation of this theorem was given by Arnold (1989)).

Suppose a given Hamiltonian system has as many as independent constants (integrals) of motion $F_i(q, p) = $ const $(i = 1, \ldots, N)$ as degrees of freedom. These integrals are called to be in *involution* if they have vanishing Poisson brackets among themselves $\{F_i, F_j\} = 0$ for all i and j. Then the orbits are confined to an N-dimensional surface in $2N$ phase space. Consider the surface (manifold) M_f determined by given values f_i $(i = 1, \ldots, N)$ of the integrals $F_i(q, p) = f_i$. Such a system is called *integrable*.

Hamiltonian motion in a finite domain of the $2N$-dimensional phase space $(q_1, \ldots, q_N, p_1, \ldots, p_N)$ defined by a manifold M_f may be viewed as a flow on the surface of the N-dimensional torus. This torus is characterized by N angular variables $(\vartheta_1, \ldots, \vartheta_N)$, modd 2π.

For instance, for the one -degree of freedom system with one constant of motion, energy, $H(p, q) = $ const, the orbits lie on the closed curves C which are contour curves on the phase plane (q, p) of constant energy $E = H(p, q)$, shown in Fig. 1.1. These curves are topologically equivalent to circles. Each point on the circle is uniquely determined by the angle ϑ (mod 2π). The latter may be viewed as an 1-D torus. From the Liouville theorem immediately follows that *one-dimensional time-independent Hamiltonian systems $H = H(q, p)$ are always integrable.*

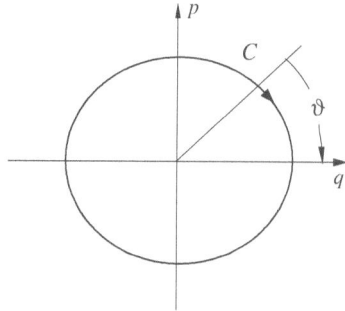

Fig. 1.1. Closed curve C in the phase plane (q, p)

Consider a two-degree of freedom Hamiltonian system $H(q_1, q_2, p_1, p_2)$ with two integrals of motion I_1, I_2 (or E, I_1). The trajectories of the system lie on a 2-D surface determined by two equations $I_1(q_1, q_2, p_1, p_2) = J_1 = \text{const}$, $I_2(q_1, q_2, p_1, p_2) = J_2 = \text{const}$. This surface is equivalent to the 2-D torus shown in Fig. 1.2.

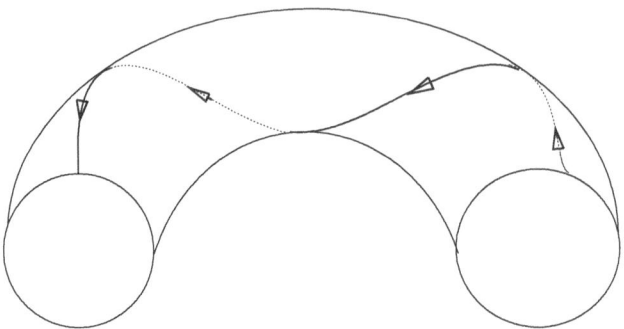

Fig. 1.2. Motion on the surface of 2-D torus

1.3.2 Hamiltonian in Action-Angle Variables

In the integrable case the Hamiltonian equations take a most convenient form if we choose the N independent integrals of motion F_i as canonical momenta conjugated to the *angle* variables ϑ_i. These canonical momenta are called *action* variables I_i $(i = 1, \ldots, N)$. The Hamiltonian function H depends on the action variables only

$$H = H(I_1, \cdots, I_N) \,. \tag{1.21}$$

Then the canonical equations become

$$\frac{d\vartheta_i}{dt} = \omega_i(I_1, \ldots, I_N), \qquad \frac{dI_i}{dt} = -\frac{\partial H}{\partial \vartheta_i} = 0 \,, \tag{1.22}$$

where

$$\omega_i(I_1, \ldots, I_N) = \frac{\partial H(I_1, \ldots, I_N)}{\partial I_i} \,, \qquad i = 1, \ldots, N \tag{1.23}$$

are the corresponding frequencies of motion along angular variables. Angular variables ϑ_i are thus linear functions of time:

$$\vartheta_i(t) = \omega_i(I_1, \cdots, I_N)t + \vartheta_i(0) \,, \qquad i = 1, \ldots, N \,. \tag{1.24}$$

1.3.3 Systems with One Degree of Freedom

First consider the simple case of an integrable system with one degree of freedom with Hamiltonian $H = H(q, p)$. Since $H = H(q, p)$ is the only one integral of motion the action variable I is a function of H, i.e., $I = I(H)$ or $H = H(I)$. In the (q, p)-plane a trajectory forms a closed curve C as shown in Fig. 1.1. The action-angle variables are introduced by a canonical transformation of the old, (q, p), variables to the new, action-angle variables (ϑ, I) and given by

$$I = \frac{1}{2\pi} \oint_C p(q; H)dq \,, \qquad \vartheta = \frac{\partial F(I, q)}{\partial I} \,, \qquad (1.25)$$

$$F(I, q) = \int^q p(q; I)dq \,.$$

The action I is equal to the area enclosed by the closed curve C divided by 2π.

Consider as an example the harmonic oscillator described Hamiltonian

$$H = H(q, p) = \frac{p^2}{2m} + k\frac{q^2}{2} \,. \qquad (1.26)$$

According to (1.25) we have the following generating function,

$$F(q, I) = I\left(\arcsin x + x\sqrt{1 - x^2}\right) \,, \qquad (1.27)$$

where $x = q(k/2H)^{1/2}$. The relation between the energy H and the action I is given by

$$H(I) = \omega I \,, \qquad \omega = \sqrt{\frac{k}{m}} \,, \qquad (1.28)$$

where ω is a frequency of harmonic oscillations. The old coordinates (q, p) are related to the action-angle variables (ϑ, I):

$$q = \sqrt{\frac{2I}{m\omega}} \sin \vartheta \,, \qquad p = \sqrt{\frac{2Im}{\omega}} \cos \vartheta \,. \qquad (1.29)$$

1.3.4 Many-Dimensional Systems with Separable Variables

In separable systems with more than one degree of freedom $(N > 1)$ action-angle variables (ϑ_i, I_i), $i = 1, \ldots, N$ are introduced as in systems with one-degree of system considered above *if* the variables in the corresponding problem are *separable*. In this case the Hamilton–Jacobi equation (1.19), a partial differential equation in N variables, can be replaced by N ordinary differential equations in a single variable. This is possible when the pair of variables (q_i, p_i), $(i = 1, \ldots, N)$ is entered in the Hamiltonian function $H(q_1, \ldots, q_N, p_1, \ldots, p_N)$ only in a combination $f_i(q_i, p_i)$. Then the solution $F(q_1, \ldots, q_N, I_1, \ldots, I_N)$ can be written as a sum of terms $F_i(q_i; I_1, \ldots, I_N)$ each depending only on one variable q_i, i.e.,

$$F(q_1, \ldots, q_N, I_1, \ldots, I_N) = \sum_{i=1}^{N} F_i(q_i; I_1, \ldots, I_N) . \tag{1.30}$$

Then the actions are defined as integrals

$$I_i = \frac{1}{2\pi} \oint_{C_i} p_i dq_i , \qquad i = 1, \ldots, N ,$$

$$p_i \equiv p_i(q_i; I_1, \ldots, I_N) = \frac{\partial F_i}{\partial q_i} , \tag{1.31}$$

taken along the closed contours C_i of constant $f_i = f_i(q_i, p_i)$ in the (q_i, p_i)-plane. The angle variables, ϑ_i, are

$$\vartheta_i = \frac{\partial F}{\partial I_i} = \sum_{i=1}^{N} \frac{\partial F_i(q_i; I_1, \ldots, I_N)}{\partial I_i} . \tag{1.32}$$

Consider as an example the *two-dimensional motion* in a polar coordinate system as an example of a separable Hamiltonian system. Let (r, θ) be the radial and angular variables, and (p_r, p_θ) be conjugated momenta, respectively. Suppose that a particle of mass $m = 1$ moves in the potential field U depending only on the radial coordinate r, $U = U(r)$. The Hamiltonian of the system is given by

$$H(r, p_r, p_\theta) = \frac{p_r^2}{2} + \frac{p_\theta^2}{2r^2} + U(r) . \tag{1.33}$$

The system has two integrals of motion: the energy $E = H(r, p_r, p_\theta)$ and the polar momentum, $p_\theta = $ const. The latter integral is due to independence of the Hamiltonian H on the polar angle θ. Therefore, the system is an integrable, and moreover it is thus reduced to two independent systems with one degree of freedom with Hamiltonians,

$$H_r(r, p_r) = \frac{p_r^2}{2} + \frac{L}{r^2} + U(r) , \qquad H_\theta(p_\theta) = \frac{p_\theta^2}{2} . \tag{1.34}$$

Consider the canonical transformation of the variables $(r, \theta, p_r, p_\theta)$ to the action-angle variables $(\vartheta_r, \vartheta_\theta, I_r, I_\theta)$. The corresponding generating function $F = F(r, \theta, I_r, I_\theta)$ is time-independent, and according to (1.19), (1.33) it satisfies the Hamilton–Jacobi equation:

$$\frac{1}{2}\left(\frac{\partial F}{\partial r}\right)^2 + \frac{1}{2r^2}\left(\frac{\partial F}{\partial \theta}\right)^2 + U(r) = \mathcal{H}(I_r, I_\theta) \ . \tag{1.35}$$

The variables r and θ in equation (1.35) are separable, and the solution F is presented as sum

$$F(r,\theta,I_r,I_\theta) = \int^r p_r(r',\mathcal{H},p_\theta)dr' + p_\theta\theta \ , \tag{1.36}$$

where

$$p_r(r,\mathcal{H},p_\theta) = \sqrt{2\left[\mathcal{H} - U(r) - \frac{p_\theta^2}{2r^2}\right]} \ .$$

The action variables (I_r, I_θ) are determined as

$$I_r = \frac{1}{2\pi}\oint_{C_r} p_r(r,\mathcal{H},p_\theta)dr \ , \qquad I_\theta = \frac{1}{2\pi}\oint_{C_\theta} p_\theta d\theta = p_\theta \ , \tag{1.37}$$

where the integrals are taken along the closed contours C_r and C_θ in the (r,p_r)– and (θ,p_θ)-planes, respectively (see Fig. 1.3). From (1.37) we obtain the dependence of the new Hamiltonian $\mathcal{H}$ on the action variables I_r and I_θ, i.e., $\mathcal{H} = \mathcal{H}(I_r, I_\theta)$, and the frequencies of motion $\omega_r(I_r, I_\theta) = \partial\mathcal{H}/\partial I_r$, $\omega_\theta(I_r, I_\theta) = \partial\mathcal{H}/\partial I_\theta$, along the radial r and the polar angle θ, respectively. According to (1.32) the angle variables $(\vartheta_r, \vartheta_\theta)$ are defined as

$$\vartheta_r = \frac{\partial F(r,\theta,I_r,I_\theta)}{\partial I_r} \ , \qquad \vartheta_\theta = \frac{\partial F(r,\theta,I_r,I_\theta)}{\partial I_\theta} \ . \tag{1.38}$$

These formulas define also the relation between the old (r,θ,p_r,p_θ) and the new action-angle variables $(\vartheta_r,\vartheta_\theta,I_r,I_\theta)$, i.e., $r = r(\vartheta_r,I_r,I_\theta)$, $\theta = \theta(\vartheta_r,\vartheta_\theta,I_r,I_\theta)$.

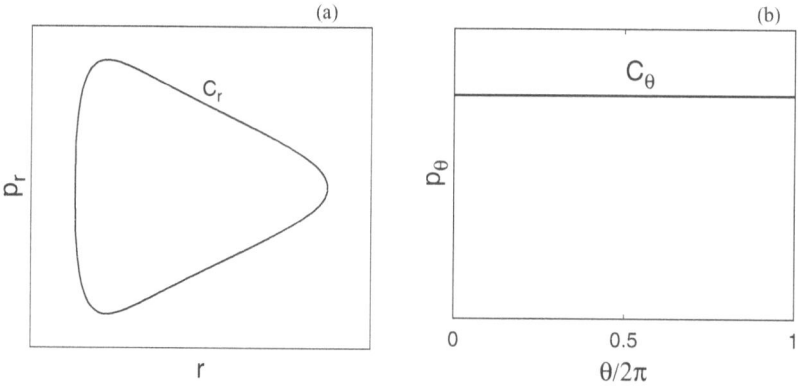

Fig. 1.3. (a) Closed contour C_r of constant $H_r(r,p_r)$ in the (r,p_r)-plane; (b) Contour C_θ of constant $H_\theta(p_\theta)$ in the (θ,p_θ)-plane

One should note that, in principle, one can introduce the action-angle variables also for integrable Hamiltonian systems with *non-separable variables*. The details of this procedure is described, e.g., in the book by Arnold (1989). However, the practical applications of this procedure in real non-separable systems is rather difficult, particularly, since in the nonseparable case no practicable test exist to decide if the system is integrable or not.

1.4 Particles in a Wave Field

We study the motion of a charged particle of mass m in a monochromatic electric wave of frequency Ω and wavelength $\lambda = 2\pi/k$. It is governed by the Newton equation:

$$m\ddot{x} = -eE_k \sin(kx - \Omega t) , \qquad (1.39)$$

where E_k is the amplitude of the wave, and e is the charge of the particle. In a coordinate system (q, t) moving with phase velocity $v_p = \Omega/k$ and normalized to the wavelength, i.e., $q = k(x - v_p t)$, the equation (1.39) may be written as $\ddot{q} + \omega_0^2 \sin q = 0$, where $\omega_0 = \sqrt{ekE_k/m}$. Introducing the canonical moment $p = \dot{q}$ the equations of motion are the Hamiltonian equations

$$\frac{dq}{dt} = \frac{\partial H}{\partial p} , \qquad \frac{dp}{dt} = -\frac{\partial H}{\partial q} , \qquad (1.40)$$

with the Hamiltonian

$$H = \frac{p^2}{2} + U(q) , \qquad U(q) = -\omega_0^2 \cos q . \qquad (1.41)$$

The Hamiltonian (1.41) describes the motion of a pendulum. It plays an important role in the study of general problems of Hamiltonian dynamics and chaos.

The system (1.40), (1.41) describes the motion of a particle of unity mass in the potential $U(q) = -\omega_0^2 \cos q$ as shown in Fig. 1.4a. The *fixed points* of the system in the (q, p)-plane defined as

$$\dot{q} = p = 0 , \qquad \dot{p} = \omega_0^2 \sin q = 0 ,$$

are $p = 0$, $q = k\pi$, $(k = 0, \pm 1, \pm 2, \ldots)$. Particularly, the points $(q_e = 2k\pi$, $p_e = 0)$ where the potential $U(q)$ has local minima, $U_{min} = -\omega_0^2$, are called *elliptic fixed points* since the orbits near them are described by the elliptic closed curves: $p^2 + \omega_0^2(q - q_e)^2 = $ const. On the other hand, the points $(q_h = (2k + 1)\pi k, p_h = 0)$ where $U(q)$ has local maxima, $U_{max} = \omega_0^2$, are called *hyperbolic fixed points*. The orbits are near these points described by hyperbolic curves, $p^2 - \omega_0^2(q - q_e)^2 = $ const.

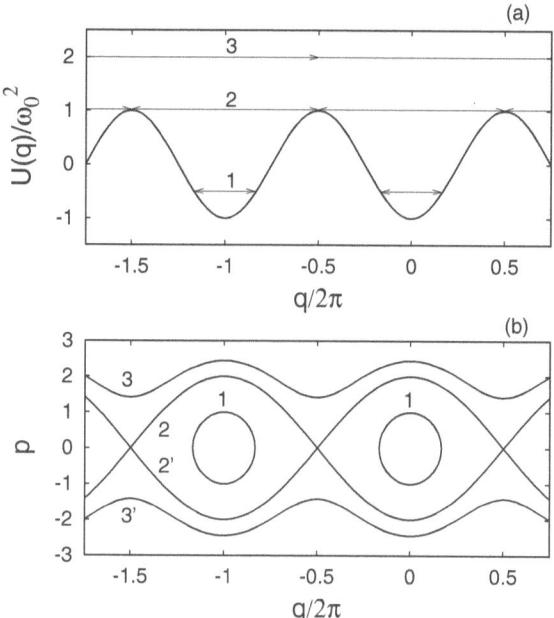

Fig. 1.4. (a) Profile of the potential $U(x)/\omega_0^2 = -\cos q$. (b) Phase space orbits of the pendulum: curves 1 correspond to the oscillatory motion $-\omega_0^2 \leq H < \omega_0^2$, curves 2 and 2' are separatrices where $H = \omega_0^2$, curves 3 and 3' correspond to rotational motion $H > \omega_0^2$

Three possible orbits of the system in the phase plane (q, p) are displayed in Fig. 1.4b. For values $-\omega_0^2 \leq H < \omega_0^2$ the motion of the pendulum is vibrational, and is described by a closed orbits curve 1. It also corresponds to the motion of a particle trapped by the wave.

For $H > \omega_0^2$ the motion is described by the unbounded curves 3 and 3'. They correspond to the rotational motion of pendulum. They also describe untrapped motion of a particle in a wave-field. The curves 2 and 2' separate the two kinds of motion, oscillatory and rotational, in the phase plane (q, p). These curves connecting the hyperbolic fixed points are known as *separatrices*. On the separatrix we have $H = \omega_0^2$.

1.4.1 Trapped Motion of Particles

Consider first the *oscillatory motion* of the pendulum, when $-\omega_0^2 \leq H < \omega_0^2$. The action variable I for this motion is determined by the integral

$$I = \frac{1}{2\pi} \oint_C p(q, H)dq = \frac{2}{\pi} \int_0^{q_m} \sqrt{2(H + \omega_0^2 \cos q)}dq \,, \qquad (1.42)$$

where q_m is a turning point of motion determined by the condition $p(q_m, H) = 0$. The integration in (1.42) gives the relation between I and H:

$$I(H) = \frac{8}{\pi} \left[E(k) - (1 - k^2)K(k) \right] , \qquad (1.43)$$

where $K(k)$ and $E(k)$ are the complete elliptic integrals of the first and the second kind, respectively, with a module $k = \sqrt{(H + \omega_0^2)/(2\omega_0^2)}$:

$$K(k) = \int_0^{\pi/2} \frac{d\varphi}{\sqrt{1 - k^2 \sin^2 \varphi}} , \qquad E(k) = \int_0^{\pi/2} \sqrt{1 - k^2 \sin^2 \varphi} d\varphi . \quad (1.44)$$

The frequency of motion $\omega(H)$ is given by

$$\omega(H) = \frac{dH(I)}{dI} = \frac{\pi \omega_0}{2K(k)} . \qquad (1.45)$$

Its dependence on H in the interval $-\omega_0^2 < H < \omega_0^2$ is shown in Fig. 1.5. At the small amplitude oscillations near the point $q = 0$, i.e., for $|H + \omega_0^2| \ll \omega_0^2$ (or $k \to 0$), it tends to the frequency of oscillations, ω_0, near the elliptic fixed points:

$$\omega(H) = \omega_0 \left(1 - \frac{k^2}{4} + O(k^4) \right) . \qquad (1.46)$$

The asymptotics of $\omega(H)$ near the separatrix when $H \to \omega_0^2$ ($k \to 1$) has the following form

$$\omega(H) = \frac{\pi \omega_0}{\ln(32/|h|)} + O(|h|) , \qquad h \to -0 . \qquad (1.47)$$

where $h = (H - \omega_0^2)/\omega_0^2$ is the energy normalized with respect to the one $H_s = \omega_0^2$ on the separatrix.

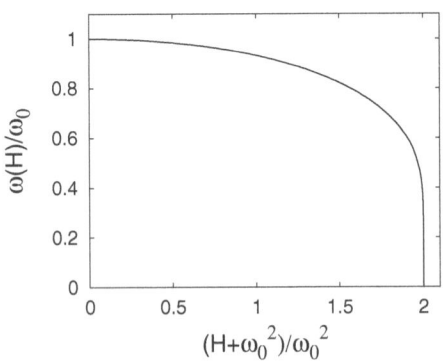

Fig. 1.5. Frequency of oscillation $\omega(H)$ versus energy H ($-\omega_0^2 < H < \omega_0^2$)

The trajectory is determined by the relation between the old coordinates (q, p) and the angle variable ϑ,

$$\sin(q/2) = k \operatorname{sn}\left(\frac{\omega_0 \vartheta}{\omega(H)}; k\right) , \qquad p = 2k\omega_0 \operatorname{cn}\left(\frac{\omega_0 \vartheta}{\omega(H)}; k\right) , \quad (1.48)$$

where $\operatorname{sn}(x; k)$ and $\operatorname{cn}(x; k)$ are the Jacobi elliptic functions with module k.

The coordinates q and p are $2\pi-$ periodic functions of the angle variable ϑ, and the relations (1.48) can be presented as Fourier series:

$$\sin(q/2) = \sum_{m=1}^{\infty} q_m \sin(2m-1)\vartheta ,$$

$$p = \sum_{m=1}^{\infty} p_m \cos(2m-1)\vartheta , \qquad (1.49)$$

where

$$q_m = \frac{2\pi}{K(k)} \frac{q^{m-1/2}}{1 - q^{2m-1}} , \qquad p_m = \frac{4\pi\omega_0}{K(k)} \frac{q^{m-1/2}}{1 + q^{2m-1}} , \qquad (1.50)$$

and $q = \exp\left(-\pi K(\sqrt{1-k^2})/K(k)\right)$. For the small amplitude oscillation, when $H \to -\omega_0^2$, $(k \to 0)$, the harmonics $m = 1$ gives the main contribution in (1.49) and the trajectory is similar to the one of harmonic oscillation given by (1.29).

When a particle approaches the separatrix, i.e., $h \to 0$ the spectrum q_m, p_m of the Fourier series (1.49) becomes broader and behave as

$$q_m \approx m^{-1} \left(1 - \pi^2/\ln(32/|h|)\right)^{m-1/2} ,$$

$$p_m \approx \left(1 - \pi^2/\ln(32/|h|)\right)^{m-1/2} . \qquad (1.51)$$

It means that in the limit $|h| \to 0$ the variables $q(\vartheta)$ and $p(\vartheta)$ become discontinuous functions of the angle variable ϑ.

1.4.2 Untrapped Motion of Particles: $H > \omega_0^2$

This corresponds to the rotational regime of pendulum (see curves 3 and 3' in Fig. 1.4). The motion is not confined along the coordinate q. Action-angle variables (ϑ, I) can be introduced using periodic boundary conditions at values q and $q + 2\pi$,

$$\vartheta = \frac{\partial}{\partial I} \int^q p(q', H)dq' = \frac{\omega(H)}{\omega_0 k} F\left(\frac{q}{2}, \frac{1}{k}\right) ,$$

$$I = \frac{1}{2\pi} \int_{-\pi}^{\pi} p(q, H)dq = \frac{4\omega_0 k}{\pi} E\left(\frac{1}{k}\right) , \qquad (1.52)$$

where $\omega(H) = dH/dI$ is the frequency of motion,

$$\omega(H) = \frac{\pi\omega_0 k}{K(1/k)}.$$ (1.53)

The trajectory is described by

$$\sin(q/2) = \mathrm{sn}\left(\frac{\omega_0 k}{\omega(H)}\vartheta\right), \qquad p = \frac{2}{k}\sqrt{1 - \frac{1}{k^2}\mathrm{sn}^2\left(\frac{\omega_0 k}{\omega(H)}\vartheta\right)}.$$ (1.54)

The asymptotics of the frequency $\omega(H)$ when an orbit approaches to the separatrix, $H \to \omega_0^2 + 0$, is determined by

$$\omega(H) = \frac{2\pi\omega_0}{\ln(32/h)} + O(|h|), \qquad h \to +0,$$ (1.55)

which is twice larger than the asymptotics (1.47).

The time-dependence of the momentum $p(t)$ for three types of orbits is plotted in Fig. 1.6: curve 1 corresponds to the small amplitude oscillations, curve 2 to the oscillatory regime close the separatrix, and curve 3 describes a rotational regime of the pendulum.

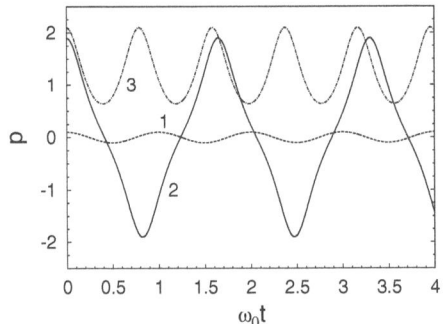

Fig. 1.6. Time-dependence of momentum $p(t)$ for three kind of motion: curve 1 corresponds to small amplitude oscillations ($|H + \omega_0^2| \ll \omega_0^2$); curve 2 — oscillations close to the separatrix ($|H - \omega_0^2| \ll \omega_0^2$); curve 3 — rotational regime ($H > \omega_0^2$)

1.4.3 Motion on Separatrix

Finally consider the orbits on the separatrix when $H = H_s = \omega_0^2$. The separatrix was shown in Fig. 1.4b by the phase space curves 2 and 2′. Using the equation $dq/dt = p(q, H_s) = \omega_0 \cos q/2$ one obtains

$$q = 4 \arctan \frac{\exp(\pm\omega_0\tau) + 1}{\exp(\pm\omega_0(t - t_0)) - 1} \ ,$$

$$p = \pm\frac{2\omega_0}{\cosh(\omega_0(t - t_0))} \ . \tag{1.56}$$

where the upper sign $(+)$ corresponds to the motion along the upper branch of the separatrix (curve 2), and lower sign (-) describes the motion along the lower brach (curve 2'), t_0 is the time instant when the orbit crosses the coordinate $q = 0$. The phase space point asymptotically approaches the saddle point:

$$q \to \mp\pi \ , \quad p \to -0 \ , \quad \text{for } t \to -\infty$$
$$q \to \pm\pi \ , \quad p \to +0 \ , \quad \text{for } t \to \infty$$

The orbits on the separatrix (1.56) will be used in Sect. 6 to obtain the mapping near the separatrix.

1.5 On Symplectic Numerical Integration of Hamiltonian Systems

Hamiltonian equations (1.1) as a set of $2N$ first order ordinary differential equations could be numerically integrated using well known standard methods, like Runge –Kutta methods, applied to the system of the first order differential equations (see, e.g., Press et al. (1992))

$$\frac{d\mathbf{x}}{dt} = \mathbf{f}(\mathbf{x}, t) \ , \tag{1.57}$$

where $\mathbf{x} = (x_1, \ldots, x_{2N})$ are M variable, and $\mathbf{f} = (f_1, \ldots, f_{2N})$ are given $2N$ force functions. However, these methods are not at all well suited for integrating Hamiltonian systems, since they do not preserve important invariants of the motion discussed in Sect. 1.1.1. Particularly, if $\mathbf{x}(t + h) = \hat{S}(h)\mathbf{x}(t)$ is the numerical solution with the time step h then the Jacobian of transformation (1.7) from the coordinates $\mathbf{x}(t)$ at time instant t to the ones $\mathbf{x}(t + h)$ at $t = t + h$ will, in general, not be exactly unity as we know it must. This means that this type of numerical integration does not preserve phase space area, and the numerically integrated Hamiltonian system becomes a dissipative system. Since the latter has attractors the numerical integration may lead to an entirely incorrect long time behavior of the solution.

To resolve this problem so called symplectic methods of integration of Hamiltonian systems have been introduced (see reviews by Sanz-Serna (1992); Sanz-Serna and Calvo (1993, 1994); Feng (1994) and references therein). The idea of the symplectic numerical integration is that each step of the integration is arranged to be a *canonical* (or *symplectic*) transformation. In the last two decades different symplectic methods of integration of Hamiltonian systems have been developed. The symplectic methods can be roughly divided into two groups. The first of them includes methods which are based on the Hamiltonian formalism using generating functions, Lie transforms, etc. The second group of methods uses well-known methods, for instance, Runge-Kutta formulas modified such, that their coefficients satisfy certain conditions. In general, both groups of methods are *implicit*, i.e., the variable x_{k+1} after a step of integration is not explicitly expressed through its previous value x_k. For instance, in the simple approximation it is determined by an equation $x_{k+1} = x_k + hf(x_{k+1})$.

Explicit numerical symplectic Runge–Kutta methods have been proposed for special Hamiltonian systems with separable variables: $H(\mathbf{p},\mathbf{q}) = T(\mathbf{p}) + V(\mathbf{q})$. Below we give formulas for the *explicit symplectic integrator*

$$(\mathbf{q}(t+h), \mathbf{p}(t+h)) = \hat{S}(\mathbf{q}(t), \mathbf{p}(t)) , \tag{1.58}$$

developed by McLachlan and Atela (1992). It is given by the following sequence of intermediate values $(\mathbf{Q}_j, \mathbf{P}_j)$, $j = 1, \ldots, s$:

$$\mathbf{P}_i = \mathbf{p}(t) + h \sum_{j=1}^{i} b_j \mathbf{f}(\mathbf{Q}_{j-1}) ,$$

$$\mathbf{Q}_i = \mathbf{q}(t) + h \sum_{j=1}^{i} a_j \mathbf{g}(\mathbf{P}_{j-1}) , \tag{1.59}$$

with $(\mathbf{Q}_0, \mathbf{P}_0) = (\mathbf{q}(t), \mathbf{p}(t))$ and $(\mathbf{q}(t+h), \mathbf{p}(t+h)) = (\mathbf{Q}_s, \mathbf{P}_s)$. In (1.59) $\mathbf{g}(\mathbf{p}) = \partial T/\partial \mathbf{p}$, $\mathbf{f}(\mathbf{q}) = -\partial V/\partial \mathbf{q}$ are the gradients. The symplectic integrator contains s coefficients a_i, b_i ($i = 1, \ldots, s$). They are listed in McLachlan and Atela (1992).

The accuracy of the symplectic integrators can be measured by the degree of energy conservation. Suppose that the symplectic map (1.58) with time step h were exact for a time -dependent Hamiltonian $K(\mathbf{p}, \mathbf{q}, t)$ which approximates the original Hamiltonian $H(\mathbf{p}, \mathbf{q})$. The deviation of H from K may serve as a good measure of the accuracy of the method. In Sect. 4.2 we will use this measure to study the accuracy of the symplectic integrator and the symplectic mappings.

In the following sections we use the the fifth order symplectic integrator (1.59) ($s = 6$) for numerical integration of Hamiltonian systems. It will be also compared with the mapping methods which will be developed later.

1.6 Bibliographic Notes

The methods of classical Hamiltonian dynamics are discussed in many textbooks. Among them one can recommend the classical books by Goldstein (1980) and by Lanczos (1962). A modern mathematical formulation of classical mechanics is given in the book by Arnold (1989). Phase space analysis of dynamical systems, particularly, Hamiltonian systems can be found in text-books by Andronov et al. (1981), Arnold (1989), Guckenheimer and Holmes (1983). The Hamilton–Jacobi method, the action-angle variable formalism are discussed in Goldstein (1980); Arnold (1989); Lanczos (1962). For more advanced readers one can recommend also books on the classical and celestial mechanics by Siegel and Moser (1995); Arnold et al. (1988).

The symplectic numerical integration methods of Hamiltonian systems are reviewed in Sanz-Serna (1992); Sanz-Serna and Calvo (1994); Feng (1994).

2 Perturbation Theory
for Nearly Integrable Systems

Real dynamical systems are generally not integrable. In many cases the deviation of a system from an integrable one is small and can be considered as perturbation of the integrable system. To study perturbed systems special theoretical methods, known as *perturbation theory*, have been developed. They are based on the assumption that the solutions of a perturbed system are close to the corresponding solutions of the unperturbed (integrable) system, and one seeks the deviation of the perturbed solution from the unperturbed one as a series in the powers of a parameter which characterizes the strength of the perturbation.

In this chapter we recall the basic ideas and methods of classical perturbation theory in Hamiltonian systems for infinite time intervals. Then we shall consider on the perturbation methods in finite time intervals since they are poorly discussed in the literature. The construction of symplectic mappings for Hamiltonian systems which will be given in Sect. 4 is mainly based on the perturbation procedure in finite time intervals.

2.1 Perturbation Methods in Infinite Time Intervals

Perturbation methods in general non-linear oscillatory systems are based on an averaging principle (see Poincaré (1892–99); Bogolyubov and Mitropol'skij (1958); Nayfeh (1973); Arnold et al. (1988); Arnold (1989)). The averaging principle or the methods of perturbation theory were first developed in the 19-th century in the problems of celestial mechanics to study the motion of planets of the Solar system. A comprehensive description of these methods is given by Poincaré in his classical book *"Les méthodes de la mécanique céleste"* (*New Methods of Celestial Mechanics*, Poincaré (1892–99)). Later the method was rediscovered by van der Pol in the study of nonlinear oscillatory systems.

Further development of the averaging procedure is connected with names of N.M. Krylov, N.N. Bogolyubov, L.I. Mandel'stam, and their followers who applied these methods to the wide-scale problems of nonlinear oscillations (see, e.g., Bogolyubov and Mitropol'skij (1958); Andronov et al. (1981)). The main idea of the averaging principle is concluded in a change of variables that allows to eliminate the fast changing phases from the equations of motion keeping only slowly varying variables. In this section we briefly recall the main

S.S. Abdullaev: *Construction of Mappings for Hamiltonian Systems and Their Applications*,
Lect. Notes Phys. **691**, 21–37 (2006)
www.springerlink.com

ideas of the averaging principle in Hamiltonian systems and its difficulties connected with small denominators.

2.1.1 A Fundamental Problem of Dynamics

Suppose an unperturbed system with N degrees of freedom is completely integrable, and one can introduce action-angle variables (I, ϑ): $(I_1, \ldots, I_N,$ $\vartheta_1, \ldots, \vartheta_N)$ to replace the physical variables (q, p). The unperturbed Hamiltonian H_0 then depends on the action variables, $H_0 = H_0(I)$, only and the unperturbed equations of motion, $\dot{I}_i = 0$, $\dot{\vartheta}_i = \partial H_0 / \partial I_i$, are integrable

$$I_i = \text{const} , \qquad \vartheta_i = \omega_i(I)(t - t_0) + \vartheta_{i0} , \qquad i = 1, \ldots, N . \qquad (2.1)$$

where

$$\omega_i(I_1, \cdots, I_N) = \frac{\partial H_0}{\partial I_i} , \qquad i = 1, \ldots, N ,$$

are the unperturbed frequencies of motion.

Suppose that the system is subjected to time-periodic perturbation. The set of perturbed Hamiltonian equations can be written as

$$\frac{d\vartheta_i}{dt} = \frac{\partial H}{\partial I_i} = \omega_i(I) + \epsilon \frac{\partial H_1}{\partial I_i} , \qquad \frac{dI_i}{dt} = -\epsilon \frac{\partial H_1}{\partial I_i} , \qquad (2.2)$$

with Hamiltonian function

$$H(I, \vartheta, t, \epsilon) = H_0(I) + \epsilon H_1(I, \vartheta, t, \epsilon) , \qquad (2.3)$$

where the dimensionless parameter ϵ stands for the perturbation strength. The perturbation Hamiltonian $H_1(I, \vartheta, t, \epsilon)$ is a $2\pi-$ periodic function in ϑ and in time with period T (or with the frequency $\Omega = 2\pi/T$). Suppose that the perturbation is small, $\epsilon \ll 1$.

The Hamiltonian problem (2.2), (2.3) describing the dynamics of integrable system affected by small perturbation has numerous applications in different areas of physics, mechanics and astronomy. Poincaré (1892–99) called this problem a *fundamental problem of dynamics*.

2.1.2 The Main Idea of Averaging Procedure

The averaging procedure is implemented by the change of fast–oscillating variables (I, ϑ) to slowly varying variables (J, ψ). In Hamiltonian systems such a change of variables ought to be canonical. The change of variables is given by a generating function $F(J, \vartheta, t, \epsilon)$ of mixed variables: old angle variables ϑ and new actions J. Since the perturbation is small, $F(J, \vartheta, t, \epsilon)$ can be presented in the form

$$F(J, \vartheta, t, \epsilon) = J\vartheta + \epsilon S(J, \vartheta, t, \epsilon) ,$$

which is close to the generating function, $F_0(J, \vartheta) = J\vartheta$, for the identical change of variables, $J = I$, $\psi = \vartheta$. Then the relation between old and new variables is given by

$$I = J + \epsilon \frac{\partial S}{\partial \vartheta}\,, \qquad \psi = \vartheta + \epsilon \frac{\partial S}{\partial J}\,. \qquad (2.4)$$

Suppose that the change of variables given by (2.4) transforms the system with Hamiltonian $H(I, \vartheta, t, \epsilon)$ (2.3) into a new one with $\mathcal{H} = \mathcal{H}(J, \epsilon)$ depending only on the action variables J. If there exists such a change of variables, then the new system is fully integrable [1], i.e.,

$$\dot{J} = -\frac{\partial \mathcal{H}(J, \epsilon)}{\partial \psi} = 0\,, \qquad \dot{\psi} = \frac{\partial \mathcal{H}(J, \epsilon)}{\partial J} = w(J, \epsilon)\,. \qquad (2.5)$$

2.1.3 Determination of the Generating Function

Below we consider the method of perturbation theory proposed by Lindstedt for problems of celestial mechanics (see Poincaré (1892–99); Arnold et al. (1988)). The generating function S determining the transformation of variables satisfies the Hamilton–Jacobi equation (1.19), which takes the following form for the Hamiltonian (2.3):

$$H_0\left(J + \epsilon \frac{\partial S}{\partial \vartheta}\right) + \epsilon H_1\left(J + \epsilon \frac{\partial S}{\partial \vartheta}, \vartheta, t, \epsilon\right) + \epsilon \frac{\partial S}{\partial t} = \mathcal{H}(J, \epsilon)\,. \qquad (2.6)$$

For small perturbation parameter, ϵ, we can seek the generating function $S(\theta, J, t; \epsilon)$ as series in powers of ϵ:

$$S(J, \vartheta, t, \epsilon) = S_1(J, \vartheta, t) + \epsilon S_2(J, \vartheta, t) + \cdots\,. \qquad (2.7)$$

We also expand the old, $H_1(I, \vartheta, t, \epsilon)$ and new, $\mathcal{H}(J; \epsilon)$, Hamiltonians in a similar series

$$\epsilon H_1(I, \vartheta, t, \epsilon) = \epsilon H_1(I, \vartheta, t) + \epsilon^2 H_2(I, \vartheta, t) + \cdots\,,$$

$$\mathcal{H}(J, \epsilon) = \mathcal{H}_0(J) + \epsilon \mathcal{H}_1(J) + \epsilon^2 \mathcal{H}_2(J) + \cdots\,. \qquad (2.8)$$

Expanding the Hamilton–Jacobi equation (2.6) in a series in powers of ϵ and equating the terms with the same power in ϵ we obtain the relation

$$\mathcal{H}_0(J) = H_0(J)\,, \qquad (2.9)$$

---

[1] The conditions at which such a transformation exists is formulated by the Kolmogorov's theorem on the existence of conditionally-periodic motion (Kolmogorov (1954)), and it constitutes the contents of the KAM theory (Arnold (1963a,b); Moser (1962)), [see Sect. 7.1 for more details.]

and the equations for the expansion coefficients $S_i \equiv S_i(\vartheta, J, t)$ of the generating function $S(\vartheta, J, t; \epsilon)$:

$$\frac{\partial S_1}{\partial t} + \frac{\partial H_0}{\partial J} \cdot \frac{\partial S_1}{\partial \vartheta} = \mathcal{H}_1(J) - H_1(\vartheta, J, t) , \tag{2.10}$$

$$\frac{\partial S_j}{\partial t} + \frac{\partial H_0}{\partial J} \cdot \frac{\partial S_j}{\partial \vartheta} = \mathcal{H}_j(J) - F_j(\vartheta, J, t) , \qquad j \geq 2 . \tag{2.11}$$

where $F_j(\vartheta, J, t)$ are the polynomial functions of derivatives $\partial S_1 / \partial \vartheta$, ..., $\partial S_{j-1} / \partial \vartheta$. Particularly, for $F_2(\vartheta, J, t)$ we have

$$F_2(\vartheta, J, t) = H_2(\vartheta, J, t) + \frac{1}{2} \frac{\partial S_1}{\partial \vartheta} \cdot \frac{\partial^2 H_0}{\partial J \partial J} \cdot \frac{\partial S_1}{\partial \vartheta} + \frac{\partial H_1}{\partial J} \cdot \frac{\partial S_1}{\partial \vartheta} . \tag{2.12}$$

In (2.10), (2.11) and (2.12) the notations $a \cdot b$, $a \cdot c \cdot b$ stand for $a \cdot b = \sum_{i=1}^{N} a_i b_i$ and $a \cdot c \cdot b = \sum_{i,j=1}^{N} a_i c_{ij} b_j$, respectively.

Using (2.1) the formal solutions of (2.10), (2.11) can be written in the form

$$S_1(\vartheta, J, t) = (\mathcal{H}_1(J) - H_{00}(J)) t - \sum_{m,n}{}' H_{mn}(J) \frac{\exp(im\vartheta - in\Omega t)}{i(m \cdot \omega(J) - n\Omega)} , \tag{2.13}$$

$$S_j(\vartheta, J, t) = \left(\mathcal{H}_j(J) - F_{00}^{(j)}(J) \right) t - \sum_{m,n}{}' F_{mn}^{(j)}(J) \frac{\exp(im \cdot \vartheta - in\Omega t)}{i(m \cdot \omega(J) - n\Omega)} , \tag{2.14}$$

where $H_{mn}(J)$ and $F_{mn}^{(j)}(J)$ are the Fourier expansion coefficients of the perturbation Hamiltonian $H_1(\vartheta, J, t)$ and the functions $F_j(I, \vartheta, t)$:

$$H_1(\vartheta, J, t) = \sum_{m,n} H_{mn}(I) \exp(im \cdot \vartheta - n\Omega t) , \tag{2.15}$$

$$F_j(\vartheta, J, t) = \sum_{m,n} F_{mn}^{(j)}(I) \exp(im \cdot \vartheta - n\Omega t) . \tag{2.16}$$

Here the following notations are used: $m = (m_1, \cdots, m_N)$, $m \cdot \vartheta = m_1 \vartheta_1 + \cdots + m_N \vartheta_N$. In (2.13), (2.14) the sum $\sum'$ means that the term $(m = 0, n = 0)$ is excluded from summation.

The series (2.14) and (2.16) contain so-called *secular terms* growing linear in time t. These terms on the right hand side of series t can be simply eliminated by choosing the correction terms $\mathcal{H}_j(J)$ in the Hamiltonian (2.8) to be equal to H_{00}, $F_{00}^{(j)}$:

$$\mathcal{H}_1(J) = H_{00} , \qquad \mathcal{H}_j(J) = F_{00}^{(j)} , \qquad j \geq 2 .$$

Suppose that we have changed the variables (I, ϑ) to new ones (J, ψ) using the generating function $F = J \cdot \vartheta + \epsilon S_1 + \cdots + \epsilon^m S_m$ with a finite number of the terms up to order of ϵ^m. Then the new Hamiltonian $\mathcal{H}(J, \psi)$ contains the terms depending on the phase ψ only in order of ϵ^{m+1} and higher, i.e., $\mathcal{H}(J, \psi) = \mathcal{H}_0(J, \epsilon) + \epsilon^{m+1} \mathcal{H}_{m+1}(J, \psi) + \cdots$. Then new variables (J, ψ) with the accuracy $O(\epsilon^{m+1})$ obey (2.5) with the simple solution

$$J_i = \text{const} , \qquad \psi_i = w_i(J, \epsilon)t + \psi_{i0} , \qquad (i = 1, \ldots, N) .$$

The original variables (I, ϑ) can be found by the canonical transformations (2.4).

However, the described procedure may fail when the series (2.7), (2.14) and (2.16) diverge. Indeed, near a set of *resonant tori* J_{mn} where the denominator $m \cdot \omega(J_{mn}) - n\Omega$ vanish, i.e.,

$$m \cdot \omega(J_{mn}) - n\Omega = 0 , \tag{2.17}$$

or negligible small the function functions S_i are not defined. As we mentioned above this problem constitutes the main difficulty of perturbation series, known as the problem of *small denominators* (or *small divisors*).

2.1.4 von Zeipel's Method

The Lindstedt's method does not allow to consider the system's behavior near resonant tori because of the divergence of the perturbation series. Von Zeipel developed a version of perturbation theory which allows to treat these cases. It is based on the same idea of eliminating of phases from Hamiltonian except slowly changing phases near the resonant tori. Below we demonstrate the method on the simple example of Hamiltonian system with one degree of freedom subjected to time-periodic perturbation.

Suppose that I_{mn} is a value of the resonant action satisfying the condition for given numbers m, n:

$$m\omega(I_{mn}) - n\Omega = 0 . \tag{2.18}$$

Consider the system near the resonant value I_{mn}. We represent the perturbed Hamiltonian in the form

$$H = H_0(I) + \epsilon H_{mn}(I) \cos(m\vartheta - n\Omega t) + \epsilon H_1'(I, \vartheta, t) , \tag{2.19}$$

where the perturbed part of Hamiltonian $H_1'(I, \vartheta, t)$ does not contain the (m, n)- resonant term. We introduce a new slowly varying variable ψ, J via the generating function $F = (m\vartheta - n\Omega t)J + \epsilon S(\vartheta, J, t, \epsilon)$. The relation between the old variables (ϑ, I) and the new ones (ψ, J) is given by

$$I = mJ + \epsilon \frac{\partial S}{\partial \vartheta} , \qquad \psi = m\vartheta - \Omega t + \epsilon \frac{\partial S}{\partial J} . \qquad (2.20)$$

The new Hamiltonian $\mathcal{H}$ is sought in the form :

$$\mathcal{H} = \mathcal{H}(J, \epsilon) + \epsilon \mathcal{H}_{mn}(J, \epsilon) \cos \psi , \qquad (2.21)$$

which retains only one term depending on the angular variables ψ. Seeking the generating function S as series of powers of ϵ (2.7) and the Hamiltonians $\mathcal{H}$ and $\mathcal{H}_{mn}$ as in the form (2.8), one can obtain a set of equations for the expansion coefficients S_i similar to (2.11). Particularly, for the generating function S_1 one obtains the equation (2.10) where the term $H_1(\vartheta, J, t)$ on the right hand side is replaced by $H'_1(\vartheta, J, t)$. For the lowest order expansion coefficients in the expansion of $\mathcal{H}$ and $\mathcal{H}_{mn}$ we obtain

$$\mathcal{H}_0(J) = H_0(mJ) - n\Omega J , \qquad \mathcal{H}_{mn}(J) = H_{mn}(mJ) . \qquad (2.22)$$

For the first order generating function S_1 one obtains

$$S_1(\vartheta, J, t) = -\mathrm{Re} \sum_{m',n'} H_{m'n'}(J) \frac{\exp(im'\vartheta - in'\Omega t)}{i(m'\omega(J) - n'\Omega)} , \qquad (2.23)$$

which does not contain the small denominator $m\omega(J) - n\Omega$ which is responsible for the divergence of the generating functions S_i ($i = 1, 2, \ldots$) near the resonant torus $J = I_{mn}$ (2.18) in the Lindstedt's method. The number of small denominators in (2.23) is reduced by one.

Suppose that we have transformed the Hamiltonian into the form (2.21) by the canonical change of variables (2.20) using the generating function S_1 (2.23) in the first order of ϵ. Then with the accuracy of ϵ^2 the new system is described by the Hamiltonian (2.21). The latter does not depend on time variable t, and it is completely integrable.

2.2 Lie Transform Methods

Lie transformation methods in Hamiltonian perturbation theory have been developed by Hori (1966); Garrido (1968); Deprit (1969); Dragt and Finn (1976); Dewar (1976) (see also book and reviews by Nayfeh (1973); Cary (1981); Lichtenberg and Lieberman (1992); Dragt (2000)). Similar to von Zeipel's method this method is also based on elimination of fast phases by the symplectic change of variables: $(I, \vartheta) \rightarrow (J, \psi)$. However, instead of the mixed variable generating function $F(\vartheta, J, t) = J\vartheta + \epsilon S(\vartheta, J, t)$ it uses a so-called a *Lie generating function* $W(\psi, J, t; \epsilon)$ which depends only on the new variables. The canonical transformation $\hat{T}$:

$$(J, \psi) = \hat{T}(I, \vartheta) , \qquad (2.24)$$

is considered as a shift along "time" ϵ, the perturbation parameter, with the "Hamiltonian" W:

$$\frac{dJ}{d\epsilon} = \frac{\partial W}{\partial \psi} , \qquad \frac{d\psi}{d\epsilon} = -\frac{\partial W}{\partial J} . \qquad (2.25)$$

At $\epsilon = 0$ the transformation $\hat{T}$ is reduced to the identical transformation $\hat{E}$: $J(\vartheta, J, t; \epsilon = 0) = I$, $\psi(I, \vartheta, \epsilon = 0) = \vartheta$. The inverse transformation $\hat{T}^{-1}$ is $(I, \vartheta) = \hat{T}^{-1}(J, \psi)$. The Hamiltonian $\mathcal{H}$ in the new variables (J, ψ) is a function of action variables J only: $\mathcal{H} = \mathcal{H}(J)$ (similar to Lindstedt's method) or it depends also on some slow varying angles (similar to von Zeipel's method). The new Hamiltonian $\mathcal{H}$ generated by the canonical transformation W at new positions (J, ψ) equals to the old Hamiltonian H at old positions (I, ϑ):

$$\mathcal{H}(J(I, \vartheta, \epsilon)) = H(I, \vartheta) . \qquad (2.26)$$

Similar to the situation in Lindstedt's method the perturbation theory with Lie transforms is based on power series expansion of the H, $\mathcal{H}$, $\hat{T}$, and $\hat{T}^{-1}$ (Deprit (1969)). They are expanded in series of powers ϵ similar to (2.8). Assuming for the Lie generating function W a slightly different expansion series,

$$W = W_1(\vartheta, I, t) + \epsilon W_2(\vartheta, I, t) + \epsilon^2 W_3(\vartheta, I, t) + \cdots , \qquad (2.27)$$

one obtains the equations for the functions W_i $(i = 1, 2, \ldots)$. For the two lowest orders they are

$$\frac{\partial W_1}{\partial t} + \frac{\partial H_0}{\partial I} \frac{\partial W_1}{\partial \vartheta} = \mathcal{H}_1 - H_1 , \qquad (2.28)$$

$$\frac{\partial W_2}{\partial t} + \frac{\partial H_0}{\partial I} \frac{\partial W_2}{\partial \vartheta} = 2[\mathcal{H}_2 - H_2] - \hat{L}(W_1)(\mathcal{H}_1 + H_1) , \qquad (2.29)$$

where $\hat{L}(f)$ is the Poisson bracket operator associated with a function f

$$\hat{L}(f)g = \{f, g\} = \frac{\partial f}{\partial \vartheta} \frac{\partial q}{\partial I} - \frac{\partial f}{\partial I} \frac{\partial q}{\partial \vartheta} .$$

The first two terms in expansion series of transformation operators $\hat{T}$, $\hat{T}^{-1}$ are given by

$$\hat{T}_1 = -\hat{L}(W_1) , \qquad \hat{T}_2 = -\frac{1}{2}\hat{L}(W_2) + \frac{1}{2}\hat{L}^2(W_1) , \qquad (2.30)$$

$$\hat{T}_1^{-1} = \hat{L}(W_1) , \qquad \hat{T}_2^{-1} = \frac{1}{2}\hat{L}(W_2) + \frac{1}{2}\hat{L}^2(W_1) . \qquad (2.31)$$

The solutions for the Lie generating functions $W_i(I, \vartheta)$ are similar to those (2.13), (2.14) for the generating functions S_i of mixed variables. Particularly, the first order solution $W_1(I, \vartheta)$ coincides with $S_1(J, \vartheta)$ (2.13) if the old variable I is replaced by the new one J.

According to (2.30) in first order of ϵ the change of variables $I, \vartheta \to J, \psi$ is given by the transformation:

$$J = I - \epsilon \frac{\partial W_1(\vartheta, I, t)}{\partial \vartheta}, \qquad \psi = \vartheta + \epsilon \frac{\partial W_1(\vartheta, I, t)}{\partial I} . \qquad (2.32)$$

This transformation coincides with the one (2.4) given by the generating function $S_1(J, \vartheta, t)$ of mixed variables when the variable J in S is replaced by I. The new variables (J, ψ) are explicitly expressed via the old variables (I, ϑ). This is one of the advantages of the Lie transform formalism over the mixed generating function formalism in which the transformation $(I, \vartheta) \to (J, \psi)$ is given by the implicit relation (2.4).

However, one should note that the transformation (2.32) is not symplectic, i.e., $|\partial(J, \psi)/\partial(I, \vartheta)| \neq 1$, unlike relation (2.4), although the Lie transform (2.24) as a whole is symplectic. When one truncates the series of type (2.27) taking a finite number of terms, which is unavoidable in computer simulations, the transformation (2.24) becomes non-symplectic. As we shall see later this property limits the application of Lie formalism methods to study the long term evolution of Hamiltonian systems.

2.3 Time-Dependent Perturbation Series

The problem with small denominators originated from the conditions imposed on the solutions, S, of the Hamilton–Jacobi equation (2.6). It is supposed that the solutions (2.13), (2.14), obtained above, work for infinite time intervals. This leads to the appearance of small denominators in the series (2.7), (2.14) and (2.16).

Since in many problems we are interested with a behavior of dynamical systems in finite time intervals it is natural to consider the Hamilton–Jacobi equation as the *Cauchy* or the *initial value problem*. To our knowledge such an approach has not been discussed systematically in the literature on perturbation theory[2]. The perturbation series for finite time intervals has been reiterated in Abdullaev (2002) to construct symplectic mappings. These series do not diverge near resonant values of J as do the perturbation series for infinite time intervals.

---

[2] A short discussion of time-dependent perturbation theory has been given in the textbook by Corben and Stehle (1974)

2.3.1 Cauchy Problem

We consider the Hamilton–Jacobi equation (2.6) as a Cauchy problem, i.e., we seek its solutions with initial condition at time $t = t_0$: $S(\vartheta, J, t = t_0, t_0) = S_0(\vartheta_0, J)$. We denote such a solution as $S(\vartheta, J, t, t_0)$. It is convenient to set $S(\vartheta, J, t = t_0, t_0) = 0$. Similar to Lindstedt's method the solution can be sought using the expansion in series of powers of a small parameter ϵ (2.7):

$$S(J, \vartheta, t, t_0, \epsilon) = S_1(J, \vartheta, t, t_0) + \epsilon S_2(J, \vartheta, t, t_0) + \cdots . \qquad (2.33)$$

All terms S_i in the expansion series (2.33) should satisfy the condition $S_i(\vartheta, J, t = t_0, t_0) = 0$. The equation for S_1 coincides with (2.10), and the equations for S_i $(i \geq 2)$ with (2.11), where $F_i \equiv F_i(\vartheta, J, t, t_0)$ are polynomial functions of derivatives $\partial S_j / \partial \vartheta$, $(1 \leq j < i)$ of lower order, generating functions satisfying the above initial conditions.

2.3.2 Generation Functions

The equations for S_i are first order partial differential equations and can be solved by the method of characteristic equations. The left hand sides of (2.10), (2.11) may be written as total time derivatives of $S_i \equiv S_i(\vartheta, J, t, t_0)$ taken along the unperturbed trajectory $(\vartheta(t), J(t))$ of the Hamiltonian $H_0(J)$ and satisfying the condition $\vartheta(t = t_0) = \vartheta_0$, i.e., $\vartheta(t) = \vartheta_0 + \omega(J)(t - t_0)$, $J(t) = $ const. The solution of (2.10) can be written as the integral

$$S_1(\vartheta, J, t, t_0) = \mathcal{H}_1(J)(t - t_0)$$

$$-\int_{t_0}^{t} H_1(\vartheta(t'), J, t')dt' = S_1(\vartheta, J, t) - S_1(\vartheta_0, J, t_0) , \qquad (2.34)$$

where integration is taken along the unperturbed trajectory $\vartheta(t') = \vartheta + \omega(J)(t - t')$. In (2.34) the functions $S_1(\vartheta, J, t)$ are defined by (2.13). Using the latter it can be rewritten in the form of Fourier series

$$S_1(\vartheta, J, t, t_0) = -(t - t_0) \sum_{m,n} c(x_{mn}) H_{mn}(J) e^{i(m \cdot \vartheta - n\Omega t)}$$

$$= -(t - t_0) \sum_{m,n} |H_{mn}(J)| \Big[a(x_{mn}) \sin(m \cdot \vartheta - n\Omega t + \chi_{mn})$$

$$+ b(x_{mn}) \cos(m \cdot \vartheta - n\Omega t + \chi_{mn}) \Big] , \qquad (2.35)$$

where $H_{mn}(J) = |H_{mn}(J)| \exp(i\chi_{mn})$, and $c(x) = b(x) - ia(x)$ is the complex function with real and imaginary parts:

$$a(x) = \frac{1 - \cos x}{x} , \qquad b(x) = \frac{\sin x}{x} . \qquad (2.36)$$

Here the argument x_{mn} of these functions is given by

$$x_{mn} = [m \cdot \omega(J) - n\Omega](t - t_0) \ .$$

The function $c(x)$ is localized near origin $x = 0$. Its real, $a(x)$, and imaginary, $b(x)$, parts have values $a(0) = 0, b(0) = 1$ at $x = 0$ and decay for large values $|x| \gg 1$ as shown in Fig. 2.1.

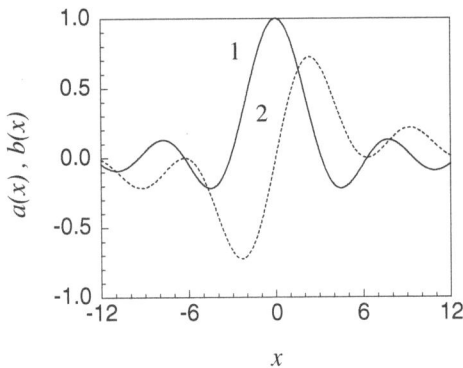

Fig. 2.1. Oscillating functions $a(x)$ (curve 2) and $b(x)$ (curve 1)

The higher order generating functions S_i $(i \geq 2)$ are determined by the integrals

$$S_i(\vartheta, J, t, t_0) = \mathcal{H}_i(J)(t - t_0) - \int_{t_0}^{t} F_i(\vartheta(t'), J, t', t_0) dt' \ , \qquad (i \geq 2) \ . \quad (2.37)$$

The determination of higher order generating functions $S_i(J, \vartheta, t)$ $(i \geq 2)$ using (2.37) is cumbersome, and requires rather complicated analytical calculations. The second order generating function $S_2(J, \vartheta, t, t_0)$ has been calculated in Abdullaev (2002) (see Eq. (A.7) in Appendix A). Here we present S_2 for the case when the perturbation Hamiltonian H_1 contains only one (m, n) term, i.e.; $H_1 = H_{mn}(I) \cos(m \cdot \vartheta - n\Omega t)$,

$$S_2(\vartheta, J, t, t_0) = -\frac{(t - t_0)^3}{4} \frac{\partial^2 H_0}{\partial J^2} m^2 H_{mn}^2(J)$$

$$\times \left\{ B_0(x_{mn}) + A_2(x_{mn}) \sin 2(m \cdot \vartheta - n\Omega t) + B_2(x_{mn}) \cos 2(m \cdot \vartheta - n\Omega t) \right\}$$

$$-\frac{(t - t_0)^2}{2} \frac{\partial H_{mn}(J)}{\partial J} m H_{mn}(J)$$

$$\times \left\{ D_0(x_{mn}) + C_2(x_{mn}) \sin 2(m \cdot \vartheta - n\Omega t) + D_2(x_{mn}) \cos 2(m \cdot \vartheta - n\Omega t) \right\} ,$$
$$(2.38)$$

where

$$B_0(x) = \frac{2}{x^2} \left(1 - \frac{\sin x}{x} \right) , \qquad D_0(x) = \frac{1}{x} \left(1 - \frac{\sin x}{x} \right) , \qquad (2.39)$$

$$A_2(x) = \frac{1}{x^2} \left[\sin 2x + \frac{1 - 4\cos x + 3\cos 2x}{2x} \right] \qquad (2.40)$$

$$B_2(x) = \frac{1}{x^2} \left[\cos 2x + \frac{4\sin x - 3\sin 2x}{2x} \right] , \qquad (2.41)$$

$$C_2(x) = \frac{1 - \cos x}{x^2} \cos x , \qquad D_2(x) = -\frac{1 - \cos x}{x^2} \sin x , \qquad (2.42)$$

are oscillating functions of x. Similar to the functions $a(x)$, $b(x)$ (2.36) in the first order generating function S_1 the functions $B_0(x)$, $D_0(x)$, $A_2(x)$, $B_2(x)$, $C_2(x)$ and $D_2(x)$ are localized near $x = 0$ and decay for large $|x| \gg 1$. They are obtained from the localized functions $U(x,y)$, $V(x,y)$, $Y(x,y)$ and $W(x,y)$ of two variables (x,y) introduced in Abdullaev (2002) (see also Appendix A) in the limits $x \to y$ and $x \to -y$, namely $A_0(x) = V(x,-x)$, $D_0(x) = W(x,-x)$ $A_2(x) = U(x,x)$, $B_2(x) = V(x,x)$, $C_2(x) = Y(x,x)$ and $D_2(x) = W(x,x)$ and are plotted in Fig. 2.2. These functions have finite values at $x = 0$, i.e., near the resonant action J_{mn} where according (2.17) to the denominator, $m\omega(J) - n\Omega$, vanishes.

2.3.3 Remarks

The main feature of the finite time perturbation series (2.33), (2.35), (2.38) is absence of divergency due to small denominators unlike the behavior in Lindstedt's perturbation series (2.13), (2.14). They are defined at all values of J including resonant values J_{mn} where the denominators $m \cdot \omega(J) - n\Omega$ vanish. This is because of cancellation of singularities in the oscillating functions $a(x_{mn})$, $b(x_{mn})$, $B_0(x_{mn})$, $A_2(x_{mn})$, $B_2(x_{mn})$, $D_0(x_{mn})$, $C_2(x_{mn})$ and $D_2(x_{mn})$ $(x_{mn} = [m \cdot \omega(J) - n\Omega](t - t_0))$ with the finite values at $(x_{mn} = 0)$.

We analyze the perturbation series (2.33) in the resonant case $m \cdot \omega(J) - n\Omega = 0$. As seen from (2.35) and (2.38), in this case the first and the second order generating functions, ϵS_1 and $\epsilon^2 S_2$, are of order of $\mu = \epsilon(t - t_0)$ and $\mu = c_1 \epsilon^2 (t - t_0)^3 + c_2 \epsilon^2 (t - t_0)^2$, respectively, where $c_1, c_2 \sim 1$. Similarly, one can expect that the higher order terms $\epsilon^k S_k$ $(k \geq 3)$ are proportional to $\epsilon^k (t - t_0)^{k\nu} c_k(J, x_{mn})$ $(\nu \geq 1)$ with finite value coefficients $c_k(J, x_{mn}) \sim 1$ at the

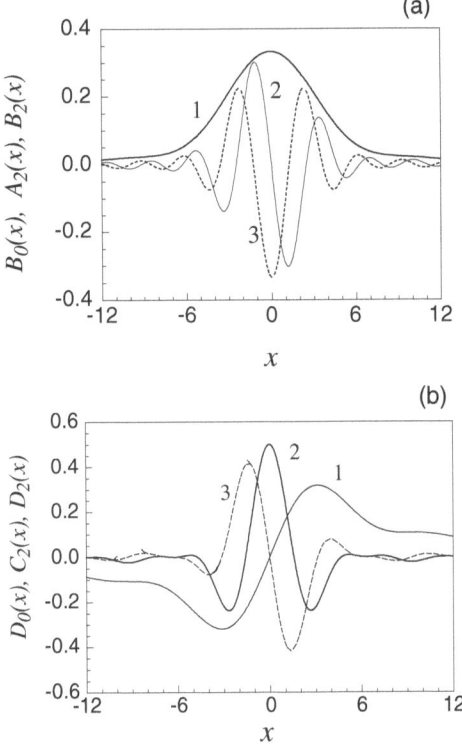

Fig. 2.2. (a) Oscillating functions $B_0(x)$ (curve 1), $A_2(x)$ (curve 2) and $B_2(x)$ (curve 3); (b) Functions $D_0(x)$ (curve 1), $C_2(x)$ (curve 2) and $D_2(x)$ (curve 3)

resonant frequencies, i.e., at $x_{mn} = 0$. It means that near the resonant values of J the *actual expansion parameter* in (2.33) the parameter, $\mu = \epsilon(t - t_0)^\nu$, plays the role of small expansion parameter, rather than the perturbation parameter ϵ. With increase in time $T = t - t_0$ this parameter μ grows. One can expect that for values of $\mu < \mu_c$, where μ_c is a certain critical value of μ, the expansion series (2.33) converges. It may diverge for $\mu > \mu_c$. The investigation of convergent properties of finite time perturbation series is difficult problem, and far beyond the scope of the book.

One should also notice that near the resonant value J_{mn} the main contribution to the perturbation series (2.33) at the given value of J comes only from terms (m, n) for which $|(m \cdot \omega(J) - n\Omega)(t - t_0)| \leq \pi$. The contributions from terms m, n with $|(m \cdot \omega(J) - n\Omega)(t - t_0)| \geq \pi$ are usually small because of the oscillatory behavior of the functions $a(x), b(x)$.

2.3.4 Time-Dependent Lie Transform Method

One can as well develop the time-dependent version of the Lie transformation method described in Sect. 2.2. For this one must impose initial conditions for all quantities of this theory. Let $\hat{T}(t, t_0, \epsilon)$ and $W(\vartheta, I, t, t_0, \epsilon)$ be the time-dependent transformation operator $\hat{T}$ and the Lie generating function W, respectively. They must satisfy the following initial conditions: $\hat{T}(t = t_0, t_0, \epsilon) = \hat{E}$, $W(\vartheta, I, t = t_0, t_0, \epsilon) = 0$. All coefficients W_i $(i = 1, 2, \cdots)$ of the expansion series in the powers of ϵ (2.27) obey the system of equations (2.28), (2.29) with initial conditions $W_i(\vartheta, I, t = t_0, t_0) = 0$. The solution for the first order Lie generating function $W_1(\vartheta, I, t, t_0)$ coincides with the first order generating function, $S_1(\vartheta, J, t, t_0)$, of mixed variables (2.35) if J is replaced by I. The higher order Lie generating functions W_i $(i \geq 2)$ can be found by integrating the right hand side of corresponding equations along the unperturbed orbits $(\vartheta(t) = \vartheta_0 + \omega(I)(t - t_0), J = \text{const.})$ similar to the integration of higher order generating functions S_i of mixed variables (2.37).

2.3.5 Example Duffing Oscillator

We illustrate the application of perturbation methods in finite time intervals described above with the example of the Duffing equation

$$\ddot{x} + \omega_0^2 x = \epsilon(\cos \Omega t + x^3) . \tag{2.43}$$

Introducing the canonical variables $(x, p \equiv \dot{x})$, the equation (2.43) is reduced to the Hamiltonian system (1.1) with Hamiltonian

$$H(x, p, t) = \frac{p^2}{2} + \frac{\omega_0^2}{2} x^2 - \epsilon \left(x \cos \Omega t + \frac{x^4}{4} \right) . \tag{2.44}$$

In action-angle variables (I, ϑ),

$$x = \sqrt{2I/\omega_0} \sin \vartheta , \qquad p = \sqrt{2I\omega_0} \cos \vartheta , \tag{2.45}$$

the Hamiltonian function (2.44) can be written in the form

$$H(I, \vartheta, t) = H_0(I, \epsilon) + \epsilon H_1(I, \vartheta, t) , \tag{2.46}$$

where

$$H_0(I, \epsilon) = \omega_0 I , \tag{2.47}$$

is unperturbed Hamiltonian, and

$$H_1(I, \vartheta, t) = -\frac{3I^2}{8\omega_0^2} - \frac{1}{2}\sqrt{\frac{I}{\omega_0}}[\sin(\vartheta - \Omega t) + \sin(\vartheta + \Omega t)]$$

$$+ \frac{I^2}{2\omega_0^2}\left[\cos 2\vartheta - \frac{1}{4}\cos 4\vartheta\right], \qquad (2.48)$$

is the perturbation Hamiltonian.

Using (2.48), the first order generating function, S_1, determined by (2.35) takes the form

$$S_1(J, \vartheta, t, t_0) = -\int_{t_0}^{t} H_1(J, \vartheta(t'), t')dt'$$

$$= (t - t_0)\sqrt{\frac{J}{4\omega_0}}\sum_{n=\pm 1}[a(x_n)\cos(\vartheta - n\Omega t) + b(x_n)\sin(\vartheta - n\Omega t)]$$

$$-(t - t_0)\frac{J^2}{2\omega_0^2}\sum_{m=2,4}\left(-\frac{2}{m}\right)^{m/2}[a(x_m)\sin m\vartheta + b(x_m)\cos m\vartheta], \quad (2.49)$$

where $x_n = (\omega_0 - n\Omega)(t - t_0)$, $x_m = m\omega_0$, and the functions $a(x)$, $b(x)$ are defined by (2.36). The first order correction $\mathcal{H}_1$ to the new Hamiltonian is chosen equal to $\mathcal{H}_1(J) = -3J^2/8\omega_0^2$ in order to avoid secular terms. The new Hamiltonian is thus

$$\mathcal{H}(J, \epsilon) = \omega_0 J - \epsilon\frac{3J^2}{8\omega_0^2}, \qquad (2.50)$$

and the perturbed frequency is

$$w(J, \epsilon) = \frac{\partial\mathcal{H}(J, \epsilon)}{\partial J} = \omega_0\left(1 - \epsilon\frac{3J}{4\omega_0^3}\right). \qquad (2.51)$$

Up to the first order of ϵ the time evolution of the new variables (ψ, J) is given by

$$J = \text{const},$$

$$\psi(t) = \psi_0 + w(J, \epsilon)(t - t_0) = \psi_0 + \omega_0\left(1 - \epsilon\frac{3J}{4\omega_0^3}\right)(t - t_0). \qquad (2.52)$$

The old variables (ϑ, I) are then determined through the generating function (2.49):

$$I(t) = J + \epsilon\frac{\partial S_1(\vartheta(t), J, t, t_0)}{\partial\vartheta}, \qquad \vartheta(t) = \psi(t) - \epsilon\frac{\partial S_1(\vartheta(t), J, t, t_0)}{\partial J} \quad (2.53)$$

At the same time (2.53), (2.52) describe the solutions of the Duffing equation (2.43) in action - angle variables (ϑ, I) and the "real world variables" (x, p) are given in terms of (ϑ, I) according to (2.45). One should note that the angle variable ϑ is determined implicitly by (2.53).

A similar perturbative solution can be obtained using the Lie generating function $W_1(J, \psi, t, t_0)$ instead of the mixed variable generating function $S_1(\vartheta, J, t, t_0)$:

$$I(t) = J + \epsilon \frac{\partial W_1(\psi, J, t, t_0)}{\partial \psi} \,, \qquad \vartheta(t) = \psi(t) - \epsilon \frac{\partial W_1(\psi, J, t, t_0)}{\partial J} \,, \quad (2.54)$$

where $W_1(\psi, J, t, t_0)$ is determined by the expression similar to (2.49) where ϑ is replaced by ψ. This equation explicitly determines the old variables (ϑ, J).

We have also integrated the Duffing equation (2.43) using the symplectic integration scheme of Mc Lachlan & Atela McLachlan and Atela (1992) (see Sect. 1.5), and compared it with the perturbative solutions (2.53), (2.52), and (2.54) for the resonant $(\omega_0 = \Omega)$ and non-resonant $(\omega_0 \neq \Omega)$ cases. (The parameters were taken as $\omega_0 = 1$, $\epsilon = 0.01$, $\Omega = 1$ for the first case and $\Omega = 1.5$ for the second case. For the resonant case $(\omega_0 = \Omega = 1)$ the solutions obtained using three different methods are plotted in Fig. 2.3: curve 1 corresponds the solution (2.53), (2.52) through the generating function $S_1(\vartheta, J, t)$ of mixed variables (2.49), curve 2 describes the same solution but through the Lie generating function $W_1(J, \psi, t)$, and curve 3 is obtained by direct integration of the Duffing equation using the symplectic integration scheme with the integration step in time $\Delta t = 2\pi/300$.

In the case of non-resonant perturbation $(\omega_0 \neq \Omega)$ both perturbation solutions are close to the numerical solutions with the relative deviation $\delta I \equiv |I - I_n|/I < 10^{-3}$ up to time intervals $T = t - t_0 \sim 10^3/\Omega$. The deviation ΔI increases linearly with T. However, the value of δI corresponding to the Lie generating function W_1 reaches the order of 1 at the time $T \sim 5 \times 10^4/\Omega$, while the magnitude of δI corresponding to the mixed variable generating function S_1 reaches the order of 1 $\delta I \sim 1$ at the later time $T \sim 2 \times 10^5/\Omega$.

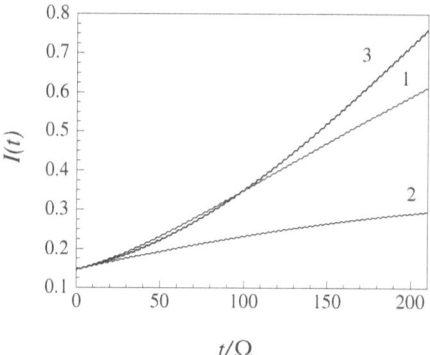

Fig. 2.3. Evolution of the action variable $I(t)$ in the resonant case $\omega_0 = \Omega$: curve 1 corresponds to transformation (2.53) with the mixed variable generating function S_1, curve 2 – to (2.54) with the Lie generating function W_1, and curve 3 – with the symplectic integration. Parameters are $\epsilon = 0.01$, $\omega_0 = \Omega = 1$

As seen from Fig. 2.3 in the *resonant case* ($\omega_0 = \Omega$) the solutions obtained by perturbation methods deviate from the one of obtained by the numerical symplectic integration at much shorter times since the perturbation series (2.7) diverges with increasing $\epsilon(t - t_0)$. However, the perturbation method with the mixed variable generating function S_1 closely reproduces the real solutions for longer time intervals than the one with the Lie generating function W_1. As was noted above this is because the transformation (2.32) does not conserve the important symplectic property of Hamiltonian systems.

Perturbation series in finite time intervals obtained in this section will be used in Chap. 4 to construct symplectic mappings for Hamiltonian systems.

2.4 Method of Successive Transformations

In this section we shortly discuss the method of successive canonical transformations in the perturbation theory. This method called *Kolmogorov's technique* has been used to prove the theorem concerning the stability of Hamiltonian systems to small perturbations known as Kolmogorov-Arnold-Moser (KAM) theory (Kolmogorov (1954); Arnold (1963a); Moser (1962), see also Arnold et al. (1988); Lichtenberg and Lieberman (1992) and Sect. 7.1). It is based on successive canonical changes of variables in a perturbation Hamiltonian system and successively eliminates terms in the Hamiltonian of higher orders in the perturbation parameter ϵ. This method is an extremely rapidly convergent with increasing numbers of transformations of variables. Below we describe the *time-dependent version of this superconvergent method*. We shall use this method in Sect. 4.6 to obtain a superconvergent version of symplectic mappings.

Consider the perturbation Hamiltonian system with Hamiltonian (2.3). Perform a change of variables $(\vartheta, I) \rightarrow (\psi_1, J_1)$ in the time interval $[t_0, t]$ with generating function $F_1(\vartheta, J_1, t, t_0, \epsilon) = \vartheta J_1 + \epsilon S_1(\vartheta, J_1, t, t_0)$, i.e.,

$$I = J_1 + \epsilon \frac{\partial S_1}{\partial \vartheta} , \qquad \psi_1 = \vartheta + \epsilon \frac{\partial S_1}{\partial J_1} , \qquad (2.55)$$

which transforms the Hamiltonian H to $\mathcal{H}^{(1)}$:

$$\mathcal{H}^{(1)}(\psi_1, J_1, t, t_0, \epsilon) = \mathcal{H}_0^{(1)}(J_1, \epsilon) + \epsilon^2 \mathcal{H}_1^{(1)}(\psi_1, J_1, t, t_0.\epsilon) . \qquad (2.56)$$

Here

$$\mathcal{H}_0^{(1)}(J_1, \epsilon) = H_0(J) + \epsilon \mathcal{H}_1(J) ,$$

$$\mathcal{H}_1^{(1)}(\psi_1, J_1, t, t_0, \epsilon) = \frac{1}{2} \frac{\partial S_1}{\partial \vartheta} \cdot \frac{\partial^2 H_0}{\partial J \partial J} \cdot \frac{\partial S_1}{\partial \vartheta} + \frac{\partial H_1}{\partial J} \cdot \frac{\partial S_1}{\partial \vartheta} + O(\epsilon) . \quad (2.57)$$

The generating function S_1 is determined by (2.34), (2.35). The new Hamiltonian (2.56) has a similar structure as (2.3) but it contains the phase

ψ_1 in the perturbed term $\epsilon^2 \mathcal{H}_1^{(1)}$ with the perturbation parameter ϵ^2. Applying a second change of variables $(\psi_1, J_1) \to (\psi_2, J_2)$ to the new system with Hamiltonian (2.56) via the generating function

$$F_1(\psi_1, J_2, t, t_0, \epsilon^2) = \psi_1 J_2 + \epsilon^2 S_2(\psi_1, J_2, t, t_0),$$

one transforms $\mathcal{H}^{(1)}$ into $\mathcal{H}^{(2)} = \mathcal{H}_0^{(2)} + \epsilon^4 \mathcal{H}_1^{(2)}$ pushing the phase-dependent perturbation terms to 4-th order in ϵ. After the m-th change of variables: $(\psi_{m-1}, J_{m-1}) \to (\psi_m, J_m)$ the Hamiltonian becomes $\mathcal{H}^{(m)} = \mathcal{H}_0^{(m)} + \epsilon^{2^m} \mathcal{H}_1^{(m)}$ with a phase-dependent term of order ϵ^{2^m}.

As was shown by Kolmogorov (1954) and Arnold (1963a) this procedure converges quadratically, similar to Newton's root-finding method (see, e.g., Lichtenberg and Lieberman (1992)). Such a quadratic convergence of perturbation series is known as *superconvergence*.

One should note that the Lindstedt's perturbation series converges linearly. After the transformation of variables with the generating function, $F = \vartheta J + \epsilon S_1 + \cdots + \epsilon^m S_m$, with the terms up to $m-$ order in ϵ the phase-dependent terms in Hamiltonian are retained in ϵ^{m+1}-th order, while in the superconvergent procedure after the m-th successive transformation they are only in ϵ^{2^m}-th order.

2.5 Bibliographic Notes

Methods of perturbation theory in celestial mechanics developed in 19-th century are thoroughly described in the treatise by Poincaré (1892–99). Modern formulations of these methods can be found in Arnold et al. (1988); Arnold (1989). Methods of perturbation theory based on the averaging principle in general dynamical systems and their applications to nonlinear oscillatory systems are discussed in Bogolyubov and Mitropol'skij (1958); Andronov et al. (1981); Nayfeh (1973); Wiggins (1990). Perturbation theory for Hamiltonian systems based on the Lie transform formalism is described in a review paper by Cary (1981) and in the books by Nayfeh (1973); Lichtenberg and Lieberman (1992); Dragt (2000).

3 Mappings for Perturbed Systems

Poincaré (1892–99) introduced a powerful tool to study dynamical systems by replacing continuous systems by discrete mappings. As was mentioned in the preface mappings significantly simplify the study of systems, by reducing the dimension of the system by one, visualizing the orbits on certain sections of phase space, and thus simplifying the formulation of many concepts of dynamical systems. In this chapter we review the traditional methods to construct symplectic maps for generic continuous Hamiltonian systems. We consider the generic Hamiltonian system (2.2) with small time-periodic perturbations. We shall discuss the main difficulties to construct mappings from Hamiltonian equations, and the shortcomings.

3.1 Poincaré Mappings

In this section we give a definition of mappings. A Poincaré *surface-to-surface map* is defined as follows. In general, trajectories lie in a $(2N + 1)$ dimensional subspace of the extended $(2N+2)$ dimensional phase space $(q_1, \ldots, q_N, p_1, \ldots, p_N, t, H)$. Consider a $2N$ dimensional cross-section Σ which transversally crosses the trajectories as shown in Fig. 3.1. Let P_k be an intersection point of the section by a trajectory. The map that associates the point P_k with the next crossing point P_{k+1} is a called *Poincaré map*:

$$P_{k+1} = \hat{M} P_k \ . \tag{3.1}$$

It is a $2N$ dimensional map. This definition of Poincaré maps includes also a *stroboscopic map*, as particular case. Suppose the Hamiltonian of a system is a periodic function of time with period $T = 2\pi/\Omega$. Let Σ_k be sections in the $(2N + 2)$ dimensional phase space at times $t = t_k = kT$, $(k = 1, 2, \ldots)$, with P_k the phase space coordinates of orbits in the section Σ_k. Then (3.1) defines a stroboscopic map relating the coordinates (I, ϑ) at the time instant $t_k = kT$ with the ones at $t_{k+1} = (k + 1)T$:

$$(I_{k+1}, \vartheta_{k+1}) = \hat{M}(I_k, \theta_k) \ , \tag{3.2}$$

S.S. Abdullaev: *Construction of Mappings for Hamiltonian Systems and Their Applications*, Lect. Notes Phys. **691**, 39–51 (2006)
www.springerlink.com
© Springer-Verlag Berlin Heidelberg 2006

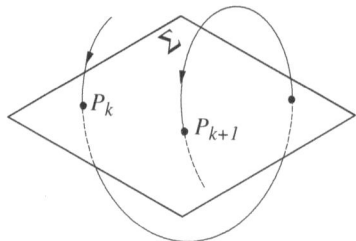

Fig. 3.1. Sketch of Poincare section and Poincaré map

where $I_k = I(t_k)$, $\vartheta_k = \vartheta(t_k)$ are $I(t) = (I_1(t), \ldots, I_N(t)$, $\vartheta(t) = (\vartheta_1(t), \ldots,$
$\vartheta_N(t))$ at the time instants $t = t_k$ $(k = 0, \pm 1, \pm 2, \cdots)$[1].

The methods of construction of Poincaré or stroboscopic maps have
been subject of many studies. For general dynamical systems including non-
Hamiltonian systems such methods were discussed in Guckenheimer and
Holmes (1983); Wiggins (1990). Special methods have been also developed
for the construction of Poincaré maps for Duffing-type oscillators with fric-
tion by Eilenberger and Schmidt (1992); Schmidt and Eilenberger (1998).
For Hamiltonian systems Poincaré maps should be presented as symplectic
maps, i.e., should satisfy the volume-preserving condition (1.7):

$$\left| \frac{\partial(I_{k+1}, \vartheta_{k+1})}{\partial(I_k, \vartheta_k)} \right| = \det \begin{pmatrix} \dfrac{\partial I_{k+1}}{\partial I_k} & \dfrac{\partial I_{k+1}}{\partial \vartheta_k} \\ \dfrac{\partial \vartheta_{k+1}}{\partial I_k} & \dfrac{\partial \vartheta_{k+1}}{\partial \vartheta_k} \end{pmatrix} = 1 . \tag{3.3}$$

There exist two main groups of methods to construct generic Hamiltonian
maps. The first which can be called as intuitive method, is based on the a
priori assumption of a special symplectic form of map, and the unknown perturbation functions associated with these maps are found from the equations
of motion. The second widely used method which can be called *method of
delta functions* is based on the assumption that a time-periodic perturbation
acting on the integrable system can be replaced by periodic delta functions.
Then the integration of the equations of motion along the delta functions
gives a mapping.

Along with these generic symplectic maps there exist also many situations for which the maps can be constructed directly (exact or approximate)
using the physical contents of the system. The map describing Fermi acceleration (Lichtenberg and Lieberman (1992); Zaslavsky (1985)) or ray maps in
a waveguide channel (Abdullaev (1993, 1994a)) are examples of such maps
(see Sects. 3.5, 13.1). In most cases these maps replace exactly a continuous
system by discrete system.

---

[1] Furthermore a subscript k will be used for the iteration steps of map. It should
not be confused with a lower index number of components of I or ϑ.

Below we present the main methods to construct generic symplectic mappings and discuss their difficulties and shortcomings.

3.2 Method of a priori Assumption

Below we present a method of constructing symplectic maps for generic Hamiltonian systems of type (2.2), (2.3). The idea of this method has been described in Lichtenberg and Lieberman (1992) for systems with two degrees of freedom. This is type of mappings has been used in various problems of plasma physics, dynamical astronomy, accelerator physics and others (see, e.g., Wobig and Fowler (1988); Wobig and Pfirsch (2001), Hadjidemetriou (1991, 1993, 1999); Šidlichovský (1997) Sándor et al. (2002) and references therein).

We demonstrate the method by constructing a stroboscopic map $\hat{M}$ (3.2) for the Hamiltonian system (2.2), (2.3) satisfying the condition (3.3). In the absence of perturbation ($\epsilon = 0$) the unperturbed Hamiltonian $H_0(I_1, \ldots, I_N)$ describes nonlinear oscillations of the system with frequencies $\omega_i(I_1, \ldots, I_N)$ (1.23). The map for this case has the form

$$I_{k+1} = I_k , \qquad \vartheta_{k+1} = \vartheta_k + \omega(I_{k+1})T . \qquad (3.4)$$

The phase plane of such a map is shown in Fig. 3.2.

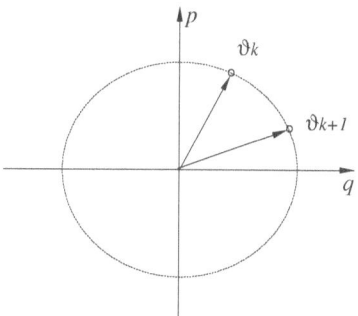

Fig. 3.2. The twist map on the phase plane

Now one assumes that in the presence of a small perturbation the action, I, and the angle, ϑ, variables acquire additional corrections proportional to ϵ thus the map takes the following form (3.4) (see, e.g., Lichtenberg and Lieberman (1992); Wobig and Fowler (1988); Wobig and Pfirsch (2001)):

$$I_{k+1} = I_k + \epsilon f(\vartheta_k, I_{k+1}) ,$$
$$\vartheta_{k+1} = \vartheta_k + \omega(I_{k+1})T + \epsilon g(\vartheta_k, I_{k+1}) , \qquad (3.5)$$

where $f(\vartheta_k, I_{k+1})$, $g(\vartheta_k, I_{k+1})$ are perturbation functions. In order to make the map symplectic according to (3.3) the functions f and g must satisfy the condition $\partial f/\partial I_{k+1} + \partial g/\partial \vartheta_k = 0$, i.e., they should be determined by a single generating function $S(\vartheta_k, I_{k+1})$:

$$f(\vartheta_k, I_{k+1}) = \frac{\partial S(\vartheta_k, I_{k+1})}{\partial \vartheta_k} , \qquad g(\vartheta_k, I_{k+1}) = -\frac{\partial S(\vartheta_k, I_{k+1})}{\partial I_{k+1}} . \quad (3.6)$$

The mapping (3.5) can be considered as a canonical transformation from the "old" variables (ϑ_k, I_k) to the "new" variables $(\vartheta_{k+1}, I_{k+1})$ given by the generating function

$$F(\vartheta_k, I_{k+1}) = \vartheta_k \cdot I_{k+1} + T H_0(I_{k+1}) + \epsilon S(\vartheta_k, I_{k+1}) , \quad (3.7)$$

and can be represented as the relation:

$$\vartheta_{k+1} = \frac{\partial F(\vartheta_k, I_{k+1})}{\partial I_{k+1}} , \qquad I_k = \frac{\partial F(\vartheta_k, I_{k+1})}{\partial \vartheta_k} . \quad (3.8)$$

The generating function S can be found from the perturbed equations of motion. For small perturbation ($\epsilon \ll 1$) one can insert the unperturbed trajectory ($I_0(t) = I_{k+1}$ const, $\vartheta_0(t) = \vartheta_k + \omega(I)(t - t_k)$) on the right hand side of the second equation (2.2) for the action I. Then integration gives the following equation for the generating function S:

$$\frac{\partial S(\vartheta, I_{k+1})}{\partial \vartheta_k} = \int\limits_{t_k}^{t_k+T} \frac{\partial H_1}{\partial \vartheta}\Big(\vartheta_k + \omega(I_{k+1})(t - t_k), I_{k+1}, t\Big) dt . \quad (3.9)$$

In (3.9) the integral is taken along unperturbed trajectories $I_0(t), \vartheta_0(t)$ over one period of perturbation. Integration of (3.9) gives

$$S(\vartheta_k, I_{k+1}) = \int d\vartheta_k \left\{ \int\limits_{t_k}^{t_k+T} \frac{\partial H_1}{\partial \vartheta}\Big(\vartheta_k + \omega(I_{k+1})(t - t_k), I_{k+1}, t\Big) dt \right\} . \tag{3.10}$$

Particularly, for the perturbation Hamiltonian H_1 in the Fourier expansion form (2.15) this yields

$$\frac{\partial H_1}{\partial \vartheta} = \mathrm{Re} \sum_{m,n} H_{mn} im \exp(im \cdot \vartheta - in\Omega t) . \quad (3.11)$$

Using (3.11), (3.10), we obtain

$$S(\vartheta_k, I_{k+1}) = T \sum_{m,n} |H_{mn}| \Big\{ - a(x_{mn}) \sin(m \cdot \vartheta_k - n\Omega t_k + \chi_{mn})$$

$$+ b(x_{mn}) \cos(m \cdot \vartheta_k - n\Omega t_k + \chi_{mn}) \Big\} , \quad (3.12)$$

where the functions $a(x)$ and $b(x)$ are defined in (2.36), and $x_{mn} = [m \cdot \omega(I_{k+1}) - n\Omega]T$.

After Moser (1962, 1973) the symplectic map (3.5) is called the *perturbed twist map* if the frequency $\omega(I)$ satisfies the so-called *twist condition*, i.e., $d\omega(I)/dI \neq 0$ (see Sect. 7.1.1). Its geometry on the torus is shown in Fig. 3.3.

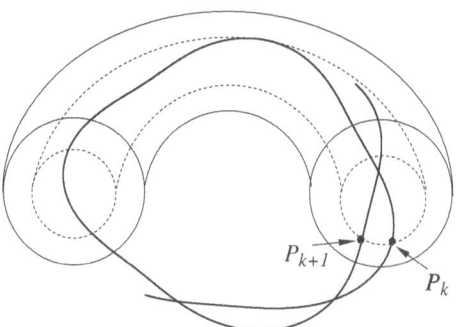

Fig. 3.3. Poincare map on the torus

Similar approach to construct symplectic mappings for autonomous Hamiltonian systems but based on the averaging procedure has been proposed by Hadjidemetriou (1991) for the applications in dynamical astronomy (see, also Sect. 13.3). The derivation of mappings similar to (3.5) for more general dynamical systems, including also Hamiltonian systems were given also in Wiggins (1990).

The described method of derivation of the symplectic maps has significant shortcomings. First of all, it restricts the possible form of the map. For instance, the following map in the symplectic form

$$I_{k+1} = I_k + \epsilon f(\vartheta_{k+1}, I_k) \,,$$
$$\vartheta_{k+1} = \vartheta_k + \omega(I_k)T + \epsilon g(\vartheta_{k+1}, I_{k+1}) \,, \tag{3.13}$$

with perturbation functions

$$f(\vartheta_{k+1}, I_k) = \frac{\partial S(\vartheta_{k+1}, I_k)}{\partial \vartheta_{k+1}} \,, \qquad g(\vartheta_{k+1}, I_k) = -\frac{\partial S(\vartheta_{k+1}, I_k)}{\partial I_k} \,,$$

equivalently describes a system as perturbed twist map in the form (3.5). The difference between the two symplectic forms (3.5), (3.6) and (3.13), (3.14) is that the action variable I_{k+1} in the map (3.5) is defined implicitly, while in (3.13) the angle variable ϑ_{k+1} is defined implicitly.

Secondly, it does not allow to obtain the higher order corrections in the perturbation parameter ϵ and therefore to estimate the accuracy of replacing the continuous system (1.1) by the map (3.5).

3.3 Method of Delta Functions

The method described below is often used to obtain symplectic maps for Hamiltonian system affected by time-periodic perturbations with broad spectrum (see, e.g., Chirikov (1979); Wisdom (1982); Zaslavsky (1985); Sagdeev et al. (1988); Zaslavsky et al. (1991); Wisdom and Holman (1991); Mendonça (1991)). The perturbation Hamiltonian of these systems can be presented in the form

$$H_1(I,\vartheta,t) = \sum_m \sum_{n=-M}^{M} H_{mn}(I)\cos(m\vartheta - n\Omega t + \chi_{mn}) , \qquad (3.14)$$

with a finite but large number of harmonics $M \gg 1$. Suppose that amplitudes $H_{mn}(I)$ and phases $\chi_{mn}(I)$ of harmonics $H_{mn}(I)$ depend only weakly on the index n. Then omitting index n from these quantities the perturbation Hamiltonian can be rewritten as

$$H_1(I,\vartheta,t) \approx \mathcal{H}_c(I,\vartheta) \sum_{n=-M}^{M} \cos(n\Omega t) + \mathcal{H}_s(I,\vartheta) \sum_{n=-M}^{M} \sin(n\Omega t) , \quad (3.15)$$

where

$$\mathcal{H}_c(I,\vartheta) = \sum_m H_m(I)\cos(m\vartheta + \chi_m) ,$$

$$\mathcal{H}_s(I,\vartheta) = \sum_m H_m(I)\sin(m\vartheta + \chi_m) .$$

Extending summation over n to $M \to \infty$ and using the Poisson summation rules

$$\sum_{n=-\infty}^{\infty} \cos 2\pi nx = \sum_{k=-\infty}^{\infty} \delta(x - k) , \qquad \sum_{n=-\infty}^{\infty} \sin 2\pi nx = 0 ,$$

one can replace the Hamiltonian (3.15) by the sum of delta functions:

$$H_1(I,\vartheta,t) = \mathcal{H}_c(I,\vartheta)\frac{2\pi}{\Omega} \sum_{k=-\infty}^{\infty} \delta\left(t - k\frac{2\pi}{\Omega}\right) . \qquad (3.16)$$

Then the equations of the perturbed motion are reduced to

$$\frac{d\vartheta}{dt} = \omega(I) + \epsilon g(I,\vartheta) \sum_{k=-\infty}^{\infty} \delta(t - k2\pi/\Omega) ,$$

$$\frac{dI}{dt} = \epsilon f(I,\vartheta) \sum_{k=-\infty}^{\infty} \delta(t - k2\pi/\Omega) , \qquad (3.17)$$

where

$$g(I, \vartheta) = \frac{\partial S(I, \vartheta)}{\partial I} , \qquad f(I, \vartheta) = -\frac{\partial S(I, \vartheta)}{\partial \vartheta} , \qquad (3.18)$$

are perturbation functions given by the generating function

$$S(I, \vartheta) = \frac{2\pi}{\Omega} \mathcal{H}_c(I, \vartheta) = \frac{2\pi}{\Omega} \sum_m H_m(I) \cos(m\vartheta + \chi_m) . \qquad (3.19)$$

The intuitive justification of the replacement of continuous perturbation functions (3.14) by series of delta functions is based on the averaging principle (Wisdom (1982); Wisdom and Holman (1991)): if high-frequency terms do not contribute significantly to the evolution, then adding these terms also does not affect the system significantly. However, one should note that such a replacement of (3.14) by (3.16) introduces artificial singularities and discontinuities to the system at periodic time instants $t_k = kT$. The orbits $I(t), \vartheta(t)$ are not defined at these times.

Integrating the equations (3.17) over one period from $t_k - 0$ to $t_k - 0 + T$ one obtains

$$\vartheta_{k+1} = \vartheta_k + \omega(I_{k+1})T + \epsilon \int_{t_k-0}^{t_k+0} g(I, \vartheta)\delta(t - kT)dt ,$$

$$I_{k+1} = I_k + \int_{t_k-0}^{t_k+0} f(I, \vartheta)\delta(t - kT)dt , \qquad (3.20)$$

where $\vartheta_k \equiv \vartheta(t_k - 0), I_k \equiv I(t_k - 0)$. In (3.20) the integrals over product of delta function with discontinuous functions are not well-defined, and it is not clear how to integrate them. In spite of these difficulties the difference equations (3.20) are often simply reduced to the form of the perturbed twist map (3.5) with the perturbation functions f and g defined in (3.18) (see, e.g., Sagdeev et al. (1988); Mendonça (1991); Abdullaev et al. (1998); da Silva et al. (2001a,b, 2002a,b)).

Difficulties related with obtaining symplectic mappings from Hamiltonian equations using the delta function formalism have been also discussed by Šidlichovský (1997); Hadjidemetriou (1998) (see also references therein). They have shown that using the formula

$$\int_{t_k-0}^{t_k+0} f(\vartheta, I)\delta(t - t_k)dt = \frac{1}{2} \left[f(\vartheta_k, I_k) + f(\vartheta_k, I_{k+1}) \right] ,$$

one obtains the map

$$I_{k+1} = I_k + \frac{\epsilon}{2} \left[f(\vartheta_k, I_k,) + f(\vartheta_k, I_{k+1}) \right] \; ,$$

$$\vartheta_{k+1} = \vartheta_k + \omega(I_{k+1})T + \frac{\epsilon}{2} \left[g(\vartheta_k, I_k) + g(\vartheta_k, I_{k+1}) \right] \; . \tag{3.21}$$

This mapping, however, is not symplectic, i.e. it does not satisfy the condition (3.3).

Eberhard (1999) has proposed the symmetric form of mapping integrating the equations of motion (3.17) from t_k to t_{k+1} using the following formula for the integration across delta function:

$$\int_{t_k}^{t_k \pm 0} f(\vartheta, I)\delta(t - t_k)dt = \pm \frac{1}{2} f(\vartheta_k, I_k) \; , \qquad (\epsilon > 0) \; .$$

where $\vartheta_k \equiv \vartheta(t_k), I_k \equiv I(t_k)$. It yielded the following symmetric form of the map

$$J_k = I_k + \frac{\epsilon}{2} f(\vartheta_k, I_k) \; ,$$

$$\Theta_k = \vartheta_k + \frac{\epsilon}{2} g(\vartheta_k, I_k) \; ,$$

$$\bar{\Theta}_k = \Theta_k + \omega(J_k)T \; ,$$

$$I_{k+1} = J_k + \frac{\epsilon}{2} f(\vartheta_{k+1}, I_{k+1}) \; ,$$

$$\vartheta_{k+1} = \bar{\Theta}_k + \frac{\epsilon}{2} g(\vartheta_{k+1}, I_{k+1}) \; . \tag{3.22}$$

The symmetric mapping (3.22) has been compared with the numerical integration of the equations (3.17) replacing the delta functions by their continuous representation $\delta_a(t) = \exp(-t^2/a^2)/a\pi^{1/2}$ $(a \to 0)$. It was found that this mapping describes the continuous system more closely than the perturbed twist map (3.5) (see Sect. 10.5 for details). However, the mapping (3.22) is not symplectic as well.

Only in the case when the function $\mathcal{H}_1$ does not depend on the action variable, i.e., $\partial \mathcal{H}_1/\partial I = g(I, \vartheta) \equiv 0$, the angle ϑ becomes continuous along time t the integration gives the map

$$\vartheta_{k+1} = \vartheta_k + \omega(I_{k+1})T \; ,$$

$$I_{k+1} = I_k + \epsilon f(\vartheta_k) \; , \tag{3.23}$$

which is known as a *radial twist map*.

One should note that in the map (3.23) the variable I_k is taken as $I_k = I(t_k - 0)$. This should be kept in mind when one compares the original continuous system with the perturbation (3.14) with the approximated one (3.16). The trajectories $\vartheta(t), I(t)$ of the original system are continuous at any time, while its replacement by (3.16) introduces discontinuities at the periodic time instants $t = t_k$ where the variables (ϑ_k, I_k) in the map (3.2) should be defined. However, for the system (3.16) they are not defined at these times. Therefore, one should not identify the variables in the map (3.23) with the ones in the original system. The presented approach to construct symplectic maps thus does not allow to find the relation between the variables of the original system and the ones in the map (3.23).

The fact that the variables in the original equations (2.2) and the ones in the mappings are not identical was noticed in Wisdom et al. (1996). In order to relate these variables, so-called symplectic correctors were introduced by means of a Lie formalism.

3.4 The Standard Map

The simplest example of the mapping (3.23), with the perturbation function $\epsilon f(\vartheta) = (K/2\pi)\sin\vartheta$, is known as the *standard map* (or the Chirikov-Taylor map) (see Chirikov (1979), Lichtenberg and Lieberman (1992)):

$$I_{k+1} = I_k + \frac{K}{2\pi}\sin\vartheta_k \,, \qquad \vartheta_{k+1} = \vartheta_k + 2\pi I_{k+1} \,. \qquad (3.24)$$

The parameter K plays role of perturbation strength. It has been extensively studied during the past two decades as one of the basic models in chaos theory. It has been widely used in many physical problems, for instance, to study particle-wave interaction, magnetic field line dynamics in magnetic confinement devices (see Sect. 10.4 and MacKay and Meiss (1987); Rechester et al. (1979); Benisti and Escande (1998); Balescu (2000a,b)). The standard map (3.24) is often assumed to be derived from the standard Hamiltonian

$$H = \frac{I^2}{2} + \frac{K}{4\pi^2}\sum_{n=-M}^{M}\cos(\vartheta - nt + \chi_n) \,, \qquad (3.25)$$

when the number of harmonics $M \to \infty$ and the phases $\chi_n = 0$, identifying the variables (ϑ_k, I_k) of the map with the variables $(\vartheta(t_k), I(t_k))$ of the continuous system (3.25) (see, for instance Chirikov (1979); Benisti and Escande (1998)). As will be shown in Sect. 4.3.2, the numerical integration of the system with Hamiltonian (3.25) with a large but finite mode number M gives a result which is significantly different from the one obtained by the standard map (3.24).

This is related with the fact that the Hamiltonian (3.25) acquires singularities at times $t = t_k$ when $M \to \infty$, and the variables I are not defined at these times. The variables (ϑ_k, I_k) in the standard map (3.24) are taken just before the instants $t = t_k$, i.e., $(\vartheta_k, I_k) = (\vartheta(t_k - 0), I(t_k - 0))$, i.e., the variables of the mapping *do not coincide* with the variables of the continuous Hamiltonian (3.25) taken at $t = t_k$.

In principle, the standard map can be obtained as well by integrating the Hamiltonian equations (3.17) with the perturbation function $\epsilon f(\vartheta) = (K/2\pi)\sin\vartheta$ in the interval $t_k + 0, t_k + 0 + T$. It gives

$$\theta_{k+1} = \theta_k + 2\pi I_k \ , \qquad I_{k+1} = I_k + \frac{K}{2\pi} \sin \theta_{k+1} \ , \qquad (3.26)$$

where $I_k = I(t = 2\pi k + 0)$. Both forms of the map (3.24), (3.26) equivalently represent the standard Hamiltonian in the limit $M \to \infty$.

From the physical point of view one can expect that the mappings (3.24), (3.26) should closely describe the Hamiltonian system (3.25) for large finite mode number M. As will be shown in Sect. 4.3.2 (see also Fig. 4.4) the standard map (3.24), however, does not reproduce Poincaré sections of the system (3.25).

One can also try to construct the standard map in the form of the perturbed twist map (3.5) for the standard Hamiltonian (3.25) using the method described in Sect. 3.2. Suppose that the perturbation parameter $\epsilon = K/4\pi^2$ is small. The unperturbed part of Hamiltonian (3.25) is $H_0(I) = I^2/2$ and the frequency of motion is $\omega(I) = dH_0(I)/dI = I$. We apply the formula (3.12) for the generating function $S(\vartheta_k, I_{k+1})$ using the perturbed part of Hamiltonian (3.25) with $H_{mn} = 1$, $\Omega = 1$, and $\chi_{mn} = 0$, and the time instants $t_k = kT = 2\pi k$, $(k = 0, 1, 2, \ldots)$. Then the generating function $S(\vartheta_k, I_{k+1})$ takes the form

$$S(\vartheta, J) = S_s(J) \sin \vartheta + S_c(J) \cos \vartheta \ , \qquad (3.27)$$

where

$$S_s(J) = - \sum_{n=-M}^{M} \frac{1 - \cos 2\pi J}{J - n} \ , \qquad S_c(J) = \sum_{n=-M}^{M} \frac{\sin 2\pi J}{J - n} \ . \qquad (3.28)$$

Extending the summation in (3.28) to $M \to \infty$ and using the formula

$$\sum_{n=-\infty}^{\infty} \frac{1}{J - n} = \frac{\pi}{\tan \pi J} \ ,$$

we obtain

$$S_s(J) = -\pi \sin 2\pi J \ , \qquad S_c(J) = 2\pi \cos^2 \pi J \ .$$

Thus, the perturbed twist map for the standard Hamiltonian (3.25) takes the form

$$I_{k+1} = I_k - \frac{K}{2\pi} \left(-\frac{\sin(2\pi I_{k+1})}{2} \cos \vartheta_k + \cos^2(\pi I_{k+1}) \sin \vartheta_k \right) \ ,$$

$$\vartheta_{k+1} = \vartheta_k + 2\pi I_{k+1} - \frac{K}{2} \left(\cos(2\pi I_{k+1}) \sin \vartheta_k + \sin(2\pi I_{k+1}) \cos \vartheta_k \right) \ .$$
$$(3.29)$$

Therefore the standard map (3.24) cannot be obtained in the form of the perturbed twist map (3.5) for the standard Hamiltonian (3.25). We will also discuss this problem in the next chapter (see Sect. 4.3.2).

3.5 Exact Mappings of Hamiltonian Systems

For some specific models of physical systems one can reduce the dynamics to exact mappings. One of the representative examples of such systems was a model proposed by Fermi to describe the acceleration of cosmic rays by momentum transfer from moving magnetic fields (Fermi (1949)). Ulam and associates considered a model consisting of a ball bouncing between two walls, one of them is fixed, and another one is oscillating. Later, different versions of the model were examined by Zaslavsky and Chirikov (1965); Brahic (1971); Lieberman and Lichtenberg (1972); Lichtenberg et al. (1980) (see, also Zaslavsky (1985); Lichtenberg and Lieberman (1992)). Ray propagation problems in inhomogeneous media is another area where dynamical equations can be exactly replaced by mappings (Abdullaev and Zaslavsky (1988); Abdullaev (1993)). This problem will be discussed in Sect. 13.1.

In this section we describe the Fermi acceleration mapping following Lichtenberg and Lieberman (1992). The geometry of the model is shown in Fig. 3.4. A particle bounces between a fixed and an oscillating wall. The fixed wall is at $x = 0$ and the oscillating wall at $x_w(t) = a + bf(t)$, where a is the average distance between the walls, b is the amplitude of oscillations, and $f(t)$ is a periodic function with the period $T = 2\pi/\Omega$: $f(t) = f(t + 2\pi/\Omega)$. Let t_k be the time of the k-th collision of particle with the fixed wall $x = 0$, v_k be the velocity after this collision. The particle collides with the moving wall at times t_c determined by $t_c = t_k + x_w(t_c)/v_k$. At collision the particle velocity changes to $v_{k+1} = v_k - 2dx_w/dt|_{t=t_c}$ and it reaches the fixed wall at the moment $t_{k+1} = t_c + x_w(t_c)/v_{k+1}$. Introducing the phases $\vartheta_k = \Omega t_k$ and $\psi_c = \Omega t_c$, and the normalized velocity $u_k = v_k/b\Omega$, the exact equations of motion can be represented in the form of a mapping $(\vartheta_k, u_k) \to (\vartheta_{k+1}, u_{k+1})$:

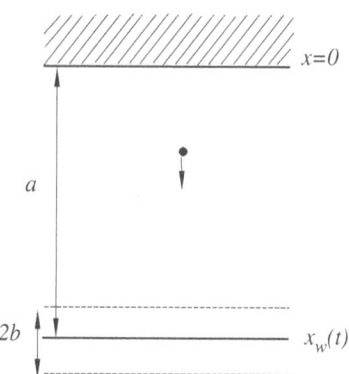

Fig. 3.4. Particle bouncing between the fixed $x = 0$ and the oscillating $x_w(t)$ walls

$$\psi_c = \vartheta_k + \frac{a/b - F(\psi_c)/2}{u_k} \, ,$$

$$u_{k+1} = u_k + F'(\psi_c) \, ,$$

$$\vartheta_{k+1} = \psi_c + \frac{a/b - F(\psi_c)/2}{u_{k+1}} \, , \qquad (3.30)$$

where $F(\psi) = -2f(\psi/\Omega)$, and $F'(\psi) \equiv dF(\psi)/d\psi$. The mapping (3.30) is written in symmetric form with respect to the variables (ϑ_s, u_s), unlike the one given in Lichtenberg and Lieberman (1992). The phase ψ_c in the first equation of (3.30) is defined implicitly. The phase ϑ is the variable conjugated to the energy $E = u^2$. By direct computation one can show that the Jacobian

$$\left| \frac{\partial(E_{k+1}, \vartheta_{k+1})}{\partial(E_k, \vartheta_k)} \right| = 1 \, .$$

The study of this map and its different simplified forms can be found in Lichtenberg and Lieberman (1992).

3.6 Difficulties in Constructing Mappings

We shall recall the main shortcomings of the two main methods to construct symplectic mappings for generic continuous Hamiltonian systems. In the first method it is a priori assumed that the corresponding map has the symplectic form (3.5). However, this form of the map is not the only possible one. Another symplectic form (3.13) of the map may equivalently represent a continuous system. The perturbation functions $f(\vartheta, I)$, $g(\vartheta, I)$ or the generating function $S(\vartheta, I)$ associated with mappings are found from the continuous equations of motion assuming smallness of the perturbation parameter ϵ. Such a method is not rigorous, and it neither allows to find the higher order corrections to the generating function nor to estimate the accuracy of replacement of the continuous system by the discrete map.

   The *method of delta functions* is usually applied to Hamiltonian systems affected by perturbations with a broad spectrum. It is based on direct replacement of the perturbation by delta functions periodic in time. This procedure introduces singularities into system at periodic time instances. Moreover, the variables of this system with delta functions are not identical to the corresponding variables of the original system. On the other hand, the integration across delta functions is defined only when the perturbation Hamiltonian H_1 depends only on the angle variables ϑ. For more general perturbation functions $H_1(\vartheta, I)$ the integration across the delta functions is not defined at all. Therefore, the derivation of symplectic maps (3.5) by the method of delta functions, as done in some publications (see, e.g., Mendonça (1991); Abdullaev et al. (1998); da Silva et al. (2001a,b, 2002a,b)) is not rigorously justified.

Another disadvantage of both methods is that the corresponding maps determine the system's phase space coordinates only at the discrete times $t_k = kT$ ($k = 0, \pm 1, \pm 2, \ldots$). They cannot recover the system's positions at arbitrary times t between the discrete times t_k. Moreover, the symplectic maps in the form (3.5) and (3.24) are not invariant with respect to the formal transformation $k \leftrightarrow k + 1$ which is equivalent to the transformation $t \rightarrow -t$, $H \leftrightarrow -H$ under which the continuous Hamiltonian system remains invariant. In the next chapter we develop rigorous methods to construct symplectic maps which does not have these shortcomings and difficulties.

4 Method of Canonical Transformation for Constructing Mappings

In this chapter we present the rigorous and systematic method to construct symplectic maps (3.2), particularly, Poincaré maps for generic Hamiltonian systems affected by perturbations. The method is based on the Hamilton–Jacobi method for integrating Hamiltonian equations and Jacobi's theorem recalled in Sect. 1.2.2. As we have seen there the idea of the Hamilton–Jacobi method consists of finding such a canonical change of variables which reduces a Hamiltonian function to a form that Hamiltonian equations are easy to integrate. The canonical transformation of variables is given by a generation function satisfying to the Hamilton–Jacobi partial differential equation. If we succeed to find a complete integral, i.e., the solutions of this equation depending N independent constants of motion, then according to Jacobi's theorem the dynamics of system is completely determined by the generating function $F(q, P, t)$. It means that the time evolution of system $(q(t), p(t))$ can be found through its initial position $(q(t_0), p(t_0))$ by the forward, $(q(t_0), p(t_0)) \rightarrow (Q(t_0), P(t_0))$, and the backward, $(Q(t), P(t)) \rightarrow (q(t), p(t))$, canonical transformations (1.20) given by the generating function $F(q, P, t)$ taken at the time instants t_0 and t, respectively. The evolution of new variables (Q, P) between these time instants is trivial and given by (1.18).

This idea can be also applied to generic perturbed Hamiltonian systems of type (2.2), (2.3) which may not possess N integrals of motion, and the system can be non-integrable (see Sect. 7.1). For these systems the complete integral of the Hamilton–Jacobi equation may not exist. The Jacobi's theorem can be also applied to these systems, supposing that there exist approximate N integrals of motion in the interval between initial, t_0, and final, t, times. Then the evolution of system in an each time interval is determined by Jacobi's theorem. By matching the orbits in neighboring time – intervals one can establish a long-term evolution of system. Such a procedure replaces the canonical equations of motion by symplectic Hamiltonian maps. Below we describe such a procedure of canonical change of variables which reduces continuous Hamiltonian systems to discrete Hamiltonian maps. Generating functions F of canonical transformation in perturbed Hamiltonian systems can be found using a perturbation series in finite time intervals (Sect. 2.3).

S.S. Abdullaev: *Construction of Mappings for Hamiltonian Systems and Their Applications*, Lect. Notes Phys. **691**, 53–81 (2006)
www.springerlink.com

4.1 Canonical Transformation and Mapping

We consider generic Hamiltonian problem described by the system of canonical equations (2.2), (2.3) namely the fully integrable system with Hamiltonian $H_0(I)$ subjected to the time-periodic perturbation $\epsilon H_1(\vartheta, I, t)$. Our task is to construct the mapping (3.2) connecting the variables $\vartheta_k = \vartheta(t_k), I_k = I(t_k)$ at the sequence of periodic time instants t_k ($k = 0, \pm 1, \pm 2, \ldots$). The mapping should be area-preserving (symplectic) (3.3).

In order to construct such a map we consider the system in a time interval $t_k < t < t_{k+1}$, and transform the variables (ϑ, I) to new ones (ψ, J), which eliminates the phases ϑ in the Hamiltonian H in this time interval. This transformation will be implemented by the time-dependent generating function $F = F(\vartheta, J, t, t_0, \epsilon) = \vartheta \cdot J + \epsilon S(\vartheta, J, t, t_0, \epsilon)$ of mixed variables satisfying the initial condition $S(\vartheta, J, t, t_0, \epsilon)|_{t=t_0} = 0$ at the time $t = t_0$ located in the interval $t_k < t_0 < t_{k+1}$. The relation between the old (ϑ, I) and the new (ψ, J) variables is given by (2.4). A new Hamiltonian $\mathcal{H}$ is $\mathcal{H}(J, \epsilon)$ and therefore

$$\psi(t) = \psi_k + w(J, \epsilon)(t - t_k), \qquad J = J_k = \text{ const}, \qquad (4.1)$$

where time t is in the interval $t_k < t < t_{k+1}$, and $w(J, \epsilon) = d\mathcal{H}(J, \epsilon)/dJ$ is a new frequency. The generating function $S(\vartheta, J, t, t_0)$ of such a transformation satisfies the Hamilton–Jacobi equation (2.6).

For the unperturbed system ($\epsilon = 0$) the generating function $F = \vartheta J$ describes the identical transformation, i.e., $\vartheta = \psi$, $I = J$, $\mathcal{H}_0(J, \epsilon) = H_0(I)$, $w(J, \epsilon) = \omega(I)$, and the map (3.2) is described by

$$I_{k+1} = I_k, \qquad \vartheta_{k+1} = \vartheta_k + \omega(I_k)(t_{k+1} - t_k). \qquad (4.2)$$

Consider the effect of non-zero perturbations ($\epsilon \neq 0$). Suppose that in the time interval $t_k < t < t_{k+1}$ there exits N constants of motion $J_k = (J_1, \ldots, J_N)$. Then according to the Jacobi's theorem the equations of motion are described as

$$I = \frac{\partial F}{\partial \vartheta} = J + \epsilon \frac{\partial S(\vartheta, J, t, t_0, \epsilon)}{\partial \vartheta},$$

$$\psi = \frac{\partial F}{\partial J} = \vartheta + \epsilon \frac{\partial S(\vartheta, J, t, t_0, \epsilon)}{\partial J}, \qquad (4.3)$$

where J and ψ are determined by equations (4.1) in the time interval $t_k \leq t \leq t_{k+1}$.

From (4.3) follows that the coordinates $(\vartheta(t), I(t))$ of trajectory at any moment of time t ($t_k \leq t \leq t_{k+1}$) with the initial coordinates (ϑ_k, I_k) at the time instant $t = t_k$ may be found by the two successive canonical transformations: the first one transforms the original variables (ϑ_k, I_k) to the new ones (ψ_k, J_k), and the second one transforms $(\psi(t), J_k)$ back to the old variables (ϑ, I) at the time instant t:

$$I_k = \frac{\partial F(\vartheta_k, J_k, t_k, t_0; \epsilon)}{\partial \vartheta_k}, \qquad \psi_k = \frac{\partial F(\vartheta_k, J_k, t_k, t_0, \epsilon)}{\partial J_k}, \qquad (4.4)$$

$$I(t) = \frac{\partial F(\vartheta(t), J_k, t, t_0, \epsilon)}{\partial \vartheta(t)}, \qquad \psi(t) = \frac{\partial F(\vartheta(t), J_k, t, t_0, \epsilon)}{\partial J_k}. \qquad (4.5)$$

Using (4.1) and the ansatz $F = \vartheta J + \epsilon S$, the map $(\vartheta_k, I_k) \to (\vartheta_{k+1}, I_{k+1})$ (3.2) can be presented in the following symplectic form

$$J_k = I_k - \epsilon \frac{\partial S_k}{\partial \vartheta_k}, \qquad \psi_k = \vartheta_k + \epsilon \frac{\partial S_k}{\partial J_k}, \qquad (4.6)$$

$$\bar{\psi}_k = \psi_k + w(J_k, \epsilon)(t_{k+1} - t_k), \qquad (4.7)$$

$$I_{k+1} = J_k + \epsilon \frac{\partial S_{k+1}}{\partial \vartheta_{k+1}}, \qquad \vartheta_{k+1} = \bar{\psi}_k - \epsilon \frac{\partial S_{k+1}}{\partial J_k}, \qquad (4.8)$$

where $S_k \equiv S(\vartheta_k, J_k, t_k, t_0, \epsilon)$, $S_{k+1} \equiv S(\vartheta_{k+1}, J_k, t_{k+1}, t_0, \epsilon)$.

The symplectic forms (4.4), (4.5) or (4.6)–(4.8) of the mapping (3.2) are general, and they are independent of assumptions of smallness of perturbation parameter ϵ.

The presented method to construct symplectic maps allows to recover the system's position at arbitrary time instant t between discrete times t_k, unlike the standard methods described in the chapter 3. Indeed, according to (4.5) the orbit $\vartheta(t), I(t)$ at the moment t located in the interval $t_k < t < t_{k+1}$ can be found by replacing the last two equations (4.7) - (4.8) in the mapping by

$$\psi(t) = \psi_k + w(J_k, \epsilon)(t - t_k),$$

$$I(t) = J_k + \epsilon \frac{\partial S(t)}{\partial \vartheta(t)}, \qquad \vartheta(t) = \psi(t) - \epsilon \frac{\partial S(t)}{\partial J_k},$$

where $S(t) \equiv S(\vartheta(t), J_k, t, t_0, \epsilon)$.

4.1.1 Nonsymmetric Forms of Maps

The generating function S associated with symplectic maps depends of the initial time t_0. By appropriate choosing this parameter in the interval $[t_k, t_{k+1}]$ one can obtain different forms of mapping. If $t_0 = t_{k+1}$ then the generating function S is identical to zero at the time instant $t = t_{k+1}$, i.e., $S(\vartheta_{k+1}, I_{k+1}, t_{k+1}, t_{k+1}, \epsilon) \equiv 0$. Then the map (4.6) - (4.8) takes the form similar to the one of the *perturbed twist map* (3.5):

$$I_{k+1} = I_k - \epsilon \frac{\partial S(\vartheta_k, I_{k+1})}{\partial \vartheta_k} \ ,$$

$$\vartheta_{k+1} = \vartheta_k + w(I_{k+1}, \epsilon)(t_{k+1} - t_k) + \epsilon \frac{\partial S(\vartheta_k, I_{k+1})}{\partial I_{k+1}} \ , \qquad (4.9)$$

where $S(\vartheta_k, I_{k+1}) \equiv S(\vartheta_k, I_{k+1}, t_k, t_{k+1}, \epsilon)$.

Similarly, taking $t_0 = t_k$ one can obtain the alternative form the perturbed twist map:

$$I_{k+1} = I_k + \epsilon \frac{\partial S(\vartheta_{k+1}, I_k)}{\partial \vartheta_{k+1}} \ ,$$

$$\vartheta_{k+1} = \vartheta_k + w(I_k, \epsilon)(t_{k+1} - t_k) - \epsilon \frac{\partial S(\vartheta_{k+1}, I_k)}{\partial I_k} \ , \qquad (4.10)$$

where $S(\vartheta_{k+1}, I_k) \equiv S(\vartheta_{k+1}, I_k, t_{k+1}, t_k, \epsilon)$.

These two alternative forms of mappings are similar to the forms (3.5) and (3.13) described in the previous section. However, the latter are only the approximate maps. They can be obtained from (4.9), (4.10) by replacing the frequency of perturbed system $w(I, \epsilon)$ by unperturbed one $\omega(I)$, and taking the first term in expansion series of the generating function S in powers of the perturbation parameter ϵ.

The orbits $(\vartheta(t), I(t))$ at arbitrary time t, $t_k < t < t_{k+1}$, can be also found by the mapping of form similar to the perturbed twist map (4.10). It is easy to show the corresponding map is determined by

$$I(t) = I_k + \epsilon \frac{\partial S(\vartheta(t), I_k)}{\partial \vartheta(t)} \ ,$$

$$\vartheta(t) = \vartheta_k + w(I_k, \epsilon)(t - t_k) - \epsilon \frac{\partial S(\vartheta(t), I_k)}{\partial I_k} \ , \qquad (4.11)$$

where $S(\vartheta(t), I) \equiv S(\vartheta(t), I_k, t, t_k, \epsilon)$.

The two forms (4.9), (4.10) of the perturbed twist map separately are not invariant with respect to the backward – forward transformations ($k \leftrightarrow k+1$). However, under this transformation the map (4.9) is transformed into (4.10), and vise-versa, if the generating function $S(\vartheta_{k+1}, I_k) = -S(\vartheta_k, I_{k+1})$. As we will see later this condition is satisfied.

4.1.2 Symmetric Map

If the initial time t_0 is located exactly in the middle of the interval, i.e., $t_0 = (t_k + t_{k+1})/2$, we call the symplectic map (4.6)-(4.8) as a *symmetric map*. This map is invariant with respect to the change of time sequences $k \leftrightarrow k+1$. The backward map $(\vartheta_{k+1}, I_{k+1}) \rightarrow (\vartheta_k, I_k)$ may be obtained from the forward map (3.2) by simple reversing sequences of canonical transformation. As will see later the accuracy of the symmetric map is higher than the perturbed twist maps (4.9), (4.10).

The stroboscopic maps (or Poincaré maps) (3.1) can be obtained from the above maps by putting the map time step $\tau = t_{k+1} - t_k$ equal to the period of perturbation $\tau = 2\pi/\Omega$.

4.1.3 The Generating Function of Mappings

Now we turn to the main task in construction of the symplectic maps, a determination of the generating function S. As was mentioned above that we should look for the solution $S(\vartheta, J, t, t_0)$ of the Hamilton–Jacobi equation (2.6) in the interval $[t_k, t_{k+1}]$ satisfying the initial condition $S = 0$ at the initial time $t = t_0$, $t_k < t_0 < t_{k+1}$. The natural approach to this problem would be the time-dependent perturbation theory developed in Sect. 2.3.

Suppose for a moment that the perturbation parameter ϵ is small. The generating function $S(\vartheta, J, t, t_0, \epsilon)$ is sought as the expansion series in powers of ϵ (2.33). For the generic Hamiltonian system (2.2), (2.3) the expansion terms S_i ($i = 1, 2, \ldots$) satisfy the (2.10), (2.11). The solutions of these equations $S_i(\vartheta, J, t, t_0)$ satisfying the initial conditions $S_i = 0$ at $t = t_0$ are determined by integrals (2.34), (2.37). Particularly, for the perturbed Hamiltonian H_1 in a Fourier expansion form (2.15) the first two terms S_1 and S_2 are given by (2.35) and (A.7).

In the first order of ϵ the mapping (4.6)–(4.8) takes the following form

$$J_k = I_k - \epsilon \frac{\partial S_1(\vartheta_k, J_k, t_k, t_0)}{\partial \vartheta_k} \ ,$$

$$\psi_k = \vartheta_k + \epsilon \frac{\partial S_1(\vartheta_k, J_k, t_k, t_0)}{\partial J_k} \ , \tag{4.12}$$

$$\bar{\psi}_k = \psi_k + \omega(J_k)(t_{k+1} - t_k) \ , \tag{4.13}$$

$$I_{k+1} = J_k + \epsilon \frac{\partial S_1(\vartheta_{k+1}, J_k, t_{k+1}, t_0)}{\partial \vartheta_{k+1}} \ ,$$

$$\vartheta_{k+1} = \psi_{k+1} - \epsilon \frac{\partial S_1(\vartheta_{k+1}, J_k, t_{k+1}, t_0)}{\partial J_k} \ , \tag{4.14}$$

where $S_1(\vartheta, J, t, t_0)$ is determined by (2.35).

Particularly if the parameter t_0 lies in the middle of the interval $[t_k, t_{k+1}]$, i.e., $t_0 = (t_k + t_{k+1})/2.0$ we have the symmetric form of the map. Taking $t_0 = t_{k+1}$ we obtain the nonsymmetric form

$$I_{k+1} = I_k - \epsilon \frac{\partial S(\vartheta_k, I_{k+1})}{\partial \vartheta_k} \ ,$$

$$\vartheta_{k+1} = \vartheta_k + \omega(I_{k+1})(t_{k+1} - t_k) + \epsilon \frac{\partial S(\vartheta_k, I_{k+1})}{\partial I_{k+1}} \ , \tag{4.15}$$

with the generating function

$$S(\vartheta, I) \equiv S_1(\vartheta, J, t_k, t_{k+1}) = (t_{k+1} - t_k) \sum_{m,n} |H_{mn}(J)|$$

$$\times \left[- a(x_{mn}) \sin(m \cdot \vartheta - n\Omega t + \chi_{mn}) \right.$$

$$\left. + b(x_{mn}) \cos(m \cdot \vartheta - n\Omega t + \chi_{mn}) \right], \tag{4.16}$$

where the argument x_{mn} of the functions $a(x)$ and $b(x)$ is given by

$$x_{mn} = [m \cdot \omega(I) - n\Omega](t_{k+1} - t_k) .$$

The mapping (4.15) coincides with the mapping defined by (3.5), (3.6), (3.12) when the time step $\Delta t = t_{k+1} - t_k$ equal to the period of perturbation $T = 2\pi/\Omega$.

We should recall the main features the expansion series (2.33), (2.34), (2.37), (2.35), (A.7) mentioned in Sect. 2.3. The main contribution to each S_i $(i = 1, 2, \ldots)$ comes at values of action J near the resonant frequencies $(\omega(J), \Omega)$ $(m \cdot \omega(J) - n\Omega = 0)$. At these resonant values of J the actual expansion parameter in the series (2.33) is not the perturbation parameter ϵ itself but its combination with the time interval $(t - t_0)$, i.e., $\mu = \epsilon(t - t_0)^\nu$ with a some exponent $\nu > 1$. As will be shown in Sect. 4.2.1 that $\mu = \epsilon(t - t_0)^2$ for a simple Hamiltonian problem, the motion of particle in the field of a wave. Such a nature of expansion allows one to apply the perturbation theory for generating functions $S(\vartheta, J, t, t_0, \epsilon)$ not only for the small values of the perturbation parameter ϵ, but also for its large values by taking the time step of mapping $\tau = t_{k+1} - t_k$ which keeps the product $\mu = \epsilon(t - t_0)^\nu$ to be small.

4.2 Accuracy of Maps

The small parameter μ for generating functions in mappings equals to $\mu = \epsilon(t_k - t_0)^\nu$. At the fixed perturbation parameter ϵ the parameter μ takes a minimal value $\mu = \epsilon(\tau/2)^\nu$ for the symmetric map (4.6)–(4.8) with $t_0 = (t_k + t_{k+1})/2$, while for the perturbed twist map μ has a maximal value $\mu = \epsilon\tau^\nu$. Therefore, the accuracy of the symmetric map is much higher than one of the perturbed twist maps (4.9), (4.10). Errors due to truncation of the power series of generating function (2.7) in μ are smaller for the symmetric map than for the perturbed twist map. As will be shown in the next sections that the symmetric map describes Hamiltonian systems much more accurate than the perturbed twist map.

One should note that from the computational point view the symplectic mappings (4.6)–(4.8) are implicit. The variable J_k in the first set of equations

(4.6) and ϑ_{k+1} in the second set of equations (4.8) are defined implicitly, and they should be found by solving corresponding algebraic equations. For this purpose one can use Newton's method (or the Newton–Raphson method) which has a high rate of quadratic convergency (Press et al. (1992), see also Sect. 2.4). As an initial value in an iterative procedure one can take

$$J_k^{(0)} = I_k - \epsilon \frac{\partial S_k^{(0)}}{\partial \vartheta_k} \,, \qquad S_k^{(0)} \equiv S(\vartheta_k, I_k, t_k, t_0, \epsilon) \,,$$

in the first equation of (4.6), and

$$\vartheta_{k+1}^{(0)} = \psi_{k+1} - \epsilon \frac{\partial S_{k+1}^{(0)}}{\partial J_k} \,, \qquad S_{k+1}^{(0)} \equiv S(\psi_{k+1}, J_k, t_{k+1}, t_0, \epsilon) \,,$$

in the second equation of (4.8). The differences $|J_k - J_k^{(0)}|$ and $|\vartheta_{k+1} - \vartheta_{k+1}^{(0)}|$ have an order of $\mu^2 = \epsilon^2 (t - t_0)^2 \ll 1$.

In Appendix F we have presented the program for the numerical implementation of the mapping (4.6)–(4.8) for the generic 1+1/2-degrees-of-freedom Hamiltonian systems. As an example of application of this program we also give a sample program for plotting Poincaré sections for a specific Hamiltonian system. The program is written in C–language and it can be easily run in personal computers with the Lunix operating system.

4.2.1 Particles in a Single-Frequency Wave Field

In order to study the accuracy of maps we consider a simple, completely integrable Hamiltonian system, namely, the motion of a particle in a single-frequency wave field studied in Sect. 1.4. The particle motion is described by the equation of motion (1.39). We consider the wave-field with the amplitude eE_k as a perturbation acting on free moving particle. Introducing angle–action variables for the free motion of particle, i.e., $\vartheta = kx$, $I = k\dot{x}$, the equation of motion can be written in Hamiltonian form with the Hamiltonian

$$H(\vartheta, I, t) = \frac{I^2}{2} - \epsilon \cos(\vartheta - \Omega t) \,, \tag{4.17}$$

where the perturbation parameter $\epsilon = ekE_k/m$. In a coordinate system running with a phase velocity of the wave, i.e., $q = \vartheta - \Omega t$, $p = I - \Omega$, it describes the pendulum with the frequency of small amplitude oscillations $\omega_0 = \sqrt{\epsilon}$. The corresponding Hamiltonian function $H_0(q, p) = H(q + \Omega t, I - \Omega, t)$ is the constant of motion (see (1.41)). For $-\omega_0^2 \leq H \leq \omega_0^2$ the motion is trapped and trajectory is oscillating near elliptic fixed points. The frequency of these nonlinear oscillations $\omega(H)$ is determined by (1.45).

We will apply mappings to study the Hamiltonian system (4.17). The unperturbed Hamiltonian $H_0(I) = I^2/2$ determines the free motion of particle with the frequency (or velocity) $\omega(I) = dH_0(I)/dI = I$: $\vartheta = \vartheta_0 + \omega(I)(t - t_0), I = \text{const.}$

The first and second order generating functions. According to (2.10), and (2.35) the first order generating function is

$$S_1(\vartheta, J, t) = (t - t_0)\Big[a(x)\sin(\vartheta - \Omega t) + b(x)\cos(\vartheta - \Omega t)\Big] , \qquad (4.18)$$

where $a(x)$ and $b(x)$ are oscillating functions (2.36), and $x = (J - \Omega)(t - t_0)$.

The second order term $S_2(\vartheta, I, t)$ is determined by (2.38). Putting $m = 1$ and $n = 1$ we obtain

$$S_2(J, \vartheta, t) = -\frac{(t - t_0)^3}{4}$$

$$\times \Big[B_0(x) + A_2(x)\sin(2\vartheta - 2\Omega t) + B_2(x)\cos(2\vartheta - 2\Omega t)\Big] , \qquad (4.19)$$

where the localized functions $B_0(x)$, $A_2(x)$ $B_2(x)$ are defined by (2.39), (2.40) and (2.41).

The map (4.6)–(4.8) for the Hamiltonian system (4.17) may be rewritten in normalized variables (x, y), $x = \vartheta - \Omega t$, $y = (J - \Omega)(t - t_0)$. Using (4.18), (4.19) the generating function S for the corresponding map $(x_k, y_k) \to (x_{k+1}, y_{k+1})$ may be reduced to

$$Y_k = y_k - \mu\frac{\partial \bar{S}_k}{\partial x_k} , \qquad X_k = x_k + \mu\frac{\partial \bar{S}_k}{\partial Y_k} , \qquad (4.20)$$

$$X_{k+1} = X_k + [w(J, \epsilon) - \Omega](t_{k+1} - t_k) , \qquad (4.21)$$

$$y_{k+1} = Y_k + \mu\frac{\partial \bar{S}_{k+1}}{\partial Y_{k+1}} , \qquad x_{k+1} = X_{k+1} - \mu\frac{\partial \bar{S}_{k+1}}{\partial x_k} , \qquad (4.22)$$

with the generating function $S_k \equiv S(x_k, Y_k, t_k, \mu)$, and

$$\mu S(x, Y, t, \mu) = \epsilon(t - t_0)S(x, Y, t) ,$$

$$S(x, Y, t, \mu) = S_1(x, Y, t) + \mu S_2(x, Y, t) + O(\mu^2) , \qquad (4.23)$$

where

$$S_1(x, Y, t) = a(Y)\sin x + b(Y)\cos x ,$$

$$S_2(x, Y, t) = -\frac{1}{4}[B_0(Y) + A_2(Y)\sin 2x + B_2(Y)\cos 2x] .$$

Here $\mu = \epsilon(t - t_0)^2$ is the expansion parameter. As was mentioned in Sect. 4.1.1 for the symmetric map the expansion parameter $\mu = \epsilon\tau^2/4$. For the perturbed twist map the corresponding expansion parameter is larger, i.e., $\mu = \epsilon \tau^2$.

Existence of such a dependence of the expansion parameter μ on the perturbation parameter ϵ and the map step τ allows one to consider also a relatively large perturbations $\epsilon \sim 1$. In these cases one can still use the first (or up to second) order term in the expansion series by choosing such a time step τ that keeps the parameter μ to be small. For instance, in the above example, the parameter μ is not changed if the perturbation ϵ is increased in two order while the time step τ is decreased only in one order. This does not increase the computational times proportional to the perturbation parameter ϵ.

Comparison of the symmetric map and the perturbed twist map We have compared the two forms of the map: the symmetric map (4.6)–(4.8) and the perturbed twist map (4.9). The several trajectories of the system in the phase plane (ϑ, I) obtained by the symmetric map and the perturbed twist map (4.9) are presented in Fig. 4.1 using the first order generating function $S_1(\vartheta, J, t, t_0)$. The value of ϵ is taken equal to 0.03, and the map step τ is 2π. The initial conditions of trajectories were $(\vartheta_0, I_0) = (0, 0.03)$, $(0, 0.2)$, $(0, 0.33)$, $(0, \pm 0.35)$ and $(0, \pm 0.45)$. Solid curves describe the exact trajectories of the pendulum, dotted curves and dashed curves correspond to the symmetric map and the perturbed twist map, respectively. As can see from Figure the symmetric map much more closely describes the trajectories than the perturbed twist map. The phase space curves obtained by the symmetric map are symmetric with respect to the axes $I = 0$ and $\vartheta = \pi$ similar to the exact trajectories, while trajectories of the perturbed twist map are slightly asymmetric.

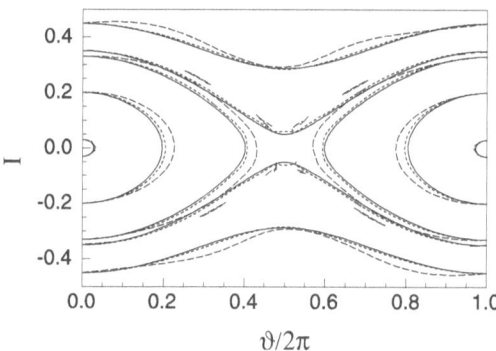

Fig. 4.1. Several orbits of the Hamiltonian system (4.17) in the (ϑ, I)-plane obtained by the symmetric map (*dotted curves*) and the perturbed twist map (*dashed curves*). *Solid curves* describe the exact orbits. Parameters are $\epsilon = 0.03$, $\Omega = 0$

4.2.2 Accuracy of the Symmetric Map

The accuracy of the symmetric map is studied on two examples: for the completely integrable Hamiltonian system (4.17) and for the standard Hamiltonian (3.25). In the first case the map solutions are compared with exact solutions. We also integrated these model Hamiltonian systems by one of the most accurate symplectic integration methods, namely the fifth –order explicit Runge–Kutta method described in Section 1.5.

First consider the integrable system (4.17). We have considered the two cases, time-independent ($\Omega = 0$) and time-dependent ($\Omega = 1$) perturbations. The amplitude ϵ is taken equal to $\epsilon = \omega_0^2 = 0.01$. We have taken a set of initial conditions ($\vartheta = 0, I = I_i$), ($\Omega \leq I_i \leq \Omega + I_s$) for trapped orbits and integrated the system up to $t = 4\pi \times 10^4$ by the symmetric map and by the symplectic integrator of McLachlan & Atela (further we abbreviate it SI). (Here $I_s = 2\omega_0$ is the value of amplitude I at the separatrix). The root-mean-square energy error $\parallel H - H_0 \parallel_2$, defined as

$$\parallel H - H_0 \parallel_2^2 = \frac{1}{N} \sum_{k=1}^{N} \left(H(\vartheta_k, I_k, t_k) - H_0 \right)^2,$$

is calculated over all instants time $t_k = 2\pi k$ ($k = 1, \ldots, N = 2 \times 10^4$). They are shown in Fig. 4.2 as a function of the oscillation amplitude I_i: a) describes the time-independent case $\Omega = 0$, and b) the time-dependent case $\Omega = 1$. The map step is $\tau = \pi$, and the integration step of the SI is $\Delta t = \pi/100$. Curve 1 corresponds to the SI, and curves 2, 3 describe the map results. The curve 2 corresponds to the map with the first order generating function S_1 (4.18), and curve 3 corresponds the case when the second order generating function S_2 (4.19) is also included, i.e., $S = S_1 + \epsilon S_2$.

As can see from Fig. 4.2, the energy errors of the SI in the time-independent case on several orders smaller than for the map. However, in the time-dependent case accuracy of the SI is significantly deteriorated whereas the accuracy of mapping has not been changed. Moreover, the energy errors of the SI become on two order higher than for maps. Inclusion of the second order of the generating function S_2 improves the accuracy of the map more than on two or three orders. Particularly, it reduces the energy error by a factor of 167 at the value of $(I - \Omega)/I_s = 0.5$. At this value of I the ratio of energy errors of the SI, the map with the first order generating function and with the second order generating function is 832:167:1 for the time-dependent case ($\Omega = 1$). Therefore, *the accuracy of the map at the fixed time-step τ does not depend on the frequency Ω*, while the accuracy of the SI significantly depends on Ω and it deteriorates with increase of Ω at the fixed integration step Δt.

For the non-integrable standard Hamiltonian system (3.25) the accuracy of the map was also tested by integrating the system forward in time up to a certain time instant t_{max} then reversing it back in time to the initial time

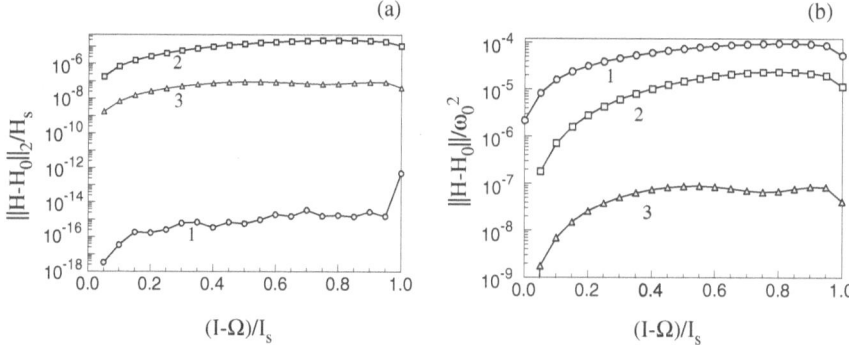

Fig. 4.2. Relative root-mean-square energy errors $\| H - H_0 \|_2 / H_s$ ($H_s = \omega_0^2$) for the system (4.17) as a function of oscillation amplitude I. Parameter $\epsilon = \omega_0^2 = 0.01$: (a) perturbation frequency $\Omega = 0$, (b) $\Omega = 1$. Curve 1 corresponds to the SI with the integration step $\Delta t = \pi/100$. The map step is $\tau = \pi$. Curve 2 corresponds to the map with the first order generating function S_1 (4.18), and curve $3 - S = S_1 + \epsilon S_2$ (4.19)

instant t_0. We checked how close the orbit comes back to the initial point. The accuracy significantly depends on whether the orbit is regular, or chaotic. For this test we have integrated the Hamiltonian system (3.25) with the initial coordinate (I_0, ϑ_0) from the time instant $t = 0$ up to $t = t_{max}$ and reversed it back in time. Let $I_f(t)$ and $I_b(t)$ be the component of the forward orbit and the backward orbit, respectively. The difference of these components $|I_f(t) - I_b(t)|$ are plotted in Fig. 4.3 as a function of $t_{max} - t$: a) describes the case of a regular orbit with the initial coordinates ($I_0 = 2, \vartheta_0 = 0.6\pi$); b) corresponds to the chaotic orbit with ($I_0 = 2, \vartheta_0 = 0.02\pi$).

In both Figures curve 1 describes the results of the SI with the integration step $\Delta t = 2\pi/4000$, and curve 2 is obtained by the map with the time step $\tau = 2\pi$. One can see from Fig. 4.3a that the accuracy of the map reversibility for the regular orbit is much more higher than one of the SI even with very small integration steps: for $t_{max} = 2\pi \times 10^3$ the regular orbit is reversed back to the initial point with accuracy less than 10^{-10} while for the SI the accuracy is of order of 10^{-2}. The error is growing linearly with time.

From Fig. 4.3b it follows that the accuracy of the map is also much higher for chaotic orbits than one of the SI. However, due to exponential growth of round – up operation errors the time t_{max} for the reversing back of a chaotic orbit to the initial point with the accuracy 10^{-10} is less than $t_{max} \approx 2\pi \times 50$. The reversibility of a chaotic orbit by the SI is much worser. The much higher reversibility of the map is mainly due to its symmetric from (4.6)–(4.8) which is invariant with respect to the map reversing transformation $k \leftrightarrow k + 1$, $H \rightarrow -H$. The latter property of the symmetric map expresses the invariance of Hamiltonian equations (1.1) with respect to the formal time reversing transformation $t \rightarrow -t$, $H \rightarrow -H$.

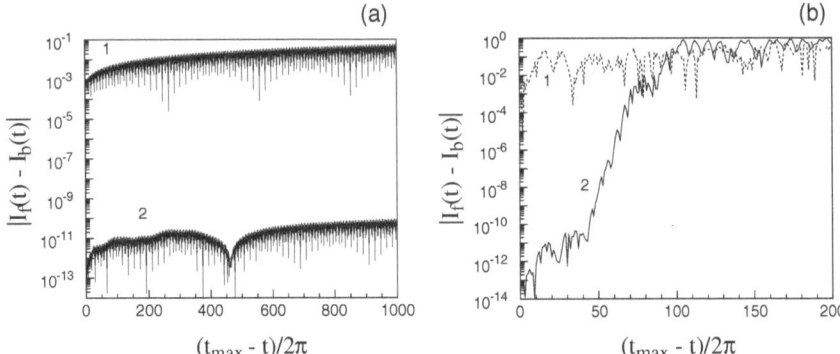

Fig. 4.3. Accuracy of time-reversing for the SI (curve 1) and the symmetric map (curve 2). **(a)** for the regular orbit; **(b)** for the chaotic orbit. Initial coordinates of the regular orbit is $(I_0 = 2, \vartheta_0 = 0.6\pi)$, and chaotic orbit $(I_0 = 2, \vartheta_0 = 0.02\pi)$. The integration step of the SI is $\Delta t = 2\pi/4000$, the map time-step $\tau = 2\pi$, $t_{max} = 2\pi \times 10^3$. The parameters are $K = 1.5$, $M = 10$

Therefore, for the exact integrable system we have shown that the symmetric map with the large time steps τ of order of perturbation period $2\pi/\Omega$, $(\tau \sim 2\pi/\Omega)$ can achieve the higher order accuracy of calculations as the symplectic integration methods with the two order smaller integration steps Δt than τ. However, the symplectic method requires at least one order shorter computational times than the symplectic integrator. The more detailed study of mapping accuracy can be found in Abdullaev (2002).

A proposed method to construct symplectic maps for generic Hamiltonian systems can be considered as a new *alternative method of symplectic integration*. It has several advantages over the standard symplectic integration methods, especially in the cases of highly oscillatory systems. For instant, the latter can be hardly applied to integrate the standard Hamiltonian system (3.25) with a finite, but large number of harmonics M since it becomes highly oscillatory. One needs to take very small integration steps Δt in order to get results with sufficient accuracy. The most convenient way to integrate such systems is to apply the symmetric maps (4.6)–(4.8). The maps with a time step of order of perturbation period have a sufficiently high accuracy. Moreover, for the fixed time step the accuracy of mapping does not depend on perturbation frequency (see Abdullaev (2002) for more details).

4.3 Mappings for Hamiltonian Systems with a Broad Perturbation Spectrum

Consider specific Hamiltonian systems for which symplectic mappings can be significantly simplified. In practical applications one encounters Hamiltonian systems subjected to perturbation with a broad spectrum. These systems can be described by the Hamiltonian in the form

$$H(\vartheta, I, t) = H_0(I) + \epsilon H_1(\vartheta, I, t) ,$$

$$H_1(\vartheta, I, t) = V(\vartheta, I) \sum_{n=-M}^{M} \cos(n\Omega t) , \tag{4.24}$$

which contains a large number $M \gg 1$ uniformly distributed harmonics in n. The dependence of perturbation on angular and action variables, (ϑ, I), is described by $V(\vartheta, I)$ which is a 2π-periodic function of angular variable ϑ. It can be also presented as a Fourier series

$$V(\vartheta, I) = \sum_m |H_m(I)| \cos(m\vartheta + \chi_m) = \text{Re} \sum_m H_m(I) e^{im\vartheta} , \tag{4.25}$$

where $H_m(I) = |H_m(I)| \exp(i\chi_m)$.

The method of delta functions to construct maps for these systems and its difficulties were discussed in Sect. 3.3. A mathematically correct and physically reasonable approach to construct symplectic maps for these systems would consist of two steps: first, to obtain a map for the finite mode number M and then to consider the limit $M \to \infty$. Such a method would avoid the uncertainty in integration along periodic delta functions discussed above.

Presenting the perturbation Hamiltonian H_1 in the form

$$H_1(\vartheta, I, t) = \epsilon \text{Re} \left\{ \sum_m \sum_{n=-M}^{M} H_m(I) e^{i(m\vartheta - n\Omega t)} \right\} , \tag{4.26}$$

the first order generating function S_1 given by (2.35) for the Hamiltonian (4.24) can be written as

$$S_1(\vartheta, J, t, t_0) = \text{Re} \sum_m i H_m(J) e^{im\vartheta}$$

$$\times \sum_{n=-M}^{M} \frac{\exp(-in\Omega t)}{m\omega(J) - n\Omega} \left(1 - e^{-i(m\omega(J) - n\Omega)(t - t_0)} \right)$$

$$= \text{Re} \sum_m i H_m(J) e^{im\vartheta} \left(\sum_{n=-M}^{M} \frac{e^{-in\Omega t}}{m\omega(J) - n\Omega} \right.$$

$$\left. - e^{-im\omega(J)(t - t_0)} \sum_{n=-M}^{M} \frac{e^{-in\Omega t_0}}{m\omega(J) - n\Omega} \right) , \tag{4.27}$$

where $\omega(I) = dH_0(I)/dI$ is the frequency of motion. Consider the asymptotics of $S_1(\vartheta, J, t)$ at the limit of large mode numbers $M \gg 1$. We use the formulas

$$\sum_{n=-\infty}^{\infty} \frac{e^{in\Omega t}}{a-n} = \frac{\pi e^{i[t]a}}{\sin \pi a} , \qquad \sum_{n=-\infty}^{\infty} \frac{1}{a-n} = \frac{\pi}{\tan \pi a} ,$$

where $[t] = t - (2k+1)\pi/\Omega$ and $2\pi k/\Omega < t < 2\pi(k+1)/\Omega$, $(k = 0, \pm 1, \pm 2, \ldots)$. For any arbitrary time-instants t, t_0 in the interval $2\pi k/\Omega < t, t_0 < 2\pi(k+1)/\Omega$ we have

$$S_1(\vartheta, J, t, t_0) = \mathrm{Re} \sum_m i H_m(J) e^{im\vartheta} \frac{\pi}{\Omega \sin(\pi\omega(J)/\Omega)}$$
$$\left(e^{-i[t]m\omega(J)/\Omega} - e^{-im\omega(J)(t-t_0)} e^{-i[t_0]m\omega(J)/\Omega} \right) + O(M^{-1}) = O(M^{-1}) ,$$

i.e., the generating function $S_1(J, \vartheta, t)$ vanish with large $M \gg 1$ (we suppose that the frequency $|\omega(J)|$ is finite, $|\omega(J)| \ll M$) as

$$S_1(\vartheta, J, t, t_0) \sim O(M^{-1}) . \tag{4.28}$$

However, at the time instants $t = t_k = 2\pi k/\Omega \pm 0$ (or $[t] = \mp\pi/\Omega$) and t_0 in the interval $2\pi k/\Omega < t_0 < 2\pi(k+1)/\Omega$ the generating function $S_1(J, \vartheta, t)$ has the finite asymptotical value for $M \gg 1$, i.e.,

$$S_1(\vartheta, J, t = 2\pi k \pm 0, t_0) = \mathrm{Re} \sum_m i H_m(J) e^{im\vartheta} \frac{\pi}{\Omega \sin(\pi\omega(J)/\Omega)}$$
$$\left(\cos(\pi\omega(J)/\Omega) - e^{\pm im\pi\omega(J)/\Omega} \right) + O(M^{-1})$$
$$= \pm \frac{\pi}{\Omega} \mathrm{Re} \sum_m H_m(J) e^{im\vartheta} + O(M^{-1}) ,$$

or

$$S_1(\vartheta, J, t = 2\pi k \pm 0, t_0) = \pm \frac{\pi}{\Omega} V(\vartheta, J)$$
$$= \pm \frac{\pi}{\Omega} \sum_m |H_m(J)| \cos(m\vartheta + \chi_m) . \tag{4.29}$$

According to (2.37) the higher order generating functions $S_i(\vartheta, J, t)$ are given by

$$S_i(\vartheta, J, t, t_0) = -\int_{t_0}^{t} F_i(\vartheta(t'), J, t') dt' . \tag{4.30}$$

The second order generating functions S_2 vanishes at the limit $M \to \infty$ at all time-instants t because of the property (4.28) and the definition of the function F_2 (2.12). Similarly, the higher-order generating functions $S_i(\vartheta, J, t)$, $(i > 2)$ vanish also, because the polynomial functions F_i of derivatives $\partial S_1/\partial\vartheta, \ldots, \partial S_{i-1}/\partial\vartheta$ on the right-hand side of equations (2.11) vanish at the limit $M \to \infty$. Therefore, the generating function $S(\vartheta, J, t)$ is determined only by the first order generating function $S_1(\vartheta, J, t)$ for arbitrary values of the perturbation parameter ϵ. The corrections to the perturbed frequency $w(J, \epsilon)$ are also vanish at the limit $M \to \infty$, i.e., $w(J, \epsilon) = \omega(J)$. Thus the symmetric map (4.6)–(4.8) can be presented in the form

$$J_k = I_k - \epsilon\frac{\partial S(\vartheta_k, J_k)}{\partial \vartheta_k} , \qquad \psi_k = \vartheta_k + \epsilon\frac{\partial S(\vartheta_k, J_k)}{\partial J_k} , \qquad (4.31)$$

$$\bar{\psi}_k = \psi_k + 2\pi\frac{\omega(J_k)}{\Omega} , \qquad (4.32)$$

$$I_{k+1} = J_k - \epsilon\frac{\partial S(\vartheta_{k+1}, J_k)}{\partial \vartheta_{k+1}} , \qquad \vartheta_{k+1} = \bar{\psi}_k + \epsilon\frac{\partial S(\vartheta_{k+1}, J_k)}{\partial J_k} , \qquad (4.33)$$

where the generating function is determined by

$$S(\vartheta, J) = \frac{\pi}{\Omega}V(\vartheta, J) = \frac{\pi}{\Omega}\sum_m |H_m(J)|\cos(m\vartheta + \chi_m) . \qquad (4.34)$$

The time step τ of the map (4.31)–(4.33) is equal to the period of perturbation $2\pi/\Omega$.

In a case when the perturbation harmonics, $H_m(J)$, does not depend on the action variable J the mapping (4.31)–(4.33) is further simplified. Indeed, this case $S(\vartheta, J) = S(\vartheta)$ and the map is reduced to

$$J_k = I_k - \epsilon\frac{\partial S(\vartheta_k)}{\partial \vartheta_k} ,$$

$$\vartheta_{k+1} = \vartheta_k + 2\pi\frac{\omega(J_k)}{\Omega} , \qquad (4.35)$$

$$I_{k+1} = J_k - \epsilon\frac{\partial S(\vartheta_{k+1})}{\partial \vartheta_{k+1}} .$$

This map can be called a *symmetric radial twist map*. The radial twist map of type (3.23) can be obtained from (4.35) for the mapping $(\vartheta_k, J_{k-1}) \to (\vartheta_{k+1}, J_k)$, i.e.,

$$J_k = J_{s-1} - 2\epsilon\frac{\partial S(\vartheta_k)}{\partial \vartheta_k} , \qquad \vartheta_{k+1} = \vartheta_k + 2\pi\frac{\omega(J_k)}{\Omega} . \qquad (4.36)$$

Therefore, the action variable I in the radial twist map (3.23) does not coincide with the original action variable in a continuous Hamiltonian system. In other words, the variables (ϑ, J) are not canonical. The mapping for the original canonical variables (ϑ, I) has a symmetric from (4.35).

4.3.1 Non-Symmetric Forms of Maps

For the Hamiltonian system (4.24) one can also construct maps in non-symmetric forms (4.9), (4.10). However, the generating functions of these maps have more complicated dependence on perturbation harmonics, $H_m(J)$, and the frequency $\omega(J)$.

To be more specific we construct the map of the form (4.9) by putting $t = t_k + 0 = 2\pi s/\Omega + 0$ and $t_0 = t_{k+1} - 0 = 2\pi(s+1)/\Omega - 0$ in (4.27). Then at the limit $M \to \infty$ the generating function S_1 is reduced to

$$S(\vartheta, J) \equiv S_1(\vartheta, J, t_k + 0, t_{k+1} - 0)$$

$$= \frac{\pi}{\Omega} \operatorname{Re} \sum_m i H_m(J) e^{im\vartheta} \frac{1 - e^{i2\pi m\omega(J)/\Omega}}{\tan(\pi m\omega(J)/\Omega)}$$

$$= \frac{2\pi}{\Omega} \sum_m |H_m(J)| \left[-\frac{1}{2} \sin\left(\frac{2\pi m\omega(J)}{\Omega} \right) \sin(m\vartheta + \chi_m) \right.$$

$$\left. + \cos^2\left(\frac{\pi m\omega(J)}{\Omega} \right) \cos(m\vartheta + \chi_m) \right]. \qquad (4.37)$$

The higher order corrections S_i $(i \geq 2)$ vanish at the limit $M \to \infty$. Thus the mapping (4.9) takes the form

$$I_{k+1} = I_k - \epsilon \frac{\partial S(\vartheta_k, I_{k+1})}{\partial \vartheta_k} ,$$

$$\vartheta_{k+1} = \vartheta_k + 2\pi \frac{\omega(I_{k+1})}{\Omega} + \epsilon \frac{\partial S(\vartheta_k, I_{k+1})}{\partial I_{k+1}} . \qquad (4.38)$$

Therefore, *one cannot obtain the perturbed twist map in the form (3.5) determined by the generating function depending only on the perturbation harmonics, $H_m(I)$, similar to the one (4.29).* The generating function (4.37) depends on the action variable I not only through the radial dependence of the harmonics, $H_m(I)$, but also because of the dependence of the frequency $\omega(I)$ on I.

The approximate nonsymmetric mappings of type (4.9) can be also obtained from the symmetric mapping (4.31)–(4.33) for the intermediate variables $(\bar{\psi}, J)$, i.e.,

$$(\bar{\psi}_k, J_k) \to (\bar{\psi}_{k+1}, J_{k+1}) , \qquad (4.39)$$

for the small values of the perturbation parameter ϵ. Indeed, using (4.31)–(4.33) the mapping for these variables can be written as

$$J_{k+1} = J_k - \frac{\epsilon}{2} \frac{\partial}{\partial \vartheta_k} [S(\vartheta_{k+1}, J_{k+1}) + S(\vartheta_{k+1}, J_k)] ,$$

$$\bar{\psi}_{k+1} = \bar{\psi}_k + 2\pi \frac{\omega(J_{k+1})}{\Omega} + \frac{\epsilon}{2} \left[\frac{\partial S(\vartheta_{k+1}, J_{k+1})}{\partial J_{k+1}} + \frac{\partial S(\vartheta_{k+1}, J_k)}{\partial J_k} \right] . \qquad (4.40)$$

According to (4.31), (4.33) for small values of ϵ we have

$$S(\vartheta_{k+1}, J_{k+1}) = S\left(\bar{\psi}_k + \epsilon\frac{\partial S(\vartheta_{k+1}, J_k)}{\partial J_k}, J_{k+1}\right)$$
$$\approx S(\bar{\psi}_k, J_{k+1}) + O(\epsilon) ,$$

$$S(\vartheta_{k+1}, J_k) = S\left(\bar{\psi}_k + \epsilon\frac{\partial S(\vartheta_{k+1}, J_k)}{\partial J_k}\right.$$
$$\left.J_{k+1} + \frac{\epsilon}{2}\frac{\partial}{\partial\vartheta_k}\left[S(\vartheta_{k+1}, J_{k+1}) + S(\vartheta_{k+1}, J_k)\right]\right) \approx S(\bar{\psi}_k, J_{k+1}) + O(\epsilon) .$$

Using these approximations and neglecting in (4.39) the small terms of order of ϵ^n, $n \geq 2$, the mapping (4.39) can be written in the non-symmetric form (4.9) with the generating function $S(\bar{\psi}_k, J_{k+1})$ given by equation (4.34).

4.3.2 Standard Hamiltonian and Corresponding Mappings

In the case $H_0(I) = I^2/2$, $\epsilon = K/4\pi^2$, $H_m(I) = \delta_{m1}$ and $\Omega = 1$ the Hamiltonian (4.24) coincides with the standard Hamiltonian (3.25) discussed in Sect. 3.4. According to (4.34) the generating function of the standard Hamiltonian is equal to $S(\vartheta) = \pi\cos\vartheta$ and the corresponding map takes the form

$$J_k = I_k + \frac{K}{4\pi}\sin\vartheta_k ,$$
$$\vartheta_{k+1} = \vartheta_k + 2\pi\omega(J_k) , \qquad (4.41)$$
$$I_{k+1} = J_k + \frac{K}{4\pi}\sin\vartheta_{k+1} .$$

We will call this map as a *symmetric standard map*. It is valid for arbitrary value of perturbation parameter K. It has been first obtained in Abdullaev (1999) for small values of K, and for the arbitrary values of K in Abdullaev (2002).

The standard map of non-symmetric forms (3.24) and (3.26) may be obtained from the symmetric radial map (4.41) for the variables (J_k, ϑ_k), i.e.,

$$J_k = J_{k-1} + \frac{K}{2\pi}\sin\vartheta_k , \qquad \vartheta_{k+1} = \vartheta_k + 2\pi J_k , \qquad (4.42)$$

for the mapping $(\vartheta_k, J_{k-1}) \rightarrow (\vartheta_{k+1}, J_k)$ or

$$\vartheta_{k+1} = \vartheta_k + 2\pi J_k, \qquad J_{k+1} = J_k + \frac{K}{2\pi}\sin\vartheta_{k+1} , \qquad (4.43)$$

for the mapping $(\vartheta_k, J_k) \rightarrow (\vartheta_{k+1}, J_{k+1})$. One can see from these mappings that the action variable J in the standard map does not coincide with the action variable I in the standard Hamiltonian (3.25). Therefore, *the variables*

(ϑ, J) *in the standard mapping are not canonical.* One should note that the canonical mapping for the standard Hamiltonian can be obtained in the non-symmetric form (4.38), (4.37). It gives the mapping (3.29) given in Sect. 3.4.

Poincaré sections, the sequence of phase space coordinates (I_k, ϑ_k), of the symmetric standard map (4.41), and the standard maps (4.42), (4.43), and the non-symmetric mapping (3.29) are shown in Fig. 4.4 for the perturbation parameter $K = 0.7$. The symmetric standard map (4.41) very well reproduces Poincaré section obtained by the integration of the standard Hamiltonian system (3.25) and by the symmetric map (4.6)–(4.8) with the generating function (4.27) with a finite number M (see Abdullaev (2002)). The non-symmetric map (3.29) also reproduces these results, but with less accuracy (Fig. 4.4d): the width of the stochastic layer (a dark area) is smaller than in the case of the symmetric map, and the invariant curves are slightly deformed.

For $K \geq 1$ the system exhibits unrestricted chaotic motion along the action variable I. We have compared the second moments of displacement $\sigma^2(t) = \langle (I(t) - \langle I \rangle)^2 \rangle$ of such a motion obtained by three different methods: by the symmetric map (4.6)–(4.8) with the generating function S (4.27) taking a

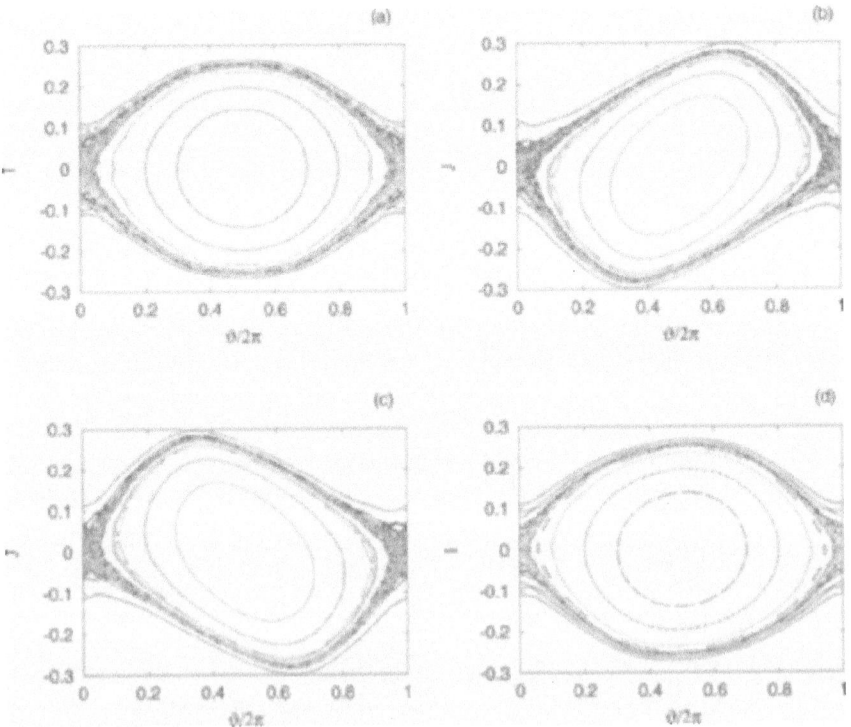

Fig. 4.4. Poincaré sections for the standard Hamiltonian in the (ϑ, I)-plane obtained by (**a**) the symmetric standard map (4.41), (**b**) the standard map (4.42), (**c**) the standard map (4.43), (**d**) the map (3.29). Parameter $K = 0.7$

finite number of harmonics $M = 20$, by the symmetric standard map (4.41), and by the standard map (4.42). The initial stage of evolution of $\sigma^2(t)$ are plotted in Fig. 4.5 for the perturbation parameter $K = 2$. As seen from Fig. 4.5 the symmetric standard map closely describes the system with a finite M (curves 1 and 2), while the standard map predicts a noisy behavior (dotted curve 3).

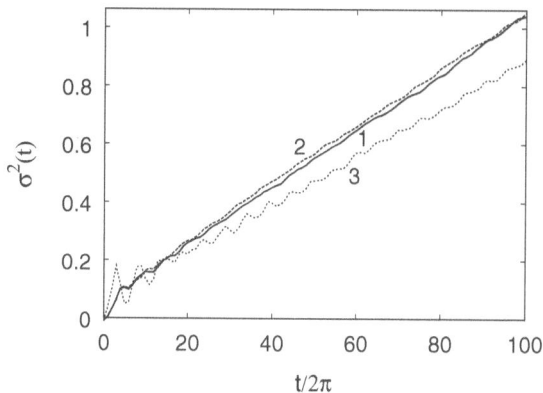

Fig. 4.5. Second moment of displacement $\sigma^2(t)$ obtained by three different methods: *solid curve* 1 corresponds to the symmetric map (4.6)–(4.8) for the standard Hamiltonian (3.25) with $M = 20$, *dashed curve* 2 – to the symmetric standard map (4.41), and *dotted curve* 3 – to the standard map (4.42). Parameter $K = 2.0$

In summary, the Hamiltonian system (2.2), (2.3), (4.24) with a broad perturbation spectrum, $M \gg 1$ and arbitrary perturbation parameter, ϵ, can be replaced by the symmetric mapping (4.31)–(4.33) with the generating function (4.34) or by the non-symmetric mapping (4.38) with the generating function (4.37). The mappings in non-symmetric forms of type (4.9) with the simple form (4.34) can be obtained only approximately for the small perturbation parameter, $\epsilon \ll 1$, in terms of the intermediate variables that are not identical to the original variables of the Hamiltonian system. The differences between the original variables and the intermediate variables are of the order of the perturbation parameter.

4.4 Mappings with Lie Generating Functions

Symplectic mappings can be also constructed using the time-dependent Lie transform method (see Sect. 2.3.4). In this case a map is determined by the time-dependent forward and inverse Lie transformation operators, $\hat{T}(t, t_0)$ and $\hat{T}^{-1}(t, t_0)$:

$$(\vartheta_{k+1}, I_{k+1}) = \hat{T}^{-1}(t_{k+1}, t_0)\hat{T}_0(t_{k+1}, t_k)\hat{T}(t_k, t_0)(\vartheta_k, I_k) ,$$

where $\hat{T}_0(t_{k+1}, t_k)$ stands for the evolution of transformed Hamiltonian system with $\mathcal{H}(J, \epsilon)$ in the time interval (t_k, t_{k+1}). Taking into account the first order terms of perturbation series for the transformation operator $\hat{T}$ the map can be presented in the form similar to the symplectic maps (4.6)–(4.8) in which the generating function of mixed variables $S(\vartheta, J, t, t_0)$ is replaced by the first order Lie generating function $W_1(J, \psi, t, t_0)$:

$$J_k = I_k - \epsilon \frac{\partial W_1(\vartheta_k, I_k, t_k, t_0)}{\partial \vartheta_k} \,,$$

$$\psi_k = \vartheta_k + \epsilon \frac{\partial W_1(\vartheta_k, I_k, t_k, t_0)}{\partial I_k} \,. \tag{4.44}$$

$$\psi_{k+1} = \psi_k + w(J_k, \epsilon)(t_{k+1} - t_k) \,, \tag{4.45}$$

$$I_{k+1} = J_k + \epsilon \frac{\partial W_1(\psi_{k+1}, J_k, t_{k+1}, t_0)}{\partial \psi_{k+1}} \,,$$

$$\vartheta_{k+1} = \psi_{k+1} - \epsilon \frac{\partial W_1(\psi_{k+1}, J_k, t_{k+1}, t_0)}{\partial J_k} \,. \tag{4.46}$$

where $W_1(\psi, J, t, t_0)$ is determined by the formula (2.34) for $S_1(\vartheta, J, t, t_0)$ replacing the variable ϑ by ψ. The main advantage of the maps (4.44)–(4.46) is that the transformation of variables are determined explicitly, unlike the mapping (4.6)–(4.8) with implicitly determined variables. It allows to integrate Hamiltonian system very fast.

However, mappings with the Lie generating functions have a serious shortcoming since they are not area–preserving. They cannot be used to study long term evolution of Hamiltonian systems, especially in non-integrable cases. As was shown in Sect. 2.3.5 in the example of the Duffing oscillator that the transformations with the Lie generating functions describes well the evolution of system in non-resonant cases over sufficiently long time intervals, but it fails in the resonant case.

Consider, as an example, the application of maps with the Lie generating function to the Duffing oscillator (2.43). The first order generating function S_1 is determined by (2.49). The Lie generating function W_1 is equal to S_1 with replacement of ϑ by ψ. The phase space of the oscillator in the (x, p)-plane obtained by the symmetric map is shown in Fig. 4.6 for the non-resonant (a) and resonant (b) perturbations. The time-step of maps is $\tau = 2\pi/\Omega$, the frequency $\omega_0 = 1$. The perturbation parameter ϵ is taken equal to 0.01 for the non-resonant case with the frequency of external perturbation $\Omega = 1.5$, and $\epsilon = 0.001$ for the resonant case $\Omega = \omega_0 = 1.5$.

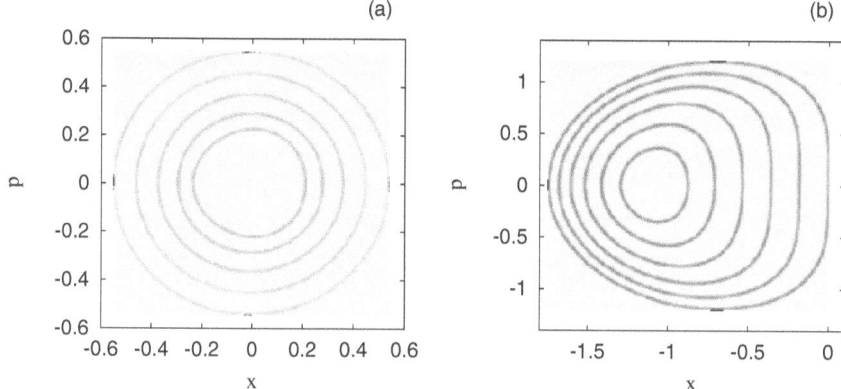

Fig. 4.6. Poincaré sections of the Duffing oscillator in the (x, p)-plane obtained by the symmetric map (4.6)–(4.8). **(a)** Non-resonant case $\Omega = 1.5$, the perturbation parameter $\epsilon = 0.01$; **(b)** Resonant case: $\Omega = 1$, the perturbation parameter $\epsilon = 0.001$. Parameter $\omega_0 = 1$

We have calculated a deviation of action variable I obtained by the maps from the one calculated by the direct integration of the Duffing equation (2.43), i.e., $\Delta I(t) = |I_{map} - I_{eqn}|$. The time – dependence of $\Delta I(t)$ for the non-resonant case is plotted in Fig. 4.7 for the orbit with the initial value of $I = 0.145$. As can see that the relative deviation for the map with the mixed variable generating function does not grow, and it does not exceed a level 5×10^{-3}. However, for the map with the Lie generating function $\Delta J(t)$ grows with time and it reaches the relative level 0.02 at the time $t = 2\pi \times 10^3$.

In the resonant case $\Omega = \omega_0$ the relative deviation $\Delta(t)/I_0$ for the map with the Lie generating function becomes of order of 1 at very short time $t \sim 200\pi$, for the map with the mixed variable generating function it reaches the level of 10% after very long time $t = 2\pi \times 10^4$.

Therefore, the maps with the Lie generating function can be used to description of short term evolution of nonresonant orbits, while maps with mixed variable generating functions are useful to study the long-term evolution of system under resonant perturbations.

4.5 Poincaré Maps at Arbitrary Sections of Phase-Space

The described method of canonical transformations to obtain symplectic maps along time can be also applied to obtain Poincaré map at the arbitrary section of phase space. Here we construct Poincaré maps for Hamiltonian systems in action-angle variables.

To be specific we consider the section Σ of phase space where the angle variable ϑ_1 is fixed (by module 2π). Since the condition $\dot{\vartheta}_1 = \partial H/\partial I_1 =$

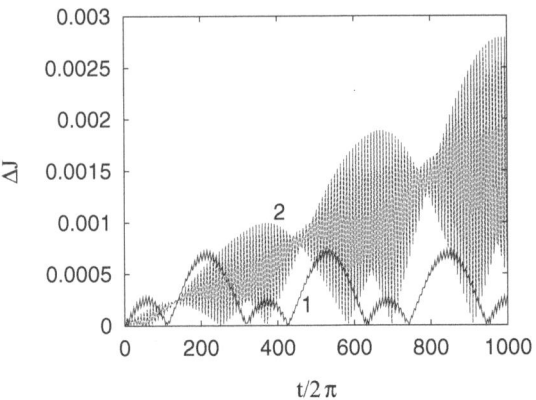

Fig. 4.7. Time-dependence of deviation of the action variable $\Delta I(t) = |I_{map}(t) - I_{eqn}(t)|$ obtained by the mappings from the one obtained the direct integration of the Duffing equation in the non-resonant case. Curve 1 corresponds to the symmetric map, curve 2 — to the map with the Lie generating function. Parameters are $\epsilon = 0.01$, $\Omega = 1.5$, $\omega_0 = 1$

$\omega_1(I) > 0$ is always satisfied one can formulate Hamiltonian equations with the angle ϑ_1 as an independent "time" -variable (see Sect. 1.1.3). We introduce notations $\bar{\vartheta} = (\vartheta_2, \cdots, \vartheta_N)$, $\bar{I} = (I_2, \cdots, I_N)$. The corresponding Hamiltonian function $K \equiv K(t, \bar{\vartheta}, p_t, \bar{I}, \vartheta_1) = -I_1$ can be obtained by inverting the original Hamiltonian $H(\vartheta, I, t)$ with respect to the action variable I_1. The time t and the energy $h = -H$ are conjugated canonical variables. The Hamiltonian equations of motion in these variables are

$$\frac{dt}{d\vartheta_1} = \frac{\partial K}{\partial h} \ , \qquad \frac{dh}{d\vartheta_1} = -\frac{\partial K}{\partial t} \ , \tag{4.47}$$

$$\frac{d\vartheta_i}{d\vartheta_1} = \frac{\partial K}{\partial I_i} \ , \qquad \frac{dI_i}{d\vartheta_1} = -\frac{\partial K}{\partial \vartheta_i} \ , \qquad (i = 2, \ldots, N) \ . \tag{4.48}$$

Suppose that the original Hamiltonian H is presented in the form (2.3). Since the perturbation parameter ϵ is small then the Hamiltonian K can be always written as an expansion series in powers of ϵ:

$$K = K_0(h, \bar{I}) + \epsilon K_1(t, \bar{\vartheta}, h, \bar{I}, \vartheta_1) + \epsilon^2 K_2(t, \bar{\vartheta}, h, \bar{I}, \vartheta_1) + \cdots , \tag{4.49}$$

where K_0 is the unperturbed Hamiltonian obtained from the original unperturbed Hamiltonian $H_0(I)$, and K_i $(i = 1, 2, \ldots)$ are the perturbed parts of Hamiltonian. The relation between the new Hamiltonian K_i and the original Hamiltonian H_0, H_1 can be found by expansion

$$K = -I_1(H_0 + \epsilon H_1) = -I_1(H_0) - \epsilon H_1 \frac{\partial I_1}{\partial H} - \epsilon^2 H_1^2 \frac{\partial^2 I_1}{\partial H^2} + \cdots . \tag{4.50}$$

Comparing the presentation (4.49) and the expansion (4.50) we obtain

$$K_0(h, \bar{I}) = -I_1(H_0),$$
$$K_1(t, \bar{\vartheta}, h, \bar{I}, \vartheta_1) = -\epsilon H_1 \frac{\partial I_1}{\partial H},$$
$$K_2(t, \bar{\vartheta}, h, \bar{I}, \vartheta_1) = -\epsilon^2 H_1^2 \frac{\partial^2 I_1}{\partial H^2}, \cdots . \tag{4.51}$$

The functions K_i ($i = 1, 2, \ldots$) are the periodic functions of the "time" variable ϑ_1 with the period 2π, and they can be presented by Fourier expansions

$$K_i(t, \bar{\vartheta}, h, \bar{I}, \vartheta_1) = \sum_m K_m^{(i)}(h, \bar{I}) e^{im \cdot \vartheta - in\Omega t} . \tag{4.52}$$

The unperturbed frequencies of the system ($\epsilon = 0$) (4.47), (4.48), (4.49) are

$$\nu_t(h, \bar{I}) = \frac{\partial K_0}{\partial h} = \frac{\partial I_1}{\partial H} \frac{\partial H_0}{\partial H} = \frac{1}{\omega_1},$$
$$\nu_i(h, \bar{I}) = \frac{\partial K_0}{\partial I_i} = -\frac{\partial I_1}{\partial H_0} \frac{\partial H}{\partial I_i} = \frac{\omega_i}{\omega_1}, \quad i = 2, \ldots, N, \tag{4.53}$$

where $\omega_i = \omega_i(I) = \partial H_0/\partial I_i$, ($i = 1, \ldots, N$) are frequencies of the original unperturbed Hamiltonian system with $H_0(I)$.

Poincaré map at the section Σ ($\vartheta_1 (\mathrm{mod}\, 2\pi) = $ const.) can be constructed similar to the time-step map (3.2). The time instants t_k should be replaced by $\vartheta_k = 2\pi k + $ const. (Furthermore we will omit the subscript "1" in the notation of the angle variable ϑ_1, and set const=0). Let $(t_k, \bar{\vartheta}_k, h_k, \bar{I}_k)$ be the k-th crossing point of the orbit $(t(\vartheta), \bar{\vartheta}(\vartheta), h(\vartheta), \bar{I}(\vartheta))$ with the section Σ.

Using a procedure described in Sect. 4.1 one can construct a return (Poincaré) map to the section Σ, i.e.,

$$(t_{k+1}, \bar{\vartheta}_{k+1}, h_{k+1}, \bar{I}_{k+1}) = \hat{M}(t_k, \bar{\vartheta}_k, h_k, \bar{I}_k) . \tag{4.54}$$

The corresponding map is similar to (4.6)–(4.8), it has the form

$$\bar{J}_k = \bar{I}_k - \epsilon \frac{\partial S_k}{\partial \bar{\vartheta}_k} , \qquad \bar{\psi}_k = \bar{\vartheta}_k + \epsilon \frac{\partial S_k}{\partial \bar{J}_k} ,$$
$$\bar{\psi}_{k+1} = \bar{\psi}_k + \Delta \vartheta \bar{\mathrm{w}}(\mathcal{H}, \bar{J}, \epsilon) , \tag{4.55}$$
$$\bar{I}_{k+1} = \bar{J}_k + \epsilon \frac{\partial S_{k+1}}{\partial \bar{\vartheta}_{k+1}} , \qquad \bar{\vartheta}_{k+1} = \bar{\psi}_{k+1} - \epsilon \frac{\partial S_{k+1}}{\partial \bar{J}_k} ,$$

for the action-angle variables $(\bar{\vartheta}, \bar{I})$, and

$$\mathcal{H}_k = h_k - \epsilon\frac{\partial S_k}{\partial t_k} , \qquad \mathcal{T}_k = t_k + \epsilon\frac{\partial S_k}{\partial \mathcal{H}_k} ,$$

$$\mathcal{T}_{k+1} = \mathcal{T}_k + \Delta\vartheta \mathbf{w}_t(\mathcal{H}, \bar{J}, \epsilon) , \qquad (4.56)$$

$$h_{k+1} = \mathcal{H}_k + \epsilon\frac{\partial S_{k+1}}{\partial t_{k+1}} , \qquad t_{k+1} = \mathcal{T}_{k+1} - \epsilon\frac{\partial S_{k+1}}{\partial \mathcal{H}_k} ,$$

for the energy–time (t, H). Here $S_k \equiv S(t_k, \bar{\vartheta}_k, \mathcal{H}_k, \bar{J}_k, \vartheta = 2\pi k, \vartheta_0, \epsilon)$, $S_{k+1} \equiv S(t_{k+1}, \bar{\vartheta}_{k+1}, \mathcal{H}_k, \bar{J}_k, \vartheta = 2\pi(k+1), \vartheta_0, \epsilon)$. The map step $\Delta\vartheta$ along the independent variable ϑ is equal to 2π. The generating function $S = S(t, \bar{\vartheta}, \mathcal{H}, \bar{J}, \vartheta)$ determines the changes of variables $(t, \bar{\vartheta}, h, \bar{I}) \rightarrow (\mathcal{T}, \bar{\psi}, \mathcal{H}, \bar{J})$ which transforms the Hamiltonian $K(t, \bar{\vartheta}, h, \bar{J}, \vartheta, \epsilon)$ (4.49) into the new one $\mathcal{K}_0 = \mathcal{K}_0(\mathcal{H}, \bar{J}, \epsilon)$. The perturbed frequencies of the system are

$$\bar{w}(\mathcal{H}, \bar{J}, \epsilon) = \frac{\partial\mathcal{K}_0(\mathcal{H}, \bar{J}, \epsilon)}{\partial\bar{J}} , \qquad \bar{w}_t(\mathcal{H}, \bar{J}, \epsilon) = \frac{\partial\mathcal{K}_0(\mathcal{H}, \bar{J}, \epsilon)}{\partial\mathcal{H}} .$$

In order to improve the accuracy one can take smaller map steps $\Delta\vartheta$. It is convenient to set $\Delta\vartheta = 2\pi/s$, where $s \geq 1$ is an integer number. Then the Poincaré map (4.54) is obtained by applying the map (4.55), (4.56) s times.

According to (2.34) the generating function S in the first order of ϵ is determined

$$S_1(t, \bar{\vartheta}, \mathcal{H}, \bar{J}, \vartheta, \vartheta_0) = -\int_{\vartheta_0}^{\vartheta} K_1(t(\vartheta'), \bar{\vartheta}(\vartheta'), \mathcal{H}, \bar{J}, \vartheta')d\vartheta' , \qquad (4.57)$$

where the integral is taken along unperturbed trajectories $\bar{\vartheta}(\vartheta') = \bar{\vartheta}(\vartheta) + \nu(\mathcal{H}, \bar{J})(\vartheta' - \vartheta)$. Using (4.51) and (4.53), (4.57) may be rewritten in the terms of unperturbed trajectory $\vartheta(t') = \vartheta(t) + \omega(J)(t' - t)$ of the original Hamiltonian $H_0(I)$:

$$S_1(t, \bar{\vartheta}, \mathcal{H}, \bar{J}, \vartheta, \vartheta_0) = \int_{t_0}^{t} H_1(t', \vartheta(t'), J)dt' . \qquad (4.58)$$

where times t and t_0 are located in the interval $t_k < t, t_0 < t_{k+1}$, where t_k is a time instant when the trajectory $\vartheta(t)$ crosses the section Σ, i.e., $\vartheta_1(t_k) = k2\pi$.

Consider the two degrees of freedom Hamiltonian system

$$H(q_1, q_2, p_1, p_2) = \frac{1}{2}(p_1^2 + p_2^2 + \omega_1^2 q_1^2 + \omega_2^2 q_2^2) + \epsilon\left[q_1 q_2 - \frac{1}{4}q_2^4\right] . \qquad (4.59)$$

It describes the interaction of two coupled oscillators. In the case $\epsilon = 1$ it coincides with the Henon–Heiles model Hénon and Heiles (1964). The oscillation along q_1 coordinate with a frequency ω_1 is linear, and the one along the q_2 coordinate with a frequency ω_2 is a weakly nonlinear. The parameter ϵ stands for the strength of interactions of oscillators, as well as for the degree of nonlinearity.

In action-angle variables (ϑ_i, I_i) introduced as

$$q_i = \sqrt{2I_i/\omega_i}\,\sin\vartheta_i\,, \qquad p_i = \sqrt{2I_i\omega_i}\,\cos\vartheta_i\,, \qquad i = 1, 2\,,$$

the Hamiltonian can be rewritten as

$$H = H_0(I_1, I_2, \epsilon) + \epsilon H_1(\vartheta_1, \vartheta_2, I_1, I_2)\,, \tag{4.60}$$

$$H_0(I_1, I_2) = \omega_1 I_1 + \omega_2 I_2\left(1 - \epsilon\frac{3I_2}{8\omega_2^3}\right)\,, \tag{4.61}$$

$$H_1(\vartheta_1, \vartheta_2, I_1, I_2) = \sqrt{\frac{I_1 I_2}{\omega_1\omega_2}}\,[\cos(\vartheta_1 - \vartheta_2) - \cos(\vartheta_1 + \vartheta_2)]$$
$$+ \frac{I_2^2}{2\omega_2^2}\left(\cos 2\vartheta_2 - \frac{1}{4}\cos 4\vartheta_2\right)\,. \tag{4.62}$$

According to (4.51), for the small perturbation parameter $\epsilon \ll 1$ the expansion terms K_0 and K_1 of the Hamiltonian K (4.50) with the angle ϑ_1 as independent "time" variable are

$$K_0(h, I_2) = \frac{h}{\omega_1} + \frac{\omega_2}{\omega_1}I_2\left(1 - \epsilon\frac{3I_2}{8\omega_2^3}\right)\,, \tag{4.63}$$

$$K_1(\vartheta_1, \vartheta_2, h, I_2) = -H_1(\vartheta_1, \vartheta_2, -h - I_2, I_2)$$
$$= \sqrt{\frac{(-h - \omega_2 I_2)I_2}{\omega_1^3\omega_2}}\,[\cos(\vartheta_1 - \vartheta_2) - \cos(\vartheta_1 + \vartheta_2)]$$
$$+ \frac{I_2^2}{2\omega_1\omega_2^2}\left(\cos 2\vartheta_2 - \frac{1}{4}\cos 4\vartheta_2\right)\,. \tag{4.64}$$

The frequencies ν_2 and ν_t determined by the unperturbed Hamiltonian (4.63) are

$$\nu_2(I_2) = \frac{\partial K_0(h, I_2)}{\partial I_2} = \frac{\omega_2}{\omega_1}\left(1 - \epsilon\frac{3I_2}{4\omega_2^3}\right)\,, \qquad \nu_t = \frac{\partial K_0(h, I_2)}{\partial h} = \frac{1}{\omega_1}\,. \tag{4.65}$$

According to (4.57) and (2.35) the first order generating function $S(t, \vartheta_2, \mathcal{H}, J_2, \vartheta, \vartheta_0)$ associated with the map (4.55), (4.56) is determined by

$$S(t, \vartheta_2, \mathcal{H}, J_2, \vartheta, \vartheta_0) = -(\vartheta - \vartheta_0)\sqrt{\frac{(-h - \omega_2 J_2)J_2}{\omega_1^3 \omega_2}}$$

$$\times \sum_{n=\pm 1} [a(x_n)\cos(\vartheta_2 - n\vartheta) + b(x_n)\sin(\vartheta_2 - n\vartheta)]$$

$$+(\vartheta - \vartheta_0)\frac{J_2^2}{2\omega_1\omega_2^2} \sum_{m=2,4} \left(-\frac{2}{m}\right)^{m/2} \left(a(x_m)\sin m\vartheta_2 + b(x_m)\cos m\vartheta_2\right),$$

$$(4.66)$$

where $x_n = (\nu_2(J_2) - n)(\vartheta - \vartheta_0)$, $x_m = m\nu_2(J_2)(\vartheta - \vartheta_0)$, and $\vartheta \equiv \vartheta_1$.

In order to obtain the Poincaré section in the (q_2, p_2)-plane with $q_1 = 0$, where $\vartheta = 0$ (or $\vartheta = \pi$) one should apply the map (4.55) $(\bar{\vartheta}_k, \bar{I}_k) \to (\bar{\vartheta}_{k+1}, \bar{I}_{k+1})$ with the map step $\Delta\vartheta = 2\pi/s$ successively s times. Figure 4.8 shows the corresponding Poincaré section for the resonant case $\omega_1 = \omega_2$ with the perturbation parameter $\epsilon = 10^{-3}$. The map step is $s = 2$ and the energy of system is taken equal to $E = 0.4$. It well reproduces the Poincaré section obtained by the direct integration of the Hamiltonian system (4.59) using the standard symplectic integration scheme.

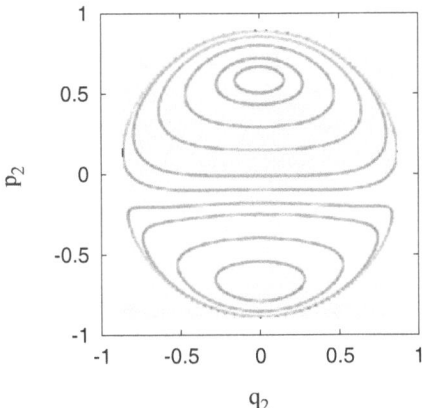

Fig. 4.8. Poincaré section of the system (4.59) in the (q_2, p_2)-plane with $q_1 = 0$ plotted by mapping (4.55). The perturbation parameter $\epsilon = 10^{-3}$, and the map step $\Delta\vartheta = \pi$

4.6 Method of Successive Canonical Transformations

The mapping method developed above uses only one canonical (forward and backward) transformation of variables, $(\vartheta, I) \to (\psi, J)$, with the generating function S sought as an expansion series in powers of perturbation parameter

ϵ: $\epsilon S = \epsilon S_1 + \epsilon^2 S_2 + \cdots + \epsilon^m S_m$. The difference between the exact generating function and the one with m terms is of order of ϵ^{m+1}.

The mapping $(\vartheta_{k+1}, I_{k+1}) = \hat{M}(\vartheta_k, I_k)$ can be also constructed using the Kolmogorov's technique of successive canonical transformations (see Sect. 2.4). Below we describe the corresponding scheme of this approach. We present the corresponding mapping in the following form

$$(\vartheta_{k+1}, I_{k+1}) = \hat{M}_- \hat{M}_0 \hat{M}_+ (\vartheta_k, I_k) , \qquad (4.67)$$

of three successive mappings $\hat{M}_+$, $\hat{M}_0$, and $\hat{M}_-$ which stand for

$$(\psi_k, J_k) = \hat{M}_+ (\vartheta_k, I_k) ,$$

$$(\bar{\psi}_k, J_{k+1}) = \hat{M}_0 (\psi_k, J_k) , \qquad (4.68)$$

$$(\vartheta_{k+1}, I_{k+1}) = \hat{M}_- (\bar{\psi}_k, J_{k+1}) ,$$

respectively. We construct the first map $\hat{M}_+$ as a sequence of maps $\hat{M}_+ = \hat{M}_+^{(m)} \cdots \hat{M}_+^{(1)}$, corresponding to the successive canonical change of variables

$$(\vartheta_k, I_k) \rightarrow (\psi_k^{(1)}, J_k^{(1)}) \rightarrow \cdots \rightarrow (\psi_k^{(m)}, J_k^{(m)}) , \qquad (4.69)$$

at the time instant $t = t_k$. This transforms the Hamiltonian as

$$H_0(I) + \epsilon H_1(\vartheta, I, t)$$
$$\rightarrow \mathcal{H}_0^{(1)}(J^{(1)}, \epsilon, t) + \epsilon^2 \mathcal{H}_1^{(1)}(\psi^{(1)}, J^{(1)}, t, \epsilon)$$
$$\rightarrow \cdots$$
$$\rightarrow \mathcal{H}_0^{(m)}(J^{(m)}, \epsilon) + \epsilon^{m^2} \mathcal{H}_1^{(m)}(\psi^{(m)}, J^{(m)}, t, \epsilon) . \qquad (4.70)$$

Each of steps in (4.69) is given by the generating functions $F_i = J^{(i)} \cdot \psi^{(i-1)} + \epsilon^{2^{(i-1)}} S^{(i)}(J^{(i)}, \psi^{(i-1)}, t, t_0)$, $(i = 1, \cdots, m)$, associated with the map $\hat{M}_+^{(i)}$ at the i-th step

$$J_k^{(i)} = J_k^{(i-1)} - \epsilon^{2^{(i-1)}} \frac{\partial S_k^{(i)}}{\partial \psi_k^{(i-1)}} , \qquad \psi_k^{(i)} = \psi_k^{(i-1)} + \epsilon^{2^{(i-1)}} \frac{\partial S_k^{(i)}}{\partial J_k^{(i)}}, \qquad (4.71)$$

where $S_k^{(i)} \equiv S^{(i)}(J^{(i)}, \psi^{(i-1)}, t = t_k, t_0)$. It is determined by the perturbation function $\mathcal{H}_1^{(i)}(\psi^{(i)}, J^{(i)}, t, \epsilon)$

$$S^{(i)}(J^{(i)}, \psi^{(i-1)}, t, t_0) = - \int_{t_0}^t \mathcal{H}_1^{(i-1)}(J^{(i)}(t'), \psi^{(i-1)}(t'), t', \epsilon) dt' . \qquad (4.72)$$

In (4.72) the integration is taken along the unperturbed orbit

$$J^{(i)} = \text{const}, \qquad \psi^{(i-1)}(t') = w_i(J^{(i)}, \epsilon)(t' - t) + \psi^{(i-1)}(t) ,$$

of the Hamiltonian $\mathcal{H}_0^{(i)}(J^{(i)}, \epsilon)$ with the frequency $w_i(J^{(i)}, \epsilon) = \partial\mathcal{H}_0^{(i)}/\partial J^{(i)}$, $(\psi^{(0)} \equiv \vartheta, \mathcal{H}_1^{(0)} \equiv H_1)$.

The map $\hat{M}_0$ simply determines evolution of variable $(\psi_k^{(m)}, J_k^{(m)})$ in the time interval:

$$J_{k+1}^{(m)} = J_k^{(m)}, \qquad \bar{\psi}_k^{(m)} = \psi_k^{(m)} + w_m(J_k^{(m)}, \epsilon)(t_{k+1} - t_k). \qquad (4.73)$$

Finally the map $\hat{M}_- = \hat{M}_-^{(1)} \cdots \hat{M}_-^{(m)}$ carries out the transformation of the new variables $(\psi_{k+1}^{(m)}, J_{k+1}^{(m)})$ at the time $t = t_{k+1}$ back to the original variables $(\vartheta_{k+1}, I_{k+1})$:

$$(\bar{\psi}_k^{(m)}, J_{k+1}^{(m)}) \to \cdots \to (\bar{\psi}_k^{(1)}, J_{k+1}^{(1)}) \to (\vartheta_{k+1}, I_{k+1}), \qquad (4.74)$$

at the time instant $t = t_{k+1}$. Each of the successive steps of the map $\hat{M}_-^{(i)}$ is given by

$$J_{k+1}^{(i-1)} = J_{k+1}^{(i)} + \epsilon^{2(i-1)} \frac{\partial S_{k+1}^{(i)}}{\partial \bar{\psi}_k^{(i-1)}}, \qquad \bar{\psi}_k^{(i-1)} = \bar{\psi}_k^{(i)} - \epsilon^{2(i-1)} \frac{\partial S_{k+1}^{(i)}}{\partial J_{k+1}^{(i)}}, \quad (4.75)$$

where $S_{k+1}^{(i)} \equiv S^{(i)}(J^{(i)}, \bar{\psi}_k^{(i-1)}, t = t_{k+1}, t_0)$, $(i = 1, \cdots, m)$, and $(\vartheta_{k+1} = \bar{\psi}_k^{(0)}, I_{k+1} = J_{k+1}^{(0)})$.

Since this technique possesses the quadratic convergence, the accuracy of mapping will be higher. However, the calculations of higher order generation functions are somehow cumbersome that makes the application of this superconvergent mapping technique to Hamiltonian problems less practical.

4.7 Summary

In summary, we have developed a general method to construct symplectic mappings for generic Hamiltonian systems. It may be applied to the general class of Hamiltonian systems which may be composed as the sum of fully integrable system and Hamiltonian perturbation. The perturbation is not required to be small. The construction of mappings is based on the application of the Hamilton–Jacobi method, particularly, the Hamilton–Jacobi equation and Jacobi's theorem to Hamiltonian systems in finite – time intervals. The generating functions of maps are solutions of the Hamilton–Jacobi equations in finite time intervals. They are found using the time-dependent perturbation series developed in Sect. 2.3 It appears that the expansion parameter near the resonant frequencies is determined by the product of the perturbation parameter and the time step of map. Particularly, it allows one to apply mapping method to systems with moderately large perturbations taking the map time-step sufficiently small.

Symplectic mappings has been also obtained for the Hamiltonian systems affected by the broad perturbation spectrum. We have considered the limit of infinite number of perturbation modes. The method to construct mappings to the arbitrary cross section of the phase space has been also developed. The superconvergent version for construction of the mappings using the Kolmogorov's technique is discussed.

We have studied the accuracy of mapping method and compared it with conventional symplectic integration methods, particularly, with the most accurate fifth-order Runge – Kutta symplectic integrator (see Sect. 1.5. It was found the map with the large time-steps comparable with the characteristic time scale of system (e.g., a perturbation period) have the same accuracy as the symplectic integrator with two or three order smaller integration steps. The most importantly that the accuracy of map does not depend on perturbation frequency, and thereby it allows one to integrate highly oscillatory Hamiltonian systems which are most challenging problem in numerical analysis (see Petzold et al. (1997)).

5 Mappings Near Separatrix. Theory

A motion near the separatrix of Hamiltonian systems has fundamental generic features. As was first shown by Poincaré (1892–99) (see also Sect. 7.1.3) any small time-periodic perturbation splits the separatrices corresponding to stable and unstable manifolds which leads to the onset of chaotic motion due to the exponential divergence of orbits with close initial conditions. This phenomenon creates the zone of phase space in the small vicinity of the unperturbed separatrix, so-called a *stochastic layer* where the motion of system is chaotic (see Sect. 7.1.3).

The dynamics of motion near the separatrix can be most conveniently described by the mappings. A mapping near the separatrix, known the *separatrix (or whisker) mapping* first introduced by Filonenko and Zaslavsky (1968); (see also Chirikov (1977, 1979); Zaslavsky et al. (1991)), and its different modifications has been widely used in various problems of physics and astronomy. It is a powerful tool to study the onset of chaotic motion due to the separatrix splitting and properties of the stochastic layer (see Sect. 6.6 for bibliographic notes.)

In this chapter we describe a systematic and rigorous method to derive symplectic mappings near the separatrix based on the Hamilton–Jacobi method to construct mappings (see Chap. 4). This method allows one to construct directly a mapping near separatrix in canonical variables. The presentation mostly follows the papers by Abdullaev (2004b, 2005).

5.1 Separatrix and Mappings

Separatrices are phase-space curves separating regions with the different type of motion. It connects the hyperbolic fixed point (or points) in phase space. Two examples of such connections are shown in Fig. 5.1. The saddle connection is called a *homoclinic orbit* if the saddle point is connected by itself (see Fig. 5.1a), or a *heteroclinic orbit*, if it connects different saddle points (see Fig. 5.1b). In typical Hamiltonian systems the separatrices are unstable to any small perturbations[1]. In particular, a time-periodic perturbation destroys the separatrix, and the motion near the unperturbed separatrix becomes chaotic.

---

[1] See Sect. 7.5.

S.S. Abdullaev: *Construction of Mappings for Hamiltonian Systems and Their Applications*,
Lect. Notes Phys. **691**, 83–104 (2006)
www.springerlink.com

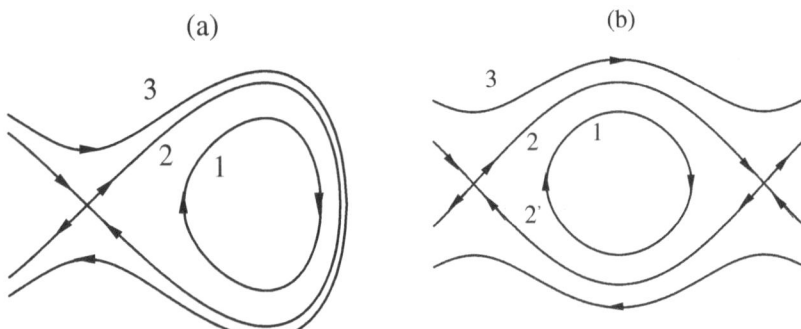

Fig. 5.1. Phase space structure of system with separatrices: (**a**) homoclinic orbits (curve 2) and (**b**) heteroclinic orbits (curves 2 and 2′) connecting different saddles points

The domain of chaotic motion, *the stochastic layer*, is formed in the small vicinity the unperturbed separatrices.

The motion near the separatrix is generic for nonlinear Hamiltonian systems. It has been studied over a long time because of its importance in numerous physical applications (see Filonenko and Zaslavsky (1968); Chirikov (1979); MacKay and Meiss (1987)). The description of motion near the separatrix by symplectic maps has own specific features. The frequency of unperturbed motion, $\omega(I)$, goes to zero when it approaches the separatrix, $\omega(I) \to 0$ for $I \to I_s$, where I_s is a value of action at the separatrix. This makes inconvenient the use the stroboscopic time step mappings (3.2) because of long calculation times near the separatrix.

As was shown in Sect. 1.4 on the example of the pendulum the spectra q_m, p_m of phase oscillations given by (1.49), (1.50) become broader while the motion approaches the separatrix, $H \to H_s$ (H_s is the energy at the separatrix) (see (1.51)). The spectrum of phase oscillations, $H_m(I)$, in the perturbed Hamiltonian $H_1(\vartheta, I, t)$, i.e.,

$$H_1(\vartheta, I, t) = \sum_m H_m(I) \cos(m\vartheta - n\Omega t) , \qquad (5.1)$$

has a similar behavior near the separatrix, $H_m \sim m^{-\beta}$. In typical cases the exponent β becomes less than 3 near the separatrix. The perturbed Hamiltonian $H_1(\vartheta, I, t)$ approaches to a singular $\delta-$like function of angle variable ϑ.

The first attempt to construct area–preserving mapping to describe the dynamics of motion near the separatrix has been taken by Filonenko and Zaslavsky (1968) for the periodically driven pendulum. They had derived a symplectic map $(\vartheta_k, I_k) \to (\vartheta_{k+1}, I_{k+1})$ for action-angle variables (ϑ, I) with the time step equal to the half of period of phase oscillations in ϑ. The mapping $(H_k, t_k) \to (H_{k+1}, t_{k+1})$ of the energy (H) and time (t) variables over half of phase oscillations in ϑ has been derived by Chirikov (1977, 1979); Zaslavsky et al. (1991). This definition of mapping is more consistent with the Hamiltonian formalism with the angle variable ϑ as an independent time-like variable (see Sect. 1.1.3). Below we recall this map known as the *separatrix* (or *whisker*) *map*.

5.1.1 The Conventional Separatrix (Whisker) Map

Consider the periodically-driven pendulum described by the Hamiltonian

$$H(q, p, t) = \frac{p^2}{2} - \omega_0^2 \cos q + 2\epsilon\omega_0^2 \cos q \cos \Omega t , \tag{5.2}$$

where ω_0 is the frequency of small oscillations, ϵ and Ω represent the amplitude and the frequency of perturbation, respectively.

We recall that the unperturbed system ($\epsilon = 0$) has elliptic fixed points at ($q = 2\pi n, p = 0$) and hyperbolic fixed points at ($q_s = 2\pi(s + 1/2), p_s = 0$) ($n, s = 0, \pm1, \pm2, \ldots$) (see Sect. 1.4 and Fig. 1.4, 5.1b). The separatrices (curve 2) connecting the saddle points q_s, p_s with $q_{s\pm1}, p_{s\pm1}$ separates the trapped orbits ($-\omega_0^2 < H < \omega_0^2$) (curve 1) from the un-trapped ones ($H > H_s = \omega_0^2$) (curve 3). The period of trapped orbits $T(H) = 2\pi/\omega(H)$ has the following asymptotics near the separatrix $H = H_s$:

$$T(H) = \frac{1}{\omega_0} \ln \frac{32\omega_0^2}{|H - \omega_0^2|} + O(|H - \omega_0^2|) , \tag{5.3}$$

$$\text{for } H \to \omega_0^2 \pm 0 .$$

The orbits on the separatrices ($H = \omega_0^2$) are

$$q_s^{(\pm)}(t) = 4 \arctan \frac{\exp[\pm\omega_0(t - t_0)] - 1}{\exp[\pm\omega_0(t - t_0)] + 1} ,$$

$$p_s^{(\pm)}(t) = \pm\frac{2\omega_0}{\cosh[\omega_0(t - t_0)]} , \tag{5.4}$$

where the signs ($\pm$) correspond to the upper (curve 2) and lower branches (curve 2') of the separatrix, respectively, and t_0 is a time instant when the orbit crosses a mid-point between two sequential saddle points.

The perturbation ($\epsilon \neq 0$) destroys the separatrix for any small amplitude of perturbation ϵ forming the stochastic layer near the unperturbed separatrix shown in Fig. 5.2. In order to describe the motion near the destroyed separatrix Chirikov (1977, 1979) introduced the map (t_k, H_k) $\to$ (t_{k+1}, H_{k+1}) as the increment of the time, $\Delta t(t_k, H_k)$, and energy, $\Delta H(t_k, H_k)$, over the half of phase rotation in phase space:

$$H_{k+1} = H_k + \Delta H(t_k, H_k) ,$$

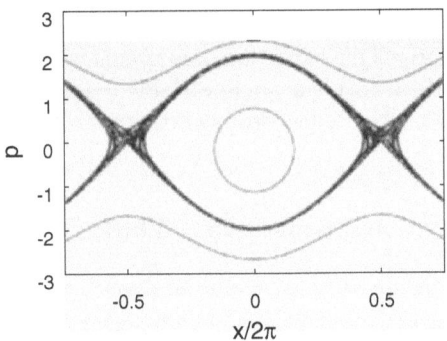

Fig. 5.2. Stochastic layer near the separatrix

$$t_{k+1} = t_k + \Delta t(t_k, H_k) \ . \tag{5.5}$$

It is required that the mapping (5.5) should be area-preserving: $|\partial(H_{k+1}, t_{k+1})/\partial(H_k, t_k)| = 1$.

The energy increment, ΔH, is found using the equation for the evolution of energy H

$$\frac{dH}{dt} = \frac{\partial H}{\partial t} = -2\epsilon\Omega\omega_0^2 \cos q \sin \Omega t \ .$$

Integration of this equation along the unperturbed orbits (5.4) on the separatrix is reduced to the Melnikov integral (Melnikov (1963), and see also Chirikov (1979); Lichtenberg and Lieberman (1992)):

$$\Delta H = -2\epsilon\Omega\omega_0^2 \int\limits_{-\infty}^{\infty} \cos q_s(t) \sin \Omega t dt \ , \tag{5.6}$$

taken along the unperturbed separatrix orbit $q_s(t)$ (5.4). For the Hamiltonian (5.2) the Melnikov integral is given by

$$\Delta H = -\omega_0^2 W \sin \Omega t_0 \ , \qquad W = \epsilon \frac{\Omega^2}{\omega_0^2} \frac{4\pi}{\sinh(\pi\Omega/2\omega_0)} \ ,$$

where t_0 is the time instant when the orbit crosses a mid-point between two sequential saddle points. The increment of time Δt is equal to the half of period of oscillation $T(H)$ (5.3).

Identifying the the time instant t_0 with time variable t_k in the mapping (5.5) the latter can be presented in the form

$$h_{k+1} = h_k - W \sin \varphi_k \ ,$$
$$\varphi_{k+1} = \varphi_k + \frac{\Omega}{\omega_0} \ln \frac{32}{|h_{k+1}|} \ , \qquad \text{mod } 2\pi \ , \tag{5.7}$$

known as *the separatrix (or whisker) map*. In (5.7) the normalized energy, $h = (H - \omega_0^2)/\omega_0^2$, and the phase, $\varphi = \Omega t_0$, variables are introduced.

5.1.2 Shortcomings of Conventional Separatrix Mappings

One of the main shortcomings of the separatrix mapping (5.7) and other similar mappings is that they are not canonical mappings, i.e., *the mapping variables* t_k, H_k *(or* φ_k, h_k*) are not canonically conjugated.* It follows from the geometrical interpretation of the separatrix mapping first given by Escande (1988) (see also Rom-Kedar (1994, 1995)) as a return map of time and energy variables defined at the different sections in phase-space (x, p) of system. As we will see in Sect. 5.2 the energy variable, H, is taken at the section near the saddle point, while the time variable, t, is in the section located in the middle between two consecutive saddle points, i.e., the sections Σ_s and Σ_c in Fig. 6.6, respectively. Therefore the separatrix map (5.7) does not coincide with the definition of the Poincaré return map where all variables are defined at the same cross section of phase space. This fact should be kept in mind when one compares the properties of original continuous Hamiltonian system with the ones of the corresponding separatrix mapping.

As an example consider the rescaling invariant property of the separatrix map (5.7) which has been found in Abdullaev and Zaslavsky (1994); Zaslavsky and Abdullaev (1995). Since the second equation in (5.7) is determined by module 2π it is easy to see that the separatrix map is invariant with respect to the following transformation of perturbation parameter ϵ:

$$\epsilon \to \epsilon' = \lambda\epsilon , \tag{5.8}$$

where the parameter $\lambda = \exp(2\pi\omega_0/\Omega)$ depending only on the perturbation frequency Ω and the frequency ω_0 of small amplitude oscillations. This interesting property of the separatrix map, however, is not revealed in the phase space (x, p) of the Hamiltonian system (5.2). As we will see in Sect. 8.1, it does not exactly coincide with the rescaling invariance property of Hamiltonian system near the hyperbolic saddle point.

We should also note an inconsistency which appears in the quantization problem of classical systems using non-canonical mappings. For instance, in several works (see Casati et al. (1990)) it has been taken attempts to quantize the conventional (non-canonical) Kepler map for studying quantum effects in the process of ionization of highly–excited hydrogen atom in a microwave. However, such a procedure of quantization of the Kepler map by presenting energy and time variables as a canonical pair of operators is not consistent with the fundamental principles of quantum mechanics since the energy and time variables in this map are not canonically conjugated.

Method to derive separatrix mappings are mainly based on the calculations of the increments of time and energy variables over phase rotation in phase space (see, for instance Chirikov (1979)). This method does not allow directly obtain canonical separatrix mappings. On the other hand this method does not allow to estimate the accuracy of the separatrix mapping.

5.2 The Hamilton–Jacobi Method
to Construct Maps Near a Separatrix

Below we present the rigorous derivation of symplectic mappings near the
separatrix of general Hamiltonian systems using the Hamilton–Jacobi method
described in Chap. 4. Furthermore we call them separatrix mappings in a
broader sense as mappings near the separatrices of arbitrary Hamiltonian
systems. This method has been first used by Abdullaev (1999) to construct
separatrix mappings for Hamiltonian systems with a single hyperbolic saddle
point. Here we consider the case of systems with arbitrary number of saddle
points. For the sake of simplicity we restrict ourselves with generic $1 + 1/2$
degrees of freedom Hamiltonian systems $H(x, p, t)$ presented in the form

$$H(x, p, t) = H_0(x, p) + \epsilon H_1(x, p, t) . \tag{5.9}$$

where $H_0(x, p)$ is the unperturbed Hamiltonian, $H_1(x, p, t)$ is the time-
dependent perturbation, and (x, p) are the canonical coordinate and momentum.

Suppose that the unperturbed system (1.1), (2.3) ($\epsilon = 0$) at certain energy
level $H_0(x, p) = $ const has a finite (or countable) number of saddle points
and corresponding number of saddle–saddle connections in phase space as
illustrated in Fig. 5.3. Let (x_s, p_s) and (x_{s+1}, p_{s+1}) be two consecutive saddle
points with a heteroclinic connection as shown in Fig. 5.4a. If the system has
only one saddle point then the points (x_s, p_s) and (x_{s+1}, p_{s+1}) coincide and
a saddle–saddle connection is a homoclinic orbit. The phase space curves 1
and 3 located on the both sides of the separatrix describes different types of
motion. We put $H = H_s = 0$ on the separatrix.

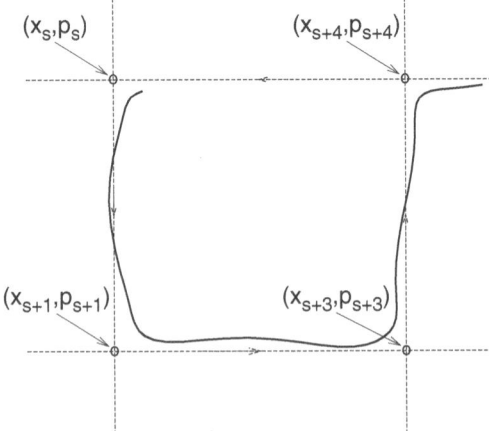

Fig. 5.3. Phase space of system with several saddle points. *Dotted curves* describe
the unperturbed. A perturbed orbit is displayed by a *solid curve*

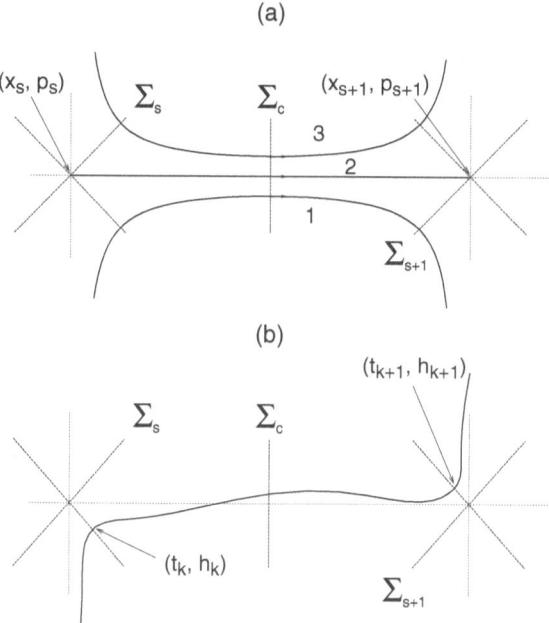

Fig. 5.4. (a) Phase curve in the neighborhood of the separatrix: curves 1 and 3 describe orbits upper and down the separatrix, curve 2 is the separatrix. (b) Schematic view of the separatrix map. *Solid curve* describes the perturbed orbit, and *dotted curve* is the unperturbed separatrix

In the phase plane (x, p) we introduce the cross sections Σ_c and Σ_s, shown in Fig. 5.4a. The section Σ_c consists of a segment perpendicular to the separatrix at the mid–point between saddle points. The section Σ_s is located near saddle points (x_s, p_s) and consists of two segments perpendicular to each other with the crossing point at (x_s, p_s). The both segments of Σ_s are also perpendicular to unperturbed phase curves.

We define action-angle variables, (I, ϑ), for the unperturbed motion in a following way. The action and angle variables are introduced as an integrals

$$I = \frac{1}{2\pi} \int_C p(x; H) dx \,, \qquad \vartheta = \frac{\partial}{\partial I} \int^x p(x'; H) dx' \,, \qquad (5.10)$$

where C is the segment of phase-space curve of constant $H = H_0(x, p)$ located between two consecutive crossing points with the sections Σ_s and Σ_{s+1}. Introduced in such a way the action variable $I(H)$ is a continuous function of energy H while it crosses the separatrix. We will set $\vartheta = 0 \pmod{2\pi}$ at the section Σ_c, $\vartheta = \mp\pi \pmod{2\pi}$ at the section Σ_s and (x_{s+1}, p_{s+1}) at Σ_{s+1}, respectively.

In typical Hamiltonian systems any small time-periodic perturbation destroys the separatrix, and orbits wobbles around the unperturbed separatrix (see Fig. 5.4b). Let t_k and h_k be a time instant and an energy at the k-th crossing point of the orbit with Σ_s. We intend to construct the map

$$(t_{k+1}, H_{k+1}) = \hat{M}_{s+1,s}(t_k, H_k) , \tag{5.11}$$

connecting the crossing point (t_k, H_k) at the section Σ_s with the corresponding point (t_{k+1}, H_{k+1}) at Σ_{s+1}. The geometric scheme of the mapping is shown in Fig. 5.4b. The change of the angular variable ϑ over one step of the map (5.11) is equal to $\Delta\vartheta \equiv \vartheta_{k+1} - \vartheta_k = 2\pi$.

Suppose that the system has N_{sep} independent saddle–saddle connections. Then there exist N_{sep} independent mappings (5.11) which completely determine the dynamics of a Hamiltonian system. The sequence of mappings $\hat{M}_{s+1,s}$ depends on the topology of saddle–saddle connections in phase space and the trajectory of motion. Below we develop a general method to construct the full set of mappings. It will be illustrated on specific examples in the next sections.

5.2.1 Mapping Along Single Saddle–Saddle Connection

Below we construct the mapping (5.11) along the single saddle–saddle connection. For this one could use the formulation of Hamiltonian equations (4.47), (4.48) with the angle variable, ϑ, as an independent variable (see Sect. 4.5). However this method fails near the separatrix where the frequency of motion $\omega(H) = dH_0/dI \to 0$ (or $dI/dH_0 \to \infty$). This singularity does not allow to invert Hamiltonian $H(I, \vartheta, t)$ in respect to the action variable I near the separatrix and to present it in the form (4.49). To avoid this difficulty we will use another approach.

We use the formulation of Hamiltonian equations in the extended phase space (t, x, p_0, p) (see Sect. 1.1). According to (1.9) and (2.3) the equations of motion in the space of action-angle and time-energy variables (t, ϑ, p_0, I) are

$$\frac{dt}{d\tau} = \frac{\partial \mathcal{H}}{\partial p_0} , \qquad \frac{dp_0}{d\tau} = -\frac{\partial \mathcal{H}}{\partial t} ,$$

$$\frac{d\vartheta}{d\tau} = \frac{\partial \mathcal{H}}{\partial I} , \qquad \frac{dI}{d\tau} = -\frac{\partial \mathcal{H}}{\partial \vartheta} , \tag{5.12}$$

with the Hamiltonian function

$$\mathcal{H}(t, \vartheta, p_0, I) = H_0(I) + p_0 + \epsilon \mathcal{H}_1(t, \vartheta, p_0) . \tag{5.13}$$

where $p_0 = -H$, and $\mathcal{H}_1(t, \vartheta, p_0) \equiv H_1(I(-p_0), \vartheta, t)$. In (5.13) the perturbation $\mathcal{H}_1$ is chosen as the function of energy H.

Suppose that the orbit crosses the section Σ_s at $\tau = \tau_k$ and the next section Σ_{s+1} at $\tau = \tau_{k+1}$. We construct the mapping

$$(t_{k+1}, \vartheta_{k+1}, h_{k+1}, I_{k+1}) = \hat{M}(t_k, \vartheta_k, h_k, I_k) \,, \tag{5.14}$$

where $(t_k, \vartheta_k, h_k, I_k) \equiv (t(\tau_k), \vartheta(\tau_k), -p_0(\tau_k), I(\tau_k))$. From the geometry of mapping illustrated in Fig. 5.4b it follows that we should impose constraints on the angle variable ϑ: $\vartheta(\tau_k) = -\pi$, and $\vartheta(\tau_{k+1}) = \pi$.

Using results obtained in Section (4.1) one can write down the mapping (5.14) in the general form (4.6)–(4.8) for the action-angle variables (ϑ, I) and for the time-energy variables (t, p_0):

$$J_k = I_k - \epsilon \frac{\partial S^{(k)}}{\partial \vartheta_k} \,,$$

$$\Theta_k = \vartheta_k + \epsilon \frac{\partial S^{(k)}}{\partial J_k} \,,$$

$$\bar{\Theta}_k = \Theta_k + w(\mathcal{H}_k, J_k, \epsilon)(\tau_{k+1} - \tau_k) \,, \tag{5.15}$$

$$I_{k+1} = J_k + \epsilon \frac{\partial S^{(k+1)}}{\partial \vartheta_{k+1}} \,,$$

$$\vartheta_{k+1} = \bar{\Theta}_{k+1} - \epsilon \frac{\partial S^{(k+1)}}{\partial J_k} \,,$$

for action-angle variables, (ϑ, I), and

$$\mathcal{H}_k = h_k + \epsilon \frac{\partial S^{(k)}}{\partial t_k} \,,$$

$$\mathcal{T}_k = t_k - \epsilon \frac{\partial S^{(k)}}{\partial \mathcal{H}_k} \,,$$

$$\bar{\mathcal{T}}_k = \mathcal{T}_k + w_t(\mathcal{H}_k, J_k, \epsilon)(\tau_{k+1} - \tau_k) \,, \tag{5.16}$$

$$h_{k+1} = \mathcal{H}_k - \epsilon \frac{\partial S^{(k+1)}}{\partial t_{k+1}} \,,$$

$$\vartheta_{k+1} = \bar{\mathcal{T}}_k + \epsilon \frac{\partial S^{(k+1)}}{\partial \mathcal{H}_k} \,,$$

for the time – energy variables $(t, h = -p_0)$. Here

$$S^{(k)} \equiv S(t_k, \vartheta_k, J_k, \mathcal{H}_k, \tau_k, \tau_0, \epsilon)$$

$$S^{(k+1)} \equiv S(t_{k+1}, \vartheta_{k+1}, J_k, \mathcal{H}_k, \tau_{k+1}, \tau_0, \epsilon)$$

are values of the generating function $S(t, \vartheta, J, \mathcal{H}, \tau, \tau_0, \epsilon)$ at $\tau = \tau_k$ and $\tau = \tau_{k+1}$, respectively. For the Hamiltonian system (5.12), (5.13) it obeys the Hamilton–Jacobi equation

$$\mathcal{H}\left(t, \vartheta, P_0 + \epsilon \frac{\partial S}{\partial t}, J + \epsilon \frac{\partial S}{\partial \vartheta}\right) + \epsilon \frac{\partial S}{\partial \tau} = \bar{H}(P_0, J, \epsilon) \tag{5.17}$$

in the time interval $\tau_k < \tau < \tau_{k+1}$ satisfying the initial condition $S|_{\tau=\tau_0} = 0$ at the time instant $\tau = \tau_0$. The time τ_0 is a free parameter lying in the interval $\tau_k < \tau_0 < \tau_{k+1}$. The new Hamiltonian $\bar{H}(P_0, J, \epsilon)$ depends only on new "action" variables (P_0, J). In (4.6), (4.56) $w(\mathcal{H}, J, \epsilon) = \partial\bar{H}(P_0, J, \epsilon)/\partial J$ and $w_t(\mathcal{H}, J, \epsilon) = \partial\bar{H}(P_0, J, \epsilon)/\partial P_0$ are the frequencies of perturbed motion. (Recall that $\mathcal{H} = -P_0$).

To solve the Hamilton–Jacobi equation (5.17) we apply the perturbation theory in finite time interval $\tau_k < \tau < \tau_{k+1}$ described in Sect. 2.3. In the first order of perturbation ϵ the generating function, S, is given by the following integral similar to (2.34):

$$S_1(t, \vartheta, \mathcal{H}, J, \tau, \tau_0) = -\int_{\tau_0}^{\tau} \mathcal{H}_1(\tau', \vartheta(\tau'), -\mathcal{H})d\tau' , \qquad (5.18)$$

where the integral over "time" τ' is taken along the unperturbed orbit $\vartheta(\tau') = \omega(\mathcal{H})(\tau' - \tau) + \vartheta(\tau)$, $t(\tau') = \tau'$. In this approximation the corresponding frequencies $w(J, \epsilon)$ and $w_t(J, \epsilon)$ can be replaced by unperturbed frequencies $\omega(\mathcal{H}) = \partial\mathcal{H}_0(I, p_0)/\partial I$ and $w_t(J) = \partial\mathcal{H}_0(I, p_0)/\partial p_0 = 1$, where $\mathcal{H}_0(I, p_0) = H_0(I) + p_0$.

Since the generating function S_1 does not depend on action J the mapping (5.14) takes the simplified form

$$J_k = I_k - \epsilon\frac{\partial S_k}{\partial\vartheta_k} ,$$

$$\vartheta_{k+1} = \vartheta_k + \omega(J_k)(\tau_{k+1} - \tau_k) , \qquad (5.19)$$

$$I_{k+1} = J_k + \epsilon\frac{\partial S_{k+1}}{\partial\vartheta_{k+1}} .$$

Recalling that $\vartheta_{k+1} - \vartheta_k = 2\pi$, and using the second equation in (5.19) we obtain $\tau_{k+1} - \tau_k = 2\pi/\omega(J_k)$. Then the mapping for time –energy (t, H) variables is reduced to

$$\mathcal{H}_k = h_k + \epsilon\frac{\partial S_k}{\partial t_k} , \qquad \mathcal{T}_k = t_k - \epsilon\frac{\partial S_k}{\partial\mathcal{H}_k} ,$$

$$\mathcal{T}_{k+1} = \mathcal{T}_k + \tau_{k+1} - \tau_k = \mathcal{T}_k + \frac{2\pi}{\omega(\mathcal{H}_k)} , \qquad (5.20)$$

$$h_{k+1} = \mathcal{H}_k - \epsilon\frac{\partial S_{k+1}}{\partial t_{k+1}} , \qquad t_{k+1} = \mathcal{T}_{k+1} + \epsilon\frac{\partial S_{k+1}}{\partial\mathcal{H}_k} ,$$

where $S_k \equiv S_1(t_k, \vartheta_k, \mathcal{H}_k, J_k, \tau_k, \tau_0)$, $\omega(\mathcal{H}_k) \equiv \omega(J_k)$.

The map (5.20) is a most general form of the mapping of time (t) and energy (H) variables to certain sections in phase space along a single saddle–saddle connection. In the first order of perturbation parameter ϵ the generating function S associated with this map is determined by (5.18). The separatrix mapping can be obtained from (5.20), (5.18) in some limiting cases. By appropriate choosing the time parameter τ_0 in (5.18) one can obtain different forms of the mapping (see Sect. 4.1.1).

5.2.2 Calculation of the Generating Function

Consider a multi-frequency perturbation with frequencies Ω_n and present the perturbed Hamiltonian $\mathcal{H}_1(\mathcal{H}, \vartheta(t), t)$ in (5.18) taken along unperturbed trajectory as a Fourier series:

$$\mathcal{H}_1(\tau, \vartheta(\tau), -\mathcal{H}) = \sum_n H_n(\mathcal{H}, \vartheta(\tau)) \cos(\Omega_n t(\tau) + \chi_n) , \qquad (5.21)$$

where χ_n are the phases of perturbation. Suppose the orbit crosses the section Σ_c at the time instant $\tau = t_c$ when the phase $\vartheta = 0$, and present the Fourier coefficients as

$$V_n(\mathcal{H}, \tau - t_c) \equiv H_n(\mathcal{H}, \vartheta(\tau)) .$$

Taking into account that the unperturbed orbit is given by $t = \tau$, $\vartheta(\tau') = \vartheta + \omega(\mathcal{H})(\tau' - \tau)$, we find that $t_c = t - \vartheta/\omega(\mathcal{H})$. Then the generating function, $S(t) \equiv S_1(t, \vartheta, \mathcal{H}, \tau, \tau_0)$ in the time interval $t_k < t, t_c < t_{k+1}$ can be reduced to

$$S(t) = -\int_{\tau_0}^{t} \sum_n V_n \left(\mathcal{H}, t' - t + \frac{\vartheta}{\omega(\mathcal{H})} \right) \cos(\Omega_n t' + \chi_n) dt' ,$$

$$= \mathrm{Re} \sum_n R_n(\mathcal{H}, \vartheta, t) \exp\left[i\Omega_n \left(t - \frac{\vartheta}{\omega(\mathcal{H})} \right) + i\chi_n \right] , \qquad (5.22)$$

where

$$R_n(\mathcal{H}, \vartheta, t) = \int_{\vartheta/\omega(\mathcal{H})}^{\tau_0 + \vartheta/\omega(\mathcal{H}) - t} V_n(\mathcal{H}, \tau) e^{i\Omega_n \tau} d\tau . \qquad (5.23)$$

At the limits $t \to t_k + 0$, $\vartheta \to -\pi$ and $t \to t_{k+1} - 0 = t_k + 2\pi/\omega(\mathcal{H})$, $\vartheta \to \pi$ we have

$$S^{(k)}(t_k, \mathcal{H}) = \sum_n \left(K_n^-(\mathcal{H}) \cos \Phi_n^+(t_k, \mathcal{H}) - L_n^-(\mathcal{H}) \sin \Phi_n^+(t_k, \mathcal{H}) \right) ,$$

$$S^{(k+1)}(t_{k+1}, \mathcal{H}) = -\sum_n \left(K_n^+(\mathcal{H}) \cos \Phi_n^-(t_{k+1}, \mathcal{H}) - L_n^+(\mathcal{H}) \sin \Phi_n^-(t_{k+1}, \mathcal{H}) \right) ,$$

$$(5.24)$$

where

$$\Phi_n^\pm(t, \mathcal{H}) = \Omega_n \left(t \pm \frac{\pi}{\omega(\mathcal{H})} \right) + \chi_n , \qquad (5.25)$$

and $K_n^\pm(\mathcal{H})$, $L_n^\pm(\mathcal{H})$ are real and imaginary parts of the integrals $R_n^\pm(\mathcal{H}) = K_n^\pm(\mathcal{H}) + iL_n^\pm(\mathcal{H})$, respectively, defined as

$$R_n^-(\mathcal{H}) \equiv R_n(\mathcal{H}, \vartheta = -\pi, t = t_k) = \int\limits_{-\pi/\omega(\mathcal{H})}^{\tau_0 - t_k - \pi/\omega(\mathcal{H})} V_n(\mathcal{H}, \tau) e^{i\Omega_n \tau} d\tau ,$$

$$R_n^+(\mathcal{H}) \equiv R_n(\mathcal{H}, \vartheta = \pi, t = t_{k+1}) = -\int\limits_{\pi/\omega(\mathcal{H})}^{\tau_0 - t_{k+1} + \pi/\omega(\mathcal{H})} V_n(\mathcal{H}, \tau) e^{i\Omega_n \tau} d\tau .$$

$$(5.26)$$

We call these integrals as Melnikov type integrals.

5.2.3 Symmetric Mappings

We call the mapping (5.20) a *symmetric map* when the free parameter τ_0 is taken exactly in the middle between τ_k and τ_{k+1}, i.e., $\tau_0 = (\tau_{k+1} + \tau_k)/2 = t_k + \pi/\omega(\mathcal{H})$. Then the integrals (5.26) take forms

$$R_n^\pm(\mathcal{H}) = K_n^\pm(\mathcal{H}) + iL_n^\pm(\mathcal{H}) = \mp \int\limits_{\pm\pi/\omega(\mathcal{H})}^{0} V_n(\mathcal{H}, \tau) e^{i\Omega_n \tau} d\tau . \qquad (5.27)$$

The Fourier integrals (5.27) are taken along the unperturbed orbits of system. In a particular case, when the orbits lie on the separatrix ($h = 0$) they coincide with the Melnikov integrals of type (5.6) (see Chirikov (1979)). Indeed, at the limit $h \to 0$ the frequency $\omega(h) \to 0$ and the integrals (5.27) are reduced to

$$R_n^+(0) = \int\limits_0^\infty V_n(0, \tau) e^{i\Omega_n \tau} d\tau ,$$

$$R_n^-(0) = \int\limits_{-\infty}^0 V_n(0, \tau) e^{i\Omega_n \tau} d\tau . \qquad (5.28)$$

It is easy to see the symmetric map conserves an invariance of Hamiltonian system with respect to time reversing, $t \to -t$, $H \to -H$, which is manifested in the invariance of mapping with respect to reversing the mapping sequence, $k \leftrightarrow k + 1$.

5.2.4 Nonsymmetric Mappings

Another forms of mappings can be obtained by setting the free time parameter τ_0 in (5.26) equal to τ_k or τ_{k+1}. They are similar to those nonsymmetric mappings presented in Sect. 4.1.1.

Consider first the case when $\tau_0 = \tau_{k+1}$ the integrals $K_n^{(-)}$ and $L_n^{(-)}$ in (5.26) vanish, and therefore $S^{(k+1)} \equiv 0$, $h_{k+1} = \mathcal{H}_k$. Then the mapping (5.20) is reduced to

$$h_{k+1} = h_k + \epsilon \frac{\partial S(t_k, h_{k+1})}{\partial t_k} ,$$
$$t_{k+1} = t_k + \frac{2\pi}{\omega(h_{k+1})} - \epsilon \frac{\partial S(t_k, h_{k+1})}{\partial h_{k+1}} , \qquad (5.29)$$

determined by only one generating function $S(t_k, h_{k+1})$:

$$S(t_k, h_{k+1}) = \sum_n \left(K_n(h_{k+1}) \cos \Phi_n^+(t_k, h_{k+1}) - L_n(h_{k+1}) \sin \Phi_n^+(t_k, h_{k+1}) \right),$$
$$(5.30)$$

where $K(h)$ and $L(h)$ are the integrals

$$K_n(h) + iL_n(h) = \int\limits_{-\pi/\omega(h)}^{\pi/\omega(h)} V_n(h, \tau) e^{i\Omega_n \tau} d\tau . \qquad (5.31)$$

On the other hand putting $\tau_0 = \tau_k$ we have $S^{(k)} \equiv 0$ and $h_k = \mathcal{H}_k$ since the integrals $K_n^{(+)} \equiv 0$ and $L_n^{(+)} \equiv 0$, and the mapping (5.20) is reduced to

$$h_{k+1} = h_k - \epsilon \frac{\partial S(t_{k+1}, h_k)}{\partial t_{k+1}} ,$$
$$t_{k+1} = t_k + \frac{2\pi}{\omega(h_k)} + \epsilon \frac{\partial S(t_{k+1}, h_k)}{\partial h_k} , \qquad (5.32)$$

where $S^{(k+1)}(t_{k+1}, h_k)$ is given by

$$S(t_{k+1}, h_k) = -\sum_n \left(K_n(h_k) \cos \Phi_n^-(t_{k+1}, h_k) - L_n(h_k) \sin \Phi_n^-(t_{k+1}, h_k) \right) . $$
$$(5.33)$$

In (5.30) and (5.33) the phases $\Phi_n^\pm(t, h)$ are defined by (5.25). At the limit $h \to 0$ we have

$$K_n(0) + iL_n(0) = \int\limits_{-\infty}^{\infty} V_n(0, \tau) e^{i\Omega_n \tau} d\tau . \qquad (5.34)$$

We call the mappings (5.29), (5.32) *nonsymmetric mappings* since they are not invariant with respect to reversing the mapping sequence, $k \leftrightarrow k+1$.

5.2.5 Properties of the Melnikov Type Integrals $K_n(h)$ and $L_n(h)$

In this section we describe some properties of the Melnikov type integrals $K_n(h)$ and $L_n(h)$ defined by (5.28), (5.31). First of all consider the relation between coefficients $H_{mn}(h)$ of the perturbation (5.21) in a Fourier series in angle variable ϑ, i.e.,

$$H_1(h, \vartheta, t) = \sum_n H_n(h, \vartheta) \cos(\Omega_n t + \chi_n) \;,$$

$$H_n(h, \vartheta) = \mathrm{Re} \sum_m H_{mn}(h) e^{im\vartheta} \;, \qquad H_{mn}^*(h) = H_{-m,n}(h) \;, \quad (5.35)$$

and the integrals $R_n(h) = K_n(h) + i L_n(h)$. According to (5.31) one obtains

$$
\begin{aligned}
R_n(h) &= \int_{-\pi/\omega(h)}^{\pi/\omega(h)} H_n(h, \vartheta) e^{i\Omega_n \tau} d\tau \\
&= \mathrm{Re} \sum_m H_{mn}(h) \int_{-\pi/\omega(h)}^{\pi/\omega(h)} e^{i(m\omega + \Omega_n)\tau} d\tau \\
&= \frac{2\pi}{\omega(h)} \sum_m \frac{\sin[\pi(m - \Omega_n/\omega(h))]}{\pi(m - \Omega_n/\omega(h))} H_{mn}^*(h) \;.
\end{aligned}
\tag{5.36}
$$

As seen from (5.36) at the values $h = h_{mn}$ of primary resonances, i.e., $m\omega(h_{mn}) = \Omega_n$, the integral $R_n(h)$ is determined by Fourier coefficients $H_{mn}^*(h)$, i.e.,

$$R_n(h_{mn}) = \frac{2\pi}{\omega(h)} H_{mn}^*(h_{mn}) = \frac{2\pi m}{\Omega_n} H_{mn}^*(h_{mn}) \;. \tag{5.37}$$

The analytical calculation of the integrals $R_n(h)$ is not straightforward. The asymptotical method to estimate these integrals is presented in Appendix B. It is shown that the integral $R_n(h)$ can be presented as a sum of regular and oscillatory parts,

$$R_n(h) = R_n^{(reg)}(h) + R_n^{(osc)}(h) \;. \tag{5.38}$$

The regular part, $R^{(reg)}(h)$, is a smooth and slowly varying function of the relative energy h. We construct this function by extending the function $R_n(h_{mn})$ (5.37) defined at discrete resonant values of h_{mn} (or m) to continuous values of h by replacing the discrete mode number m by the continuous one $m = \Omega_n/\omega(h)$, i.e.,

$$R_n^{(reg)}(h) = \frac{2\pi}{\omega(h)} H_{\Omega_n/\omega(h),n}^*(h) \;. \tag{5.39}$$

At the limit $|h| \to 0$ it tends to the value $R(0)$, i.e., to the Melnikov-Arnold integrals (5.28), (5.34). Analytical and numerical calculations of $R^{(reg)}(h)$

for typical Hamiltonian systems presented in Chap. 6 show that $R^{(reg)}(h)$ is sufficiently close to $R(0)$ at certain small region near the separatrix $h = 0$, i.e., $R^{(reg)}(h) \approx R(0)$.

The oscillatory part, $R^{(osc)}(h)$, is a fast–oscillating function of h with a vanishing amplitude at the limit $|h| \to 0$: $R^{(osc)}(h) \to 0$. The asymptotical formulae for $R^{(osc)}(h)$ at small values of h are given in Appendix B. For Hamiltonian systems with hyperbolic saddle points they have generic features near the separatrix. It was shown that for the perturbation Hamiltonian, $H_1(x, p, t)$, with the non-vanishing first derivative at the saddle point, (x_s, p_s), i.e., $\partial H_1(x, p, t)/\partial x \neq 0$ (or $\partial H_1(x, p, t)/\partial p \neq 0$), the leading term of $R^{(osc)}(h)$ has the following asymptotics like

$$
R^{(osc)}(h) \sim \sqrt{|h|}
\begin{cases}
\sin\left(\frac{\pi \Omega_n}{\omega(h)}\right), & \text{for } h < 0 \ , \\[2mm]
\cos\left(\frac{\pi \Omega_n}{\omega(h)}\right), & \text{for } h > 0 \ .
\end{cases}
\tag{5.40}
$$

Since $\omega(h) \to 0$ when $|h| \to 0$ the frequency of oscillations of $R^{(osc)}(h)$ in h increases with approaching the separatrix. According to the definitions of $R^{(reg)}(h)$ and $R^{(osc)}(h)$, given by (5.38), (5.39) and the property (5.37) the function $R^{(osc)}(h)$ has zeros at the primary resonance values of $h = h_{mn}$ when $m\omega(h_{mn}) = \Omega_n$. As we will see later (see Sect. 5.3) this fact is important to understanding the applicability and the justification of the separatrix mapping.

Another important feature of the integrals $R^{(osc)}(h)$ is its rescaling invariance near the separatrix. As we see later the frequency of motion, $\omega(h)$, near the separatrix of Hamiltonian systems with hyperbolic saddle points has an universal asymptotics, $\omega(h) \sim 2\pi\gamma/\ln|h|$, where γ is a growth increment of orbits (see Sects. 6.1, 6.2, 8.1, 8.5). Then from (5.40) follows that

$$
R^{(osc)}(\lambda^2 h) = \lambda R^{(osc)}(h) \ ,
\tag{5.41}
$$

where $\lambda = \exp(2\pi\gamma/\Omega_n)$ is the universal rescaling parameter mentioned in Sect. 5.1.1 (see Eq. (5.8)).

Complete asymptotical formulae for the integrals $R^{(osc)}(h)$ for generic Hamiltonian systems are derived in Appendix B. We shall also study the properties of these integrals in Chap. 6 for the specific Hamiltonian systems.

5.3 Simplification of Mappings

The symmetric mapping (5.20) with the generating functions (5.24), as well as the nonsymmetric mappings (5.29), (5.32) determined by generating functions (5.30), (5.33), respectively, have a rather complicated structure due to presence of oscillatory parts of integrals $R_n(h)$ in the generating functions $S(h, t)$. This may cause some difficulties in numerical solutions of implicit equations in the mappings, especially when h approaches to 0. For this reason it is desirable to simplify mappings. Below we consider such an approximation which would not only simplify mappings, but also justifies the separatrix mappings.

5.3.1 "Primary Resonant" Approximation

As have been noted above the oscillatory parts of the integrals $R_n(h)$ have zeros at primary resonant values of $h = h_{mn}$, where $R_n(h_{mn})$ proportional to Fourier coefficients H_{mn} of the perturbation Hamiltonian $H_1(I, \vartheta, t)$ (see Eq. (5.37)). Since the primary resonant perturbation, $H_{mn} \cos(m\vartheta - \Omega_n t)$, affects significantly on the system near the resonant values of h_{mn}, $(m\omega(h_{mn}) = \Omega_n)$, while the effect of other nonresonant terms is negligible. Then near the resonant values of h_{mn} the oscillatory part $R_n^{(osc)}(h)$ is significantly small than the regular part $R_n^{(reg)}(h)$:

$$|R_n^{(osc)}(h)| \ll |R_n^{(reg)}(h)|, \qquad \text{for } h \approx h_{mn}. \tag{5.42}$$

Then one can neglect the oscillatory parts, $R_n^{(osc)}(h)$, in the generating functions, $S(h,t)$, replacing the integrals $R_n(h)$ by their regular parts $R_n^{(reg)}(h)$. Furthermore we shall call this approximation as a *"primary resonant"* ap*proximation*.

5.3.2 Simplified Form of Mappings

Further simplification of the mappings can be done using the smallness of perturbation parameter ϵ. Eliminating the intermediate variables, $\mathcal{H}$, $\mathcal{T}$, we transform a set of (5.20) into

$$h_{k+1} = h_k - \epsilon \left(\frac{\partial S_{k+1}}{\partial t_{k+1}} - \frac{\partial S_k}{\partial t_k} \right),$$

$$t_{k+1} = t_k + \frac{2\pi}{\omega(\mathcal{H}_k)} + \epsilon \left(\frac{\partial S_{k+1}}{\partial \mathcal{H}_k} - \frac{\partial S_k}{\partial \mathcal{H}_k} \right). \tag{5.43}$$

Using (5.20) and (5.24), one can show that

$$\frac{2\pi}{\omega(\mathcal{H}_k)} + \epsilon \left(\frac{\partial S_{k+1}}{\partial \mathcal{H}_k} - \frac{\partial S_k}{\partial \mathcal{H}_k} \right) = \frac{\pi}{\omega(h_k)} + \frac{\pi}{\omega(h_{k+1})} + G(t_k, h_{k+1}, h_k) + O(\epsilon^2),$$

$$\epsilon \left(\frac{\partial S_{k+1}}{\partial t_{k+1}} - \frac{\partial S_k}{\partial t_k} \right) = \epsilon F(t_k, h_{k+1}, h_k) + O(\epsilon^2),$$

where

$$F(t_k, h_{k+1}, h_k) = \sum_n \Omega_n \left(K_n^{(reg)}(h_{k+1}) \sin \Phi_n^+(t_k, h_k) \right.$$

$$\left. + L_n^{(reg)}(h_k) \cos \Phi_n^+(t_k, h_k) \right) ,$$

$$G(t_k, h_{k+1}, h_k) = \sum_n \left(\frac{dK_n^{(reg)}(h_{k+1})}{dh_{k+1}} \cos \Phi_n^+(t_k, h_k) \right.$$

$$\left. - \frac{dL_n^{(reg)}(h_{k+1})}{dh_{k+1}} \sin \Phi_n^+(t_k, h_k) \right) . \tag{5.44}$$

with the regular parts of the integrals $K_n(h)$, $L_n(h)$ defined by (5.31). Neglecting the terms of order ϵ^2 one obtains

$$h_{k+1} = h_k - \epsilon F(t_k, h_{k+1}, h_k) ,$$

$$t_{k+1} = t_k + \frac{\pi}{\omega(h_k)} + \frac{\pi}{\omega(h_{k+1})} + \epsilon G(t_k, h_{k+1}, h_k) . \tag{5.45}$$

A straightforward calculation shows that $\det|\partial(h_{k+1}, t_{k+1})/\partial(h_k, t_k)| = 1$, i.e., the mapping (5.45) is a area–preserving. It is also invariant with respect to the time reversing transformation, $k \leftrightarrow k + 1$.

The mapping (5.45) can be also obtained from the nonsymmetric forms of the mappings (5.29), (5.32) using a similar procedure.

5.3.3 Separatrix Mapping Approximation

For typical Hamiltonian systems the regular part, $R_n^{(reg)}(h)$, is a smooth function of h and its deviation from $R_n^{(reg)}(0)$ is small. Then the integrals (5.27) in the generating functions (5.24) can be replaced by the Melnikov type integrals (5.28), i.e., $K_n(h) = K_n(0)$, $L_n(h) = L_n(0)$[2]. Then the mapping (5.45) can be further simplified to

$$h_{k+1} = h_k - \epsilon \sum_n \Omega_n \left\{ K_n(0) \sin \left[\Omega_n \left(t_k + \frac{\pi}{\omega(h_k)} \right) + \chi_n \right] \right.$$

$$\left. + L_n(0) \cos \left[\Omega_n \left(t_k + \frac{\pi}{\omega(h_k)} \right) + \chi_n \right] \right\} ,$$

$$t_{k+1} = t_k + \frac{\pi}{\omega(h_k)} + \frac{\pi}{\omega(h_{k+1})} , \tag{5.46}$$

where $K_n(0)$ and $L_n(0)$ are the integrals defined by (5.34).

---

[2] For specific examples see Sections 6.1 and 6.2 where we shall construct the separatrix mapping for the perturbed double–well potential and the periodically–driven pendulum.

The mapping (5.46) has been first introduced in Abdullaev and Zaslavsky (1995, 1996) to study magnetic field lines in plasmas (see Sect. 10). It is called as a *shifted separatrix mapping* since as it was obtained from the conventional separatrix mapping by shifting the phases Ωt_k (see below Sect. 5.4.1).

Since both variables (t, h) in the mappings (5.45), (5.46) are defined in the neighborhood of the saddle points, they become important to study the dynamics and statistical properties of chaotic motion in a system. This is because of the fact that trajectories spend most of time near the saddle point, and therefore, the whole dynamics is mainly determined by the phase space structure of system in the neighborhood of saddle points. We will discuss this subject in Chap. 8.

5.4 Mapping at Arbitrary Sections of Phase Space

In some applications it is necessary to construct mappings with variables defined at the arbitrary sections of phase space. These mappings can be also constructed similar to the ones presented above. However, in general, a construction of such mappings is not straightforward as the mappings to the sections Σ_s along the single saddle–saddle connection (see Fig. 5.4). In the latter case the mapping (5.11) is determined only by orbits between sections Σ_s and Σ_{s+1}. In order to obtain a mapping (5.11) where the variables (t, H) are defined, for instance, at the sections Σ_c (see Fig. 5.4) one needs to know the topology of all saddle–saddle connections.

Figure 5.5 illustrates the example of the mapping to the sections Σ_c. Suppose the orbit crosses the section Σ_c at (t_k, h_k). Because of a sensitive dependence of orbits near the separatrix on their initial condition the next crossing point (t_{k+1}, h_{k+1}) may lie the section Σ_c either on the left side for

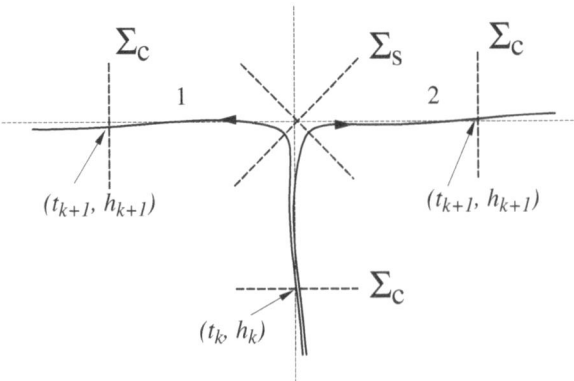

Fig. 5.5. Geometrical illustration of the mapping to the sections Σ_c. *Solid curves* 1 and 2 describe the perturbed orbits, and *dotted curves* are the unperturbed separatrices

the orbits of type 1 or on the right side for the orbits of type 2. The direction
of orbits is determined by the value of energy variable H at the crossing point
of the orbit with Σ_s. Depending on the condition $H > H_s = 0$ or $H < H_s = 0$
the orbit crosses the section Σ_c on the left hand side or on the right hand side.
In this sense the mapping, in general, should be constructed in algorithmic
way. Below we construct corresponding mapping for the system with a single
saddle point and only one a saddle–saddle connection. The construction of
mappings in a system with more than one saddle–saddle connections will be
considered in Sect. 6.2.3 for the periodically–driven pendulum.

5.4.1 Mapping to a Section Σ_c

The phase space of such a system is shown in Fig. 5.6. The hyperbolic saddle
point is located at $(x_s = 0, p_s = 0)$. The dotted curve describes the unperturbed separatrix. The orbit reflects from the rigid border $x = 0$ changing
the sign of momentum p. Then the mapping $(t_k, h_k) \rightarrow (t_{k+1}, h_{k+1})$ defines
the Poincaré return map to the section Σ_c. It has a general form (5.20) with
the generating function (5.22), (5.23). We should put $\vartheta = 0 \pmod{2\pi}$ at both
time instants $t = t_k$ and at $t = t_{k+1}$. For the generating functions S_k and
S_{k+1} of the mapping (5.20) we have

$$S_k(t_k, \mathcal{H}) = \sum_n \left[K_n^+(\mathcal{H}) \cos(\Omega_n t_k + \chi_n) - L_n^+(\mathcal{H}) \sin(\Omega_n t_k + \chi_n) \right] ,$$

$$S_{k+1}(t_{k+1}, \mathcal{H}) = \sum_n \left[K_n^-(\mathcal{H}) \cos(\Omega_n t_{k+1} + \chi_n) - L_n^-(\mathcal{H}) \sin(\Omega_n t_{k+1} + \chi_n) \right] ,$$

$$(5.47)$$

where $K_n^{\pm}(\mathcal{H})$, $L_n^{\pm}(\mathcal{H})$ are the integrals

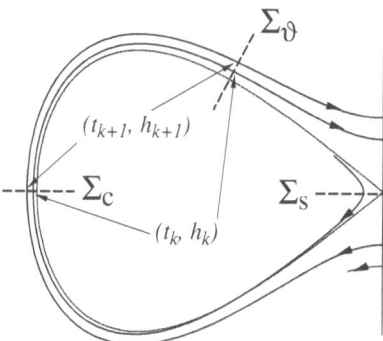

Fig. 5.6. Geometry of the mapping to the sections Σ_c and Σ_ϑ in the system with
the single saddle point

$$K_n^\pm(\mathcal{H}) + iL_n^\pm(\mathcal{H}) = \int\limits_0^{\pm\pi/\omega(\mathcal{H})} V_n(\mathcal{H},\tau)e^{i\Omega_n\tau}d\tau \ . \tag{5.48}$$

Equations (5.20) with the generating functions (5.47) determine the corresponding map. The first set of this mapping is implicit in energy variable $\mathcal{H}$, and the last set is implicit in time t_{k+1}.

The mapping can be significantly simplified if the stochastic layer is sufficiently thin. In this case one can replace the integrals (5.48) by their values at $\mathcal{H} = 0$: $K_n^\pm \equiv K_n^\pm(0)$, $L_n^\pm \equiv L_n^\pm(0)$:

$$K_n^\pm + iL_n^\pm = \int\limits_0^{\pm\infty} V_n(0,\tau)e^{i\Omega_n\tau}d\tau \ . \tag{5.49}$$

In this approximation the generating function (5.47) does not depend on energy variable $\mathcal{H}$, and the mapping (5.20) takes the simplified form

$$\begin{aligned}
\mathcal{H}_k &= h_k + \epsilon\frac{\partial S_k}{\partial t_k} \ , \\
t_{k+1} &= t_k + \frac{2\pi}{\omega(\mathcal{H}_k)} \ , \\
h_{k+1} &= \mathcal{H}_k - \epsilon\frac{\partial S_{k+1}}{\partial t_{k+1}} \ ,
\end{aligned} \tag{5.50}$$

or

$$\begin{aligned}
\mathcal{H}_k &= h_k - \epsilon\sum_n \Omega_n\left[K_n^+ \sin(\Omega_n t_k + \chi_n) + L_n^+ \cos(\Omega_n t_k + \chi_n)\right] \ , \\
t_{k+1} &= t_k + \frac{2\pi}{\omega(\mathcal{H}_k)} \ , \\
h_{k+1} &= \mathcal{H}_k + \epsilon\sum_n \Omega_n\left[K_n^- \sin(\Omega_n t_{k+1} + \chi_n) + L_n^- \cos(\Omega_n t_{k+1} + \chi_n)\right] \ .
\end{aligned} \tag{5.51}$$

This map determines the Poincaré return map of energy (H) and time (t) variables to the section Σ_c.

The map (5.51) can be also written in the form of mapping $(t_k, \mathcal{H}_{k-1}) \to (t_{k+1}, \mathcal{H}_k)$:

$$\begin{aligned}
\mathcal{H}_k &= \mathcal{H}_{k-1} + \epsilon\sum_n \Omega_n\left[K_n \sin(\Omega_n t_k + \chi_n) + L_n \cos(\Omega_n t_k + \chi_n)\right] \\
t_{k+1} &= t_k + \frac{2\pi}{\omega(\mathcal{H}_k)} \ ,
\end{aligned} \tag{5.52}$$

where the coefficients $K_n = K_n^+ - K_n^-$ and $L_n = L_n^+ - L_n^-$ are determined by the Melnikov type integrals (5.34).

Equations (5.52) coincides with the conventional form of the separatrix map in which the energy (H) and time (t) are defined at the different sections of phase space. Indeed, the intermediate variable $\mathcal{H}$ coincides with the energy taken at the section Σ_s, while t is on the section Σ_c.

One should also note that the mapping (5.46) at the section Σ_s can be formally derived from the mapping (5.52) by shifting the time variable t_k from the section Σ_c to the section Σ_s. Since the phase difference between these sections is π the time difference along the unperturbed orbit equals to $\pi/\omega(\mathcal{H})$. Taking into account that the energy variable $\mathcal{H}_k$ in the map (5.52) coincides with one at the section Σ_s and replacing $\mathcal{H}_{k-1} \to h_k$, $t_k \to t_k + \pi/\omega(h_k)$ we obtain the mapping (5.46).

5.4.2 Mapping to Section $\Sigma_{\vartheta=\eta}$

Now we consider the mapping to the arbitrary section Σ_η of phase space with constant phase $\vartheta = \eta$. The schematic view of the section Σ_ϑ is shown in Fig. 5.6. It is specified by the phase $\vartheta = \eta = \text{const}$, $(-\pi < \eta < \pi)$, and consists of the segment of a straight line that can be reached from the section Σ_c in time $\Delta t(h) = |\eta|/\omega(h)$ along unperturbed phase curves about the separatrix. The return map $(t_k, h_k) \to (t_{k+1}, h_{k+1})$, where the variables (t_k, h_k) are at the section Σ_η, is given by (5.20). According to (5.22), (5.23) the generating functions S_k, S_{k+1} are

$$S_k = \sum_n \left(K_n^+(\mathcal{H}, \eta) \cos \Phi_n(t_k, \mathcal{H}) - L_n^+(\mathcal{H}, \eta) \sin \Phi_n(t_k, \mathcal{H}) \right) ,$$

$$S_{k+1} = \sum_n \left(K_n^-(\mathcal{H}, \eta) \cos \Phi_n(t_{k+1}, \mathcal{H}) - L_n^-(\mathcal{H}, \eta) \sin \Phi_n(t_{k+1}, \mathcal{H}) \right) ,$$

$$(5.53)$$

where

$$\Phi_n(t, H) = \Omega_n \left(t - \frac{\eta}{\omega(H)} \right) + \chi_n , \tag{5.54}$$

and $K_n^\pm(\mathcal{H}) + iL_n^\pm(\mathcal{H}) = R_n^\pm(\mathcal{H})$:

$$R_n^+(\mathcal{H}, \eta) \equiv R_n^+(\mathcal{H}, \eta, t = t_k) = \int\limits_{\eta/\omega(\mathcal{H})}^{\pi/\omega(\mathcal{H})} V_n(\mathcal{H}, \tau) e^{i\Omega_n \tau} d\tau ,$$

$$R_n^-(\mathcal{H}, \eta) \equiv R_n^-(\mathcal{H}, \eta, t = t_{k+1}) = \int\limits_{\eta/\omega(\mathcal{H})}^{-\pi/\omega(\mathcal{H})} V_n(\mathcal{H}, \tau) e^{i\Omega_n \tau} d\tau . \quad (5.55)$$

Consider the case of the thin stochastic layer taking the limiting case $\mathcal{H} \to 0$. Then the integrals (5.55) has the following limits. For the value of the phase η in the interval $-\pi < \eta < 0$ we have $R_n^+(0) = K_n(0) + iL_n(0)$ determined by the integral (5.34) and $R_n^- = 0$. Therefore, the generating function $S_{k+1} = 0$ and the mapping (5.20) is reduced to

$$h_{k+1} = h_k + \epsilon \frac{\partial S(t_k, h_{k+1})}{\partial t_k} \ ,$$
$$t_{k+1} = t_k + \frac{2\pi}{\omega(h_{k+1})} - \epsilon \frac{\partial S(t_k, h_{k+1})}{\partial h_{k+1}} \ , \tag{5.56}$$

with the generating function

$$S(t, h) = \sum_n \left(K_n(0) \cos \Phi_n(t, h) - L_n(0) \sin \Phi_n(t, h) \right) \ . \tag{5.57}$$

Similarly, for $0 < \eta < \pi$ we obtain $R_n^+ = 0$ and $R_n^-(0) = -K_n(0) - iL_n(0)$. The mapping becomes

$$h_{k+1} = h_k + \epsilon \frac{\partial S(t_{k+1}, h_k)}{\partial t_{k+1}} \ ,$$
$$t_{k+1} = t_k + \frac{2\pi}{\omega(h_k)} - \epsilon \frac{\partial S(t_{k+1}, h_k)}{\partial h_k} \ . \tag{5.58}$$

The mapping (5.56) is implicit with respect to the variable h_{k+1}, while the mapping (5.58) is implicit in the variable t_{k+1}. One should note that only the phases $\Phi_n(t, h)$ of the generating function $S(t, h)$ depends on the energy h.

5.5 Conclusion

In this Chapter we have described the general method to construct canonical mappings near the separatrix of Hamiltonian systems. The application of these methods to the specific Hamiltonian system will be considered in the next chapter. There we shall present some critical comments on existing mapping methods near the separatrix and give bibliographic notes.

6 Mappings Near Separatrix. Examples

In this chapter we shall apply the methods developed in the previous section to construct mappings near separatrix for some specific problems. They include a motion in a double–well potential, dynamics of the periodically-driven oscillator. We consider also the dynamics of particles near the separatrix of long–range potential field. These problems are a motion of particle in a periodically–driven Morse potential and the Kepler problem.

6.1 Motion in a Perturbed Double–Well Potential

As an example we consider a motion of particle in a double − well potential under external time−periodic perturbation. The system is described by Hamiltonian

$$H = H_0(x, p) + \epsilon H_1(x, p, t) \ ,$$

$$H_0(x, p) = \frac{p^2}{2} - \frac{x^2}{2} + \frac{x^4}{4} \ ,$$

$$H_1(x, p, t) = \epsilon x \cos(\Omega t + \chi) \ . \tag{6.1}$$

The potential function $U(x) = -x^2/2 + x^4/4$ and the phase space of unperturbed motion ($\epsilon \equiv 0$) are in Fig. 6.1. The unperturbed system has a single hyperbolic fixed point at $(x = 0, p = 0)$ and two elliptic fixed points at $(x = \pm 1, p = 0)$ (see Fig. 6.1b). For $-1/4 < H = H_0(x, p) < 0$ a motion is trapped in potential wells (curves 1), and for $H > 0$ a motion is un-trapped (curve 3), and the separatrix ($H = 0$) is described by the curve 2. The action variable I for the trapped motion ($H < 0$) is given by

$$I = \frac{1}{2\pi} \oint p(x; H) dx = \frac{1}{\pi} \int_{-a}^{-b} \sqrt{2(H + x^2/2 - x^4/4)} dx$$

$$= \frac{2a}{3\pi} [E(k) - b^2 K(k)] \ , \tag{6.2}$$

where $(a, b) = \sqrt{1 \pm \sqrt{1 + 4H}}$, $K(k)$ and $E(k)$ are the complete elliptic integrals of the first kind and the second kind, respectively, with a module k:

S.S. Abdullaev: *Construction of Mappings for Hamiltonian Systems and Their Applications*,
Lect. Notes Phys. **691**, 105–138 (2006)
www.springerlink.com

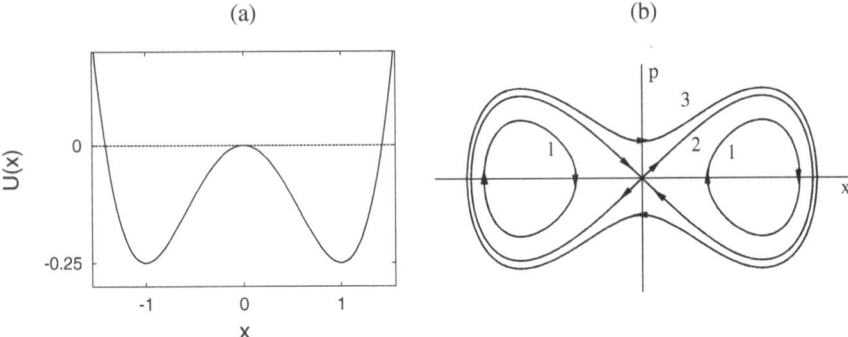

Fig. 6.1. (a) Double–well potential potential $U(x) = -x^2/2 + x^4/4$; (b) Phase space of motion

$$k = \frac{\sqrt{2}(1 + 4H)^{1/4}}{(1 + \sqrt{1 + 4H})^{1/2}} \, .$$

For the untrapped motion $(H > 0)$ we introduce the action variable I according to the definition given above:

$$I = \frac{1}{\pi} \int_{-a}^{0} \sqrt{2(H + x^2/2 - x^4/4)} \, dx$$

$$= \frac{\sqrt{2}}{3\pi} [(a^2 - b^2) E\left(k^{-1}\right) + b^2 K\left(k^{-1}\right)] \, . \tag{6.3}$$

The action $I(H)$ is a continuous function of energy H at the separatrix $H = 0$.

The unperturbed trajectory $x(t), p(t)$ can be determined by the second relation in (5.10). For the trapped motion $(H < 0)$ we have

$$x(\vartheta) = \pm a \sqrt{1 - k^2 \operatorname{sn}^2\left(K(k)\frac{\vartheta}{\pi}, k\right)} = \pm a \operatorname{dn}\left(K(k)\frac{\vartheta}{\pi}, k\right) \, ,$$

$$p = \mp \frac{a^2 k^2}{\sqrt{2}} \operatorname{sn}\left(K(k)\frac{\vartheta}{\pi}, k\right) \operatorname{cn}\left(K(k)\frac{\vartheta}{\pi}, k\right) \, , \tag{6.4}$$

where $\operatorname{sn}(u; k), \operatorname{cn}(u; k), \operatorname{dn}(u; k)$ are the Jacobi elliptic functions. The solution (6.4) is chosen to order to have the orbit which crosses the section Σ_c ($x = \pm a$) when $\vartheta = 0$, and the section Σ_s at $\vartheta = \pm \pi$. The unperturbed frequency, $\omega(H) = dH_0(I)/dI$, of this motion is

$$\omega(H) = \frac{\pi a}{\sqrt{2} K(k)} \, . \tag{6.5}$$

Outside the potential wells $(H > 0)$ the orbit is described by

$$x(\vartheta) = \pm a \, \mathrm{cn}\left(u; k^{-1}\right) \, ,$$

$$p(\vartheta) = \mp a(1 + 4H)^{1/4} \, \mathrm{sn}\left(u; k^{-1}\right) \sqrt{1 - k^{-2} \, \mathrm{sn}^2\left(u; k^{-1}\right)} \, ,$$

$$u = K\left(k^{-1}\right) \frac{\vartheta}{\pi} \, . \tag{6.6}$$

Note that at $\vartheta = \pm \pi$, $p(\pm \pi) = \pm \sqrt{2H}$. The frequency is given by

$$\omega(H) = \frac{\pi k a}{\sqrt{2} K\left(k^{-1}\right)} \, . \tag{6.7}$$

In Fig. 6.2 the dependence of the frequency of motion, $\omega(h)$, given by (6.5), (6.7) on the energy h is plotted.

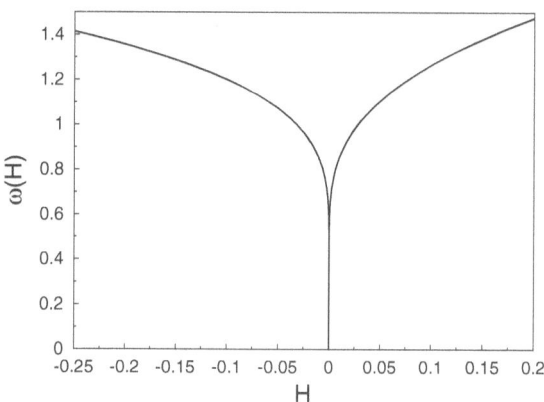

Fig. 6.2. Dependence of frequency of motion, $\omega(h)$ on energy h

Near the separatrix $(H \to 0)$ the frequency, $\omega(h)$, goes to zero according to the following asymptotics:

$$\omega(H) = \frac{1}{\ln \frac{16}{|H|}} + O(H) \, , \qquad |H| \to 0 \, . \tag{6.8}$$

The trajectory on the unperturbed separatrix $(H = 0)$ is described by

$$x_s^{(\pm)}(t) = \pm \frac{\sqrt{2}}{\cosh(t - t_c)}, \qquad p_s^{(\pm)}(t) = \mp \frac{\sqrt{2} \sinh(t - t_c)}{\cosh^2(t - t_c)} \, , \tag{6.9}$$

where t_c is a time instant when the orbit crosses the section Σ_c.

Using (6.4), (6.6) the perturbation Hamiltonian $H_1(x, p, t)$ in (6.1) can be expanded in Fourier series in the angle variable ϑ:

$$H_1(I, \vartheta, t) = \sum_{s=-1,1} \sum_{m=1}^{\infty} H_m(H) \begin{cases} \cos(m\vartheta - s\Omega t), & \text{for } H < 0 \, , \\ \cos([m - 1/2]\vartheta - s\Omega t), & \text{for } H > 0 \, , \end{cases} \tag{6.10}$$

$$H_m(H) = \pm a \begin{cases} \frac{\pi}{K(k)} \frac{q_-^m}{1+q_-^{2m}}, & \text{for } H < 0 \\[2ex] \frac{\pi}{K(k^{-1})} \frac{q_+^{m-1/2}}{1+q_+^{2m-1}}, & \text{for } H > 0 \ , \end{cases} \qquad (6.11)$$

where

$$q_- = \exp\left[-\pi K(\sqrt{1-k^2})/K(k)\right] \ ,$$

$$q_+ = \exp\left[-\pi K\left(\sqrt{1-k^{-2}}\right)/K\left(k^{-1}\right)\right] \ .$$

From (6.10) follows that the primary resonance conditions are $m\omega(H) = \Omega$ ($m = 0, 1, 2, \ldots$) for the trapped motion ($H < 0$). For untrapped motion ($H > 0$) the corresponding conditions are $(2m - 1)\omega(H) = 2\Omega$.

The geometry of the separatrix map for the Hamiltonian system (6.1) is shown in Fig. 6.3. The cross sections Σ_c consist of the segments of the x axis located near the farthest crossing points of the unperturbed separatrix with the x axis. The section Σ_s is located near the saddle point (x_s, p_s) and consists of two perpendicular to each other segments of the x and p axes with the center at $(x_s = 0, p_s = 0)$. There are two types of saddle–saddle connections. For the sake of simplicity we construct the nonsymmetric form of the separatrix mapping (5.29) presented in Sect. 5.2.4. The generating function $S(t_k, h_{k+1})$ of this mapping is given by (5.30) which takes the following form for the Hamiltonian (6.1)

$$S(t_k, h_{k+1}) = K(h_{k+1}) \cos \Phi^+(t_k, h_{k+1}) - L_n(h_{k+1}) \sin \Phi^+(t_k; h_{k+1}) \ .$$

From (6.1) follows that the perturbation function $V_n(\mathcal{H}, \tau)$ is equal to $x(\tau)$, and according to the definition (5.34) the integral $K(h)$ is equal to:

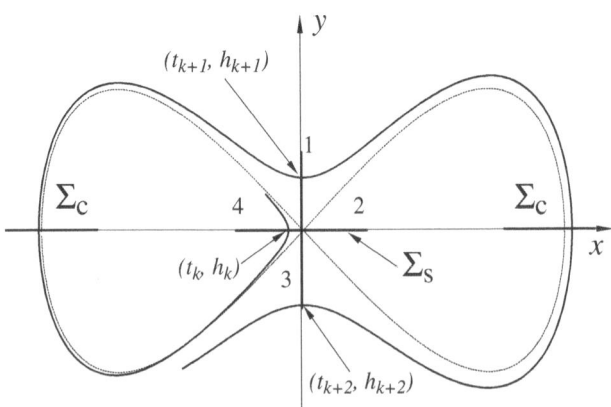

Fig. 6.3. Geometry of the separatrix map. *Solid curve* describes the perturbed orbit, and *dotted curve* is the unperturbed separatrix

$$K(h) = \int_{-\pi/\omega(h)}^{\pi/\omega(h)} x^{(\pm)}(\tau)\cos(\Omega\tau)d\tau = \pm F(h) \ ,$$

$$F(h) = \frac{a\pi}{\omega(h)} \int_{-1}^{1} \begin{cases} \mathrm{dn}\left(K(k)\tau; k\right)\cos(\Omega\pi\tau/\omega)d\tau, & \text{for } h < 0 \ , \\ \mathrm{cn}\left(K(k^{-1})\tau; k^{-1}\right)\cos(\Omega\pi\tau/\omega)d\tau, & \text{for } h > 0 \ , \end{cases} \tag{6.12}$$

for the right ($x > 0$) and left ($x < 0$) halves of phase space, respectively. At the separatrix $h = 0$ we have

$$K(0) = \int_{-\infty}^{\infty} x_s^{(\pm)}(t)\cos(\Omega\tau)d\tau$$

$$= \pm\sqrt{2} \int_{-\infty}^{\infty} \frac{\cos(\Omega\tau)d\tau}{\cosh\tau} = \frac{\pm\sqrt{2\pi}}{\cosh(\pi\Omega/2)} \ . \tag{6.13}$$

One can also show that

$$L(h) = \int_{-\pi/\omega(h)}^{\pi/\omega(h)} x^{(\pm)}(\tau)\sin(\Omega\tau)d\tau = 0 \ . \tag{6.14}$$

As was shown in Sect. 5.2.5 the integral $K(h)$ consists of regular and oscillatory parts: $K(h) = K^{(reg)}(h) + K^{(osc)}(h)$. According to (5.39) the regular part can be expressed through the Fourier components $H_m(H)$ (6.11) of the perturbation Hamiltonian. Using (6.5), (6.7) we have

$$K^{(reg)}(h) = \frac{2\pi}{\omega(h)}H_{\Omega/\omega(h)}(h) = \pm\frac{2\pi a}{\sqrt{1 + \sqrt{1 + 4h}}}\frac{q_-^{\Omega/\omega(h)}}{1 + q_-^{2\Omega/\omega(h)}}$$

$$= \pm\frac{\pi\sqrt{2}}{\cosh\left(\sqrt{2}\Omega K\left(\sqrt{1 - k^2}\right)/a\right)} \ , \tag{6.15}$$

for the case $h < 0$, and

$$K^{(reg)}(h) = \frac{2\pi}{\omega(h)}H_{\Omega/\omega(h)+1/2}(h) = \pm\frac{2\pi a}{(1 + 4h)^{1/4}}\frac{q_+^{\Omega/\omega(h)}}{1 + q_+^{2\Omega/\omega(h)}}$$

$$= \pm\frac{\pi\sqrt{2}}{k\cosh\left(\sqrt{2}\Omega K\left(\sqrt{1 - k^{-2}}\right)/ka\right)} \ , \tag{6.16}$$

for the case $h > 0$. In the limit $|h| \to 0$ the both expressions of $K^{(reg)}(h)$ coincide with $K(0)$ (6.13) obtained by the direct integration.

The asymptotical formula for $K^{(osc)}(h)$ can be found using the general asymptotical formulae for the Melnikov type integrals, $R^{(osc)}(h)$, near the separatrix obtained in Appendix B. Our problem corresponds to the cases (ii) and (iii) considered in Sect. B.1.4, and the expression for $K^{(osc)}(h)$ is given by (B.35). For the Hamiltonian system (6.1) we have the following parameters $\alpha = \gamma = 1$, $a_\xi = 1$, $a_\eta = b_{\xi\xi} = b_{\xi\eta} = b_{\eta\eta} = 0$. Therefore, one obtains

$$K^{(osc)}(h) = \mp \frac{2\sqrt{2|h|}}{\Omega^2 + 1} \begin{cases} \Omega\sin[\pi\Omega/\omega(h)], & \text{for } h < 0\ , \\ -\cos[\pi\Omega/\omega(h)], & \text{for } h > 0\ , \end{cases} \qquad (6.17)$$

where the upper sign $(-)$ corresponds to the right side of the phase space $(x > 0)$, and the lower sign $(+)$ – to the left side of the phase space $(x < 0)$.

Figure 6.4 shows the dependence of the integral $K(h)$ on h obtained by the direct numerical integration of (6.12), as well as by the analytical formulae (6.15), (6.16), and (6.17): solid curve 1 corresponds to the numerical calculations, dashed curve 2 – to the analytical result, and dotted curve 3 – to the regular part $K^{(reg)}(h)$ given by (6.15), (6.16). The perturbation frequency is taken equal to $\Omega = 4.53236$. The corresponding rescaling parameter $\lambda = \exp(2\pi\gamma/\Omega)$ is equal to 4.

As seen from Figs. 6.4a-c analytical formulae (6.15), (6.16) the regular part $K^{(reg)}(h)$ and the asymptotical formula (6.17) for the oscillatory part, $K^{(osc)}(h)$, well describe the behavior of the integral $K(h)$. The accuracy of approximation increases with approaching the separatrix. On the other hand the numerical calculations confirm also the following rescaling property,

$$K^{(osc)}(\lambda^2 h) = \lambda K^{(osc)}(h)\ , \qquad (6.18)$$

of the oscillatory part of the integral $K(h)$ following from the asymptotical formula (6.17).

One can easily see that zeros of $K^{(osc)}(h)$ coincide with the primary resonant values of h_{mn}: $m\omega(h) = \Omega$ $(m = 0, 1, 2, \ldots)$ at $h < 0$, and $(2m - 1)\omega(h) = 2\Omega$ for $h > 0$. Therefore, according to the primary resonant approximation (see Sect. 5.3.1) one can neglect oscillatory parts, $K^{(osc)}(h)$, in the mapping retaining only smooth regular parts $K^{(reg)}(h)$.

The dynamics of the system is described by two mappings, $(t_{k+1}, h_{k+1}) = \hat{M}^{(\pm)}(t_k, h_k)$, (5.30), corresponding to the two different saddle–saddle connections:

$$h_{k+1} = h_k \mp \epsilon\Omega F(h_{k+1})\sin\left(\varphi_k + \frac{\pi\Omega}{\omega(h_{k+1})} + \chi\right)\ ,$$

$$\varphi_{k+1} = \varphi_k + \frac{2\pi\Omega}{\omega(h_{k+1})} + \epsilon\frac{dF(h_{k+1})}{dh_{k+1}}\cos\left(\varphi_k + \frac{\pi\Omega}{\omega(h_{k+1})} + \chi\right)$$

$$\pm\epsilon F(h_{k+1})\frac{d}{dh_{k+1}}\frac{\pi\Omega}{\omega(h_{k+1})}\sin\left(\varphi_k + \frac{\pi\Omega}{\omega(h_{k+1})} + \chi\right)\ , \qquad (6.19)$$

where $F(h_{k+1}) \equiv K^{(reg)}(h)$ and the phase variable $\varphi = \Omega t$ is introduced. The map with the $(+)$ sign describes the right side of phase space $(x > 0)$, while one with the (-) sign corresponds to $(x < 0)$.

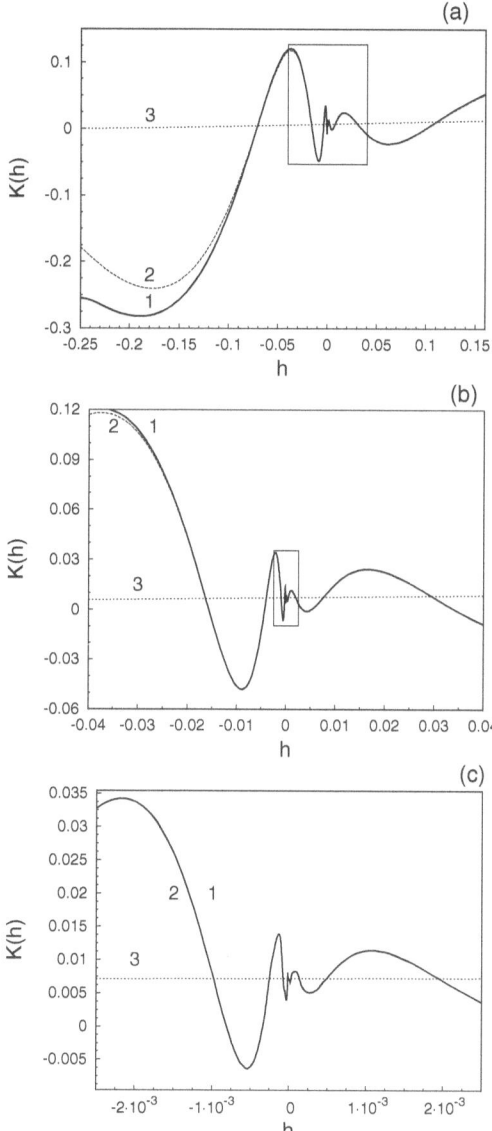

Fig. 6.4. Dependence of the integral $K(h)$ on the relative energy h: *solid curve* 1 describes $K(h)$ obtained by the numerical integration of the integral (6.12), and *dashed curve* 2 describes $K(h)$ obtained by the analytical formulae (6.15), (6.16), and (6.17), and *dotted curve* 3 corresponds to the regular part $K^{(reg)}(h)$ given by (6.15), (6.16). (**a**) in the interval $-0.25 < h < 0.16$: (**b**) Expanded view of $K(h)$ in the rectangular box region shown in (a); (**c**) Expanded view of the rectangular box region shown in (b). The perturbation frequency $\Omega = 4.53236$, and the rescaling parameter $\lambda = \exp(2\pi\gamma/\Omega) = 4$

The simplified form of the mapping given (5.45) in our case can be written as

$$h_{k+1} = h_k \mp \epsilon \Omega K^{(reg)}(h_{k+1}) \sin \left(\varphi_k + \frac{\pi \Omega}{\omega(h_k)} + \chi \right) ,$$

$$\varphi_{k+1} = \varphi_k + \frac{\pi \Omega}{\omega(h_{k+1})} + \frac{\pi \Omega}{\omega(h_k)}$$
$$\mp \epsilon \frac{dK^{(reg)}(h_{k+1})}{dh_{k+1}} \cos \left(\varphi_k + \frac{\pi \Omega}{\omega(h_k)} + \chi \right) . \qquad (6.20)$$

Further simplification of the mapping can be done near the small neighborhood of the separatrix. Using the asymptotics of the frequency $\omega(h)$ (6.8) $|h| \to 0$ and replacing $K^{(reg)}(h)$ by $K(0)$ we obtain

$$h_{k+1} = h_k \mp \epsilon \Omega K(0) \sin \left(\varphi_k + \frac{\Omega}{2} \ln \frac{16}{|h_k|} + \chi \right) ,$$

$$\varphi_{k+1} = \varphi_k + \frac{\Omega}{2} \left(\ln \frac{16}{|h_k|} + \ln \frac{16}{|h_{k+1}|} \right) . \qquad (6.21)$$

The map (6.21) is an example of the *algorithmic separatrix map*, a term which is introduced by Shevchenko (1999). The dynamics of system near the separatrix is described by the sequence of iterations of the maps $\hat{M}^{(\pm)}$. This sequence is determined by a certain rule. Let $S^{(+)}$ and $S^{(-)}$ be domains of phase space (x, p) in the right $(x > 0)$ and the left $(x < 0)$ half planes, respectively. Then

$$\hat{M}_{k+1} = \begin{cases} \hat{M}^{(+)} & \text{if } z_k \in S^{(+)} \text{ and } h_k < 0 , \\ \hat{M}^{(-)} & \text{if } z_k \in S^{(+)} \text{ and } h_k > 0 , \\ \hat{M}^{(+)} & \text{if } z_k \in S^{(-)} \text{ and } h_k > 0 , \\ \hat{M}^{(-)} & \text{if } z_k \in S^{(-)} \text{ and } h_k < 0 , \end{cases} \qquad (6.22)$$

where $z_k = (x_k, p_k)$. Applications of the separatrix mapping (6.21) to study the rescaling properties of Hamiltonian systems near the saddle points will be given in Sect. 8.2, and the statistics of a residence time and a Poincaré recurrence − in Sect. 9.2.

The mappings (6.20), (6.21) are valid for small values of perturbation, $\epsilon \ll 1$. Note, that the second mapping (6.21) is applicable only the area close to the separatrix, while the first mapping (6.20) can be applied also far from the separatrix.

For illustration we have applied the mapping (6.20) to obtain Poincaré sections of system in the (φ, h) plane of the 1-th $(h > 0)$ and 4-th branches $(h < 0)$ of the section Σ_s shown in Fig. 6.3a. It has been also compared with the small step numerical integration of the Hamiltonian system (6.1) using the symplectic integrator (see Sect. 1.5). The mapping result is shown in Fig. 6.5a, and the results obtained from the numerical integration is presented Fig. 6.5b. Calculations in both cases were performed with a set of identical

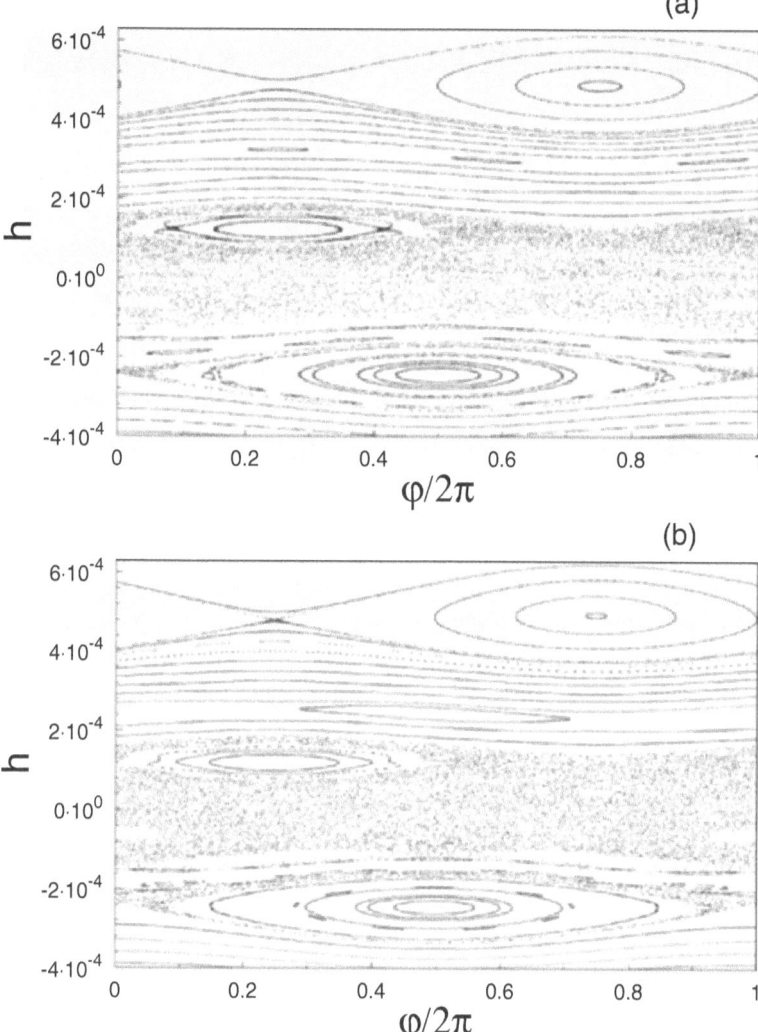

Fig. 6.5. Comparison of the mapping (6.20) with the direct integration of the Hamiltonian system (6.1). The perturbation frequency Ω is the same as in Fig. 6.4, and the perturbation amplitude $\epsilon = 10^{-3}$. (**a**) corresponds to the mapping; (**b**) − to the equations

initial coordinates. As seen from Fig. 6.5 the mapping well reproduces the structure of phase space. For the time-step of integration, Δt, of the equation, equal to $4\pi \times 10^{-3}/\Omega$, the mapping runs two order faster than the small-step symplectic integrator.

6.2 Mapping for the Periodically Driven Pendulum

Consider the periodically–driven pendulum given the following Hamiltonian

$$H(x, p, t) = H_0(x, p) + \epsilon H_1(x, p, t) \, , \tag{6.23}$$

$$H_0(x, p) = \frac{p^2}{2} - \omega_0^2 \cos x \, , \tag{6.24}$$

$$\epsilon H_1(x, p, t) = \epsilon \omega_0^2 \left[A \cos(x - \Omega t - \chi) + B \cos(x + \Omega t + \chi) \right] \, . \tag{6.25}$$

The quantities A and B describe amplitudes of waves propagating in positive and negative directions of the x-axis.

The geometry of the separatrix map $(t_k, h_k) \to (t_{k+1}, h_{k+1})$ is shown in Fig. 6.6. The sections Σ_s on the (x, p)-plane consist of two perpendicular segments of x and p axes with the center at the hyperbolic fixed points $(x_s = 2\pi(s+1/2), p_s = 0)$ $(s = 0, \pm 1, \pm 2, \ldots)$. Sections Σ_c consist of segments perpendicular to the unperturbed separatrices $(x_s(t), p_s(t))$ at the midpoint between two consecutive saddle points, $x_c = 2\pi s$.

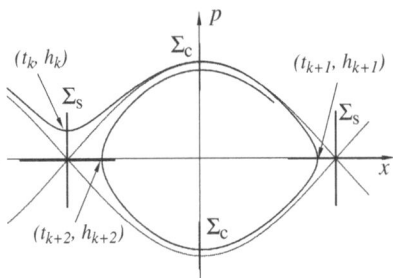

Fig. 6.6. Geometry of the separatrix map to the section Σ_s for the periodically driven pendulum

The system is described by the Hamiltonian (6.23). Changing the Hamiltonian to $H \to h = (H - \omega_0^2)/\omega_0^2$, $\omega_0 t \to t$, $p/\omega_0 \to p$ we write the Hamiltonian in the form

$$H = \frac{p^2}{2} - \cos x - 1 + \epsilon \left[A \cos(x - \Lambda t - \chi) + B \cos(x + \Lambda t + \chi) \right] \, , \tag{6.26}$$

where $\Lambda = \Omega/\omega_0$. The unperturbed motion ($\epsilon = 0$) is trapped for $H < 0$ and un-trapped for $H > 0$ (curves 1 and 3 in Fig. 5.1). The action-angle variables (I, ϑ) for the unperturbed Hamiltonian ($\epsilon = 0$) (see Sect. 1.4) should be introduced in such a way that they should be continuous at the separatrix

$H = 0$. For this we define the action variable I for the trapped motion as an integral taken along the segment of the orbit on the upper, $p > 0$, (lower, $p < 0$) half of the phase space (x, p):

$$I = \frac{1}{2\pi} \int_b^a p(x, H)dx = \frac{1}{2\pi} \int_b^a \sqrt{2(H + 1 + \cos x)}dx$$

$$= \frac{4}{\pi} \left[E(k) - (1 - k^2)K(k) \right] , \qquad (6.27)$$

where $K(k), E(k)$ are the complete elliptic integrals with a module

$$k = \sqrt{1 + H/2} ,$$

and a, b are the roots of the equation $p(x, H) = 0$ $(2\pi(s - 1/2) < a, b < 2\pi(s + 1/2))$. The corresponding angle variable ϑ is introduced as

$$\vartheta = \frac{\partial}{\partial I} \int_{2\pi s}^x p(x, H)dx = \omega(H) \int_0^x \frac{dx}{\sqrt{2(H + 1 + \cos x)}}$$

$$= \omega(H)F(k^{-1}\arcsin(x/2); k) , \qquad (6.28)$$

with the conditions that $\vartheta = 0$ at the sections Σ_c $(x = 2\pi s)$ and $\vartheta(\mathrm{mod}\ 2\pi) = \pm\pi$ at Σ_s, $(x = 2\pi(s + 1/2)$. From (6.28) follows that

$$x(\vartheta; H) = 2\arcsin[k\,\mathrm{sn}(\vartheta/\omega(H); k)] . \qquad (6.29)$$

The frequency of motion $\omega(H) = dH(I)/dI = \pi/K(k)$ has the following asymptotics near the separatrix:

$$\omega(H) = \frac{2\pi}{\ln(32/|H|)} , \qquad \text{for } H \to -0 . \qquad (6.30)$$

For the untrapped motion $(H > 0)$ the action-angle variables are introduced as in Sect. 1.4.

$$I = \frac{1}{2\pi} \int_{-2\pi(s-1/2)}^{2\pi(s+1/2)} p(x, H)dx = \frac{4k}{\pi} E\left(k^{-1}\right) ,$$

$$\vartheta = \frac{\partial}{\partial I} \int_{2\pi s}^x p(x', H)dx' == \omega(H)k^{-1}F\left(x/2, k^{-1}\right) ,$$

$$\sin(x/2) = \mathrm{sn}\left(k\vartheta/\omega(H); k^{-1}\right) = \mathrm{sn}\left(k(t - t_0); k^{-1}\right) . \qquad (6.31)$$

The frequency $\omega(H) = \pi k/K\left(k^{-1}\right)$ has the same asymptotics (6.30) at $H \to +0$.

The orbits on the upper (lower) branches of the separatrix $(H = 0)$ are

$$\sin x^{\pm}(t) = \pm\frac{\sinh(t - t_c)}{\cosh^2(t - t_c)} , \qquad \cos x^{\pm}(t) = \frac{2}{\cosh^2(t - t_c)} - 1 . \ (6.32)$$

For the system with the Hamiltonian (6.26) the perturbation Hamiltonian $\mathcal{H}_1(t, \vartheta, p_0)$ in (5.13)

$$\mathcal{H}_1(t, \vartheta, p_0) = \epsilon \left\{ A \cos \left[x(\vartheta, p_0) - \Lambda t - \chi \right] + B \cos \left[x(\vartheta, p_0) + \Lambda t + \chi \right] \right\} ,$$

can be presented in the form (5.21) with

$$V_1(\mathcal{H}, t - t_c) = (A + B) \cos x(\vartheta; \mathcal{H}), \qquad \Omega_1 = \Lambda , \qquad \chi_1 = \chi ,$$

$$V_2(\mathcal{H}, t - t_c) = (A - B) \sin x(\vartheta; \mathcal{H}) , \qquad \Omega_2 = \Lambda , \qquad \chi_2 = \chi - \frac{\pi}{2} . \,(6.33)$$

Remind that $V_n(\mathcal{H}, t - t_c) \equiv H_n(\mathcal{H}, \vartheta)$. On the separatrix we have

$$V_1(0, \tau) = (A + B) \left(\frac{2}{\cosh^2 \tau} - 1 \right) ,$$

$$V_2(0, \tau) = \pm(A - B) \frac{\sinh \tau}{\cosh^2 \tau} . \tag{6.34}$$

6.2.1 Behavior of Integrals $K_n(h)$ and $L_n(h)$ (5.31)

We evaluate the integrals $K_n(h)$ and $L_n(h)$ (5.31) which appear in the simplified form of the mapping (5.45). Using the relations (6.33), (6.29), (6.31) these integrals can be reduced to

$$K_1(h) = \int\limits_{-\pi/\omega(h)}^{\pi/\omega(h)} V_1(h, \tau) \cos(\Lambda\tau) d\tau = 4(A + B)$$

$$\times \begin{cases} k^2 \int\limits_0^{\pi/\omega(h)} \mathrm{cn}^2 \left(\tau, k \right) \cos \left(\Lambda\tau \right) d\tau, & \text{for } h < 0 , \\ \int\limits_0^{\pi/\omega(h)} \mathrm{cn}^2 \left(k\tau, \frac{1}{k} \right) \cos \left(\Lambda\tau \right) d\tau, & \text{for } h > 0 , \end{cases} \tag{6.35}$$

$$L_2(h) = \int\limits_{-\pi/\omega(h)}^{\pi/\omega(h)} V_2(h, \tau) \sin \Lambda\tau d\tau = \pm 4(A - B)$$

$$\times \begin{cases} k \int\limits_0^{\pi/\omega(h)} \mathrm{sn} \left(\tau, k \right) \sqrt{1 - k^2 \mathrm{sn}^2 \left(\tau, k \right)} \sin \left(\Lambda\tau \right) d\tau, & \text{for } h < 0 , \\ \int\limits_0^{\pi/\omega(h)} \mathrm{sn} \left(k\tau, \frac{1}{k} \right) \mathrm{cn} \left(k\tau, \frac{1}{k} \right) \sin \left(\Lambda\tau \right) d\tau, & \text{for } h > 0 , \end{cases} \tag{6.36}$$

$$K_2(h) = \int\limits_{-\pi/\omega(h)}^{\pi/\omega(h)} V_2(h, \tau) \cos \Lambda\tau d\tau = 0 ,$$

$$L_1(h) = \int\limits_{-\pi/\omega(h)}^{\pi/\omega(h)} V_1(h,\tau) \sin \Lambda\tau d\tau = 0 , \qquad (6.37)$$

where

$$\omega(h) = \pi \begin{cases} 1/K(k), & \text{for } h < 0 , \\ k/K(k^{-1}), & \text{for } h > 0 . \end{cases} \qquad (6.38)$$

According to Section (5.2.5) the integrals $K_1(h)$, $L_2(h)$ consist of regular and oscillatory parts:

$$K_1(h) = K_1^{(reg)}(h) + K_1^{(osc)}(h) ,$$

$$L_2(h) = L_2^{(reg)}(h) + L_2^{(osc)}(h) . \qquad (6.39)$$

The asymptotical behavior of oscillatory parts $K_1^{(osc)}(h)$, $L_2^{(osc)}(h)$ in the limit $|h| \to 0$ are found in Appendix B.2. Putting in (B.39), (B.40) the parameter values $\alpha = 1$, $\gamma = 1$ and $\Omega = \Lambda$ we have

$$K_1^{(osc)}(h) = (A+B)\frac{|h|}{\Lambda} \begin{cases} \left(1 + \frac{\Lambda^2}{4+\Lambda^2}\right) \sin\left(\frac{\pi\Lambda}{\omega(h)}\right) , & \text{for } h < 0 \\ -\left(1 - \frac{\Lambda^2}{4+\Lambda^2}\right) \sin\left(\frac{\pi\Lambda}{\omega(h)}\right) , & \text{for } h > 0 , \end{cases} \qquad (6.40)$$

$$L_2^{(osc)}(h) = \mp(A-B)\frac{2\sqrt{2|h|}}{1+\Lambda^2}$$
$$\times \begin{cases} \Lambda\left(1 + \frac{|h|}{12}\left[3 + \frac{1+\Lambda^2}{9+\Lambda^2}\right]\right) \cos\left(\frac{\pi\Lambda}{\omega(h)}\right) , & \text{for } h < 0 , \\ \left(1 + \frac{|h|}{4}\left[-1 + \frac{1+\Lambda^2}{9+\Lambda^2}\right]\right) \sin\left(\frac{\pi\Lambda}{\omega(h)}\right) , & \text{for } h > 0 , \end{cases} \qquad (6.41)$$

where the sign $(-)$ corresponds to the upper branch of the separatrix, and the sign $(+)$ corresponds to the lower branch of the separatrix. In (6.41) for $L_2^{(osc)}(h)$ we have also presented the next expansion term proportional to $|h|^{3/2}$.

For $h = 0$, when the functions $V_n(h,\tau)$ are given by (6.34) the integrals (6.35), (6.36) can be exactly integrated

$$K_1(0) = 2(A+B) \int\limits_{-\infty}^{\infty} \frac{\cos(\Lambda\tau)d\tau}{\cosh^2 \tau} = \frac{2\pi\Lambda(A+B)}{\sinh(\pi\Lambda/2)} ,$$

$$L_2(0) = \pm(A-B) \int\limits_{-\infty}^{\infty} \frac{\sinh\tau \sin(\Lambda\tau)d\tau}{\cosh^2 \tau} = \pm\frac{2\pi\Lambda(A-B)}{\cosh(\pi\Lambda/2)} . \qquad (6.42)$$

For arbitrary values of h we have numerically integrated the integrals (6.35), (6.36). The dependencies of the integrals $K(h)$, $L(h)$ on the energy h

in the interval $[-2, 1]$ are presented in Figs. 6.7a and 6.8a, where solid lines 1 correspond to $K(h)$ and $L(h)$, while dashed lines 2 describe the regular parts, $K_1^{(reg)}(h)$ and $L_2^{(reg)}(h)$, respectively. The latter are found by subtracting the oscillatory parts, $K_1^{(osc)}(h)$ and $L_2^{(osc)}(h)$ given by (6.40), (6.41) from $K(h)$ and $L(h)$. Figures 6.7(b)–(c), 6.8, (b)–(c) show the consecutive expanded views of the dependencies $K_1(h)$ and $L_2(h)$ vs h at its small values. They confirm the rescaling properties of the oscillatory parts, $K_1^{(osc)}(h)$ and $L_2^{(osc)}(h)$, for small values of h described by the asymptotical formulae (6.40), (6.41):

$$K_1^{(osc)}(\lambda^2 h) = \lambda^2 K_1^{(osc)}(h) \ ,$$

$$L_2^{(osc)}(\lambda^2 h) = \lambda L_2^{(osc)}(h) \ , \tag{6.43}$$

where $\lambda = \exp(2\pi/\Lambda)$ is a rescaling parameter coinciding with the corresponding parameter in (5.8). Such a rescaling behavior of $K_1(h)$ and $L_2(h)$ is related with the logarithmic asymptotics (6.30) of the frequency of oscillations $\omega(h)$ near the separatrix.

One should note that the integrals $K_n(0)$, $L_n(0)$ are exponentially small at large frequency of perturbation $\Lambda = \Omega/\omega_0 \gg 1$, i.e.,

$$K_n(0), L_n(0) \sim e^{-\pi\Lambda/2} \ll 1 \ . \tag{6.44}$$

In this case the oscillatory parts $K_n^{(osc)}(h)$, $L_n^{(osc)}(h)$ become dominant.

6.2.2 Mapping to Sections Σ_s

The general form of this mapping is given by (5.29). The corresponding generating function (5.30) in our case, according to (6.33), is determined by

$$S(t, h) = K(h) \cos \Phi(t, h) \ , \tag{6.45}$$

where

$$K(h) = \Lambda \left[K_1(h) + L_2(h) \right] \ ,$$

$$\Phi(t, h) = t\Lambda + \frac{\pi\Lambda}{\omega(h)} + \chi \ . \tag{6.46}$$

One can show that the oscillatory parts of the integrals $K_1(h)$, $L_2(h)$ have zeros at the primary resonant values of h determined by $(2m - 1)\omega(h) = 2\Omega$ for $h < 0$ and $m\omega(h) = \Omega$ for $h > 0$. Then according to the primary resonance approximation (Section 5.3.1) the integrals $K_1(h)$, $L_2(h)$ can be approximated only by their regular parts $K_1^{(reg)}(h)$, $L_2^{(reg)}(h)$.

Below we present the simplified form of mapping at the sections Σ_s given by (5.45). Using (6.45) one obtains the following expressions for perturbation functions

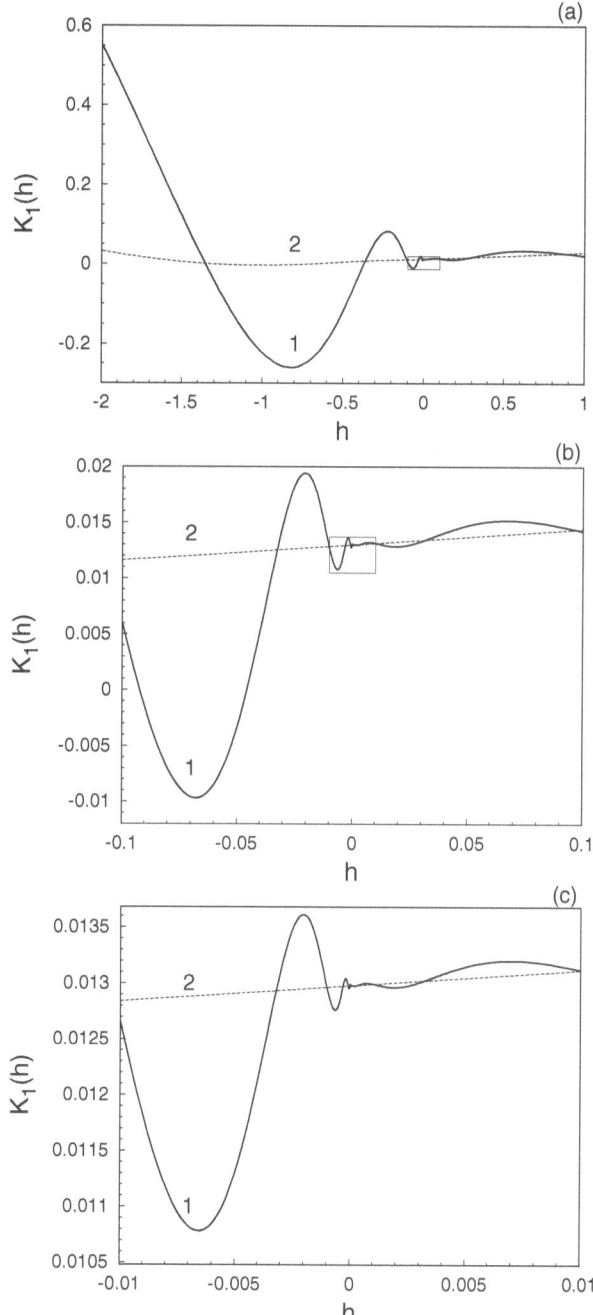

Fig. 6.7. Dependence of integrals $K_1(h)$ (normalized to $A + B$) on h (*solid curve* 1). *Dashed lines* 2 correspond to $K_1^{(reg)}(h)$; (b) and (c) show the expanded view of boxed areas in (a) and (b), respectively. The value $\Lambda = 5.4575$, $\lambda^2 = \exp(4\pi/\Lambda) = 10$

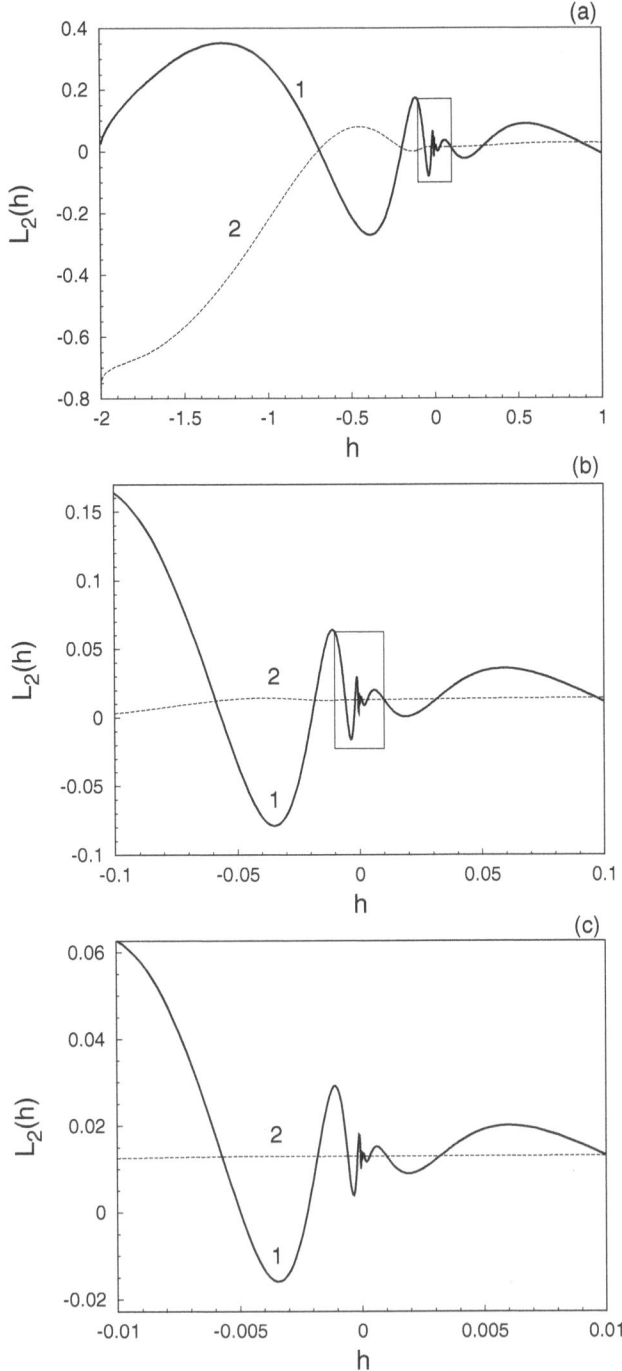

Fig. 6.8. The same as in Fig. 6.7, but for $L_2(h)$ (normalized to $A - B$): (b) and (c) show the expanded view of boxed areas in (**a**) and (**b**), respectively

$$F(t_k, h_{k+1}, h_k) = K^{(reg)}(h_{k+1}) \sin \Phi(t_k, h_k) \ ,$$

$$G(t_k, h_{k+1}, h_k) = \frac{1}{\Lambda} \frac{dK^{(reg)}(h_{k+1})}{dh_{k+1}} \cos \Phi(t_k, h_k) \ , \tag{6.47}$$

and the mapping (5.45) takes the form

$$h_{k+1} = h_k - \epsilon K^{(reg)}(h_{k+1}) \sin \left(\varphi_k + \frac{\pi \Lambda}{\omega(h_k)} + \chi \right) \ ,$$

$$\varphi_{k+1} = \varphi_k + \frac{\pi \Lambda}{\omega(h_k)} + \frac{\pi \Lambda}{\omega(h_{k+1})} + \epsilon \frac{dK^{(reg)}(h_{k+1})}{dh_{k+1}} \cos \left(\varphi_k + \frac{\pi \Lambda}{\omega(h_k)} + \chi \right) \ , \tag{6.48}$$

where $\varphi = \Lambda t$.

As seen from Figs. 6.7, 6.8 the regular parts, $K_1^{(reg)}(h)$, $L_2^{(reg)}(h)$, are smooth functions h. In the small neighborhood of the separatrix $h = 0$ their deviations from $K_1(0)$, $L_2(0)$ are small, i.e., $|K_1^{(reg)}(h) - K_1(0)| \ll 1$, $|L_2^{(reg)}(h) - L_2(0)| \ll 1$. Then the mapping (6.48) can be further simplified by replacing $K_1^{(reg)}(h)$ and $L_2^{(reg)}(h)$ by $K_1(0)$, $L_2(0)$. Using (6.42) and the asymptotics of $\omega(H)$ (6.30) near the separatrix the mapping (6.48) is reduced to

$$h_{k+1} = h_k - \epsilon K^{\pm}(0) \sin \left(\varphi_k + \frac{\Lambda}{2} \ln \frac{32}{|h_k|} + \chi \right) \ ,$$

$$\varphi_{k+1} = \varphi_k + \frac{\Lambda}{2} \left(\ln \frac{32}{|h_{k+1}|} + \ln \frac{32}{|h_k|} \right) \ , \tag{6.49}$$

where

$$K^{\pm}(0) = \Lambda(K_1(0) + L_2(0)) = \frac{4\pi \Lambda^2}{\sinh(\pi \Lambda)} \left[A e^{\pm \pi \Lambda / 2} + B e^{\mp \pi \Lambda / 2} \right] \ . \tag{6.50}$$

The sign ($\pm$) corresponds to the integral taken along the separatrix on the upper (lower) half phase space, $p > 0$, ($p < 0$), respectively.

Let (φ_k, h_k) be the phase and the energy at the k-th mapping step. Suppose also, that (x_k, p_k) are the corresponding phase space coordinates. The sequence of the mapping iteration $\hat{M}_k$: ($\varphi_{k+1}, h_{k+1}) = \hat{M}_{k+1}(\varphi_k, h_k)$, and the coordinates ($x_{k+1}, p_{k+1}$) after one map iteration are determined by the following algorithm

$$\hat{M}_{k+1} = \begin{cases} \hat{M}^{(+)}, & \text{if } \hat{M}_k = \hat{M}^{(+)} \text{ and } h_k > 0 \ , \\ \hat{M}^{(-)}, & \text{if } \hat{M}_k = \hat{M}^{(-)} \text{ and } h_k > 0 \ , \\ \hat{M}^{(-)}, & \text{if } \hat{M}_k = \hat{M}^{(+)} \text{ and } h_k < 0 \ , \\ \hat{M}^{(+)}, & \text{if } \hat{M}_k = \hat{M}^{(-)} \text{ and } h_k < 0 \ , \end{cases} \tag{6.51}$$

where $\hat{M}^{(\pm)}$ are mappings (6.49) along upper and lower branches, respectively.

The separatrix mapping (6.49) will be used in Sect. 8.1 to analyze the rescaling properties of Hamiltonian system near the hyperbolic saddle point, and in Sect. 9.3 to study a chaotic transport along the stochastic layer.

6.2.3 Mapping to Sections Σ_c

For the problem under consideration there are two different types of sections Σ_c: Σ_c^+ and Σ_c^- corresponding to the upper and lower branches of the separatrix. The geometry of the separatrix mapping to the cross sections $\Sigma_c^\pm$ is schematically shown in Fig. 6.9. In general, there are four independent mappings of the sections $\Sigma_c^\pm$ to $\Sigma_c^\pm$ which fully determine the dynamics of the system. These mapping should be constructed in two steps: in the first step one should find the map from the section $\Sigma_c^\pm$ to Σ_s along a certain saddle–saddle connection, and in the second step one should map Σ_s to $\Sigma_c^\pm$ along another saddle–saddle connection which depends on the sign energy on the section Σ_s. Therefore the dynamics of system is then fully determined, in general, by four independent mappings, $\hat{T}_1^{(\pm)}$ and $\hat{T}_2^{(\pm)}$, where $\hat{T}_1^{(\pm)}$ stands for the mapping of variables $(t_k, h_k) \in \Sigma_c^\pm$ to $(\mathcal{T}_k, \mathcal{H}_k) \in \Sigma_s$ along upper $(+)$ or lower $(-)$ branches of the separatrix, and $\hat{T}_2^{(\pm)}$ stands for the mapping of variables $(\mathcal{T}_k, \mathcal{H}_k) \in \Sigma_s$ to $(t_{k+1}, h_{k+1}) \in \Sigma_c^\pm$ along upper $(+)$ or lower $(-)$ branches of the separatrix, respectively.

These mappings can be constructed using the general mapping form (5.16) described in Sect. 5.2. Below we derive nonsymmetric forms of these mappings similar to ones (5.29), (5.32) obtained in the subsection 5.2.4.

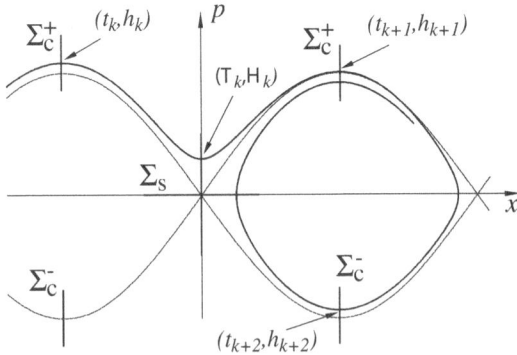

Fig. 6.9. Geometry of the separatrix map to sections $\Sigma_c^\pm$ for the periodically–driven pendulum

Mapping $\Sigma_c^\pm$ to Σ_s

For the sake of simplicity we choose $\tau_0 = \tau_{k+1}$. Recall that $\tau_{k+1} - t_k = \pi/\omega(\mathcal{H}_k)$. Taking $t = t_k$, $\vartheta = 0$ at the section $\Sigma_c^\pm$ the generating function (5.22) is reduced to

$$S_k(t_k, \mathcal{H}_k) = \sum_n [K_n(\mathcal{H}_k)\cos(\Omega_n t_k + \chi_n) - L_n(\mathcal{H}_k)\sin(\Omega_n t_k + \chi_n)],$$

$$(6.52)$$

where

$$K_n(\mathcal{H}) = \int_0^{\pi/\omega(\mathcal{H})} V_n(\mathcal{H}, \tau)\cos(\Omega_n \tau)d\tau,$$

$$L_n(\mathcal{H}) = \int_0^{\pi/\omega(\mathcal{H})} V_n(\mathcal{H}, \tau)\sin(\Omega_n \tau)d\tau, \qquad (6.53)$$

Since $\tau_0 = t_{k+1}$ we have $S_k(t_{k+1}, \mathcal{H}_k) \equiv 0$.

Further for the sake of simplicity we consider the case of thin stochastic layer by taking the limit $\mathcal{H} \to 0$. Using formulae for the quantities $V_n(\mathcal{H}, \tau)$ at the separatrix given by (6.33), (6.34) we obtain the following integrals

$$K_1 \equiv K_1(0) = 2(A + B)\int_0^\infty \frac{\cos \Lambda\tau}{\cosh^2 \tau}d\tau = \frac{\pi\Lambda(A + B)}{\sinh(\pi\Lambda/2)},$$

$$K_2 \equiv K_2(0) = \pm(A - B)\int_0^\infty \frac{\sinh\tau\cos\Lambda\tau}{\cosh^2 \tau}d\tau,$$

$$L_1 \equiv L_1(0) = 2(A + B)\int_0^\infty \frac{\sin\Lambda\tau}{\cosh^2 \tau}d\tau,$$

$$L_2 \equiv L_2(0) = \pm(A - B)\int_0^\infty \frac{\sinh\tau\sin\Lambda\tau}{\cosh^2 \tau}d\tau = \pm\frac{2\pi\Lambda(A - B)}{\cosh(\pi\Lambda/2)}, \quad (6.54)$$

and the generating function takes the form

$$S_k(t_k) = \frac{1}{2\Lambda}\epsilon\left[K^\pm\cos(\Lambda t_k + \chi) + L^\pm\sin(\Lambda t_k + \chi)\right], \qquad (6.55)$$

where the following notations are introduced: $K^\pm = 2\Lambda(K_1 + L_2)$ and $L^\pm = 2\Lambda(K_2 - L_1)$. Notice that the coefficient $K^\pm$ coincides with (6.50). Taking into account that $S_k(t_{k+1}, \mathcal{H}_k) \equiv 0$ and using (5.16), (6.55), the mapping $(\mathcal{T}_k, \mathcal{H}_k) = \hat{T}_1^{(\pm)}(t_k, h_k)$ can be written as

$$\mathcal{H}_k = h_k - \frac{1}{2}\epsilon \left[K^{\pm} \sin(\varphi_k + \chi) + L^{\pm} \cos(\varphi_k + \chi) \right] \ ,$$

$$\Phi_k = \varphi_k + \frac{\Lambda}{2} \ln \frac{32}{|\mathcal{H}_k|}, \qquad \Phi_k = \Omega \mathcal{T}_k \ . \tag{6.56}$$

Mapping Σ_s to $\Sigma_c^{\pm}$

To construct this map suppose we choose $\tau_0 = \tau_k$ in (5.22), which gives $S_k(t_k) \equiv 0$. Then taking $t = t_{k+1}$, $\vartheta = 0$ at the section $\Sigma_c^{\pm}$ and performing similar procedure as above we obtain the following generating function

$$S_{k+1}(t_{k+1}) = -\frac{1}{2\Lambda}\epsilon \left[K^{\pm} \cos(\Lambda t_k + \chi) - L^{\pm} \sin(\Lambda t_k + \chi) \right], \tag{6.57}$$

Therefore the mapping $(t_{k+1}, h_{k+1}) = \hat{T}_2^{(\pm)}(\mathcal{T}_k, \mathcal{H}_k)$ can be presented as

$$\varphi_{k+1} = \Phi_k + \frac{\Lambda}{2} \ln \frac{32}{|\mathcal{H}_k|} \ ,$$

$$h_{k+1} = \mathcal{H}_k - \frac{1}{2}\epsilon \left[K^{\pm} \sin(\varphi_{k+1} + \chi) - L^{\pm} \cos(\varphi_{k+1} + \chi) \right] \ . \tag{6.58}$$

According to scheme shown in Fig. 6.9 the mapping

$$(t_{k+1}, h_{k+1}) = \hat{M}(t_k, h_k) \tag{6.59}$$

of the $\Sigma_c^{\pm}$ to $\Sigma_c^{\pm}$ is presented by two consecutive mappings $\hat{T}_1^{\pm}$, $\hat{T}_2^{\pm}$, given by the following rules

$$\hat{M} = \begin{cases} \hat{T}_2^+ \hat{T}_1^+, & \text{if } (t_k, h_k) \in \Sigma_c^+ \text{ and } \mathcal{H}_k > 0 \ , \\ \hat{T}_2^- \hat{T}_1^+, & \text{if } (t_k, h_k) \in \Sigma_c^+ \text{ and } \mathcal{H}_k < 0 \ , \\ \hat{T}_2^+ \hat{T}_1^-, & \text{if } (t_k, h_k) \in \Sigma_c^- \text{ and } \mathcal{H}_k < 0 \ , \\ \hat{T}_2^- \hat{T}_1^-, & \text{if } (t_k, h_k) \in \Sigma_c^- \text{ and } \mathcal{H}_k > 0 \ . \end{cases} \tag{6.60}$$

Consider a particular case $A = B = 1$ when the problem coincides with the one considered in Sect. 5.1.1. Then the coefficient $K^{\pm} = W/\epsilon$, where W is defined by (5.6), $L^{\pm} = -2\Lambda L_1$. The mapping (6.59) can be written in the following simplified form as a mapping $(t_k, \mathcal{H}_{k-1}) \to (t_{k+1}, \mathcal{H}_k)$ for the non-canonical variables $(t, \mathcal{H})$:

$$\mathcal{H}_k = \mathcal{H}_{k-1} - W \sin(\varphi_k + \chi) \ ,$$

$$\varphi_{k+1} = \varphi_k + \Lambda \ln \frac{32}{|\mathcal{H}_k|} \ , \tag{6.61}$$

with the energy, $\mathcal{H}_k$, defined at the section Σ_s and time (or phase), φ_k defined at the sections $\Sigma_c^{\pm}$. The mapping (6.61) formally coincides with the conventional separatrix mapping (5.7) (supposing $\chi = 0$). The latter can be also obtained from the mapping (6.49) by replacement of the phase, $\varphi_k + \Lambda/2\ln(32/|h_k|) \to \varphi_k$. The mapping (6.61) clarifies the meaning of variables (t, h) in the conventional separatrix mapping (5.7).

In spite of this coincidence, however, there is a fundamental difference between these mappings. The canonical mappings (6.49), (6.56), (6.58) are supplemented with the corresponding rules (6.51), (6.60) of their application which fully determine the evolution of the system in phase space. The formal derivation of the conventional separatrix mapping (5.7) by calculating the increments does not give any rules to apply this mapping to study the dynamics of the system. For this reason, it has been mostly employed to estimate the width of the stochastic layer.

6.3 Mapping for the Periodic–Driven Morse Oscillator

Consider the example of a Hamiltonian system with the saddle point located at infinity, namely the classical Morse oscillator driven by time-periodic force. This system has been widely used as the main model in the studies of stochastic excitation and dissociation of diatomic molecules in a microwave field and associated with the onset of chaos (see Davis and Wyatt (1982); Goggin and Milonni (1988)). The model is described by the Hamiltonian

$$H = \frac{p^2}{2m} + D\left(1 - e^{-x/a}\right)^2 + xE_0 d\cos(\Omega t + \chi) , \qquad (6.62)$$

where D is the depth of potential well, d is a molecule's dipole moment, Ω and E_0 are the frequency and the amplitude of a microwave field, the parameter a is the effective width of the unperturbed potential function $U(x) = D(1 - e^{-x/a})^2$, shown in Fig. 6.10. The latter has a minimum at $x = 0$.

The phase-space structure of unperturbed motion ($E_0 = 0$) is shown in Fig. 6.10b. The unperturbed motion is trapped when $H < D$ (curve 1), and it is unbounded when $H > D$ (curve 3). There are two fixed points of unperturbed motion: the elliptic fixed point at $(x = 0, p = 0)$ and the non-hyperbolic saddle point at $(x = \infty, p = 0)$. The oscillations frequency near the fixed point $(x = 0, p = 0)$ is $\omega_0 = (2D/m)^{1/2}a^{-1}$.

Introducing the normalized energy, $h = H/D - 1$, one can show that the unperturbed orbit of trapped motion, $h < 0$, is described by

$$\exp(x/a) = |h|^{-1}\left[1 - \sqrt{1 - |h|}\cos\vartheta\right] ,$$

$$p = \frac{ma\omega(h)\sqrt{1 - |h|}\sin\vartheta}{1 - \sqrt{1 - |h|}\cos\vartheta} , \qquad (6.63)$$

where $\vartheta = \vartheta_0 + \omega(h)(t - t_c)$ is the angle variable, t_c is a time instant when the orbit crosses the point $(x_1, p = 0)$, x_1 is a left turning point of motion

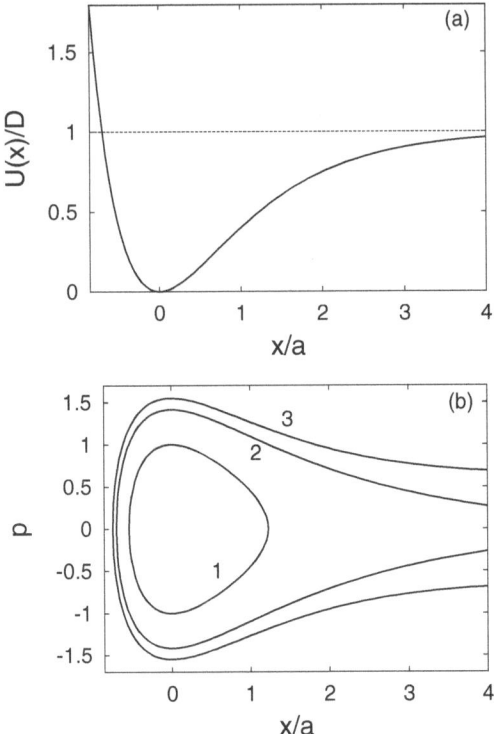

Fig. 6.10. (a) Morse potential $U(x) = D(1 - e^{-x/a})^2$; (b) Phase space of the unperturbed Hamiltonian (6.62) ($E_0 = 0$): curve 1 corresponds to the trapped motion ($H < 1$), curve 2 — to the separatrix ($H = 1$), and curve 3 — to the unbounded motion ($H > 1$)

($p(x_1) = 0$). We set $\vartheta_0 = 0$ in order to have $\vartheta(t_c) = 0$. The relation between the action variable (I) and the energy h is given by

$$I = \frac{1}{2\pi} \oint p dx = \frac{1}{\pi} \int_{x_1}^{x_2} p dx = a\sqrt{2mD} \left(1 - |h|^{1/2}\right) ,$$

$$H = D\left[1 - (1 - I/I_0)^2\right] , \qquad (6.64)$$

where $I_0 = a\sqrt{2mD}$, x_1, x_2 are two turning points of motion: $x_{1,2} = a \times \ln\left[\left(1 \mp \sqrt{1 - |h|}\right)/|h|\right]$, ($x_1 < x_2$). The frequency of oscillations $\omega(h) = dH(I)/dI$ is

$$\omega(h) = \omega_0 |h|^{1/2} . \qquad (6.65)$$

According to quasi-classical quantization rules $I = \hbar(m + 1/2)$, ($m = 0, 1, 2, \ldots$), where $\hbar$ is Planck's constant, one can obtain the discrete energetic levels of the Morse oscillator

$$H_m = D\left[1 - \left(1 - \hbar(m + 1/2)/I_0\right)^2\right] . \tag{6.66}$$

For the lowest energetic levels, $m\hbar \ll I_0$, we have the energy levels of the harmonic oscillator: $H_m \approx \hbar\omega_0(m + 1/2)$.

The motion on the separatrix $h = 0$ is given the formula

$$x_s(t) = a \ln \frac{1 + \omega_0^2(t - t_c)^2}{2} , \qquad p_s(t) = 2p_0 \frac{\omega_0(t - t_c)}{1 + \omega_0^2(t - t_c)^2} , \tag{6.67}$$

where $p_0 = ma\omega_0 = \sqrt{2Dm}$. The orbits of unbounded motion, $h > 0$, are given

$$\exp(x/a) = h^{-1}\left[-1 + \sqrt{1 + h}\cosh\left(\omega(h)(t - t_c)\right)\right] . \tag{6.68}$$

6.3.1 Mapping

We formulate the Hamiltonian system (6.62) in the extended phase space of the action-angle (I, ϑ) and the time-energy (t, p_0) variables in the form (5.12) with the Hamiltonian (5.13):

$$H_0(I) = -(1 - I/I_0)^2 ,$$
$$\epsilon H_1(t, \vartheta, p_0) = \epsilon \frac{x(\vartheta, p_0)}{a} \cos(\Omega t + \chi) , \tag{6.69}$$

where $\epsilon = E_0 ad/D$ is the dimensionless perturbation parameter. We intend to construct the Poincaré return map $(t_k, h_k) \to (t_{k+1}, h_{k+1})$ near the separatrix to the cross sections Σ_s and Σ_c on the phase space. The geometry of this mapping is plotted in Fig. 6.11. The cross sections Σ_c and Σ_s consist of the segments on the x-axis covering the left, x_1, and the right, x_2 turning points of unperturbed motion, respectively. The general form of the corresponding mapping in the first order of ϵ is given by (5.20) with the generating function (5.18). Non-symmetric forms of the mappings are given by (5.29) or (5.32) with the generating function (5.30).

The stochastic layer formed near the separatrix of the Morse oscillator is sufficiently large even for small perturbations. The variation of energy in the stochastic layer may be large enough that the deviation of the generating function $S(H, t)$ (5.30), (5.27) from its value $S(H = 0, t)$ at the unperturbed separatrix $H = 0$ would be not negligible. In this case the dependence of the generating function $S(H, t)$ on the energy variable H becomes important.

6.3.2 The Symmetric Mapping

According to the relations (6.69), (6.63) the perturbation function $V_n(\mathcal{H}, \tau)$ is equal to $V_n = x(\vartheta, H)/a$. Using (6.65) for the frequency $\omega(h)$ the integrals (5.27) can be reduced to:

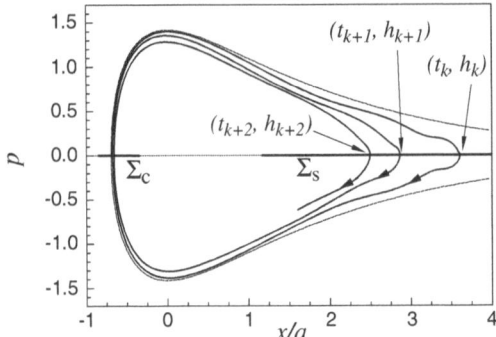

Fig. 6.11. Geometry of the separatrix map to the section Σ_s for the periodically driven Morse oscillator. A *thin curve* describes unperturbed separatrix

$$R^{\pm}(h, \vartheta = \pm\pi) = a^{-1} \int\limits_{\mp\pi/\omega(h)}^{0} x(\tau, h) e^{i\Omega\tau} d\tau = \frac{1}{i\Omega a} \int\limits_{0}^{\mp\pi/\omega(h)} e^{i\Omega\tau} \frac{dx}{dt} d\tau$$

$$= \frac{\sqrt{1 - |h|}}{i\Omega} \int\limits_{0}^{\mp\pi} \frac{e^{i\Lambda|h|^{-1/2}\eta} \sin\eta}{1 - \sqrt{1 - |h|}\cos\eta} d\eta \,, \tag{6.70}$$

where $\Lambda = \Omega/\omega_0$. Then according to (5.24), the generating function takes the form

$$S(t_k \pm 0, \mathcal{H}) = K^{\pm}(\mathcal{H})\Omega^{-1} \cos\left(\Omega t_k \pm \pi\Lambda|\mathcal{H}|^{-1/2} + \chi\right)$$
$$- L^{\pm}(\mathcal{H})\Omega^{-1} \sin\left(\Omega t_k \pm \pi\Lambda|\mathcal{H}|^{-1/2} + \chi\right) \,, \tag{6.71}$$

where $K^{+}(h) = -K^{-}(h) = -K(h)$, $L^{+}(h) = L^{-}(h) = -L(h)$, and

$$K(h) = \sqrt{1 - |h|} \int_0^\pi \frac{\sin\tau \sin(\Lambda\tau/|h|^{1/2})}{1 - \sqrt{1 - |h|}\cos\tau} d\tau \,,$$

$$L(h) = \sqrt{1 - |h|} \int_0^\pi \frac{\sin\tau \cos(\Lambda\tau/|h|^{1/2})}{1 - \sqrt{1 - |h|}\cos\tau} d\tau \,. \tag{6.72}$$

These integrals also can be also presented as a sum of regular and oscillatory parts: $K(h) = K^{(reg)}(h) + K^{(osc)}(h)$. The asymptotical estimations of both parts of $K(h)$ are given in Sect. B.3 of Appendix B.3. The dependence of the integral $K(h)$ on h is shown in Fig. 6.12 at fixed value of the parameter $\lambda = 4$.

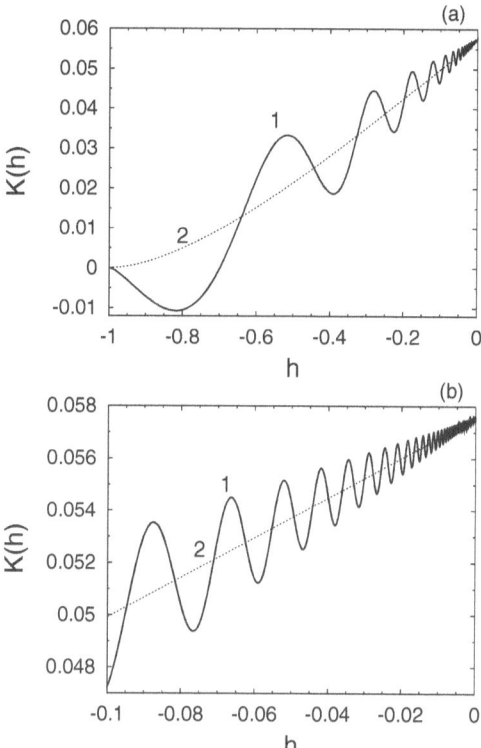

Fig. 6.12. Dependence of the integral $K(h)$ on h: (**a**) in the whole interval of h. (**b**) in the interval $-0.1 < h < 0$. *Solid curve* 1 corresponds $K(h)$, and *dashed curve* 2 describes its regular part $K^{(reg)}(h)$. Parameter $\lambda = 4$

Using the generating function (6.71) the symmetric mapping (5.20) can be written as

$$\mathcal{H} = h_k + \epsilon \left(K(\mathcal{H}) \sin \alpha_k + L(\mathcal{H}) \cos \alpha_k \right) ,$$

$$\Phi_k = \Omega t_k + \epsilon \left(\frac{dK(\mathcal{H})}{d\mathcal{H}} \cos \alpha_k - \frac{dL(\mathcal{H})}{d\mathcal{H}} \sin \alpha_k \right)$$
$$+ \epsilon \frac{\pi \Lambda}{2|\mathcal{H}|^{3/2}} \left(K(\mathcal{H}) \sin \alpha_k + L(\mathcal{H}) \cos \alpha_k \right) , \qquad (6.73)$$

$$h_{k+1} = \mathcal{H} + \epsilon \left(K(\mathcal{H}) \sin \alpha_{k+1} - L(\mathcal{H}) \cos \alpha_{k+1} \right) ,$$

$$\Omega t_{k+1} = \Phi_k + \frac{2\pi \Lambda}{|\mathcal{H}|^{1/2}} + \epsilon \left(\frac{dK(\mathcal{H})}{d\mathcal{H}} \cos \alpha_{k+1} + \frac{dL(\mathcal{H})}{d\mathcal{H}} \sin \alpha_{k+1} \right)$$
$$- \epsilon \frac{\pi \Lambda}{2|\mathcal{H}|^{3/2}} \left(K(\mathcal{H}) \sin \alpha_{k+1} - L(\mathcal{H}) \cos \alpha_{k+1} \right) , \quad (6.74)$$

where

$$\alpha_k = \Omega t_k + \frac{\pi\Lambda}{|\mathcal{H}|^{1/2}} + \chi\,, \qquad \alpha_{k+1} = \Omega t_{k+1} - \frac{\pi\Lambda}{|\mathcal{H}|^{1/2}} + \chi\,.$$

The first set of equations (6.73) is implicit with respect to the energy variable $\mathcal{H}$, and the second set (6.74) is implicit with respect to the time t_{k+1}.

The mapping (6.73), (6.74) can be simplified using a smallness of the perturbation parameter ϵ. Carrying out the transformations similar to ones made in Sect. 5.3.2 and neglecting the terms of order ϵ^2 one obtains

$$h_{k+1} = h_k - 2\epsilon K(h_{k+1})\sin\left(\varphi_k + \frac{\pi\Lambda}{|h_k|^{1/2}} + \chi\right)\,,$$

$$\varphi_{k+1} = \varphi_k + \frac{\pi\Lambda}{|h_k|^{1/2}} + \frac{\pi\Lambda}{|h_{k+1}|^{1/2}} - 2\epsilon\frac{dK(h_{k+1})}{dh_{k+1}}\cos\left(\varphi_k + \frac{\pi\Lambda}{|h_k|^{1/2}} + \chi\right)\,. \tag{6.75}$$

A straightforward calculation shows that $\det|\partial(h_{k+1}, t_{k+1})/\partial(h_k, t_k)| = 1$, i.e., the mapping (6.75) is a area–preserving. It is also invariant with respect to the time reversing: $k \leftrightarrow k + 1$.

6.3.3 A Nonsymmetric Mapping

We present also the nonsymmetric form of the separatrix mapping. Particularly, we consider the mapping in the form (5.29). Putting $\vartheta = -\pi$ in (5.27), one can show that $L(h) = 0$ and $K(h)$ is determined by (6.72). Then the generating function (5.30) becomes

$$S(t_k, h_{k+1}) = -2K(h_{k+1})\Omega^{-1}\cos\left(\Omega t_k + \frac{\pi\Lambda}{|h_{k+1}|^{1/2}} + \chi\right)\,.$$

Then the separatrix mapping (5.29) takes the following form:

$$h_{k+1} = h_k - 2\epsilon K(h_{k+1})\sin\left(\varphi_k + \frac{\pi\Lambda}{|h_{k+1}|^{1/2}} + \chi\right)\,,$$

$$\varphi_{k+1} = \varphi_k + \frac{2\pi\Lambda}{|h_{k+1}|^{1/2}} - 2\epsilon\frac{dK(h_{k+1})}{dh_{k+1}}\cos\left(\varphi_k + \frac{\pi\Lambda}{|h_{k+1}|^{1/2}} + \chi\right)$$

$$- \frac{\pi\Lambda}{|h_{k+1}|^{3/2}}\epsilon K(h_{k+1})\sin\left(\varphi_k + \frac{\pi\Lambda}{|h_{k+1}|^{1/2}} + \chi\right)\,. \tag{6.76}$$

This map can be also transformed into the form (6.75) by eliminating the last term in the second equation (6.76) using the first equation, and neglecting the terms of order ϵ^2.

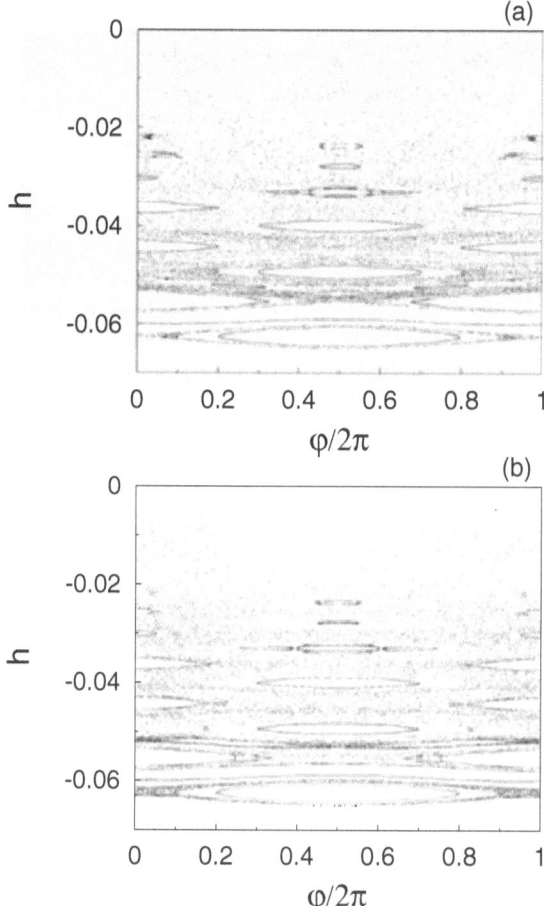

Fig. 6.13. (a) Phase space of the separatrix map (6.75); (b) A corresponding section obtained by a direct numerical integration of Hamiltonian system (6.62). Perturbation parameter $\epsilon = 0.01$, normalized frequency $\varLambda = 4$

6.3.4 Comparison with a Numerical Integration

The phase space of the separatrix mapping (6.13) near the separatrix region is plotted in Fig. 6.13a for the perturbation parameter $\epsilon = 0.01$ and the normalized frequency $\varLambda = \Omega/\omega_0 = 4$. It is supposed that the orbit leaves the system when the energy h exceeds the zero, $h > 0$, which corresponds to the unbounded motion. We have compared the separatrix map with the direct numerical integration of Hamiltonian system with (6.62) using the symplectic integrator presented in Sect. 1.5. The results are shown in Fig. 6.13b. As seen from Figs. 6.13a,b the separatrix map quantitatively well reproduces all features of the regular and chaotic motion of the system: locations of KAM islands and their widths.

6.4 The Kepler Map

The developed method can be directly applied to construct mappings near the separatrix of Hamiltonian systems with a saddle point located at infinity. Particularly, the separatrix mapping of type (6.75) can be directly obtained for one-dimensional hydrogen atom in the field of a monochromatic electromagnetic wave. The latter problem has been subject of the numerous studies related to the chaotic ionization of highly excited hydrogen atom in a microwave field (see a review by Jensen et al. (1991); Sanders and Jensen (1996)). This problem is similar to the mentioned above problem of dissociation of molecules in a microwave field. In several publications by Casati et al. (1987); Gontis and Kaulakys (1987); Casati et al. (1988); Jensen et al. (1988); Kaulakys and Vilutis (1999) the so called *Kepler map* has been introduced to study this problem. Here we present the simplified form of mapping which describes the classical motion of an electron in a one dimensional model of hydrogen atom in the field of monochromatic electromagnetic field similar to the map (6.75) for the driven Morse oscillator.

In atomic units ($m_e = \hbar = e = 1$) the system is described by Hamiltonian

$$H = H(x, p, t) = \frac{p^2}{2} - \frac{1}{x} + xF \cos(\Omega t + \chi) , \qquad x \geq 0 , \qquad (6.77)$$

where Ω and F is the microwave frequency and amplitude, respectively. In the absence of microwave field ($F = 0$) the classical orbit, $x(t - t_0, H)$ of bounded electron ($H < 0$) is given by

$$\vartheta = \omega(H)(t - t_0) = \arcsin \sqrt{\frac{x}{x_c}} - \sqrt{\frac{x}{x_c} \left(1 - \frac{x}{x_c} \right)} , \qquad (6.78)$$

where t_0 is the moment of time when electron reflects from boundary $x = 0$ (*perihelion*), and $x_c = 1/|H|$ is the turning point of classical motion (*aphelion*). The frequency of motion $\omega(H)$ is determined by the relation between the action (I) and energy (H):

$$\omega(H) = \frac{dH_0(I)}{dI} = \frac{1}{I^3} = (2|H|)^{3/2} , \qquad H_0(I) = -\frac{1}{2I^2} . \qquad (6.79)$$

From the perturbed Hamiltonian $\epsilon H_1 = xF \cos(\Omega t + \chi)$ in (6.77) we have $V_n(H, \tau = x(\tau, H)$, and using the relation (6.78) one obtains the Melnikov type integral $K_n(H)$ (5.31)

$$K(H) = \int_{-\pi/\omega(H)}^{\pi/\omega(H)} x(\tau, H)\cos\Omega\tau d\tau = -\frac{2}{\Omega}\int_0^{x_c}\sin\Omega\tau(x)dx$$

$$= -\frac{2\pi}{|H|\Omega}\mathbf{J}'_\nu(\nu) = -\frac{4\pi}{\Omega^{5/3}}\nu^{2/3}\mathbf{J}'_\nu(\nu) ,$$

$$L(H) = \int_{-\pi/\omega(H)}^{\pi/\omega(H)} x(\tau, H)\sin\Omega\tau d\tau = 0 , \tag{6.80}$$

where $\nu = \Omega/\omega(H)$, and $\mathbf{J}'_\nu(z) \equiv d\mathbf{J}_\nu(z)/dz$ is the derivative of the Anger function

$$\mathbf{J}_\nu(z) = \frac{1}{\pi}\int_0^\pi \cos(\nu x - z\sin x)dx . \tag{6.81}$$

For $\nu \leq 1$ (or $|H| \geq \Omega^{3/2}/2$) the function $\mathbf{J}'_\nu(\nu)$ is approximated by a polynomial function

$$\mathbf{J}'_\nu(\nu) = \nu\sum_{k=0}^3 a_k\nu^k ,$$
$$a_0 = 0.49819 , \quad a_1 = 0.0183892 ,$$
$$a_2 = -0.280396 , \quad a_3 = 0.0888974 . \tag{6.82}$$

For $\nu \geq 1$ (or $|H| \leq \Omega^{3/2}/2$), it has the following asymptotics

$$\mathbf{J}'_\nu(\nu) \approx \frac{a}{\nu^{2/3}}\left[1 + \sum_{k=1}^\infty \frac{\gamma_k}{\nu^{2k}}\right] - \frac{b}{\nu^{4/3}}\sum_{k=0}^\infty \frac{\delta_k}{\nu^{2k}} - \frac{\sin\nu\pi}{4\pi\nu^2}\sum_{k=0}^\infty \frac{s_k}{\nu^{2k}} , \tag{6.83}$$

where the coefficients a, b, and the first three coefficients γ_k, δ_k, s_k are

$$a = \frac{2^{2/3}}{3^{1/3}\Gamma(1/3)} , \qquad b = \frac{2^{1/3}}{3^{2/3}\Gamma(2/3)} ,$$
$$\gamma_1 = \frac{23}{3150}, \qquad \gamma_2 = -9.373\times10^{-4}, \qquad \gamma_3 = 4.44\times10^{-4} ,$$
$$\delta_0 = 0.2, \qquad \delta_1 = -\frac{947}{346500} , \qquad \delta_2 = 6.047\times10^{-4} ,$$
$$\delta_4 = -3.8\times10^{-4} , \qquad s_0 = 1 , \qquad s_1 = -1/8 , \qquad s_2 = 1/32 .$$

The maximal relative deviation of the asymptotical formula (6.83) from the value of $\mathbf{J}'_\nu(\nu)$ obtained by numerical integration of the integral (6.81) is 6.3×10^{-3} at $\nu = 1.34$.

At $|H| \leq \Omega^{3/2}/2$ ($\nu \geq 1$) the integral $K(h)$ can be presented as a sum

$$K(H) = K^{(reg)}(H) + K^{(osc)}(H) , \tag{6.84}$$

where

$$K^{(reg)}(H) = K_0 \left\{ a \left[1 + \sum_{k=1}^{\infty} \frac{\gamma_k}{\nu^{2k}} \right] - \frac{b}{\nu^{2/3}} \sum_{k=0}^{\infty} \frac{\delta_k}{\nu^{2k}} \right\},$$

$$K^{(osc)}(H) = -K_0 \frac{\sin \pi \nu}{4\pi \nu^{2/3}} \sum_{k=0}^{\infty} \frac{s_k}{\nu^{2k}}, \qquad (6.85)$$

$$K_0 = -\frac{4\pi}{\Omega^{5/3}}, \qquad \nu = \frac{\Omega}{(2|H|)^{3/2}}.$$

The oscillatory part $K^{(osc)}(H)$ has zeros at the primary resonant values of H, determined by $m\omega(H_m) = \Omega$, i.e., $K^{(osc)}(H_m) = 0$. From (6.80) follows that the normalized integral $K(h)/K_0$ depends on H only through the quantity ν. The dependence $K(h)/K_0$ on ν is plotted in Fig. 6.14 by the solid curve 1, the dashed curve 2 corresponds to the normalized regular part, $K^{(reg)}(H)/K_0$. The dependence of the normalized integral $K(H)/K_0$ on the energy H is shown in Fig. 6.15 by the solid curve 1: a) in the interval $-1 \leq H < 0$; b) in the interval $-0.1 \leq H < 0$. Dashed curve 2 describes the normalized regular part, $K^{(reg)}(H)/K_0$.

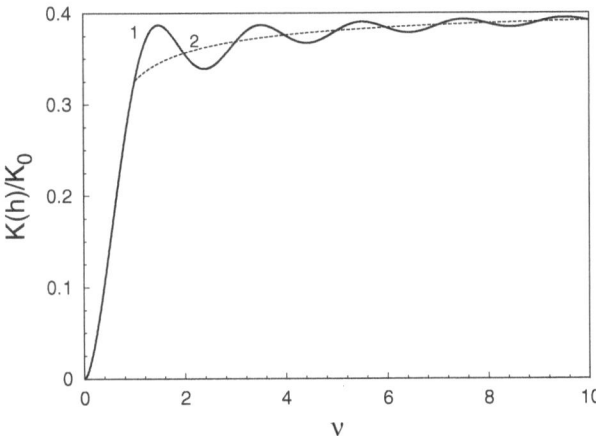

Fig. 6.14. Normalized integral $K(H)/K_0$ versus the parameter ν (curve 1); curve 2 describes the normalized regular part, $K^{(reg)}(H)$, for $\nu \geq 1$

The simplified form of the map given (5.45) takes the following form

$$h_{k+1} = h_k - F\Omega K^{(reg)}(h_{k+1}) \sin\left(\varphi_k + \frac{\pi\Omega}{|h_k|^{3/2}} + \chi \right),$$

$$\varphi_{k+1} = \varphi_k + \frac{\pi\Omega}{|h_k|^{3/2}} + \frac{\pi\Omega}{|h_{k+1}|^{3/2}}$$

$$-F\Omega \frac{dK^{(reg)}(h_{k+1})}{dh_{k+1}} \cos\left(\varphi_k + \frac{\pi\Omega}{|h_k|^{3/2}} + \chi \right), \qquad (6.86)$$

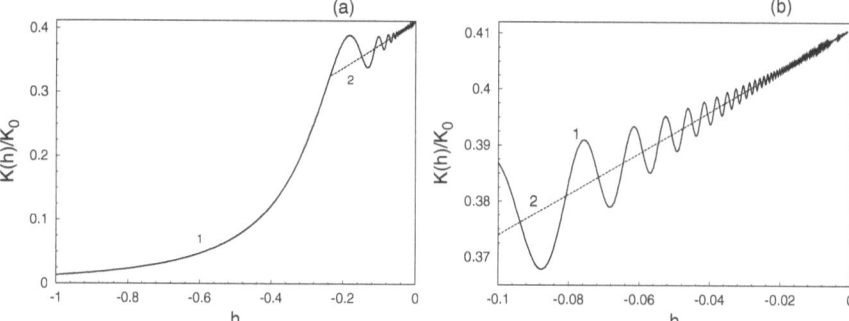

Fig. 6.15. (a) Dependence of the normalized integral $K(h)/K_0$ on the energy h (curve 1) for the frequency $\Omega = 0.325$; (b) Its expanded view in the interval $-0.1 \leq h < 0$. Curve 2 describes the normalized regular part, $K^{(reg)}(H)$

In this map the both variables $\varphi_k = \Omega t_k$, $h_k = H(t_k)$ are defined at the same section Σ_s of phase space (x, p) located at the maximum distance from the center $x = 0$ (aphelion).

In the region very close to the separatrix $H = 0$ the quantity $K^{(reg)}(h)$ can be approximated by $aK_0 = K^{(reg)}(0)$. Then the mapping is reduced to

$$h_{k+1} = h_k - F\Omega aK_0 \sin\left(\varphi_k + \frac{\pi\Omega}{|h_k|^{3/2}} + \chi\right) ,$$
$$\varphi_{k+1} = \varphi_k + \frac{\pi\Omega}{|h_k|^{3/2}} + \frac{\pi\Omega}{|h_{k+1}|^{3/2}} . \tag{6.87}$$

The mapping obtained by Gontis and Kaulakys (1987); Kaulakys and Vilutis (1999) can be recovered from (6.86) shifting the time t (or the phase φ) by the half period of unperturbed motion, $\pi/\omega(H)$ (or $\pi\Omega/\omega(H)$), i.e.,

$$\varphi_k + \frac{\pi\Omega}{|h_k|^{3/2}} = \bar{\varphi}_k \to \varphi_k .$$

In the number works by Petrosky (1986); Casati et al. (1987); Gontis and Kaulakys (1987); Casati et al. (1988); Jensen et al. (1988); Sagdeev and Zaslavsky (1987); Petrosky and Broucke (1988); Chirikov and Vecheslavov (1989) the conventional Kepler map has been derived calculating the increments of energy H and phase φ over one phase rotation in phase space. It has the following form (in our notations)

$$h_{k+1} = h_k - F\Omega K_0 \sin \bar{\varphi}_k ,$$
$$\bar{\varphi}_{k+1} = \bar{\varphi}_k + \frac{2\pi\Omega}{|h_k|^{3/2}} . \tag{6.88}$$

This mapping can be formally obtained from the canonical mapping (6.87) by shifting the phase $\varphi_k \to \bar{\varphi}_k$.

We should note that the variables in the map obtained by Gontis and Kaulakys (1987); Kaulakys and Vilutis (1999) as well as in the Kepler map (6.88) are defined at the different sections of the phase space: the energy H is at the maximum distance from the center (aphelion), and the phase, $\bar{\varphi}$ (or time t) at the minimum distance (perihelion). Because of this the variables H and φ are not canonically conjugated. The Kepler map in canonical variables has been constructed by Nauenberg (1990) and later by Pakoński and Zakrzewski (2001) by integrating Hamiltonian equations in extended phase space. However, the map obtained in such a way has a complicated form although it is in a good agreement with direct numerical integrations. On the other hand, the Kepler map is valid only for the small perturbations F and in the area close to the separatrix.

Casati et al. (1987); Gontis and Kaulakys (1987); Casati et al. (1988); Jensen et al. (1988) applied the Kepler map in the form (6.88) to study the classical chaotic ionization of hydrogen atoms in a microwave field which has been investigated experimentally (see Jensen et al. (1991) and references therein). The map obtained by Gontis and Kaulakys (1987); Kaulakys and Vilutis (1999) which is equivalent to the map (6.86) allows one to analyze the frequency dependence of ionization process and to study the adiabatic and chaotic regimes of ionization.

The Kepler map (6.88) has been also proposed by Petrosky (1986); Sagdeev and Zaslavsky (1987); Petrosky and Broucke (1988); Chirikov and Vecheslavov (1989) to study the chaotic motion of comets near parabolic orbits in the Solar system. Particularly, Chirikov and Vecheslavov (1989) have shown that the motion of Halley's comet is chaotic.

6.5 Comments on Separatrix Map Methods

During the last decade there were a number of studies devoted to the derivation of the separatrix mapping and its generalization (see Chirikov (1979); Zaslavsky et al. (1991); Ahn et al. (1996); Abdullaev and Zaslavsky (1995, 1996); Shevchenko (2000); Luo and Han (2000, 2001); Luo (2002)). These studies were mainly based on the calculations of the increments of the time, t, and the energy, h, over phase rotation in phase space, similar to the derivation of the perturbed twist map, described in Sect. 3.2. This approach leads to the mappings near the separatrix with time and energy variables defined at the different sections of phase space (Escande (1988)). Therefore, the variables of these separatrix maps are not canonical which makes it difficult, in principle, to compare the phase space structure of a system with ones of the separatrix mapping. On the other hand the these approaches, in general, does not allow estimate the accuracy of the separatrix mapping.

For this reason there were several studies to construct the separatrix maps with variables defined at the same sections of phase space. Particularly, in

Abdullaev and Zaslavsky (1995, 1996) the so-called shifted separatrix mapping has been proposed where both time and energy variables are defined at the sections near the saddle points. The separatrix mapping to the section Σ_c (in our terminology) for the periodically driven-pendulum (5.2) has been also obtained in Shevchenko (1998, 2000). These mappings has been obtained from the conventional form of the separatrix map (6.61) by the calculation of the increment of the energy to the corresponding sections. However, the map proposed by Shevchenko (1998, 2000) unlike the one (6.59) obtained here, is implicit in both, time and energy variables, which makes it difficult for practical calculations.

Nauenberg (1990) and later Pakoński and Zakrzewski (2001) constructed the implicit symplectic map for canonical variables defined at the same sections of phase space in the perturbed Kepler problem. Since they employed the method of calculations of increments in time and energy over one phase rotation in phase space, it requires to set additional assumptions on dependence of these increments on time and energy variables.

Ahn et al. (1996); Shevchenko (1998, 2000) have proposed the so-called *exact separatrix mappings* to describe the dynamics of system for large perturbation parameter ϵ. For instance, for the periodically driven pendulum Shevchenko (1998, 2000) obtained the exact map from the conventional separatrix mapping (6.61) by replacing the asymptotics of frequency of motion $\omega(H) = 2\pi/\ln(32/|H|)$ by its exact expression, $\omega(H) = 2\pi k/K\left(k^{-1}\right)$ for $H > 0$ and $\omega(H) = 2\pi/K(k)$ for $H < 0$, while keeping the same integrals $K^{\pm}$ taken along the unperturbed separatrix. However, these maps cannot exactly describe Hamiltonian system, at least, because of two reasons. Firstly, since for the large energy variations one should calculate the energy increments not only along unperturbed separatrix, but along the unperturbed closed orbits inside and outside the separatrix. Secondly, the construction of maps by calculating the increments of time and energy along the unperturbed orbits is valid only in the first order of the perturbation theory ϵ (see Sect. 5.2.1).

The Hamilton–Jacobi method discussed in this chapter gives a systematic and rigorous way to construct mappings near the separatrix. It allows to construct mappings to the arbitrary sections of phase space which is important in different applications.

6.6 Bibliographic Notes

The separatrix mapping has been first proposed by Filonenko and Zaslavsky (1968) and later by Chirikov (1979). The geometrical interpretation was first given by Escande (1988). In early works it has been used to estimate the width of the stochastic layer near the separatrix (see Filonenko and Zaslavsky (1968); Chirikov (1979); Zaslavsky et al. (1991)). Later the application of the separatrix mapping has been extended not only to study the mixing and transport processes in the stochastic layer in the abstract Hamiltonian systems Rom-Kedar (1994, 1995); Vecheslavov (1996); Treschev (1998); Vecheslavov (1999) but also in specific physical systems: fluid dynamics, plasma physics, dynamical astronomy.

A number of studies by Ahn et al. (1996); Abdullaev (1999); Treschev (2002); Abdullaev (2004b) were devoted to the rigorous derivation of the separatrix mapping. The different forms of the separatrix map or its generalizations have been discussed in Abdullaev and Zaslavsky (1995, 1996); Vecheslavov (1996, 1999); Luo and Han (2000); Shevchenko (2000); Luo and Han (2001); Luo (2002); Vecheslavov (2002). Among the mathematical studies to construct separatrix maps one should mention the works by Treschev (2002, 2004). He has constructed a multidimensional analog of the separatrix map which determines the chaotic dynamics in the vicinity of asymptotic (separatrix) surfaces of a hyperbolic torus. The systematic and rigorous method to construct mappings near separatrix is developed by Abdullaev (2004b, 2005).

Rom-Kedar (1994, 1995) has applied the separatrix mapping to study the mixing and transport processes in chaotic Hamiltonian systems. The separatrix mapping has been used by Treschev (2004) to obtain the rigorous estimations for the diffusion rates in Hamiltonian systems with many degrees of freedom. The separatrix mappings were also played an instrumental role in a study of the rescaling invariant properties of Hamiltonian systems near the separatrix and the chaotic transport in the stochastic layer (Abdullaev and Zaslavsky (1994); Zaslavsky and Abdullaev (1995); Abdullaev (1997); Kuznetsov and Zaslavsky (1997); Abdullaev (2000); Kuznetsov and Zaslavsky (2002)).

The separatrix mappings have been widely used to study a chaotic transport of passive particles in structured fluids in Weiss and Knobloch (1989); Ahn and Kim (1994); Latka and West (1995); Weeks et al. (1996); Ahn and Kim (1997); del Castillo-Negreto (1998); Kuznetsov and Zaslavsky (1998), to study magnetic field lines in tokamaks in Yamagishi (1995); Abdullaev and Zaslavsky (1995, 1996); Abdullaev and Finken (1998), and transport in plasmas (Escande (1988)), to describe a particle motion in electromagnetic fields in Lichtenberg and Wood (1989); Zaslavsky et al. (1991). In Shevchenko and Scholl (1997); Shevchenko (1998, 1999) the separatrix map has been used to study the dynamics small planetary bodies, asteroids, in the Solar system, and rotational motion of a satellite.

The Kepler map , which is a specific form of the separatrix map in the systems with along–range interaction, has been introduced in several publications by Casati et al. (1987); Gontis and Kaulakys (1987); Casati et al. (1988); Jensen et al. (1988); Casati et al. (1990); Kaulakys and Vilutis (1999); Petrosky (1986); Sagdeev and Zaslavsky (1987); Petrosky and Broucke (1988); Chirikov and Vecheslavov (1989). It has been used to study a classical ionization of hydrogen atoms in a microwave field and the motion of comets in the Solar system.

7 The KAM Theory Chaos Nontwist and Nonsmooth Maps

In this chapter we will discuss some problems of dynamics and chaos in non-standard Hamiltonian systems. First of these problems is the dynamics of Hamiltonian systems in which a so-called *twist condition* is violated. Particular example of these system is a one–degree-of-freedom system with a non-monotonic dependence of the frequency of oscillations on action variable. The second problem is a study of systems subjected to non-smooth perturbations. These problems have been mainly investigated in the last decade, and they are less discussed in monographes and reviews. Regular and chaotic dynamics of these Hamiltonian systems are outside the scope of the Kolmogorov–Arnold–Moser (KAM) theory. Before discussing these problems we recall the KAM theory and the onset of chaotic dynamics in standard Hamiltonian systems. We will use the mapping approach to study these problems.

7.1 Conservation of Conditionally Periodic Motions. The KAM Theory

In this section we recall the main ideas of the Kolmogorov-Arnold-Moser (KAM) theory emphasizing on conditions of its applicability. This theory concerns the conservation of conditionally –periodic motion of Hamiltonian systems subjected to small perturbations in *infinite time interval*. The corresponding theorem has been first formulated by Kolmogorov (1954) with the sketch of its proof. The theorem has been later proven by Arnold (1963a,b) and Moser (1962) (see, e.g., Arnold et al. (1988), Arnold (1989)).

Consider an integrable Hamiltonian system described by Hamiltonian $H_0(I_1, \ldots, I_N)$. The orbits in phase space lie on the surface of the invariant torus determined by the actions $(I_i = \text{const})$ $(i = 1, \ldots, N)$. The invariant torus is called *irrational* (or *non-resonant*), if the frequencies of motion, $\omega_i = \partial H_0(I)/\partial I_i$, $(i = 1, \ldots, N)$ are rational independent,

$$k_1\omega_1 + k_2\omega_2 + \cdots + k_N\omega_N \neq 0 \,,$$

for all integer numbers k_i not all of which are zero. If the frequencies ω_i $(i = 1, \ldots, N)$ are rationally dependent then the tori are rational (or resonant).

S.S. Abdullaev: *Construction of Mappings for Hamiltonian Systems and Their Applications*,
Lect. Notes Phys. **691**, 139–174 (2006)
www.springerlink.com

The unperturbed system is *nondegenerate* if the frequencies are functionally independent:

$$\det \left(\frac{\partial^2 H_0}{\partial I^2} \right) \neq 0 \, . \tag{7.1}$$

Suppose that the integrable system is subjected to a small perturbation. The perturbed system is described by Hamiltonian (2.3). The Kolmogorov's theorem (Kolmogorov (1954)) states that under sufficiently small perturbation ϵ the majority of *non-resonant tori* are not destroyed, but slightly deformed. For the unperturbed system with H_0 satisfying the nondegenerate condition (7.1) the motion on these tori is still conditionally-periodic with the perturbed frequencies slightly different from the unperturbed ones $w(J)$. Orbits densely fill these invariant tori.

Formally, it means that the dynamical system with the Hamiltonian (2.3) can be transformed to the integrable Hamiltonian system (2.5) with the Hamiltonian $\mathcal{H}(J, \epsilon)$ by the canonical change of variables $(\vartheta, I) \rightarrow (\psi, J)$ (2.4). Then the invariant tori is determined by new action variable $J_i = \text{const}$ $(i = 1, \ldots, N)$, and the motion on tori is conditionally−periodic with the frequencies $w(J, \epsilon) = \partial \mathcal{H}(J, \epsilon)/\partial J$. The proof of the Kolmogorov's theorem is based on the convergence of perturbation series by the successive canonical changes of variables which eliminate phases in higher orders of ϵ. (see Sect. 2.4).

7.1.1 Invariant Tori for Mapping

Below we present the variant of the theorem on invariant tori for the mapping of $2n$ dimensional annulus to itself $(\vartheta, I) \rightarrow (\bar{\vartheta}, \bar{I})$ (see Moser (1973); Arnold et al. (1988)):

$$\bar{I} = I - \epsilon \frac{\partial S(\vartheta, \bar{I}, \epsilon)}{\partial \vartheta} \, ,$$

$$\bar{\vartheta} = \vartheta + \alpha(\bar{I}) + \epsilon \frac{\partial S(\vartheta, \bar{I}, \epsilon)}{\partial \bar{I}} \, , \tag{7.2}$$

satisfying to the so-called *twist condition*,

$$\det \frac{\partial \alpha(I)}{\partial I} \neq 0 \, . \tag{7.3}$$

where $\alpha(I) = (\alpha_1, \ldots, \alpha_N)$ is rotation angles (or transforms). This condition corresponds to the nondegenerate condition (7.1). The map (7.2) coincides with the perturbed twist map (3.5) (see Sect. 3) with the frequency $w(I) = \alpha(I)/T$.

For the case $N = 1$ (7.2) describes the mapping of the annulus on the plane to itself. In the unperturbed case $(\epsilon = 0)$ the orbits lie on circles $I =$

const. If the rotation angle $\alpha(I)$ is incommensurable with 2π, i.e., the ratio $\alpha(I)/2\pi$ is irrational, then orbits completely fill the circle. The corresponding circle is said to be *nonresonant*. When $\alpha(I)$ is commensurable with 2π, i.e., $\alpha(I_{mn})/2\pi = n/m$ with n and m are integer numbers, the circle is resonant. The orbit consists of m periodic points on the circle $I = I_{mn}$. The twist condition (7.3) means that the rotational angle monotonically changes from one circle to another one.

The nonresonant and resonant orbits in the (ϑ, I) plane are shown in Fig. 7.1a[1]. The horizontal lines correspond to the circles.

In the presence of perturbation ($\epsilon \neq 0$) the nonresonant circles satisfying the irrationality condition

$$\left| \alpha(I) - 2\pi \frac{n}{m} \right| > c\sqrt{\epsilon} m^{-\nu} , \tag{7.4}$$

are not destroyed, but they are slightly deformed. (Here $\nu = 2.5$, and c is a constant). The invariant curves have the form

$$I = J + \epsilon f(J, \psi; \epsilon) , \qquad \vartheta = \psi + \epsilon g(J, \psi; \epsilon) , \tag{7.5}$$

where f, g are continously differentiable functions of period 2π in ψ: $f(J, \psi; \epsilon) = f(J, \psi + 2\pi; \epsilon)$, $g(J, \psi; \epsilon) = g(J, \psi + 2\pi; \epsilon)$, and of order of $f, g \sim 1$. The variable J is a constant, and the angle variable ψ is determined by the mapping

$$J = \text{const} , \qquad \bar{\psi} = \psi + w(J; \epsilon)T , \tag{7.6}$$

where $w(J; \epsilon)$ is a perturbed frequency on the invariant curve.

To illustrate this theorem consider the example of a Hamiltonian system given by Hamiltonian

$$H(\vartheta, I, t) = \int \omega(I) dI + \epsilon \sum_{m=4}^{8} \cos(m\vartheta - t) , \qquad \omega(I) = I^{-1} . \tag{7.7}$$

The rotation angle $\alpha(I) = 2\pi w(I) = 2\pi/I$, and the resonant circles are $I_{mn} = m/n$. Several nonresonant and resonant orbits on the (ϑ, I) plane are shown in Fig. 7.1a in the absence of perturbation ($\epsilon = 0$). The horizontal lines correspond to the circles, and the periodic points describe the resonant orbits.

Figure 7.1b shows the corresponding orbits in the presence of perturbation with $\epsilon = 4 \times 10^{-4}$. As seen from Figure the non-resonant orbits are deformed about the unperturbed ones.

---

[1] Furthermore all calculations throughout the book will be performed using the mappings (4.6)–(4.8) with the first order generating function (2.35) unless it is specially noted.

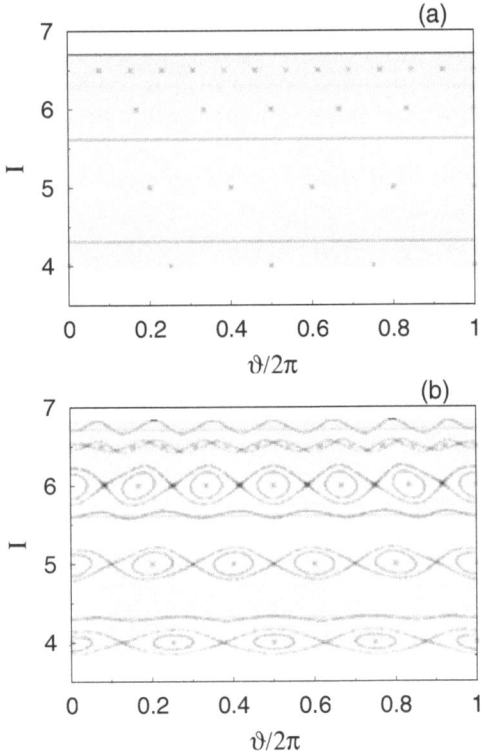

Fig. 7.1. (a) Non-resonant and resonant orbits for $m : n = 5{:}1$, $6{:}1$ and $13{:}2$ in the absence of perturbation ($\epsilon = 0$). (b) Corresponding to these orbits invariant curves in the presence of perturbation ($\epsilon \neq 0$). The resonant orbits are destroyed. *Dashed straight* lines correspond unperturbed orbits

7.1.2 Destruction of Resonant Orbits: Nonlinear Resonance

The resonant orbits as well as orbits nearby are completely modified in the presence of perturbation. The behavior of system in this case is described by the Poincaré–Birkhoff theorem (Poincaré (1892–99); Birkhoff (1927)). In the physical literature this phenomenon is known as a *nonlinear resonance* (see Zaslavsky and Chirikov (1971); Chirikov (1979); Zaslavsky (1985); Sagdeev et al. (1988)). Here we consider the dynamics of resonant orbits using von Zeipel's method of the perturbation theory (see Sect. 2.1.4).

Consider for simplicity the Hamiltonian system (2.19) with $N = 1$ near the resonant value I_{mn}, $m\omega(I_{mn}) - n\Omega = 0$. We introduce a canonical change of variables $(I, \vartheta) \rightarrow (J, \psi) = (I/m, m\vartheta - n\Omega t)$ in the Hamiltonian system (2.19) by means of the generating function $F = (m\vartheta - n\Omega t)J + \epsilon S(\vartheta, J, t, \epsilon)$. Hamiltonian H is transformed to the new one $\mathcal{H}(\psi, J, t) = H(\vartheta(\psi), I(J), t) + \partial F/\partial t$, which can be written as

$$\mathcal{H}(J, \psi, t) = H_{res}(\psi, J) + \epsilon \mathcal{H}_1'(\psi, J, t, \epsilon) ,\tag{7.8}$$

where $\mathcal{H}'(\psi, J, t) \equiv H_1'(\vartheta(\psi), I(J), t)$ does not contain the resonant (m, n) term, and

$$H_{res}(\psi, J) = H_0(mJ) - n\Omega J + \epsilon H_{mn}(mJ) \cos \psi,\tag{7.9}$$

is an integrable resonant Hamiltonian.

First we consider the behavior of the resonant Hamiltonian system (7.9). Suppose that the deviation of action J from the resonance action I_{mn}/m is small, i.e., $|\Delta J| = |J - I_{mn}/m| \ll I_{mn}/m$. We expand the Hamiltonian (7.9) in series of powers of ΔJ. As we will see below that $\max|\Delta J| \propto \sqrt{\epsilon}$. Then retaining only the terms up to the first order of ϵ and taking into account the resonant condition $m\omega(I_{mn}) - n\Omega = 0$, one obtains

$$\bar{H}_{res} = \frac{m^2 \omega'(I_{mn})}{2}(\Delta J)^2 + (k + k'\Delta J) \cos \psi ,\tag{7.10}$$

where $\bar{H}_{res}(J, \psi) = H_{res}(\psi, J) - h_0 - n\Omega I_{mn}/m$ and

$$h_0 = H_0(I_{mn}) , \quad k = \epsilon H_{mn}(I_{mn}) , \quad k' = \epsilon m \frac{dH_{mn}}{dI}\bigg|_{I=I_{mn}} ,$$

$$\omega(I_{mn}) = \frac{dH_0}{dI}\bigg|_{I=I_{mn}} , \quad \omega'(I_{mn}) = \frac{d^2 H_0}{dI^2}\bigg|_{I=I_{mn}} .$$

According to the twist condition (7.3) the quantity $\omega'(I_{mn})$ does not vanish, i.e., $d\omega(I)/dI \neq 0$. For a certainty suppose that $\omega'(I_{m,n})H_{mn}(I_{m,n}) < 0$. When the derivative k' is neglected the Hamiltonian (7.10) coincides with Hamiltonian describing the pendulum studied in Sect. 1.4. Introducing the momentum $y = m^2\omega'(I_{m,n})\Delta J$ and the normalized Hamiltonian $H = \bar{H}_{res}m^2\omega'(I_{m,n})$ the Hamiltonian (7.10) can be reduced to the standard form (1.41)

$$H = y^2/2 - \omega_0^2 \cos \psi ,$$

with the frequency

$$\omega_0 = m\left|\epsilon H_{mn}(I_{m,n})\omega'(I_{m,n})\right|^{1/2} .\tag{7.11}$$

The phase space of the pendulum was shown in Fig. 7.2a. In the (ψ, y) plane the system has the elliptic fixed point at $(\psi = 2\pi, y = 0)$, and the hyperbolic fixed point at $(\psi = \pi, y = 0)$. Since $\psi = m\vartheta - n\Omega t$ there exist m elliptic points and m hyperbolic points in the original (ϑ, I) plane shown in Fig. 7.1b. The hyperbolic points are connected by heteroclinic orbits. They sharply divide the trapped motion with $H < \omega_0^2$ from the un-trapped motion with $H > \omega_0^2$. The maximum variation of the trapped motion in momentum y is equal $\Delta y = 4\omega_0$ which defines a *width of the nonlinear resonance*. The width of the resonance in the action variable I is

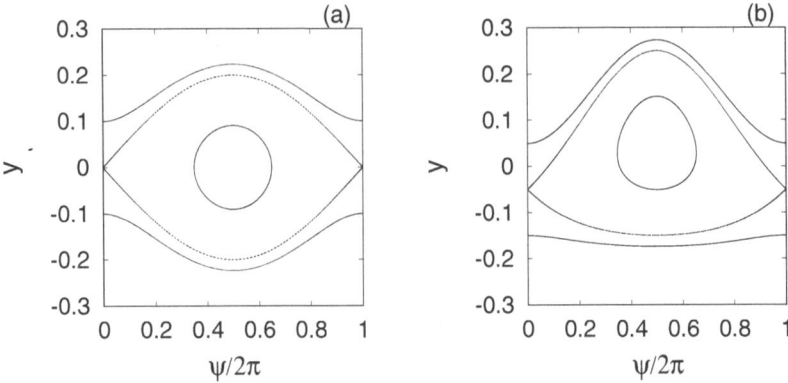

Fig. 7.2. Phase space of Hamiltonian (7.2): (a) $k' = 0$; (b) $k' = 0.05$. The parameter $\omega_0 = 0.1$

$$W_{mn} = 4 \left| \frac{\epsilon H_{mn}(I_{m,n})}{\omega'(I_{m,n})} \right|^{1/2} . \tag{7.12}$$

When k' is not small enough that could be neglected, the Hamiltonian (7.10) is reduced to

$$H = y^2/2 - (\omega_0^2 + k'y)\cos\psi . \tag{7.13}$$

This system has the elliptic point ($\psi = \pi, y = k'$), and the hyperbolic fixed point at ($\psi = 0, y = -k'$). The separatrix dividing the regions of trapped and un-trapped motion becomes asymmetric with respect to the resonant line $y = 0$. The topology of phase space of the Hamiltonian (7.13) in this case is plotted in Fig. 7.2b.

Such asymmetric nonlinear resonances appears when the perturbation, $H_1(\vartheta, I, t)$, has a strong dependence on action variable I. It occurs, for instance, in the study of magnetic field lines in a so called ergodic divertor tokamaks (see Chap. 11).

7.1.3 Chaotic Layer Near a Separatrix

Now we consider the influence of the nonresonant perturbation $\mathcal{H}_1(\psi, J, t)$ in (7.8) near the separatrix of system. Under the time-periodic perturbations the majority of invariant curves are preserved: they may be only slightly deformed. However, the orbits on the separatrix which asymptotically ($t \to \pm\infty$) approach the hyperbolic points, i.e., *stable and unstable manifolds*, do not coincide any more. This phenomenon known as *splitting of separatrices* was discovered by Poincaré (1892–99) (for more details a reader may consult Lichtenberg and Lieberman (1992); Guckenheimer and Holmes (1983)).

A behavior of orbits sufficiently close to the separatrix becomes very sensitive to the small change of initial conditions. As was noted by Poincaré "... *it may happen that small differences in the initial conditions produce very great ones in the final conditions. A small error [change] in the former will produce an enormous error [change] in the latter. Prediction becomes impossible ...*". In general, any small error in initial conditions will be growing exponentially, that the prediction of results will be practically impossible. In other words the distance, $d(t)$, between two orbits with close initial conditions grows exponentially,

$$d(t) = d(0)e^{\sigma t} , \tag{7.14}$$

where $d(0)$ is an initial distance, and the exponent σ, $(\sigma > 0)$, is the measure of divergency of orbits. An example of such exponential divergence of two orbits near the separatrix with very close initial conditions is shown in Fig. 7.3. This phenomenon is known as a *dynamical chaos* or simply *chaos*.

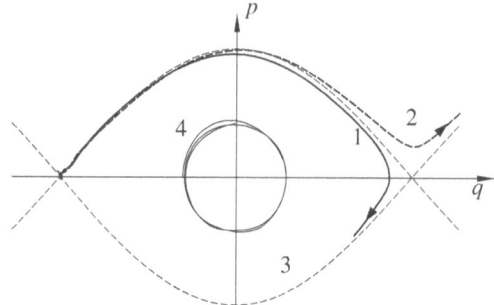

Fig. 7.3. Exponential divergence of orbits near the unperturbed separatrix with very small differences in the initial conditions (curves 1 and 2). Dashed curve 3 describe the unperturbed separatrix, curve 4 — stable orbit

The region near the separatrix with chaotic orbits is known as a *stochastic layer*. The stochastic layer is formed for any small magnitude of perturbation[2]. It constitutes a seed or as called by Chirikov an *embryo* of chaos (Chirikov (1979)). For sufficiently small perturbation ϵ the width of the stochastic layer is exponentially small and the chaotic motion is confined in this very small region. The chaotic orbits cannot diffuse far from its initial positions because of existence of invariant curves.

In the Chaps. 5, 6 we have developed mapping methods to study a motion near the separatrix. The structure of the stochastic layer, its properties and chaotic transport in a stochastic layer will be also studied in Chaps. 8, 9.

---

[2] As was recently shown by Vecheslavov (2001); Vecheslavov and Chirikov (2002) that for a certain class of Hamiltonian systems the resonance separatrices are not destroyed even for moderately strong perturbation; see Sect. 7.5

7.1.4 Lyapunov Exponents

The *Lyapunov exponent* is the important characteristic of chaotic systems. It gives a precise quantitative measure of exponential divergence of nearby orbits. Below we give a definition of the Lyapunov exponent and the method of its calculation for symplectic mappings (3.2). For simplicity we consider the system described by the two-dimensional map. Let

$$d\mathbf{I}_k = \begin{pmatrix} dI_k \\ d\vartheta_k \end{pmatrix} \tag{7.15}$$

be an infinitesimal vector separating neighboring orbits at the k-th step. The evolution of the vector $d\mathbf{I}_k$ for one map iteration is described by the following equation

$$d\mathbf{I}_{k+1} = \mathbf{J}_k d\mathbf{I}_k \,, \tag{7.16}$$

where $\mathbf{J}_k$ is the Jacobian matrix of the mapping (3.2):

$$\mathbf{J}_k = \begin{pmatrix} \frac{\partial I_{k+1}}{\partial I_k} & \frac{\partial I_{k+1}}{\partial \vartheta_k} \\ \frac{\partial \vartheta_{k+1}}{\partial I_k} & \frac{\partial \vartheta_{k+1}}{\partial \vartheta_k} \end{pmatrix} \,. \tag{7.17}$$

We recall that $\det \mathbf{J}_k = 1$. Consider the evolution of the distance $ds_k = \sqrt{dI_k^2 + d\vartheta_k^2}$ between orbits in the (ϑ, I) plane. According to (7.16) its square, ds_k^2, can be written as

$$\begin{aligned} ds_k^2 &= d\mathbf{I}_k^T d\mathbf{I}_k = d\mathbf{I}_{k-1}^T \mathbf{J}_k^T \mathbf{J}_k d\mathbf{I}_{k-1} \\ &= d\mathbf{I}_0^T \mathbf{J}_1^T \cdots \mathbf{J}_k^T \mathbf{J}_k \cdots \mathbf{J}_1 d\mathbf{I}_0 \,, \end{aligned} \tag{7.18}$$

Let $\lambda_1^{(k)}, \lambda_2^{(k)}$ be the eigenvalues of the matrix $\mathbf{J}_k$ which satisfy the eigenvalue problem,

$$\mathbf{J}_k \mathbf{A} = \lambda \mathbf{A} \,,$$

where $\mathbf{A}$ is the corresponding eigenvector. The eigenvalues are found as solutions of the equation

$$\det \begin{pmatrix} \frac{\partial I_{k+1}}{\partial I_k} - \lambda & \frac{\partial I_{k+1}}{\partial \vartheta_k} \\ \frac{\partial \vartheta_{k+1}}{\partial I_k} & \frac{\partial \vartheta_{k+1}}{\partial \vartheta_k} - \lambda \end{pmatrix} = 0 \,,$$

from which it follows that

$$\lambda_{1,2}^{(k)} = D \pm \sqrt{D-1} \,, \tag{7.19}$$

where

$$D = \frac{1}{2} \left(\frac{\partial I_{k+1}}{\partial I_k} + \frac{\partial \vartheta_{k+1}}{\partial \vartheta_k} \right) \,.$$

In (7.19) we have taken into account that according to the volume-preserving condition (3.3) the determinant of the Jacobi matrix $\mathbf{J}_k$ is unity.

In general, the eigenvalues $\lambda_1^{(k)}, \lambda_2^{(k)}$ are functions of local coordinates (ϑ_k, I_k). In the case $D > 1$ the eigenvalues $\lambda_{1k}, \lambda_{2k}$ are real, and satisfy the condition

$$\lambda_1^{(k)} > 1, \qquad \lambda_2^{(k)} < 1, \qquad \lambda_1^{(k)} \lambda_2^{(k)} = 1 .$$

In this case the orbits is locally unstable. If $D < 1$ the eigenvalues λ_{1k} and λ_{2k} are complex numbers, with the unity modules, $|\lambda_{1k}| = |\lambda_{2k}| = 1$. The orbits in this case are locally stable.

The *Lyapunov exponent*, σ, is defined as

$$\sigma = \lim_{N \to \infty} \frac{1}{N} \ln \frac{ds_N}{ds_0} , \tag{7.20}$$

or $ds_n = \exp(N\sigma)ds_0$, i.e., it characterizes the degree of exponential divergency of orbit per one map iteration. According to (7.18), it is determined by the largest eigenvalue, $\lambda^{(k)} = \max(\lambda_1^{(k)}, \lambda_2^{(k)})$, of the Jacobian matrix $\mathbf{J}_k$:

$$\sigma = \lim_{N \to \infty} \frac{1}{N} \ln \prod_{k=1}^{N} \lambda^{(k)} . \tag{7.21}$$

When the eigenvalue $\lambda^{(k)}$ does not depend on the local coordinates (ϑ_k, I_k), i.e., $\lambda^{(k)} = \lambda_1$, then $\sigma = \ln \lambda_1$. The Lyapunov exponent, σ, is positive for the unstable orbits, and it vanishes, $\sigma = 0$, for the stable orbit.

We conclude this section with calculation of the Jacobi matrix of the mapping (4.12)–(4.14) determined by the generating function (2.35). Use the presentation of the mapping in the form (4.67) of three successive mappings (4.68) each of them are given by (4.12), (4.13) and (4.14), respectively. Then the Jacobian matrix (7.17) can be written as a product of three Jacobian matrices, corresponding to three successive mappings,

$$\mathbf{J}_k = \hat{M}_{k+1} \hat{M}_0 \hat{M}_k , \tag{7.22}$$

where

$$\hat{M}_k = \begin{pmatrix} \frac{\partial J_k}{\partial I_k} & \frac{\partial J_k}{\partial \vartheta_k} \\ \frac{\partial \psi_k}{\partial I_k} & \frac{\partial \psi_k}{\partial \vartheta_k} \end{pmatrix} , \tag{7.23}$$

$$\hat{M}_0 = \begin{pmatrix} \frac{\partial J_k}{\partial J_k} & \frac{\partial J_k}{\partial \psi_k} \\ \frac{\partial \psi_k}{\partial J_k} & \frac{\partial \psi_k}{\partial \psi_k} \end{pmatrix} = \begin{pmatrix} 1 & 0 \\ \omega'(J_k)(t_{k+1} - t_k) & 1 \end{pmatrix} , \tag{7.24}$$

$$\hat{M}_{k+1} = \begin{pmatrix} \frac{\partial I_{k+1}}{\partial J_k} & \frac{\partial I_{k+1}}{\partial \psi_k} \\ \frac{\partial \vartheta_{k+1}}{\partial J_k} & \frac{\partial \vartheta_{k+1}}{\partial \psi_k} \end{pmatrix} . \tag{7.25}$$

The derivatives in the matrices (7.23), (7.25) are easily calculated from the mappings given by (4.12) and (4.14)):

$$\frac{\partial J_k}{\partial I_k} = \frac{1}{1+\epsilon A_{12}(t_k)}, \quad \frac{\partial J_k}{\partial \vartheta_k} = -\frac{\epsilon A_{22}(t_k)}{1+\epsilon A_{12}(t_k)},$$

$$\frac{\partial \psi_k}{\partial I_k} = \frac{\epsilon A_{11}(t_k)}{1+\epsilon A_{12}(t_k)}, \quad \frac{\partial \psi_k}{\partial \vartheta_k} = 1+\epsilon A_{12}(t_k) - \frac{\epsilon^2 A_{11}(t_k)A_{22}(t_k)}{1+\epsilon A_{12}(t_k)}, \quad (7.26)$$

$$\frac{\partial I_{k+1}}{\partial J_k} = 1+\epsilon A_{12}(t_{k+1}) - \frac{\epsilon^2 A_{11}(t_{k+1})A_{22}(t_{k+1})}{1+\epsilon A_{12}(t_{k+1})},$$

$$\frac{\partial I_{k+1}}{\partial \bar{\psi}_k} = \frac{\epsilon A_{22}(t_{k+1})}{1+\epsilon A_{12}(t_{k+1})},$$

$$\frac{\partial \vartheta_{k+1}}{\partial J_k} = -\frac{\epsilon A_{11}(t_{k+1})}{1+\epsilon A_{12}(t_{k+1})}, \quad \frac{\partial \vartheta_{k+1}}{\partial \bar{\psi}_k} = \frac{1}{1+\epsilon A_{12}(t_{k+1})}, \quad (7.27)$$

where

$$A_{11}(t) = \frac{\partial^2 S_1(\vartheta, J, t, t_0)}{\partial J^2}, \quad A_{12}(t) = \frac{\partial^2 S_1(\vartheta, J, t, t_0)}{\partial J \partial \vartheta},$$

$$A_{22}(t) = \frac{\partial^2 S_1(\vartheta, J, t, t_0)}{\partial \vartheta^2}. \quad (7.28)$$

Obtained formulae can be used to calculate the Lyapunov exponents for generic Hamiltonian systems (2.2), (2.3), (2.15) using the mapping procedure (4.12)–(4.14).

7.2 Applicability of KAM Theory

Conditionally periodic motion of Hamiltonian systems subjected to small perturbations is preserved only when the certain conditions on the unperturbed Hamiltonian and perturbation are satisfied. We list these conditions of applicability of the KAM theory on invariant tori. They are following:

1. Perturbation ϵ is sufficiently small.
2. Perturbations $H_1(I, \vartheta, t)$ or $S(I, \vartheta)$ are sufficiently smooth functions.
3. Rotation angles $\alpha(I)$ satisfy the twist condition (7.3).
4. Rotation angles $\alpha(I)$ are sufficiently irrational (7.4).

The case of violation of twist condition will be considered in Sect. 7.3. Here we discuss the first two conditions.

7.2.1 On the Smallness of Perturbations

The *smallness of the perturbation parameter* ϵ is an important condition for the existence of invariant tori. With the increase of ϵ the width of destroyed

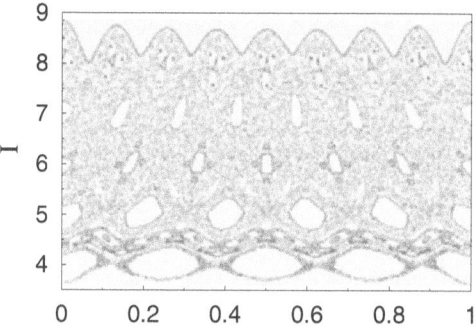

Fig. 7.4. A case of global chaos in the Hamiltonian system (7.7) for $\epsilon = 1.8 \times 10^{-3} >$ ϵ_c

rational tori (or nonlinear resonances) located between invariant tori grows. At certain level of $\epsilon > \epsilon_c$ neighboring resonances start to overlap destroying all invariant tori. Orbits are no longer confined near the resonant tori but they may diffuse far from its initial positions (see, e.g., Chirikov (1979); Lichtenberg and Lieberman (1992). The system becomes *globally chaotic*. It is illustrated in Fig. 7.4 for the Hamiltonian system (7.7) at the perturbation $\epsilon = 7.6 \times 10^{-3}$ which exceeds the critical perturbation ϵ_c. The case $(\epsilon < \epsilon_c)$ was shown in Fig. 7.1b.

The qualitative condition for the onset of global chaos is described by the resonance overlapping criteria or the *Chirikov criteria* (Chirikov (1979)). It states that the global chaos occurs when a sum of half widths of neighboring resonances, $(W_{m,n} + W_{m+1,n})/2$ exceeds the distance between them, $\delta I = |I_{m+1,n} - I_{m,n}|$,

$$\sigma = \frac{W_{m,n} + W_{m+1,n}}{2(I_{m+1,n} - I_{m,n})} \geq 1 .$$

For simple Hamiltonian systems the exact condition of global chaos can be found using the renormalization group theory (see, e.g., MacKay (1983, 1993)). For instance, the critical perturbation threshold K_c for the global chaos in the standard map is $K_c = 0.98$. In more complex systems the exact critical perturbation ϵ_c may be found by numerical calculations.

In the case of systems with more than two degrees of freedom, $N > 2$, a global chaotic motion may happen at any small perturbations $\epsilon > 0$. In these systems trajectories could wander arbitrarily far from their point of departure. However, such a chaotic motion known as *Arnold diffusion* occurs only at special initial conditions (Arnold (1964), see also Arnold (1989); Lichtenberg and Lieberman (1992)). The diffusion rate of such a chaotic motion is very small and decays with ϵ as $\exp(-C/\epsilon)$ (C is a constant).

The prediction of particle orbits in the stochastic zone becomes practically impossible because of the exponential growth of small errors in initial conditions. For this reason we need a statistical approach to study the dynamics of system. The elements of the statistical description of dynamically chaotic systems will be recalled in Sect. 9.1.

7.2.2 On the Smoothness of Perturbations

The term "smooth" means that the function has a finite number of continuous derivatives. One introduces a smoothness parameter, β, which is related with the dependence of Fourier coefficients, $H_m(I)$, of the perturbed Hamiltonian $H_1(\vartheta, I, t)$ (2.15) on harmonics number m. For the analytical functions H_1 the coefficients $H_m(I)$ decay exponentially, $H_m(I) \propto \exp(-Cm)$. For the smooth Hamiltonian the coefficients $H_m(I)$ has the power–law dependence, $H_m(I) \propto |m|^{-(\beta+1)}$ (Chirikov (1991)).

The original proof of the Kolmogorov's theorem given by Arnold was based on the analyticity of perturbation function (Arnold (1963a)). Moser (1962) has first proven the theorem for $\beta > \beta_c = 333$. Later it was improved by Rüssman to $\beta > \beta_c = 4$ (see, Moser (1973)). At these cases there exists a threshold value of perturbation $\epsilon_c(\beta)$ that a global chaos takes place only for perturbation $\epsilon > \epsilon_c(\beta)$. The critical perturbation $\epsilon_c(\beta)$ goes to zero as $\beta \to \beta_c$. As was shown by Chirikov (1991) (see also Vecheslavov and Chirikov (2002)) for simplest two dimensional maps there exists the non zero threshold $\epsilon_c(\beta)$ of onset of global chaos if $\beta > \beta_c = 3$.

On the other hand it has been proven that in the case of Hamiltonian systems with $\beta = 1$ the invariant tori do not longer exist for any small ϵ (Takens (1971)). However, the similar statement has not been proven for the case $\beta = 2$. Nevertheless, the existence of the global invariant curves in Hamiltonian systems with $\beta < \beta_c$ were shown in a number mathematical studies by Wojtkowski (1981); Hénon and Wisdom (1983); Bullet (1986). Recent extensive numerical and analytical studies of Hamiltonian systems with $\beta = 2$ revealed the existence of invariant resonance structures (Vecheslavov (2001); Vecheslavov and Chirikov (2002); Chirikov and Vecheslavov (2002)). These structures act as barriers preventing a global chaotic motion in phase space. These problems will be shortly mentioned in Sect. 7.5 for simplest two dimensional maps.

7.3 Non-Twist Maps

A behavior of the system described above may be significantly changed in the case when the twist condition (7.3) is violated. It takes place, for instance, when the frequency of nonlinear oscillations $\omega(I)$ is a non-monotonic function of action I and it has local maximum or minimum. At these points

$$\det\left(\frac{\partial \omega(I)}{\partial I}\right) = 0 \,. \tag{7.29}$$

This case occurs in many problems, for instance, in the problem of passive advection in shear flows (del Castillo-Negreto and Morrison (1993a,b); Morrison (2000)), in condensed matter physics (Soskin (1994); Soskin et al. (2003)), rays dynamics in waveguides (Abdullaev (1994b)), in tokamak magnetic fusion devices with reversed magnetic shear (see, e.g., a review Wolf (2003) and references therein), in the study of $\mathbf{E} \times \mathbf{B}$ transport in magnetized plasmas (Horton (1990)) and others. More detailed investigation of these problems has been started at the beginning of 90s. The recent review by Soskin et al. (2003) summarizes achievements of these activities called by authors as "zero-dispersion phenomena".

In this section we study this phenomenon in Hamiltonian systems from the perspective of mappings. We restrict ourselves with one-degree-of-freedom systems. In this case the non-twist condition (7.29) determines local extremal points of the frequency $\omega(I)$, where $\omega'(I_0) = 0$. For one-degree-of-freedom system ($N=1$) the curve $I = I_0$ is called a *shearless curve* (del Castillo-Negreto et al. (1996)). These extremal points may be local maximum, minimum or bending points. At the local maximum point $I = I_0$, $\omega''(I_0) < 0$, for local minimum $\omega''(I_0) > 0$, and the bending point $\omega''(I_0) = 0$.

7.3.1 Dynamics of Systems with a Non-Monotonic Frequency

To be specific we consider a Hamiltonian system similar to (7.7) but with a non-monotonic frequency $\omega(I)$

$$H(\vartheta, I, t) = \int \omega(I)dI + \epsilon \sum_{m=1}^{10} \cos(m\vartheta - t) , \qquad \omega(I) = \frac{I}{I^2 + I_0^2} \cdot \quad (7.30)$$

The profile of $\omega(I)$ is plotted in Fig. 7.5. The unperturbed frequency has maximum at $I = I_0$. The resonant values of I_{mn} ($\omega(I_{mn}) = n/m$),

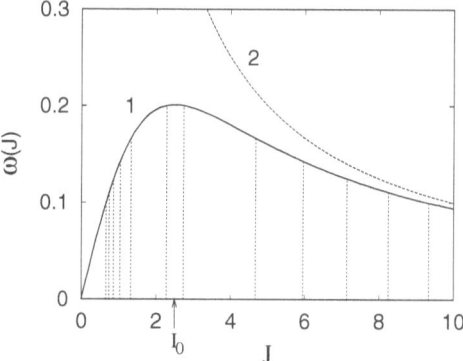

Fig. 7.5. Non-monotonic dependence of $\omega(I)$ on I (curve 1). Curve 2 describes the case $\omega(I) = I^{-1}$. A value $I_0 = 2.49$

$$I_{mn} = \frac{m}{2n} \pm \left(\frac{m^2}{4n^2} - I_0^2 \right)^{1/2},$$

with the same numbers (m, n) are located *up* and *down* the shearless curve $I = I_0$. The resonant actions I_{mn} for the primary resonances (m, n), $(m = 5, \ldots, 10; n = 1)$ are shown in Fig. 7.5 by vertical lines. The distance between neighboring resonant curves is relatively large near the shearless curve. Below the shearless curve it becomes more dense, while above the shearless the corresponding distance is large. There is a pair of resonances $(m, n)=(5:1)$ just about the shearless curve $I = I_0$.

Consider the dynamics of the system under influence of perturbation. Poincaré section of the Hamiltonian system with (7.30) is shown in Fig. 7.6 for three values of the perturbation parameter ϵ: (a) $\epsilon = 5.97 \times 10^{-4}$, (b) $\epsilon = 5 \times 10^{-3}$, (c) $\epsilon = 10^{-2}$. As seen from Fig. 7.6 for sufficiently small perturbation ($\epsilon = 5.97 \times 10^{-4}$) most non-resonant curves located far from the shearless curve are survived in accordance with the KAM theorem. The resonant curves are destroyed. The motion below the shearless curve becomes chaotic because of interaction of closely located resonances. However, the dynamics of pair of resonances $(m : n = 5 : 1)$ near the shearless curve is different from one predicted by the KAM theory. When $\epsilon < \epsilon_c = 5.97 \times 10^{-4}$ they are separated. At $\epsilon = 5.97 \times 10^{-4}$ the separatrices of resonances of reconnected as shown in Fig. 7.6a.

With increase the perturbation ϵ the resonances above the shearless curve start to overlap breaking invariant curve. It is shown in Fig. 7.6b when $\epsilon = 5 \times 10^{-3}$. The pair of resonances near the shearless curve is modified. However, for this value of perturbation there are still invariant curves just above the these resonances. These invariant curves are "hard" to break. These curves are destroyed for larger perturbation. This case is shown in Fig. 7.6c when $\epsilon = 10^{-2}$. Nevertheless, even for this relatively large perturbation there exists an invariant curve between pair of resonance about the shearless curve. This invariant curve prevents a diffusion of orbits from the highly chaotic bottom region of the system to the another chaotic region above the shearless curve. Described picture of transition to global chaos in the system with the non-monotonic frequency is completely different from the one with monotonic frequency considered in the previous section.

In the following two subsections we analyze specific features of system near the shearless curves. A study will be based on analysis of the Hamiltonian (7.9) with only one resonant term located near the extremal point $I = I_0$. The effect of non-resonant terms $\epsilon \mathcal{H}'_1(\psi, J, t)$ will be studied by constructing symplectic mappings.

7.3.2 Behavior Near Local Maxima or Minima

First consider the case when a local minimum or maximum point I_0 close to the resonant action I_{mn}. Dynamics of the system near this point can

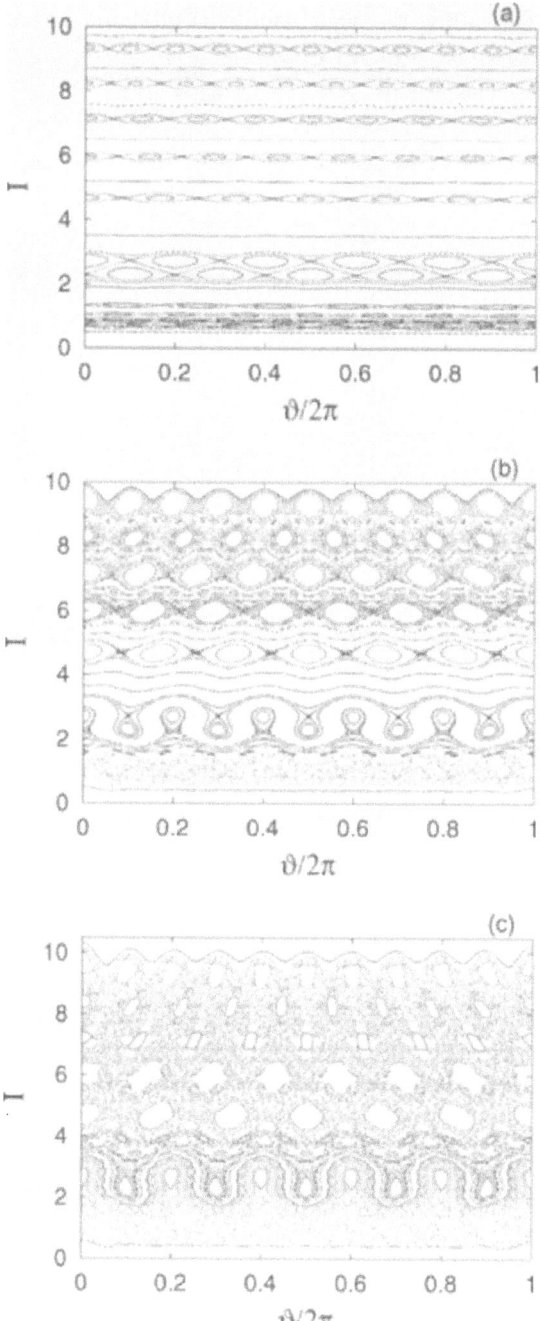

Fig. 7.6. Poincaré section of the Hamiltonian system (7.30). (a) $\epsilon = 5.97 \times 10^{-4}$, (b) $\epsilon = 5 \times 10^{-3}$, (c) $\epsilon = 10^{-2}$. A value $I_0 = 2.49$

be studied by expanding the right hand side of Hamiltonian (7.9) near the extremal point I_0 up to third order term $\Delta J = J - I_0/m$. Taking into account that $\omega'(I_0) = 0$ we obtain

$$H_{res}(\psi, J) = H_0(I_0) - \frac{n\Omega}{m}I_0 + (m\omega_0 - n\Omega)\Delta J$$
$$+ \frac{m^3\omega_0''}{6}(\Delta J)^3 + (k_0 + k'\Delta J)\cos\psi , \qquad (7.31)$$

where

$$k_0 = \epsilon H_{mn}(I_0) , \qquad k' = m\epsilon\frac{dH_{mn}}{dI}\bigg|_{I=I_0} , \qquad \omega_0'' = \frac{d^3 H_0}{dI^3}\bigg|_{I=I_0} .$$

Introducing a new canonical variable $y = \sqrt{m^3|\omega_0''|/2}\Delta J$ and $H = (H_{res} - H_0(I_0) - n\Omega I_0/m)\sqrt{m^3|\omega_0''|/2}$ the Hamiltonian (7.31) is reduced to

$$H(\psi, y) = ay + \sigma\frac{y^3}{3} + (b + k'y)\cos\psi , \qquad (7.32)$$

where $a = m\omega_0 - n\Omega$, $b = k_0\sqrt{m^3|\omega_0''|/2}$, and $\sigma = \mathrm{sgn}(\omega_0'')$. The quantity a describes mismatch between the frequency ω_0 at the extremal point I_0 and the resonant frequency: $a = m(\omega_0 - \omega(I_{mn}))$. The number $\sigma = 1$ (or $\omega_0'' > 0$) when the frequency $\omega(I)$ has a minimum at I_0, and therefore $a < 0$. The value $\sigma = -1$ (or $\omega''(I_0) < 0$) when the frequency has a maximum at I_0 and $a > 0$. Furthermore we put $k' = 0$ for the sake of simplicity.

The fixed points of the Hamiltonian system with (7.31) are determined by

$$\dot{\psi} = \frac{\partial H}{\partial y} = a + \sigma y^2 = 0 ,$$

$$\dot{y} = -\frac{\partial H}{\partial \psi} = b\sin\psi = 0 .$$

For $\sigma = 1$ ($a < 0$) there are two kind of fixed points: elliptic fixed points at $(\psi_e^{(+)} = \pi, y_e^{(+)} = \sqrt{|a|})$ and $(\psi_e^{(-)} = 0, y_e^{(-)} = -\sqrt{|a|})$. The up and down hyperbolic fixed points are $(\psi_h^{(+)} = 0, y_h^{(+)} = \sqrt{|a|})$ and $(\psi_h^{(-)} = \pi, y_h^{(-)} = -\sqrt{|a|})$, respectively. Similarly, one can find fixed points for the case $\sigma = -1$. The phase space of Hamiltonian (7.32) is plotted in Fig. 7.7 for the different values of the perturbation parameter b.

For small values of b the up and down hyperbolic points are connected only with themselves by heteroclinic orbits similar to the case of the standard nonlinear resonance. At the certain level of perturbation b_c the up and the down hyperbolic points are reconnected. It occurs when

$$H(\psi = \pi, y = y_h^{(-)}) = H(\psi = 0, y = y_h^{(+)}) .$$

According to the Hamiltonian (7.32) we obtain the reconnection threshold

$$b_c = 2|a|^{3/2}/3 \ . \tag{7.33}$$

When $b > b_c$ the lower hyperbolic point starts to connect itself homoclinically.

The maximum variation of the action variable J near resonant orbits is proportional to $\epsilon^{1/3}$, which is larger than for the conventional nonlinear resonance where it is proportional to $\epsilon^{1/2}$ (see Eq. (7.12)).

Under influence of the perturbation term $\epsilon\mathcal{H}'_1(\psi, J, t)$ (7.8) the stable and unstable homoclinic orbits are split. The motion near the separatrix becomes chaotic. This and other features of dynamics of system near extremal points of non-monotonic frequency will be discussed in the next section.

7.3.3 Behavior Near a Bending Point

Near the bending point where $\omega''(I_{mn}) = 0$ one can expand Hamiltonian (7.9) near the resonant action I_{mn} up to term of order of $(\Delta J)^4$. Supposing $k' = 0$ one obtains

$$\bar{H}_{res} = H_0(I_0) - n\Omega I_0 + a\Delta J + \frac{m^4\omega_0'''}{24}(\Delta J)^4 + k_0 \cos\psi \ , \tag{7.34}$$

where $\omega_0''' = d^3\omega(I)/dI^3\big|_{I=I_0}$. In a new canonical variable $y = (m^4\omega_0'''/6)^{1/3}\Delta J$ the Hamiltonian (7.34) is reduced to

$$H(\psi, y) = ay + \frac{y^4}{4} + b\cos\psi \ , \tag{7.35}$$

where $b = k(m^4\omega'''/6)^{1/3}$, and $H = (H_{res} - H_0(I_0) - n\Omega I_0)(m^4\omega_0'''/6)^{1/3}$. When $a \neq 0$ the elliptic point of Hamiltonian (7.35) is $\psi_e = \pi, y_e = (-a)^{1/3}$, and the hyperbolic point is $\psi_h = 2\pi, y_h = (-a)^{1/3}$. In the case $a = 0$ the latter becomes a non-hyperbolic fixed point $\psi_h = 2\pi, y_h = 0$. At this point the orbit on the separatrix approaches the fixed point parallel to the y axis.

The phase space of the Hamiltonian (7.35) is plotted in Fig. 7.8 for the perturbation parameter $b = 0.01$ and for different values of a: a) $a = 0.09$; b) $a = 0$. When $a = 0$ the phase space is symmetric with respect to the resonance line $y = 0$. The maximum variation of variable y for the trapped motion is larger than in cases of the standard nonlinear resonance and of the local maximum (or minimum), and it has the order of $\epsilon^{1/4}$. When $a \neq 0$ the phase space becomes asymmetric with respect to the line $y = y_h$.

7.3.4 Non-Twist Standard Maps

Described a behavior of the system near extremal points of frequency is not valid when the system is subjected to any small time-periodic perturbation with more than one harmonics. Then homoclinic (or heteroclinic) connections

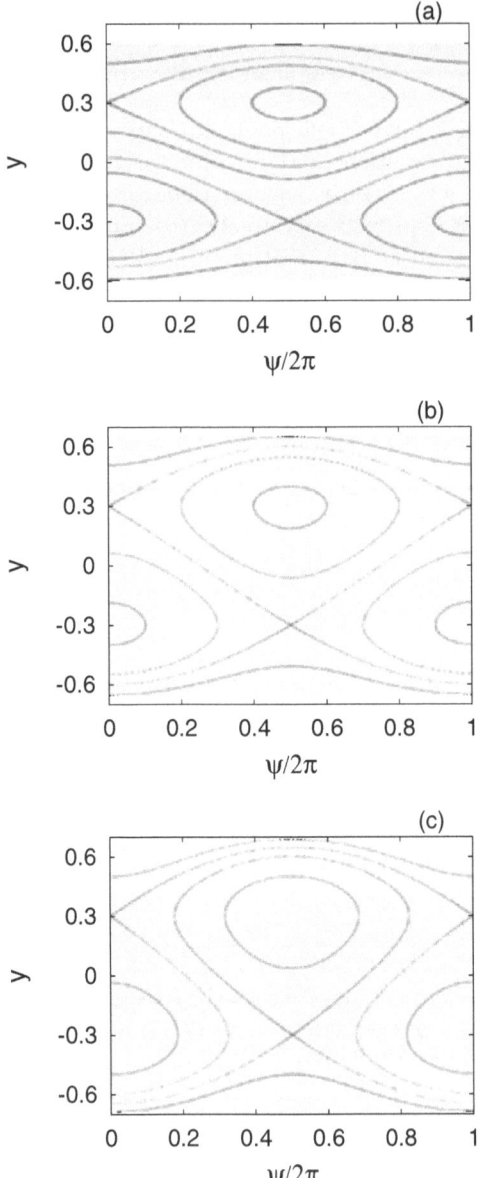

Fig. 7.7. Phase space of Hamiltonian (7.32) for the three different perturbation parameter b: (**a**) $b = 0.01$; (**b**) $b = 0.018$; (**c**) $b = 0.025$. The parameter $a = -0.09$

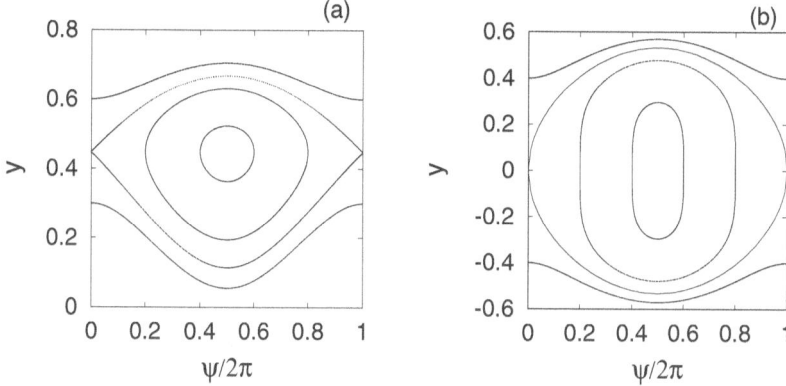

Fig. 7.8. Phase space of Hamiltonian (7.35): (**a**) $a = 0.09$; (**b**) $a = 0$. The parameter $b = 0.01$

are split, and the motion near the separatrix becomes chaotic. The most effi-cient way to study this phenomenon is to use symplectic maps. Near extremal point these maps has a specific non-twist property, since the twist condition (7.3) is violated. For this reason the corresponding maps are called *non-twist maps*. Formally, we will define these maps corresponding to the Hamiltonian systems (2.3) for which the *nondegenaracy condition* (7.1) is violated, i.e., $\det (\partial^2 H_0/\partial I^2) = 0$. In general, the non-twist maps have the forms of sym-metric maps (4.6)–(4.8) with the unperturbed frequencies $\omega(I)$ satisfying the nontwist condition (7.29) at certain values of action variables I_0. If the per-turbation has a broad spectrum the maps can be reduced to the simpler form (4.31)–(4.33) with generating functions of type (4.34).

These mapping can be simplified near the extremal point I_0 ($\omega'_0 = 0$) by expanding the frequency $\omega(I)$ in series of powers of $I - I_0$:

$$\omega(I) = \omega_0 + \frac{1}{2}\omega''_0(I - I_0)^2 + \frac{1}{6}\omega'''_0(I - I_0)^3 + \cdots . \qquad (7.36)$$

and approximating the generating function $S(\vartheta, J)$ by

$$S(\vartheta, J) \approx S_0(\vartheta) = \frac{\pi}{\Omega} \sum_m |H_m(I_0)| \cos(m\vartheta + \chi_m) . \qquad (7.37)$$

When the frequency has a local maximum (or minimum), $\omega''_0 \neq 0$, one can neglect the last term (7.36). Then introducing a dimensionless variable $u = \sqrt{|\omega''_0|/2\omega_0}(I - I_0)$ the map (4.31) - (4.33) can be reduced to

$$U_k = u_k + \frac{\beta}{2} \sum_m mH_m(I_0) \sin(m\vartheta_k + \chi_m) ,$$

$$\vartheta_{k+1} = \vartheta_k + \alpha(1 - \sigma U_k^2) , \qquad (7.38)$$

$$u_{k+1} = U_k + \frac{\beta}{2} \sum_m m H_m(I_0) \sin(m\vartheta_{k+1} + \chi_m),$$

where $\sigma = \mathrm{sgn}(\omega_0'') = \pm 1$,

$$\alpha = 2\pi\omega_0/\Omega, \qquad \beta = 2\pi\epsilon\sqrt{|\omega_0''|/2\omega_0}/\Omega, \qquad U = \sqrt{|\omega_0''|/2\omega_0}(J - I_0).$$

If the perturbation contains only one harmonics with $m = 1$, $H_m(I_0) = -1$ and $\chi_m = 0$, we obtain the following map

$$U_k = u_k - \frac{\beta}{2} \sin\vartheta_k,$$

$$\vartheta_{k+1} = \vartheta_k + \alpha(1 - \sigma U_k^2), \tag{7.39}$$

$$u_{k+1} = U_k - \frac{\beta}{2} \sin\vartheta_{k+1},$$

This map can be rewritten in a nonsymmetric form as the map $(\vartheta_k, U_{k-1}) \to (\vartheta_{k+1}, U_k)$ of the variables (ϑ, U):

$$U_k = U_{s-1} - \beta\sin\vartheta_k, \qquad \vartheta_{k+1} = \vartheta_k + \alpha(1 - \sigma U_k^2), \tag{7.40}$$

When $\sigma = 1$ this map coincides with the map introduced in del Castillo-Negreto and Morrison (1993a), which was called a *non-twist standard map*. This map as well as its symmetric form (7.39) describes dynamics of system near a local maximum or minimum of frequency. The properties of the non-twist standard map (7.40) have been discussed in detail in del Castillo-Negreto et al. (1996, 1997). Particularly, the periodic points and transition to chaos have been studied. In del Castillo-Negreto and Morrison (1993a) obtained the map (7.40) from the Hamiltonian

$$H(\vartheta, u, t) = -\alpha u + \frac{\alpha u^3}{3} + \beta \sum_{n=-\infty}^{\infty} \cos(\vartheta - n\Omega t)$$

$$= -\alpha u + \frac{\alpha u^3}{3} + \beta \frac{2\pi}{\Omega} \cos\vartheta \sum_{k=-\infty}^{\infty} \delta\left(t - k\frac{2\pi}{\Omega}\right), \tag{7.41}$$

identifying $U_k = u(t_k - 0)$ and $\vartheta_k = \vartheta(t_k - 0)$, where $t_k = k(2\pi/\Omega)$.

The map parameters α, β, and the variable u are related with the corresponding parameters a, b, and the variable y in the Hamiltonian (7.32) near the primary resonances $(m = 1, n)$:

$$\alpha = \frac{2\pi}{\Omega}(a + n\Omega), \qquad \beta = \frac{2\pi}{\Omega}\frac{b}{\sqrt{a + n\Omega}}, \qquad u = \frac{y}{\sqrt{a + n\Omega}}. \tag{7.42}$$

The nontwist standard maps (7.39) and (7.40) describe all features of the separatrix reconnection and chaotic motion near a maximum (or minimum) point I_0 of frequency $\omega(I)$.

Symmetry properties. The symmetric nontwist standard map (7.39) has a symmetry with respect to transform

$$k \leftrightarrow k+1 \,, \qquad \vartheta_k \leftrightarrow 2\pi - \vartheta_{k+1} \,, \qquad (7.43)$$

which corresponds to the time-reversal symmetry of continuous Hamiltonian equations, $t \to -t$, $H \to -H$.

7.3.5 Fixed Points and Transit to Chaos

The *periodic point of order* n , $\mathbf{x} = (\vartheta, u)$ is defined as $\hat{M}^n \mathbf{x} = \mathbf{x}$, where $\hat{M}$ stands for the map $(\vartheta_{k+1}, u_{k+1}) = \hat{M}(\vartheta_k, u_k)$. In general, a determination of periodic points is a difficult two–dimensional root finding problem. In del Castillo-Negreto et al. (1996) this problem for the nontwist standard map (7.40) has been reduced to a one–dimensional root finding problem. For instance, the periodic points with odd n are found as zeros of the function $F(u) = \sin\left[2\pi\left(\bar{\vartheta} - \alpha(1 - \bar{U}^2)/2\right)\right] = 0$, where $\bar{\vartheta}(u)$, $\bar{U}(u)$ are functions of u: $(\bar{\vartheta}(u), \bar{U}(u)) = \hat{M}^{(n+1)/2}(0, u)$.

The periodic points of the nontwist standard map come in pairs due to violation of twist condition: there are two periodic points with the same rotation number α. With the variation of map parameters the periodic up and down points may collide. This bifurcation phenomenon accompanying with the separatrix reconnection has been studied in del Castillo-Negreto et al. (1996). For instance, the one-period periodic points $(m : n = 1 : 0)$ are located at $(\vartheta = 0, u = \pm 1)$ and $(\vartheta = \pi, u = \pm 1)$. The reconnection occurs when the orbit emerging from the up hyperbolic point $(0, 1)$ joins the down hyperbolic point $(\pi, -1)$. The threshold, $\beta_c = 2\alpha/3$, of such a reconnection found from the condition $H(\vartheta = 0, u = 1) = H(\vartheta = \pi, u = -1)$, where $H(\vartheta, u)$ is the Hamiltonian (7.41) containing only $n = 0$ term in the sum. Using the relations (7.42) for $n=0$ is easy to see that it coincides with the reconnection threshold (7.33).

Periodic points of the symmetric nontwist standard map (7.39) can be expressed through the ones of the map (7.40) using the relation between corresponding variables u and U: $u = U + \beta/2 \sin \vartheta$. Due to the symmetry property (7.43) they are symmetric with respect to the axis $\vartheta = \pi$ unlike the periodic points of the nontwist standard map (7.40) (see Figures 7.10a, b). One should note that the phase space locations of periodic orbits of the continuous Hamiltonian system (7.41) coincide with the periodic points of the symmetric map (7.40) rather than the nontwist standard map (7.40). This can be easily confirmed by direct integration of corresponding Hamiltonian equations with a finite number of harmonics in (7.41).

The phase space of the symmetric nontwist map (7.39) for the primary resonance $(m : n) = (1 : 0)$ are plotted in Fig. 7.9 for the two values of the perturbation parameter β: a) $\beta = 0.02094$; b) for the reconnection threshold $\beta_c = 0.37699$, and $\alpha = 0.5655$. They correspond to the parameter $b = 0.01$,

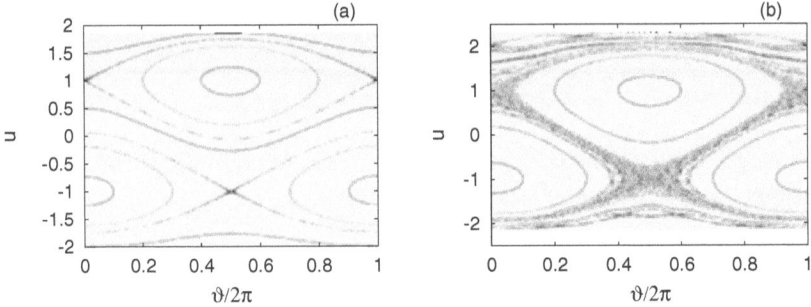

Fig. 7.9. Phase space of the symmetric nontwist map (7.39) for the resonance $(m : n) = (1 : 0)$: **(a)** $\beta = 0.02094$; **(b)** $b = \beta_c = 0.37699$. The parameter $\alpha = 0.5655$

$b_c = 0.018$ and $|a| = 0.09$ in the Hamiltonian (7.32). As seen from Figures 7.9 a, b the symmetric nontwist map well reproduces the phase space and the separatrix reconnection in the integrable Hamiltonian (7.32) near the extremal point (see Figs. 7.7a, b). Figures 7.9 b shows the destruction of the reconnected separatrix and the formation of the chaotic layer due to the high frequency perturbations in the Hamiltonian (7.41).

The phase space of the nontwist standard maps (7.39) and (7.40) in the case of secondary higher order resonances $(m : n = 5 : 2)$ are shown in Fig. 7.10 for the parameters taken in del Castillo-Negreto and Morrison (1993a): (a) and (b) $\alpha = 2.565$, $\beta = 0.664$, and (c) $\alpha = 2.8$, $\beta = 0.7$. The cases (a) and (c) correspond to the symmetric nontwist standard map, while (b) − to the nontwist standard map.

These figures show that even at these relatively large perturbations there are still barriers near the shearless curve which prevents a diffusion between the highly developed stochastic zones located on the upper and down regions. In order to break these barriers one needs to significantly increase the perturbation ϵ.

7.4 Non-Smooth Mappings

As was mentioned in Sect. 7.2 the invariant curves of non smooth Hamiltonians with $\beta = 1$ do not survive for any small perturbation. In the latter case a Hamiltonian is the continuous function of angle variable but it has a discontinuous first derivative. Such Hamiltonian systems appears in problems of ray propagations in waveguides (Abdullaev and Zaslavsky (1988); Tappert et al. (1991); Abdullaev (1993, 1994b)), in models of the Fermi mechanism of acceleration of cosmic rays (Zaslavsky and Chirikov (1965), and see Lichtenberg and Lieberman (1992)).

In this section we study some new features of dynamics of Hamiltonian systems subjected to non-smooth perturbations which have been found

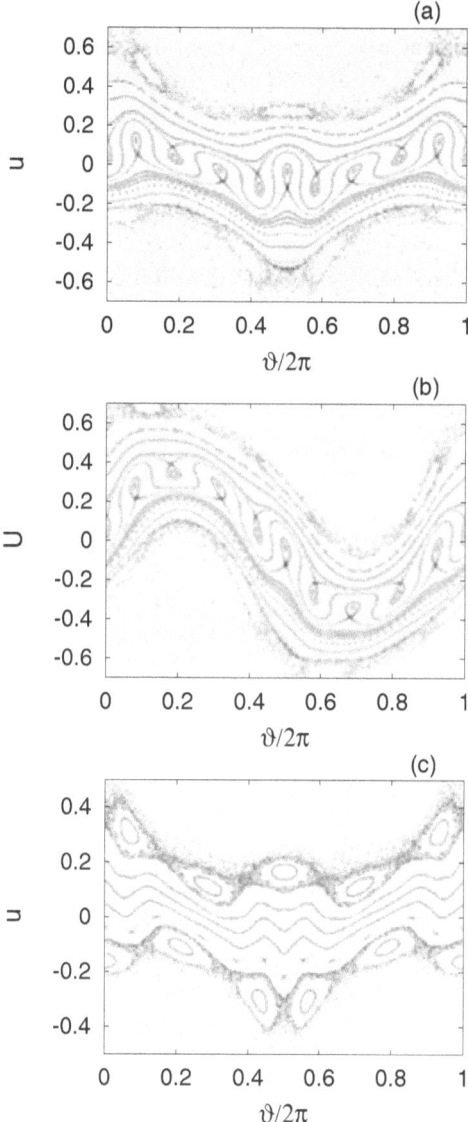

Fig. 7.10. Separatrix reconnection and chaotic motion for the higher order resonance $m : n = 5 : 2$ obtained by the symmetric nontwist map (7.39) (a), (c), and the nonsymmetric nontwist map (7.40) (b). Parameters are (a), (b) $- \alpha = 2.656$, $\beta = 0.664$; (c) $\alpha = 2.8$, $\beta = 0.7$

recently. It takes place in systems when the twist condition is violated. Although, the global chaos in system occurs for any small perturbation, the diffusion through the layer formed near the shearless curve is significantly reduced due to an intermittent nature of motion.

7.4.1 Intermittence in Nontwist Systems

In this section we present a specific example of system in which the two conditions of the KAM theory applicability, a twist condition (7.3) and a sufficiently smoothness $\beta > \beta_c = 3$, are violated (Abdullaev (1994b)). Consider a motion of particle of unit mass $m = 1$ in the potential well subjected to time-periodic perturbation with a broad spectrum:

$$H(x, p, t) = \frac{p^2}{2} + U(x) + \epsilon H_1(x, p, t) , \tag{7.44}$$

where ϵ is a dimensionless perturbation parameter and the potential function $U(r)$ is given by (see Fig. 7.11)

$$U(x, t) = \begin{cases} L/x^2, & \text{for } 0 < x < 1 , \\ U_0, & \text{for } x > 1 . \end{cases} \tag{7.45}$$

We present the perturbation H_1 in the form

$$H_1(x, p, t) = \begin{cases} x^2/2 \sum_{n=-M}^{M} \cos n\Omega t, & \text{for } 0 < x < 1 , \\ 0, & \text{for } x > 1, \end{cases} \tag{7.46}$$

where M is the largest harmonics number. In (7.45) L is constant parameter. When $M \to \infty$ the sum in (7.46) can be replaced by the periodic delta function $\delta_1(t) = \sum_{k=-\infty}^{\infty} \delta(t - k2\pi/\Omega)$. Then the problem is reduced to the

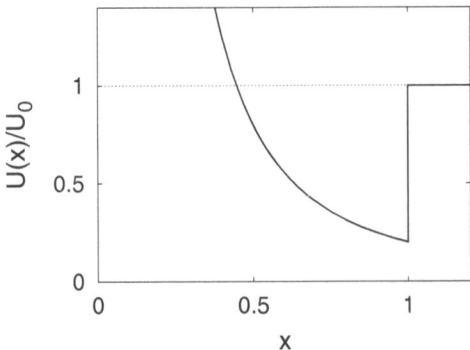

Fig. 7.11. Profile of the potential $U(r)$ for $L = 0.2$

dynamics of a particle in a potential well under action of periodic kicks. In this case and when $L = 0$ the model (7.44) is known as a "kicked oscillator" which is pertinent to problems in classical and quantum chaos in Hamiltonian systems (see, e.g., Dana et al. (1989); Schwägerl and Krug (1991); Casati et al. (1979)). As was shown in Abdullaev (1994b) the problem of ray propagation in a waveguide with the circular cross section and with a periodic array of lenses along its axis is reduced to the Hamiltonian problem (7.44).

In the absence of perturbation ($\epsilon = 0$) the system is integrable with energy E as the integral of motion. The orbit of particle can be found using the action-angle variables (ϑ, I). For the action I we have

$$I = \frac{1}{2\pi} \oint p(x)dx = \frac{L^{1/2}}{\pi}(q - \arctan q) , \tag{7.47}$$

where

$$p(x, I) = \left(2E(I) - \frac{L}{x^2}\right)^{1/2} , \qquad q = \sqrt{\frac{2E - L}{L}} .$$

The permissible energy E of particle in the potential well is $E > E_{min} = L/2$, $E < U_0$. The relation between the angle variable ϑ and the coordinate x is

$$\vartheta = \frac{\partial}{\partial I} \int^x p(x', I)dx' = \frac{\omega}{\sqrt{2E}} \begin{cases} \sqrt{x^2 - b^2}, & \text{for } 0 < \vartheta < \pi , \\ 2\sqrt{1 - b^2} - \sqrt{x^2 - b^2}, & \text{for } \pi < \vartheta < 2\pi , \end{cases}$$

where $b^2 = L/2E$, and $\omega(I)$ is the frequency of oscillations

$$\omega(I) = \frac{\partial H_0(I)}{\partial I} = \frac{2\pi E(I)}{\sqrt{2E(I) - L}} . \tag{7.48}$$

Inverting the relation (7.48), we obtain

$$x^2(\vartheta, I) = b^2 + (1 - b^2)F(\vartheta) ,$$

$$p = \left(2E(I) - \frac{L}{x^2(\vartheta, I)}\right)^{1/2} \begin{cases} 1, & \text{for } 2\pi n < \vartheta < (2n+1)\pi, \\ -1, & \text{for } (2n+1)\pi < \vartheta < 2(n+1)\pi , \end{cases} \tag{7.49}$$

where n is an integer number, $n = 0, 1, 2, \cdots$, and

$$F(\vartheta) = \begin{cases} (\vartheta - 2\pi n)^2/\pi^2, & \text{for } 2\pi n < \vartheta < (2n+1)\pi, \\ (2\pi - \vartheta + 2\pi n)^2/\pi^2, & \text{for } (2n+1)\pi < \vartheta < 2(n+1)\pi . \end{cases} \tag{7.50}$$

The profile of the function $F(\vartheta)$ is shown in Fig. 7.12a. It is the continuous function of ϑ with the discontinuous derivative $dF(\vartheta)/d\vartheta$ at $\vartheta = \pi$.

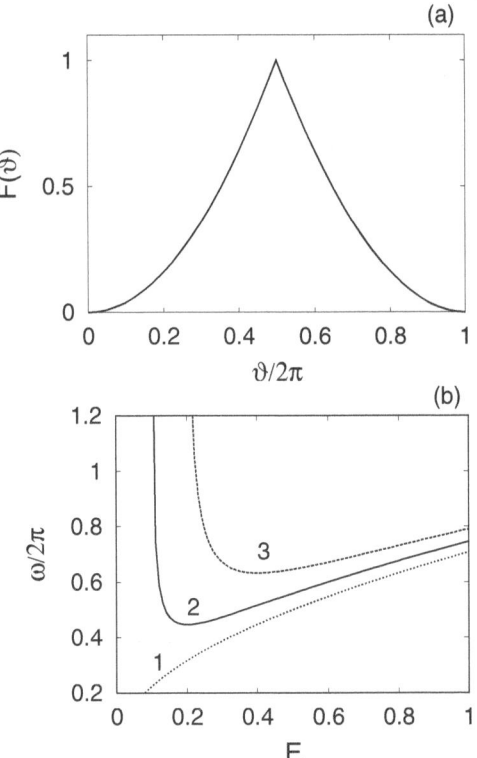

Fig. 7.12. (a) Profile of the function $F(\vartheta)$ (7.50); (b) Frequency of oscillation $\omega(E)$ versus the energy E: curve 1 corresponds to $L = 0$, curve $2 - L = 0.2$, curve $2 - L = 0.4$

When $L \neq 0$ the frequency of oscillations $\omega(I)$ is the non-monotonic function of energy E (or action I) as shown in Fig. 7.12b. It has a minimum at $E = L$.

In the presence of the perturbation ($\epsilon \neq 0$) the full Hamiltonian (7.44) in the action-angle (ϑ, I) takes the form

$$H(\vartheta, I, t) = H_0(I) + \epsilon \frac{x^2(\vartheta, I)}{2} \sum_{n=-M}^{M} \cos n\Omega t \ . \tag{7.51}$$

In the limit $M \to \infty$ the evolution of the perturbed system (7.51) can be reduced to the mapping $(\vartheta_k, I_k) \to (\vartheta_{k+1}, I_{k+1})$ in the symmetric form (4.31) - (4.33) with the generating function

$$S(\vartheta, J) = \frac{\pi}{2\Omega} x^2(\vartheta, J) = \frac{\pi}{2\Omega} \left[b^2 + (1 - b^2)F(\vartheta) \right] \ . \tag{7.52}$$

[We recall that $(\vartheta_k, I_k) = (\vartheta_r(t_k), I_r(t_k))$, $t_k = s2\pi/\Omega$, $s = 0, \pm 1, \pm 2, \ldots$].

The Hamiltonian H (7.51) and the generating function $S(\vartheta, J)$ (7.52) are continuous functions, but their first derivative with respect to ϑ are discontinuous (see Fig. 7.12a), i.e., the smoothness parameter $\beta = 1$. Therefore the KAM theory is not applicable to this case. The invariant curves do not exist for any small perturbations, and the orbits of system may diffuse far from its initial state.

First we consider the case when $L = 0$ describing a monotonic frequency $\omega(I)$ (curve 1 in Fig. 7.12b). The phase space of system obtained by the mapping (4.31)–(4.33) is shown in Fig. 7.13. As seen from Figure that the phase space consists of stability islands of various sizes surrounded by thin chaotic layers formed by destroyed resonant separatrices. The chaotic layers of neighboring resonances are connected and form a so-called *stochastic web* which extends to the whole phase space of the system. Orbits with initial conditions located on the stochastic web can diffuse far from their initial positions. The orbits with initial coordinates inside the islands are always confined and do not diffuse. Described behavior of Hamiltonian systems is known as *a weak chaos*.

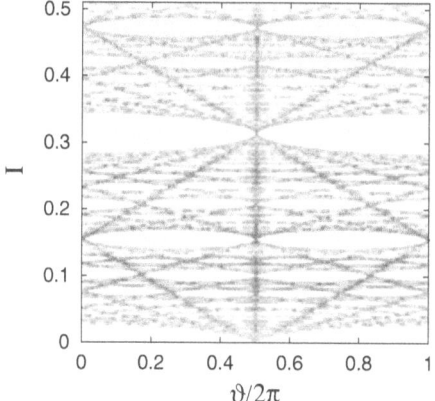

Fig. 7.13. Phase space of the Hamiltonian system with (7.51) for the case $L = 0$. The perturbation parameter $\epsilon = 0.01$, and the perturbation frequency $\Omega = \pi$

Let us turn to the case $L \neq 0$ when the frequency $\omega(I)$ becomes non-monotonic, and the twist condition $d\omega(I)/dI \neq 0$ is violated at the shearless curve $I = I_0$, $\omega'(I_0) = 0$ (curves 2, 3 in Fig. 7.12b). The dynamics of such a system under smooth perturbations has been studied in Sect. 7.3. Here, we consider the case of non-smooth perturbations.

From the relations (7.48) and (7.47) it follows that $I_0 = \sqrt{L}(1 - \pi/4)/\pi$. The phase space structure of system for the value $L = 0.6$ is plotted Fig. 7.14. One can see that the phase space consists of two clearly distinct zones of the weak chaos and the *strong chaos* separated by the region along the shearless

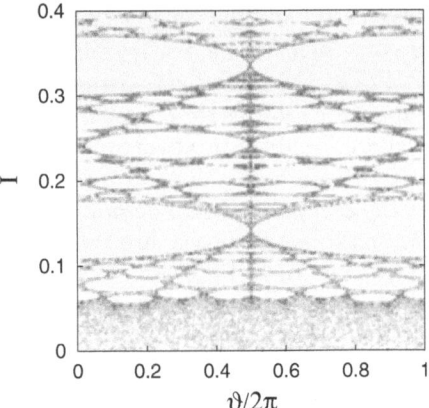

Fig. 7.14. Phase space of the Hamiltonian system (7.51) on the (ϑ, I)-plane for the case $L = 0.6$. The perturbation parameter $\epsilon = 0.0146$, and the perturbation frequency $\Omega = 1.795$

line $I = I_0 \approx 0.053$ (for $L = 0.6$). The zone of strong chaos is located down the shearless curve, $0 < I < I_0$ (or $L/2 < E < L$). In this region all orbits are completely chaotic without any stability islands. The region of weak chaos is on the upper side of the shearless curve, $I > I_0$ (or $E > L$). There is a very thin *transitional layer* near the shearless curve $I = I_0$ connecting the areas of weak and strong chaos. The width of this layer depends on resonances near I_0. As we will see below the dynamics of the system in this transitional layer has an *intermittent nature*.

When the shearless action I_0 coincides with the resonant action I_{mn} $(\omega(I_{mn})/\Omega = n/m)$ with the low resonance numbers (m, n) the transitional layer becomes fairly large. The case of $m : n = 3/2$ resonance is shown in Fig. 7.15 for the same values of parameters L and ϵ as in Fig. 7.14. The perturbation frequency Ω that creates this resonance is $\Omega = 3.244$. As seen from Fig. 7.15 the transitional layer surrounds stability islands whose center coincide with two periodic fixed points. The motion over a considerably long time interval exceeding the perturbation period $T = 2\pi/\Omega$ occurs along a regular phase curve which loops the stability island. After one loop of the island the motion switches to another regular phase curve in an irregular way and the process may be repeated, or it may transit into the region of strong chaos. Such a behavior of dynamical systems is known as *intermittence*. The evolution of energy $E(t)$ with time shown in Fig. 7.16 clearly demonstrates the intermittent nature of motion in the transitional layer. A more detailed analysis of this phenomenon will be given below in Sect. 7.4.3 using simplified non-smooth mappings near a shearless curve.

The term "intermittency" originated in the theory of turbulence is usually understood a weak chaotic dynamics with a certain quasiregular structure in

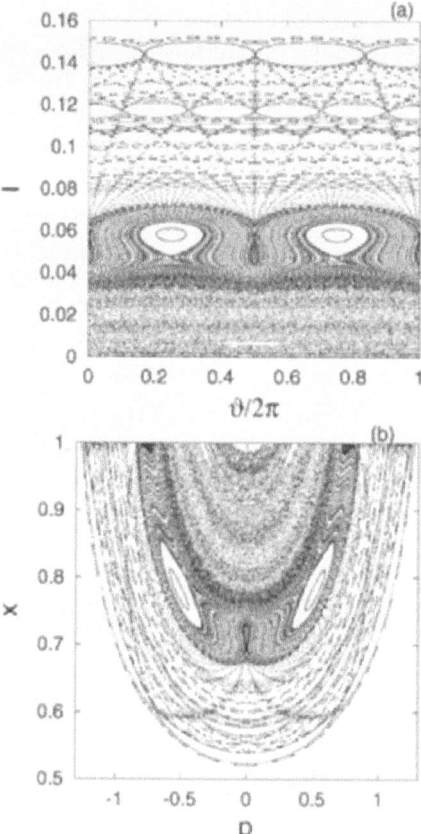

Fig. 7.15. The same as in Fig. 7.14 but for the perturbation frequency $\Omega = 3.244$. Other parameters are the same. (a) On the (ϑ, I)-plane; (b) on the (x, p)-plane

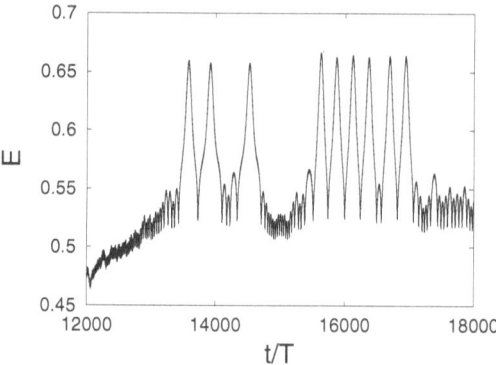

Fig. 7.16. Time evolution of the energy E along the single orbit near the transition region

space or time (see, for example Paladin and Vulpiani (1987); Stanley and Meakin (1988)). The examples of intermittent dynamics in Hamiltonian systems can be found in Zaslavsky et al. (1991).

7.4.2 Simplified Non-Smooth Mappings

In order to study a local behavior of system near the certain resonance (m, n) we simplify the mapping (4.31)–(4.33) by expanding the frequency $\omega(I)$ near the resonant action I_{mn}. Consider first the case when the resonant action I_{mn} does not lie on the shearless curve I_0. Then $\omega(I) = \omega(I_{mn}) + \omega'(I_{mn})(I - I_{mn})$. Approximating the generating function (7.52) at the resonant action I_{mn}: $S(\vartheta, J) \approx S(\vartheta, I_{mn})$ and introducing notations

$$x = \vartheta/2\pi , \qquad y = \omega'(I_{mn})(I - I_{mn})/\Omega ,$$

$$K = 4\frac{\epsilon\omega'(I_{mn})}{\Omega^2} \left[1 - b^2(I_{mn})\right] , \qquad (7.53)$$

the mapping (4.31)–(4.33) can be reduced to

$$Y_k = y_k - \frac{K}{2} f(x_k) ,$$
$$x_{k+1} = x_k + m/s + Y_k , \qquad (7.54)$$
$$y_{k+1} = Y_k - \frac{K}{2} f(x_{k+1}) ,$$

where $f(x)$ is the discontinuous function

$$f(x) = \begin{cases} x, & \text{for } 0 < x < 1/2 , \\ -1 + x, & \text{for } 1/2 < x < 1 . \end{cases} \qquad (7.55)$$

The map (7.54) can be also written in the form

$$Y_k = Y_{k-1} - Kf(x_k) , \qquad x_{k+1} = x_k + m/s + Y_k , \qquad (7.56)$$

for the variables (x, Y). It is known that mappings of type (7.56) with a discontinuous perturbation function $f(x)$ are chaotic for the arbitrary small perturbation K (Rokhlin (1961); Chirikov (1969); Aubry (1978); Percival (1979)). The dynamics of these maps either is completely chaotic (without stability island), or their phase space is covered by a stochastic web with stability islands of various sizes. To determine the regime of motion one should find the Lyapunov exponent, σ, (7.20), determined by the maximum eigenvalue, λ_k, of the Jacobian matrix (7.17). For the map (7.54) we obtain

$$\mathbf{J}_k = \begin{pmatrix} 1 - K/2 & -K(1 - K/4) \\ 1 & 1 - K/2 \end{pmatrix} .$$

The eigenvalues of the matrix are

$$\lambda_{1,2}^k = 1 + \frac{K}{2}\left(1 \pm \sqrt{1 - \frac{4}{K}}\right) . \tag{7.57}$$

When the parameter $K > 4$ or $K < 0$ the eigenvalues $\lambda_{1,2}$ are real, $\lambda_1 > 1, \lambda_2 < 1$, ($\lambda_1 \lambda_2 = 1$), and the Lyapunov exponent $\sigma = \ln \lambda_1$ is a positive everywhere on the phase space (x, y). In these cases all orbits are chaotic. On the other hand when $0 < K < 4$ the eigenvalues $\lambda_{1,2}$ are complex, $\lambda_1^* = \lambda_2$, ($|\lambda_1| = 1$), and the Lyapunov exponent $\sigma = 0$. For this case there are stable orbits inside the resonance structures.

According to (7.53) the condition $K < 0$ is satisfied in the region below the shearless action, i.e., $I < I_0$, where $\omega'(I_{mn}) < 0$. Similar, the parameter $K > 0$ in the region $I > I_0$ ($\omega'(I_{mn}) > 0$). This explains the formation of regions of strong chaos for $I < I_0$, and of weak chaos for $I > I_0$ (see Fig. 7.14).

7.4.3 A Nontwist Map and Intermittency

Consider now the dynamics of system near the shearless action I_0. The simplified map in this case can be obtained in a similar way. It has the same form as the nontwist map (7.39) but with the discontinuous perturbation function $f(x)$:

$$\begin{aligned} Y_k &= y_k - \frac{K}{2} f(x_k) , \\ x_{k+1} &= x_k + \alpha_0 + Y_k^2, \\ y_{k+1} &= Y_k - \frac{K}{2} f(x_{k+1}) , \end{aligned} \tag{7.58}$$

where the following notations are introduced

$$y = \sqrt{\frac{\omega''(I_0)}{2\Omega}}(I - I_0) , \qquad \alpha_0 = \frac{\omega(I_0)}{\Omega} , \qquad K = 4 \frac{\epsilon \sqrt{2\omega''(I_0)}}{\Omega^{3/2}} .$$

For the non-canonical variables (x, Y) the map (7.58) can be written in the nonsymmetric form

$$Y_k = Y_{k-1} - K f(x_k), \qquad x_{k+1} = x_k + \alpha_0 + Y_k^2 , \tag{7.59}$$

which has been obtained by Abdullaev (1994b).

The phase space of the mapping (7.58) is shown in Fig. 7.17a for the resonant case $\alpha_0 = n : m = 1:1$. As seen from Fig. 7.17 that over fairly long time interval of order of $10^2 T$ the orbit near the curve $y = 0$ follows the regular curve. During this interval there exists the following approximate integral of motion along this curve:

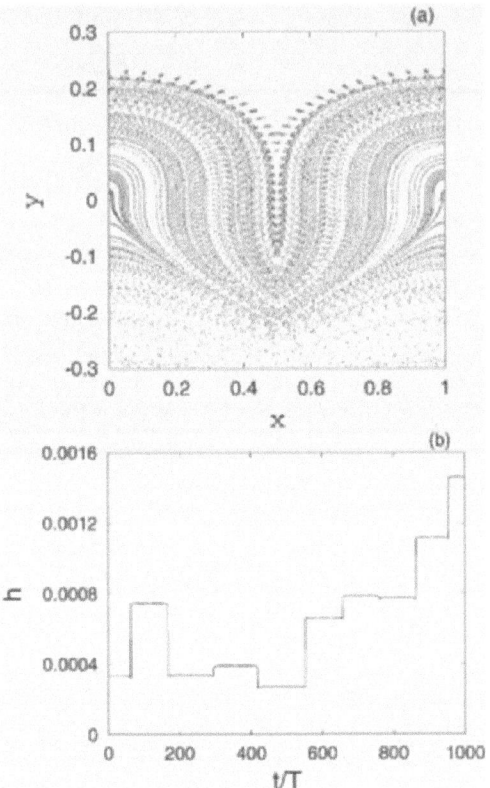

Fig. 7.17. (a) Phase space of the nontwist map (7.58) for the resonance m : $n = 1:1$. (b) Evolution of the approximate integral of motion $h = y^3/3 + KF(x)$. Perturbation parameter $K = 0.03$, $\alpha_0 = 1$

$$h = y^3/3 + KV(x) \,, \qquad (7.60)$$

$$V(x) = \int f(x)dx = \begin{cases} x^2/2, & \text{for } 0 < x < 1/2 \,, \\ (1-x)^2/2, & \text{for } 1/2 < x < 1 \,. \end{cases}$$

The evolution of h is plotted in Fig. 7.17b. After passing the discontinuity point $x = 1/2$ the motion randomly jumps to another regular curve with the different value of h, and the process repeats itself. This constitutes the intermittent behavior of the system. Below following Abdullaev (1994b) we give a qualitative analysis of this phenomenon.

For the sake of simplicity we consider the case $\alpha_0 = 1$. Since the function $f(x)$ (7.55) is a periodic with a period 1, $f(x) = f(x+1)$, the second equation in the mapping (7.58) can be replaced by $x_{k+1} = x_k + Y_k^2$. Then according to results of Sect. 4.3 Hamiltonian function, [see Eqs. (4.24), (4.35) and (4.34)], corresponding to the mapping (7.58) has the following form:

$$H(x,y,t) = \frac{y^3}{3} + KV(x) \sum_{n=-M}^{M} \cos(2\pi nt) \,, \qquad (7.61)$$

The generating function of the mapping (7.58) is $\epsilon S(x) = (K/2)\bar{F}(x)$. The canonical equations of motion can be written in the form

$$\frac{dx}{dt} = \omega(y) \,, \qquad \frac{dy}{dt} = \epsilon Y(x,t) \,, \qquad (7.62)$$

where

$$\omega(y) = \frac{\partial H}{\partial y} = y^2 \,,$$

$$\epsilon Y(x,t) = -\frac{\partial H}{\partial x} = -Kf(x) \sum_{n=-M}^{M} \cos(2\pi nt) \,, \qquad (7.63)$$

For the small values of y the period of motion along the coordinate x is of order of $T_x = 2\pi/\omega(y) = 2\pi/y^2$ which is much greater than the period of perturbation $T = 1$, $T_x \gg 1$. Thus (7.62) describes the effect of small amplitude fast–oscillating perturbations on a nonlinear system. Such a system can be effectively analyzed using the method of averaging of perturbations (see Arnold (1989); Arnold et al. (1988); Bogolyubov and Mitropol'skij (1958)).

Consider the domain D of the phase space (x,y): $|y| < C < 1$, $2\pi x \in S$, where S is a circle, $(0 \leq x \pmod 1) < 1$. The functions $\omega(y)$ and $Y(x,t)$ in the domain are smooth functions, except a discontinuity line at $x = 1/2$. We also suppose that the perturbation parameter $\epsilon \equiv K$ is small. Then in the domain D where the line $x = 1/2$ is excluded the system of equations (7.62) can be replaced by the averaged equations

$$\frac{d\xi}{dt} = \Omega(\zeta) \,, \qquad \frac{d\zeta}{dt} = \epsilon Y_0(\xi,t) \,, \qquad (7.64)$$

where

$$\Omega(\zeta) = \frac{1}{T} \int_0^T \omega(y)dt = \omega(\zeta) \,,$$

$$Y_0(\xi) = \frac{1}{T} \int_0^T Y(x,t)dt = -Kf(\xi) \,. \qquad (7.65)$$

According to Bogolyubov (1945) (see, also Bogolyubov and Mitropol'skij (1958); Arnold et al. (1988)) the solutions, $x(t), y(t)$, of the exact equations (7.62) are close to the solutions, $\xi(t), \zeta(t)$, of the averaged equations (7.64) with the identical initial coordinates at $t = 0$: $x(0) = \xi(0), y(0) = \zeta(0)$, i.e., $|x(t) - \xi(t)|, |y(t) - \zeta(t)| < \epsilon$, on the interval $0 \leq t \leq 1/\epsilon$. Particularly, the exact integral

$$\bar{h} = \zeta^3/3 + KV(\xi) \,, \tag{7.66}$$

of the averaged equations (7.64) will be also close to the numerically found approximate integral of motion h (7.60) of the exact equations (7.62) (or the corresponding mapping (7.58)). Phase space curves of the averaged system of equations are shown in Fig. 7.18. They are very close to those shown in Fig. 7.17b for the same perturbation parameter K.

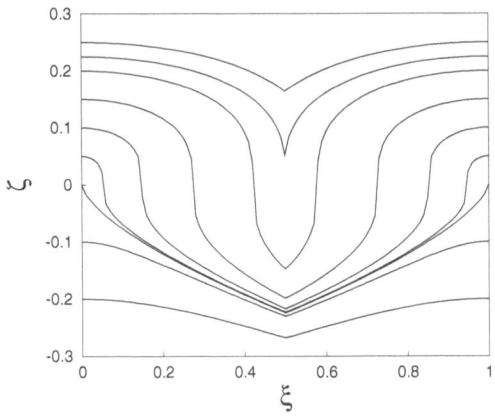

Fig. 7.18. Phase space curve of the averaged system (7.64) for perturbation parameter $K = 0.03$

The averaging procedure is not applicable near the line $x = 1/2$ of the phase space domain D where the smoothness of the function $Y(x, t)$ is violated, i.e., the following condition (Bogolyubov (1945), see also Bogolyubov and Mitropol'skij (1958))

$$|Y(x, t) - Y(x', t)| \le C|x - x'| \,,$$

where C is a some positive number, is violated because of the presence of the discontinuous function $f(x)$. Therefore, the averaged equations (7.64) does not describe the original system (7.62) (or the mapping (7.58)) near the line $x = 1/2$. A numerical analysis shows that the system evolves along a smooth orbit $(x(t), y(t))$ until it crosses the lines $x = 1/2$. After crossing this line it jumps to another smooth orbit with the different constant of motion h. Direction of a jump occurs along the y-axis is random. However, the rigorous analytical proof of this numerical finding has not yet been given.

The existence of a transitional layer between the regions of weak and strong chaos with intermittent nature significantly reduces the global diffusion. It acts as a partial barrier to the global chaotic motion in Hamiltonian systems.

7.5 Suppression of Chaos in Smooth Hamiltonian Systems

As was mentioned in Sect. 7.2 global invariant curves exist in Hamiltonian systems with sufficiently smooth perturbations whose Fourier coefficients, $H_m(I)$, decay as $H_m(I) \propto m^{-(\beta+1)}$ with the exponent $\beta > \beta_c = 3$. However, the opposite statement that for $\beta < \beta_c = 3$ there are no global invariant curves suppressing a global diffusion for any small perturbation has not been proven (except the case $\beta = 1$) (see Moser (1973)). For the case $\beta = 2$ a specific example of a Hamiltonian system has been found where there exists a so called *invariant resonant structure* suppressing the global chaos for a certain set of perturbation parameter (see, Vecheslavov (2001) and references therein). Below we shortly discuss this example.

Consider the mapping (Vecheslavov and Chirikov (2001))

$$y_{k+1} = y_k - K f(x_k), \qquad x_{k+1} = x_k + y_{k+1}, \qquad (\text{mod } 1) \,, \qquad (7.67)$$

where $f(x)$ is the periodic, $f(x) = f(x+1)$, and antisymmetric, $f(-x) = -f(x)$, sawtooth perturbation function

$$f(x) = \begin{cases} 2x/(1-d), & \text{for } |x| \le (1-d)/2 \,, \\ \\ (1-2x)/d, & \text{for } |1/2 - x| \le d/2 \,, \end{cases} \qquad (7.68)$$

Here d is the distance between the sawteeth $|f(x)| = 1$. The function $f(x)$ and the corresponding potential function $U(x) = -\int f(x)dx$ are plotted in Fig. 7.19. In the limit $d \to 0$ we obtain the discontinuous function (7.55).

Bullet (1986) and recently, Ovsyannikov[3] considered the mapping (7.67) in the case $d = 1/2$. They have shown that for a certain countable set of special values K_m of the perturbation parameter K the resonance separatrices of integer number m of Hamiltonian system are un-split, and it acts as a barrier to the global chaos along the momentum variable y.

The extensive numerical experiments by Vecheslavov (2000, 2001) have shown that not only integer but also fractional resonances $m : n$ of arbitrary orders have critical numbers $K_{m,n}$ whose separatrices do not split. The theory of the separatrix conversation based on the Melnikov method has been developed by Vecheslavov (2001); Vecheslavov and Chirikov (2001). Previously, this approach has been used by Hénon and Wisdom (1983) to study the separatrix conversation in the oval billiard problem. The diffusion processes in the mapping (7.67) are studied recently by Vecheslavov and Chirikov (2002); Chirikov and Vecheslavov (2002).

---

[3] The Ovsyannikov's theorem has not been published. It is given in Vecheslavov (2001)

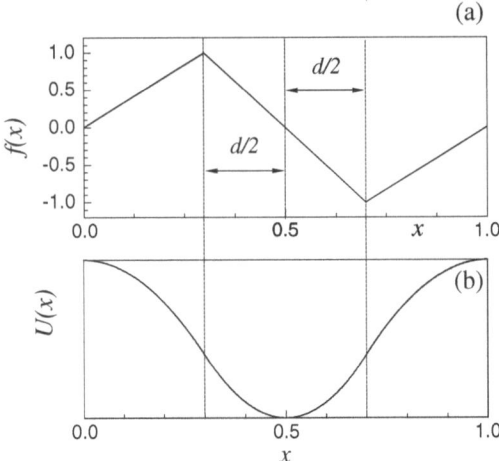

Fig. 7.19. Sawtooth perturbation function $f(x)$ (7.68) and the potential $U(x)$

7.6 Bibliographic Notes

A phenomenon of stochastic instability or chaos in Hamiltonian systems is discussed in many review papers and monographs. Introduction to this phenomenon in dynamical systems including also Hamiltonian systems intended for general audience can be found in a review by Jensen (1992) and in a book by Hilborn (2000). The first comprehensive review of chaos theory has been given by Chirikov (1979). A systematic description of regular and chaotic motion in Hamiltonian systems is given in the books by Lichtenberg and Lieberman (1992); Zaslavsky (1985); Sagdeev et al. (1988). More rigorous mathematical approach to study dynamical systems is discussed in the books by Guckenheimer and Holmes (1983); Wiggins (1990); Ott (1993). Renormalization theory in Hamiltonian dynamics is discussed in a review paper by Escande (1985) and in a monography by MacKay (1993). Different aspects of Hamiltonian chaos theory can be found in the collection of papers selected by MacKay and Meiss (1987). In the review paper by Meiss (1992) a mapping approach to Hamiltonian dynamics and chaos is discussed.

8 Rescaling Invariance
of Hamiltonian Systems Near Saddle Points

In typical Hamiltonian systems chaotic motion appears because of the separatrix destruction by any small time-periodic perturbation (see Sect. 7.1.3). It forms a stochastic layer, a zone, of chaotically unstable motion near the unperturbed separatrix. In this section we study important properties of the stochastic layer, namely, a rescaling invariance of phase space of systems near the saddle points. This property of motion is generic for typical Hamiltonian systems subjected to time-periodic perturbations. The rescaling invariance consists of that the phase space (x, p) of system near hyperbolic saddle points is invariant with respect to the scaling transformation of the perturbation parameter $\epsilon \to \lambda\epsilon$, the shift of perturbation phase $\chi \to \chi + \pi$, and the phase space coordinates $(x, p) \to (\lambda^{1/2}x, \lambda^{1/2}p)$. The rescaling parameter λ depends only on the frequency of perturbation, Ω, and the divergence exponent γ of unperturbed orbits near the saddle point, $\lambda = \exp(2\pi\gamma/\Omega)$. It means that the topology of phase space near the saddle point is a periodic function of $\log \epsilon$ with the certain period, $\log \lambda$.

The rescaling invariance property of motion has been first found in Abdullaev and Zaslavsky (1994); Zaslavsky and Abdullaev (1995) in numerical simulations, and later it has been explained by Abdullaev and Zaslavsky (1995, 1996) using the separatrix mapping method. The analytical proof of the rescaling invariance has been given in Abdullaev (1997) for some class of Hamiltonian systems. Existence of additional rescaling properties near the saddle points due to the symmetry of Hamiltonian system in phase space has been found in Abdullaev (2000). Rescaling invariance properties in autonomous two-degrees-of-freedom Hamiltonian systems has been studied in Kuznetsov and Zaslavsky (1997). Recently Kuznetsov and Zaslavsky (2002) have studied a condition at which the phase space structures near the saddle points of different systems are similar.

This property of motion plays important role in understanding of the chaotic transport in the stochastic layer. Particularly, it predicts to the universal periodic $\log \epsilon-$ dependence of chaotic transport in the stochastic layer (Abdullaev and Spatschek (1999); Abdullaev (2000)).

Below we study these generic rescaling properties of motion near the hyperbolic saddle points in typical Hamiltonian systems. The study is mainly based on using the separatrix mappings approach described in the previous Chaps. 5, 6.

S.S. Abdullaev: *Construction of Mappings for Hamiltonian Systems and Their Applications*,
Lect. Notes Phys. **691**, 175–195 (2006)
www.springerlink.com

8.1 Rescaling Invariance Near Saddle Points and Separatrix Maps

In this section we consider one degree of freedom Hamiltonian system subjected to time-periodic perturbation, and discuss its rescaling invariant properties near hyperbolic saddle points. Existence of this property is proven by the direct numerical integration of Hamiltonian equations and using the separatrix mappings. One should note that the conventional separatrix mappings of type (5.7), (5.52) are not suitable for these analysis. It is because a time and energy variables in these mappings are defined at the different sections of phase space, and therefore they cannot be compared with Hamiltonian equations. For this reason we will use the separatrix mappings at the section Σ_s (5.46) located near hyperbolic saddle points.

First we consider the universal rescaling invariance property of system, and then the rescaling invariance due to the symmetry of Hamiltonian system in phase space.

8.1.1 Structure of Phase Space Near Saddle Points

Consider the Hamiltonian system given by Hamiltonian

$$H = H_0(q,p) + \epsilon H_1(q,p,t+t_0) , \qquad (8.1)$$

describing the effect of time dependent perturbation on one degree of freedom Hamiltonian system $H_0(q,p)$. The perturbation $H_1(q,p,t)$ is a periodic in time with the frequency Ω: $H_1(q,p,t) = H_1(q,p,t+2\pi/\Omega)$. A constant t_0 is related to the phase of perturbation, $\chi = \Omega t_0$.

Suppose that the unperturbed Hamiltonian system $H_0(q,p)$ has hyperbolic fixed points. Let $(q_s = 0, p_s = 0)$ be coordinates of one saddle point. The $H_0(q,p)$ may be presented as a power series near the saddle point:

$$H_0(q,p) = H_0(q_s,p_s) \pm \frac{\alpha_s^2}{2}q^2 \mp \frac{\beta_s^2}{2}p^2 + O(\delta^3) , \qquad (8.2)$$

where α_s and β_s are the expansion coefficients, $O(\delta^3)$ is higher order expansion terms $(\delta \sim x, p)$. By a linear rotational transformation

$$x = \frac{1}{\sqrt{2\alpha_s\beta_s}}(\alpha_s q + \beta_s p) , \qquad y = \frac{1}{\sqrt{2\alpha_s\beta_s}}(-\alpha_s q + \beta_s p) , \qquad (8.3)$$

the Hamiltonian (8.2) can be reduced to

$$H_0(q,p) = H_0(q_s,p_s) \pm \gamma_s xy + O(\delta^3) , \qquad (8.4)$$

where $\gamma_s = \alpha_s\beta_s$ is a coefficient determining the exponential growth (or decrease) of coordinates near the saddle points: $x \sim \exp(\pm\gamma_s t), y \sim \exp(\mp\gamma_s t)$.

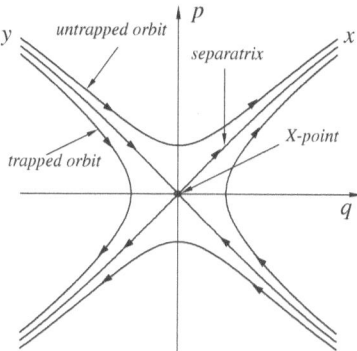

Fig. 8.1. Phase space structure near the hyperbolic fixed point

The phase space of the unperturbed Hamiltonian near the hyperbolic saddle point is shown in Fig. 8.1. Further suppose that all hyperbolic saddle points which lie on the same energy surface $H(q_s, p_s) = $ constant have the same increments $\gamma \equiv \gamma_s$. These saddle points are connected heteroclinically (or homoclinically in the case of one saddle point). The examples of homoclinic and heteroclinic connections of saddle points were shown in Fig. 5.1a, and Fig. 5.1b, respectively.

The frequency of closed motion $\omega(H)$ has the universal asymptotics when orbit approaches to the separatrix:

$$\omega(H) = \frac{2\pi\gamma}{\ln \frac{A}{|H - H_s|}} \ , \qquad H \to H_s \ , \qquad (8.5)$$

which depends on the increment, γ, and a certain constant parameter, A. In (8.5) H_s is then energy on the separatrix.

In typical Hamiltonian systems any small time-periodic perturbation destroys the separatrices, and motion near the unperturbed separatrices becomes chaotic (see Fig. 5.2). However the stochastic layer formed in the small vicinity of the unperturbed separatrices is not uniform. There are regions inside the stochastic layer with regular motions (KAM–stability islands). The examples of stochastic zone are shown in Fig. 8.2. The mutual positions of these islands and their relative sizes determine the topology of the stochastic layer. As we will show later that it plays main crucial role in chaotic transport along the stochastic layer. Particularly, the structure of the stochastic layer near saddle points mainly determines the statistical properties of chaotic motion because particles spend relatively large times in regions near saddle points.

8.1.2 Universal Rescaling Invariance

By numerous simulations of different Hamiltonian systems of type (8.1) it has been found that the phase-space, (x, y), of the perturbed motion near saddle points is invariant with respect to the rescaling transformation

$$\epsilon \to \epsilon' = \lambda\epsilon \, , \qquad\qquad \chi \to \chi' = \chi + \pi \, ,$$

$$x \to x' \approx \lambda^{1/2}x \, , \qquad\qquad y \to y' \approx \lambda^{1/2}y \, , \qquad (8.6)$$

with the rescaling parameter,

$$\lambda = \exp(2\pi\gamma/\Omega) \, , \qquad (8.7)$$

depending only on the perturbation frequency Ω and the coefficient γ describing the behavior of the unperturbed Hamiltonian $H_0(x,y)$ near saddle point (8.4). Here $\chi = t_0\Omega/2\pi$ is the initial phase of perturbation.

This rescaling property is illustrated in Fig. 8.2 by plotting Poincaré sections of the Hamiltonian system (6.23) of the periodically driven pendulum near the saddle point $(x = \pi \pmod 1), p = 0)$ for the two set of parameters: (a) $\epsilon_a = 0.02$, $\chi_a = 0$, and (b) $\epsilon_b = \lambda\epsilon_a = 0.08$, $\chi_b = \pi$. The perturbation frequency Ω is chosen equal to $\Omega = 4.53236\omega_0$ in order to have the rescaling parameter $\lambda = \exp(2\pi\gamma/\Omega) = 4$. One can see that the phase-space topologies of the stochastic layer near hyperbolic saddle points (q_s, p_s) corresponding to the two set of parameters (ϵ_a, χ_a) and (ϵ_b, χ_b) are similar, i.e., mutual positions of islands of types 1, ..., 6 are similar. The coordinates of their elliptic fixed points, (x_e, p_e), are related according to (8.6). Numerically calculated ratios of these coordinates are presented in Table 8.1. (In Table (x_a, y_a) and (x_b, y_b) stand for the coordinates of fixed points for the case (a) and (b), respectively, $x_s = \pi$ is a hyperbolic fixed point of the unperturbed system.) As seen from the Table these ratios are approximately equal to $\lambda^{1/2} \approx 2$, which confirms the rescaling law (8.6).

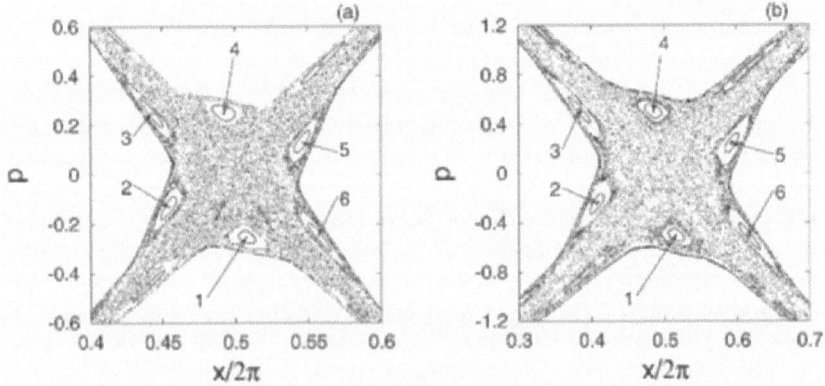

Fig. 8.2. Poincaré sections of the Hamiltonian system (5.2) near the saddle point $(q = \pi, p = 0)$: (a) $\epsilon_a = 0.02$, $\chi_a = \pi + 1$, (b) $\epsilon_b = \lambda\epsilon_a = 0.08$, $\chi_b = \chi_a - \pi$. The rescaling parameter is $\lambda = 4$, $A_+ = B_- = 1$

Table 8.1. Ratio of relative x- and p-coordinates corresponding the two set of perturbation parameters (ϵ_a, χ_a) and (ϵ_b, χ_b)

	1	2	3	4	5	6
$\frac{x_b - x_s}{x_a - x_s}$	2.000	2.033	2.026	2.000	2.033	2.028
$\frac{p_b}{p_a}$	1.995	1.855	2.052	1.995	1.848	2.048

Due to stickiness of orbits to the islands of types 2, 3, 5, 6 particles may be trapped for a long time while the stickiness to the island 1 (or 4) may lead long distance flight along positive (or negative) direction of the x-axis. The variation of perturbation amplitude, ϵ, periodically changes the topology of phase space of perturbed motion near saddle points. One expects that this property can lead to the quasi-periodic dependencies of statistical properties of chaotic transport in the stochastic layer Abdullaev (2000) (see Chap. 9).

8.1.3 Proof of the Rescaling Invariance of Hamiltonian Equations

Below we present an analytical proof the rescaling invariance of Hamiltonian equations near saddle points following by Abdullaev (1997). Consider an arbitrary Hamiltonian system subjected by time-periodic perturbation,

$$\frac{dx}{dt} = \frac{\partial H_0(x,y)}{\partial p} + \epsilon g_1(x,y,t+t_0) ,$$

$$\frac{dy}{dt} = -\frac{\partial H_0(x,y)}{\partial x} + \epsilon g_2(x,y,t+t_0) . \qquad (8.8)$$

The canonical variables (x,y) are chosen in such a way that the unperturbed Hamiltonian $H_0(x,y)$ near the saddle point $(x=0, y=0)$ has the form (8.4). The perturbation is given by the time-periodic functions $g_i(x,p,t)$, $(i=1,2)$, with period $T = 2\pi/\Omega$:

$$g_i(x,y,t) = g_i(x,y,t+T) , \qquad (i=1,2) , \qquad (8.9)$$

We does not require that the perturbations to be Hamiltonian, when the functions $g_i(x,p,t)$, $(i=1,2)$ are determined by the perturbation Hamiltonian: $g_1(x,y,t) = \partial H_1/\partial y$, $g_2(x,y,t) = -\partial H_1/\partial x$. Suppose also that the average value of the perturbation functions, $g_i(x,p,t)$, over one period, T, are zero

$$\int_0^T g_i(x,p,t)dt = 0, \qquad (i=1,2) . \qquad (8.10)$$

The dimensionless perturbation parameter ϵ is small, $0 < \epsilon \ll 1$.

Consider the behavior of the system (8.8) near the hyperbolic saddle point $(x=0, y=0)$ by expanding its right hand side in a series of powers of x, y.

$$g_i(x,y,t) = a_i(t) + c_{i1}(t)x + c_{i2}(t)y + O(x^2, y^2, xy) , \qquad (i=1,2) , \quad (8.11)$$

where

$$a_i(t) = g_i(0, 0, t) ,$$

$$c_{i1}(t) = \left.\frac{\partial g_i(x, y, t)}{\partial x}\right|_{x=y=0} , \qquad c_{i2}(t) = \left.\frac{\partial g_i(x, y, t)}{\partial y}\right|_{x=y=0} .$$

The coefficients $a_i(t)$, $c_{ij}(t)$ $(i, j = 1, 2)$ are periodic functions of time with the period $T = 2\pi/\Omega$, $c_{ij}(t) = c_{ij}(t + T)$, and

$$\int_0^T a_i(t)dt = \int_0^T c_{i,k}(t)dt = 0 , \qquad (i = 1, 2) . \qquad (8.12)$$

Neglecting the small terms of order of $O(x^2, y^2, xy)$ the equations (8.8) are reduced to

$$\frac{dx}{dt} = -\gamma x + \epsilon \left[a_1(t + t_0) + c_{11}(t + t_0)x + c_{12}(t + t_0)y\right] ,$$

$$\frac{dy}{dt} = \gamma y + \epsilon \left[a_2(t + t_0) + c_{21}(t + t_0)x + c_{22}(t + t_0)y\right] , \qquad (8.13)$$

Let $(x_a(t), y_a(t))$ be a solution of (8.13) for a certain small value of $\epsilon = \epsilon_a$, and the phase $\chi_a = \Omega t_0$, $t_0 = t_a$. Suppose that the solution $(x_b(t), y_b(t))$ of the same equation (8.13) corresponds to the transformed parameters

$$\epsilon = \epsilon_b = \epsilon_a/\lambda , \qquad \chi = \chi_b = \chi_a - \pi , \qquad (8.14)$$

where the parameter $\lambda = \exp(2\pi\gamma/\Omega)$. Then one can show (see Appendix C) that there exists the following relation between solutions $(x_a(t), y_a(t))$ and $(x_b(t), y_b(t))$ corresponding to (ϵ_a, t_a) and (ϵ_b, t_b), respectively:

$$x_b = \lambda^{-1/2}x_a - \epsilon_b \left(\lambda^{-1/2}c_{12}^0 y_a + p_1^0\right) + O(\epsilon_b^2) ,$$

$$y_b = \lambda^{-1/2}y_a - \epsilon_b \left(\lambda^{-1/2}c_{21}^0 y_a + p_2^0\right) + O(\epsilon_b^2) . \qquad (8.15)$$

where the coefficients

$$c_{12}^0 = 2 \int_0^{T/2} e^{2\gamma t}c_{12}(t - T/2)dt , \qquad p_1^0 = \int_0^{T/2} e^{\gamma t}p_1(t - T/2)dt ,$$

$$c_{21}^0 = 2 \int_{-T/2}^0 e^{-2\gamma t}c_{21}(t - T/2)dt , \qquad p_2^0 = \int_{-T/2}^0 e^{-\gamma t}p_2(t - T/2)dt ,$$

are constants of order of $O(1)$. The relation (8.15) confirms the numerically found rescaling law (8.6). It also shows that the rescaling invariance may be violated in a small neighborhood of the saddle point $(x = 0, y = 0)$ with a radius of the order of ϵ.

Consider now the case when the $(n-1)$, $(n \geq 2)$, derivatives of the perturbation functions $g_i(x, y, t)$, i.e., $\partial^j g_i(x, y, t)/\partial x^k \partial y^{j-k}$, $(0 \leq k \leq j, 1 \leq j \leq n-1)$, at the saddle point $(x = y = 0)$ are zeros. Then the lowest order non-zero terms in expansions of g_i in powers of x, y are given by the homogeneous polynomial functions of order n,

$$g_i(x, y, t) = \sum_{k=0}^{n} c_{ik}^{(n)}(t) x^k y^{n-k}, \qquad (i = 1, 2), \qquad (8.16)$$

where the coefficients

$$c_{ik}^{(n)}(t) = \frac{n!}{k!(n-k)!} \frac{\partial^n g_i(x, y, t)}{\partial x^k \partial y^{n-k}} \bigg|_{x=y=0},$$

are periodic functions of time t with a period T. In this case the equations of motion (8.8) take the form

$$\frac{dx}{dt} = -\gamma x + \epsilon \sum_{k=0}^{n} c_{1k}(t) x^k y^{n-k},$$

$$\frac{dy}{dt} = \gamma y + \epsilon \sum_{k=0}^{n} c_{2k}(t) x^k y^{n-k}. \qquad (8.17)$$

Analysis of the equations of motion (8.17) with the perturbation functions (8.16) is provided in Appendix C.2. It was shown that the relations between solutions $(x_a(t), y_a(t))$ and $(x_b(t), y_b(t))$ of these equations corresponding to the parameters (ϵ_a, χ_a) and (ϵ_b, χ_b) related according to (8.14) are given by

$$x_b = \lambda^{-1/2} x_a + O(\epsilon_b^2) + \epsilon O(\lambda^{-2}),$$

$$y_b = \lambda^{-1/2} y_a + O(\epsilon_b^2) + \epsilon O(\lambda^{-2}). \qquad (8.18)$$

It follows from (8.18), that the rescaling law (8.6) is valid for large values of the rescaling parameter λ, that $\lambda^2 \gg 1$. On the other hand, the rescaled perturbation parameter ϵ_a and $\epsilon_a \lambda$ should be small.

8.1.4 Separatrix Mapping Approach

The rescaling invariance of system near saddle points immediately follows from the property of the separatrix mapping for time and energy variables defined at the section of phase space near saddle points. Consider, for example, the system with a several hyperbolic saddle points shown in Fig. 8.3. The orbit crosses the branches 1, 2, ..., 4 of the cross sections Σ_s near the saddle points. The each step of the map connecting the branches of the sections Σ_s is described by the separatrix mapping (5.46). Suppose the perturbation is a periodic with a single frequency Ω. Then the frequencies Ω_n of higher harmonics n are simply related to Ω: $\Omega_n = n\Omega$. Using the asymptotics (8.5) of

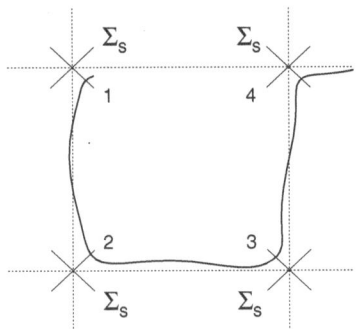

Fig. 8.3. Scheme of the separatrix mapping in the system with a several saddle points

the frequency of unperturbed motion, $\omega(H)$, near the separatrix the mapping (5.46) can be presented as

$$
h_{k+1} = h_k + \epsilon \sum_n n\Omega \left[K_n \sin \left(n\varphi_k + n\frac{\Omega}{2\gamma} \ln \frac{A}{|h_k|} + \chi_n \right) \right.
$$
$$
\left. + L_n \cos \left(n\varphi_k + n\frac{\Omega}{2\gamma} \ln \frac{A}{|h_k|} + \chi_n \right) \right] ,
$$
$$
\varphi_{k+1} = \varphi_k + \frac{\Omega}{2\gamma} \left(\ln \frac{A}{|h_k|} + \ln \frac{A}{|h_{k+1}|} \right) . \tag{8.19}
$$

Since the phase $\varphi = \Omega t$ is defined by module 2π, the mapping (8.20) is invariant with respect to the transformation

$$
\epsilon \rightarrow \epsilon\lambda \,, \qquad \chi_n \rightarrow \chi_n + n\pi \,, \qquad h \rightarrow h\lambda \,, \tag{8.20}
$$

where the rescaling parameter λ is determined by (8.7). This rescaling invariance property of the separatrix mapping corresponds to the rescaling law (8.6).

8.1.5 Rescaling Invariance in Parameter Space

The separatrix mapping (8.19) uniquely determines the dynamics of system near saddle points. It depends on a few parameters: γ and A in the asymptotics of the frequency of oscillations $\omega(H)$ (8.5), the perturbation frequency Ω, and the Melnikov-Arnold type integrals K_n, L_n (5.28). We analyze the rescaling invariance properties of the mapping in the parameter space $\gamma, A, \Omega, K_n, L_n$. It is easy to check that the separatrix map is invariant with respect to the transformation of the parameters $(\gamma, A, \Omega, K_n, L_n) \rightarrow (\bar{\gamma}, \bar{A}, \bar{\Omega}, \bar{K}_n, \bar{L}_n)$:

$$A \to \bar{A} = A\delta\lambda^{-s} , \qquad (s = 0, 1, 2 \cdots) ,$$

$$K_n \to \bar{K}_n = K_n\delta, \qquad L_n \to \bar{L}_n = L_n\delta ,$$

$$\chi_n \to \bar{\chi}_n = \chi_n + sn\pi , \qquad h \to \bar{h} = h\delta , \qquad (8.21)$$

at the fixed perturbation parameter ϵ and the ratio $\Omega/\gamma = \bar{\Omega}/\bar{\gamma}$. The parameter δ is arbitrary, and λ is the rescaling parameter (8.7). One should note, that the transformations $K_n \to \bar{K}_n = K_n\delta$, $L_n \to \bar{L}_n = L_n\delta$ are equivalent to the transformation of the perturbation parameter $\epsilon \to \bar{\epsilon} = \epsilon\delta$, at the fixed parameters K_n, L_n.

The rescaling invariance (8.21) means that the two different systems have the similar structure of phase space in the neighborhood of saddle points with the rescaling law

$$x \to \bar{x} = \delta^{1/2}x , \qquad p \to \bar{p} = \delta^{1/2}p ,$$

if their parameters $(\gamma, A, \Omega, K_n, L_n)$ and $(\bar{\gamma}, \bar{A}, \bar{\Omega}, \bar{K}_n, \bar{L}_n)$ are related according to (8.21).

In a particular case, $s = 0$, this rescaling invariant property of system coincides with the one which has been recently found in Kuznetsov and Zaslavsky (2002). They have shown that the phase space of two different systems near saddle points has a similar structure if the separatrix mappings corresponding to these systems are equivalent, i.e., the separatrix mapping of the first system can be transformed to the one of the second system by the rescaling transformation of mapping parameters.

8.2 Rescaling Invariance due to the Symmetry of Hamiltonians

If Hamiltonian system (1.1) has some symmetries in the phase-space of canonical variables (x, p) then there exists additional rescaling invariance of system near saddle points with respect to the transformation of perturbation amplitude and its phase. Consider, for example, the motion of particle in the perturbed double – well potential studied in Sect. 6.1. The hyperbolic saddle point is located at $(x = 0, p = 0)$ and the parameter γ is equal to $\gamma = 1$. Therefore the rescaling parameter is $\lambda = \exp(2\pi/\Omega)$. The unperturbed Hamiltonian, $H_0(x, p)$, in (6.1) is symmetric with respect to $x \to -x$, $p \to -p$, i.e., $H_0(-x, -p) = H_0(x, p)$, and but the perturbed Hamiltonian $H_1(x, p, t)$ is antisymmetric, i.e., $H_1(-x, -p, t) = -H_1(x, p, t)$.

The time–periodic perturbation destroys the separatrix, and the motion near becomes chaotic. Poincaré section of the system in the (x, p) plane obtained by direct symplectic numerical integration is shown in Fig. 8.4 for the

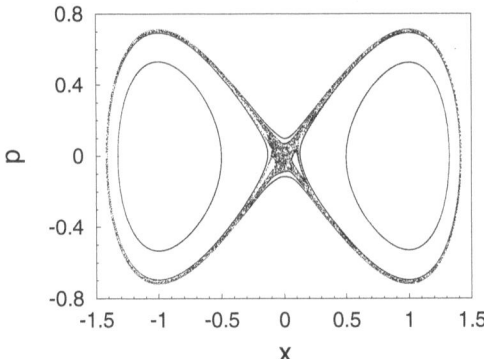

Fig. 8.4. Poincaré sections of the Hamiltonian (6.1) for $\epsilon_a = 0.0025$, $\chi_a = \pi - 1$. The frequency $\Omega = 2\pi/\ln\lambda$ (the rescaling parameter is $\lambda = 16$)

perturbation parameter $\epsilon = 0.0025$, and phase $\chi = \pi - 1$. The frequency of perturbation Ω is chosen to have the value of the rescaling parameter $\lambda = \exp(2\pi/\Omega)$ equal to 16, i.e., $\Omega = 2\pi/\ln\lambda = 2.2662$. The perturbed motion near the saddle point beside the universal rescaling property (8.6) has an additional rescaling property due to the symmetry properties of Hamiltonian (6.1) above mentioned. By direct numerical integration of the equations of motion we have found that the system near the saddle point $(x = 0, p = 0)$ is invariant with respect to the following transformation

$$\epsilon \to \epsilon' = \lambda^{1/2}\epsilon , \qquad\qquad \chi \to \chi' = \chi \pm \pi/2 ,$$

$$x \to x' \approx \pm\lambda^{1/4}p , \qquad\qquad p \to p' \approx \pm\lambda^{1/4}x . \qquad (8.22)$$

Poincaré sections of the system (6.1) near the saddle point are shown in Fig. 8.5 for the two values of perturbation amplitude ϵ and phase χ: (a) $\epsilon_a = 0.0025$, $\chi_a = \pi - 1$ and (b) $\epsilon_b = \lambda^{1/2}\epsilon_a = 0.01$, $\chi_b = \chi_a - \pi/2$. (The rescaling parameter $\lambda = \exp(2\pi/\Omega) = 16$). Note that in Fig. 8.5b the coordinate q is along the vertical axis, and the momentum p is along the horizontal axis. As can see from Fig. 8.5 the q- and p-axes are rescaled according to (8.22).

8.2.1 Separatrix Mapping Analysis

The existence of the rescaling invariance (8.22) can be proven by the separatrix mapping. To prove this we will use the separatrix mapping (6.21) obtained in Sect. 6.1. The geometry of the mapping was shown in Fig. 5.4. The rescaling transformation (8.22) may be also formulated as

$$\epsilon \to \lambda^{1/2}\epsilon , \qquad \chi \to \chi \pm \pi/2 , \qquad h \to -\lambda^{1/2}h . \qquad (8.23)$$

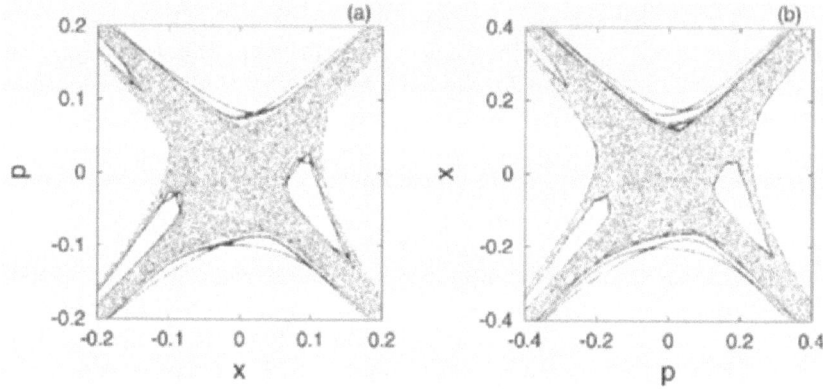

Fig. 8.5. Poincaré sections of the Hamiltonian (6.1) near the saddle point ($x = 0, p = 0$): (a) $\epsilon_a = 0.0025$, $\chi_a = \pi - 1$, (b) $\epsilon_b = \lambda^{1/2}\epsilon_a = 0.01$, $\chi_b = \chi_a - \pi/2$. The rescaling parameter is $\lambda = 16$

We study the rescaling invariance property of fixed points of motion near saddle points. Consider the cross-sections of orbits at the branches $p = 1, 2, 3, 4$ of the section Σ_s shown in Fig. 6.3. Let $(\varphi_{q,s}^{(p)}, h_{q,s}^{(p)})$ be the (q, s) fixed point at the p-th branch of Σ_s defined as

$$\left(\varphi_{q,s}^{(p)} + 2\pi s, h_{q,s}^{(p)}\right) = \left(\hat{F}_p\right)^q \left(\varphi_{q,s}^{(p)}, h_{q,s}^{(p)}\right), \tag{8.24}$$

where $q, s = 1, 2, \ldots$ are integer numbers. The maps $\hat{F}_p$, ($p = 1, 2, 3, 4$) are composed by the consecutive application of the separatrix maps $\hat{M}^{(\pm)}$:

$$\hat{F}_1 = \left(\hat{M}^{(-)}\hat{M}^{(+)}\right)^q, \qquad \hat{F}_2 = \left(\hat{M}^{(+)}\right)^q,$$
$$\hat{F}_3 = \left(\hat{M}^{(+)}\hat{M}^{(-)}\right)^q, \qquad \hat{F}_4 = \left(\hat{M}^{(-)}\right)^q. \tag{8.25}$$

Although each of the separatrix maps $\hat{M}^{(\pm)}$ is not invariant with respect to the transformation (8.23), but their combinations of type $\hat{M}^{(\mp)}\hat{M}^{(\pm)}$, $\left(\hat{M}^{(\pm)}\right)^2$ are transformed as

$$\left(\hat{M}^{(+)}\right)^2 \to \hat{M}^{(\pm)}\hat{M}^{(\mp)}, \qquad \hat{M}^{(-)}\hat{M}^{(+)} \to \left(\hat{M}^{(\mp)}\right)^2,$$
$$\left(\hat{M}^{(-)}\right)^2 \to \hat{M}^{(\mp)}\hat{M}^{(\pm)}, \qquad \hat{M}^{(+)}\hat{M}^{(-)} \to \left(\hat{M}^{(\pm)}\right)^2, \tag{8.26}$$

which can be easily proved by direct calculations. Therefore the rescaling transformations (8.23) transform the maps $\hat{F}_p$ as follow

$$\hat{F}_1 \to \hat{F}_4, \qquad \hat{F}_2 \to \hat{F}_3, \qquad \hat{F}_3 \to \hat{F}_2, \qquad \hat{F}_4 \to \hat{F}_4 \tag{8.27}$$

for $\chi \to \chi + \pi/2$, and

$$\hat{F}_1 \to \hat{F}_2, \qquad \hat{F}_2 \to \hat{F}_1, \qquad \hat{F}_3 \to \hat{F}_4, \qquad \hat{F}_4 \to \hat{F}_3 \tag{8.28}$$

for $\chi \to \chi - \pi/2$. One can see that the transformations (8.23), (8.27), (8.28) are equivalent to the rescaling invariance of perturbed motion equations near the saddle point (8.22) found by the numerical integration of the equations of motion.

The rescaling property (8.23) is demonstrated in Fig. 8.6 by plotting Poincaré sections of orbits at the section Σ_s by the separatrix map for the same parameters as in Fig. 8.5: (a) $\epsilon_a = 0.025$, $\chi_a = \pi - 1$; (b) $\epsilon_b = \lambda^{1/2}\epsilon_a = 0.01$, $\chi_b = \chi_a - \pi/2$. They are obtained using the rules of application of the separatrix mapping (6.22). The region $h > 0$ corresponds to the 1-st branch of the section Σ_s, and $h < 0$ corresponds to its 4-th branch, respectively (see Fig. 6.3). Note that the axis h in Fig. 8.6b is inverted. One can clearly see that the rescaling transformations (8.23) indeed conserves the topology of phase space with the rescaling law $h \to -\lambda^{1/2}h$.

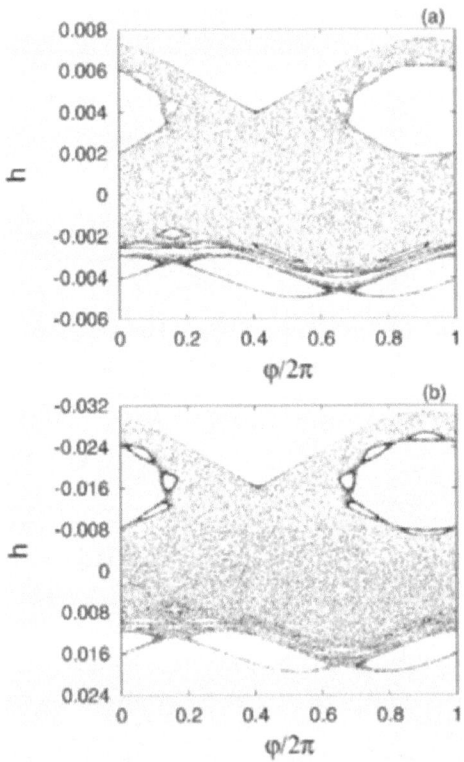

Fig. 8.6. Poincaré sections of orbits at section Σ_s obtained by the separatrix mapping (6.21) for the same parameters as in Fig. 8.5: (a) $\epsilon_a = 0.0025$, $\chi_a = \pi - 1$, (b) $\epsilon_b = \lambda^{1/2}\epsilon_a = 0.01$, $\chi_b = \chi_a - \pi/2$. The rescaling parameter is $\lambda = 16$

Such a rescaling invariance near saddle points with respect to the transformation (8.22) occurs only due to a specific symmetry of Hamiltonian system. The periodically driven pendulum studied in Sect. 6.2 also has such a property for certain type of perturbation. Indeed, one can show that the motion near the saddle points described by Hamiltonian (6.26) is invariant with respect to the transformations (8.22) if the amplitudes of perturbation waves A and B, propagating in opposite directions are $A = B$. In this case the perturbation parameter $K^\pm$ (6.50) in the mapping (6.49) is reduced to

$$K^\pm = \pm|A_+| \frac{4\pi A^2}{\cosh(\pi A/2)} \ ,$$

where the sign $(+)$ corresponds to the map along the upper half of phase space, $p > 0$, and the sign $(-)$ corresponds to the one along the down half of phase space. Then the separatrix map (6.49) takes the form similar to (6.21). For the latter we have proven the existence of the rescaling invariance (8.23).

8.3 2D-Periodic Vortical Flow

In this section we consider the rescaling invariance of motion in Hamiltonian systems with a several types of hyperbolic saddle points. This system is a two–dimensional periodic vortical flow. It is well known that the Lagrangian trajectories of fluid elements in a plane are given by the solution of the Hamiltonian equations of motion

$$\frac{dx}{dt} = -\frac{\partial \psi}{\partial y} \ , \qquad \frac{dy}{dt} = \frac{\partial \psi}{\partial x} \ ,$$

with the stream-function ψ playing role of Hamiltonian, H and the spatial coordinates (x, y) as canonical variables in Hamiltonian dynamics (see Ottino (1989); Pedlosky (1982)).

8.3.1 Model

Consider two dimensional periodic vortical flow subjected to small time–periodic perturbation. The system is determined by Hamiltonian function (see Bertozzi (1988))

$$H = H_0(x, y) + \epsilon H_1(x, y, t) \ ,$$

$$H_0(x, y) = \frac{1}{2\pi} \cos(2\pi x)\cos(2\pi y) \ . \tag{8.29}$$

For convenience we have chosen the unperturbed Hamiltonian $H_0(x, y)$ in (8.29) with the x-coordinate shifted by a half of spatial period in comparison with one given in Bertozzi (1988); Ahn and Kim (1994). Hamiltonian (8.29) is a good model for many convective flows, including the axisymmetric Taylor vortex, as well as the Rossby waves in geophysical fluid dynamics (Pedlosky (1982)).

The phase–space of the unperturbed flow is shown in Fig. 8.7. It has elliptic fixed points at $(x_m^{(e)} = (m-1)/2, y_n^{(e)} = (n-1)/2)$ and hyperbolic fixed points at $[x_m = (m-1/2)/2, y_n = (n-1/2)/2]$, $(n, m = 0, \pm 1, \pm 2, \ldots)$. There are four different types of saddle points: (x_m, y_n), (x_m, y_{n+1}), (x_{m+1}, y_n), (x_{m+1}, y_{n+1}). Because of periodicity of the system along $x-$ and $y-$ axes with the period of 1 all other hyperbolic fixed points (x_{m+2k}, y_{n+2p}), $(k, p = 0, \pm 1, \pm 2, \ldots)$ whose coordinates are shifted on integer numbers belong to the same type. Therefore there are only eight independent saddle–saddle connections.

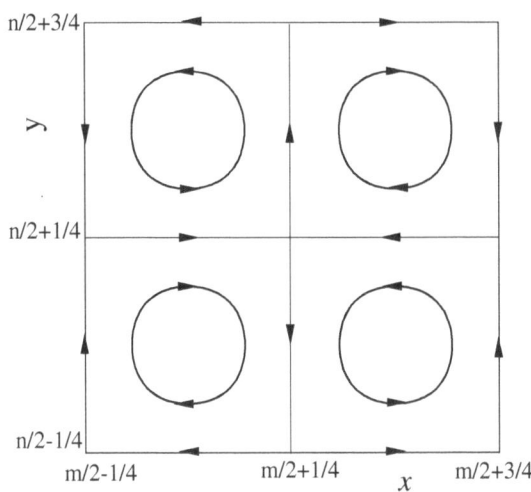

Fig. 8.7. Phase space of the system (8.29)

Near the saddle points the unperturbed Hamiltonian $H_0(x, y)$ in (8.29) has the following expansion in powers of $(x - x_m), (y - y_n)$:

$$H_0(x, y) = (-1)^{m+n} 2\pi (x - x_m), (y - y_n) , \qquad (8.30)$$

and according to the expansion (8.4) the parameter γ is equal to 2π.

For $H = H_0(x, y) = 0$ the saddle points are connected along horizontal $x-$ and vertical $y-$ axes. These saddle connections are described by orbits

$$\cos(2\pi \alpha_s(t)) = \pm 1/\cosh[2\pi(t - t_0)] , \qquad \alpha = x, y , \qquad (8.31)$$

where t_0 is a time instant when a trajectory passes a midpoint between two adjacent saddle points.

Inside of each cell the trajectories are closed with the frequency of oscillations

$$\omega(H) == \frac{\pi^2}{K(k)} \,, \qquad k^2 = 1 - 4\pi^2 H^2 \,. \tag{8.32}$$

Near the separatrices $H \rightarrow 0$ the frequency $\omega(H)$ has the following asymptotics:

$$\omega(H) = \frac{\pi^2}{\ln(2/\pi|H|)} + O(H) \,. \tag{8.33}$$

8.3.2 Rescaling Invariance Property

Any time–periodic perturbation destroys the separatrices. The motion near the unperturbed separatrices becomes chaotic forming a stochastic web along unperturbed separatrices. The structure of the stochastic web near the saddle points are invariant with respect to the universal rescaling transformation (8.6) with the rescaling parameter $\lambda = \exp(2\pi\gamma/\Omega) = \exp(4\pi^2/\Omega)$ for arbitrary small time–periodic perturbation $H_1(x, y, t)$.

For some wide class of perturbations $H_1(x, y, t)$ it may also exist a rescaling invariance with respect to the rescaling transformations of type (8.22). Specifically, we consider the time-periodic perturbation of the flow (8.29) in the form of traveling waves with the same spatial periods as the unperturbed flow and the phase velocity Ω coinciding with the perturbation frequency:

$$H_1(x, y, t) = \frac{\epsilon}{2\pi} [a_y \sin(2\pi y - \Omega t - \chi) - a_x \sin(2\pi x - \Omega t - \chi)] \,, \tag{8.34}$$

where a_x and a_y are the relative amplitudes of traveling-waves along the x- and y-axes, respectively. The perturbed Hamiltonian has a following symmetry property in the (x, y) space:

$$H_1(x + 1/2, y + 1/2, t) = -H_1(x, y, t) \,. \tag{8.35}$$

To integrate the Hamiltonian system (8.29), (8.34) we used a fifth order Bulirsch–Stoer Runge–Kutta method with adaptive step size control, and 10^{-7} accuracy (Press et al. (1992)). Poincaré sections of orbits near the saddle points are presented in Fig. 8.8 for the two different amplitudes, ϵ, and phases χ of the perturbation related with the rescaling parameter $\lambda = \exp(4\pi^2/\Omega) = 16$: (a) $\epsilon_a = 0.0208$, $\chi_a = 0$; (b) $\epsilon_b = \lambda^{-1/2}\epsilon_a = 0.0052$, $\chi_b = \chi_a + \pi/2$. The relative amplitudes of waves are chosen as $a_x = 1$ and $a_y = 0.5$. Figure 8.8a shows Poincaré sections near the saddle points $(x_{m=0}, y_{n=0}) = (-1/4, -1/4)$ and $(x_{m=0}, y_{n=1}) = (-1/4, 1/4)$. Corresponding plots near the points $(x_{m=1}, y_{n=1}) = (1/4, 1/4)$ and $(x_{m=1}, y_{n=0}) = (1/4, -1/4)$ are similar to those near $(x_{m=0}, y_{n=0})$ and $(x_{m=0}, y_{n=1})$, and they may be obtained from the latter by rotating plots by $180°$ around the corresponding points. Figure 8.8b presents Poincaré sections near points $(1/4, -1/4)$ and $(-1/4, -1/4)$

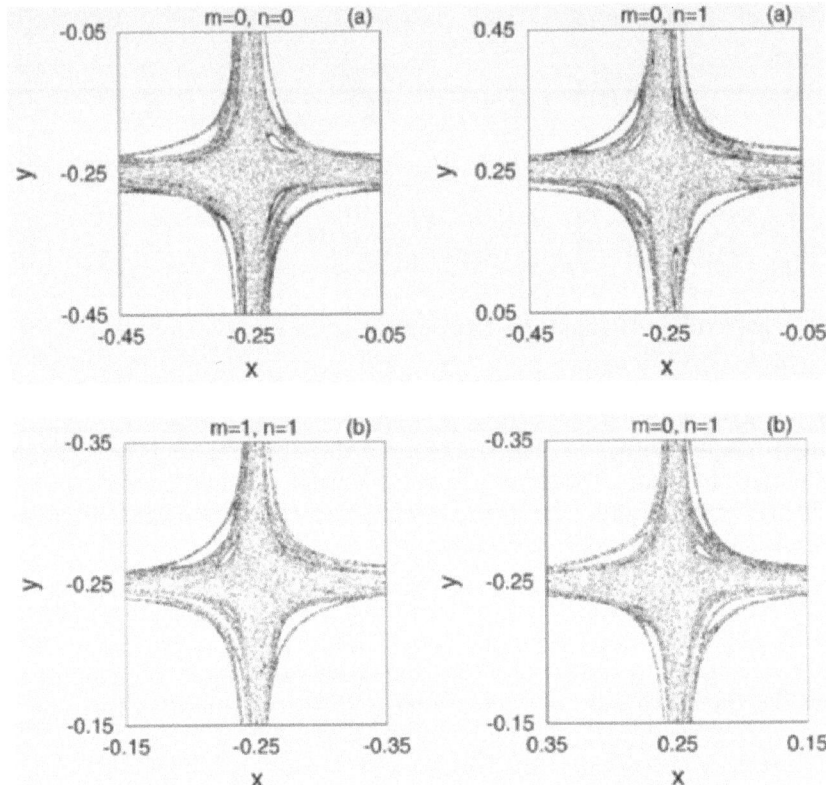

Fig. 8.8. Poincaré sections of orbits in perturbed 2D lattice flow (8.29), (8.34) near the four different saddle points for the perturbation amplitudes: (**a**) $\epsilon_a = 0.0208$, phase $\chi_a = 0$; (**b**) $\epsilon_b = \lambda^{-1/2}\epsilon_a = 0.0052$, phase $\chi_b = \chi_a + \pi/2$. The rescaling parameter $\lambda = \exp(4\pi^2/\Omega) = 16$. Other parameters are $a_x = 1$ and $a_y = 0.5$

with the inverted coordinates $(x \to -x,\ y \to -y)$. In the inverted coordinates (x, y) they correspond to the points $(x_{m=0}, y_{n=1})$ and $(x_{m=1}, y_{n=1})$.

As can see from Fig. 8.8. that the topology of phase space near the saddle points are conserved with respect to the rescaling transformation of parameters

$$\epsilon \to \lambda^{-1/2}\epsilon\,, \qquad \chi \to \chi + \pi/2\,,$$
$$(x - x_m) \to -\lambda^{-1/4}(x - x_{m'})\,,$$
$$(y - y_n) \to -\lambda^{-1/4}(y - y_{n'})\,. \qquad (8.36)$$

One should note that unlike the rescaling law (8.22) for systems with a single saddle point, in this case the structure of phase space near the saddle point (x_m, y_n) is transformed to the one near the other saddle point $(x_{m'}, y_{n'})$. For the even sum $m + n$ the transformation $(m, n) \to (m', n')$ is the following

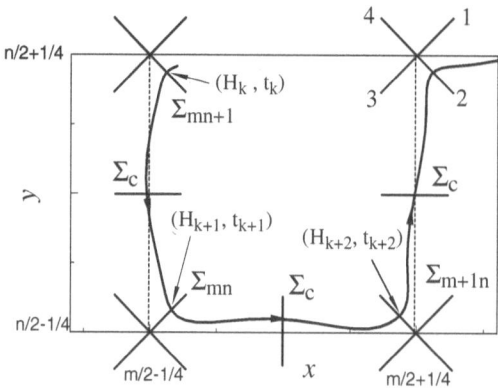

Fig. 8.9. Geometry of the separatrix map for the periodic vortical flow (8.29)

$$(m, n) \to (m, n + 1),$$
$$(m, n + 1) \to (m + 1, n + 1),$$
$$(m + 1, n + 1) \to (m + 1, n),$$
$$(m + 1, n) \to (m, n). \tag{8.37}$$

The rescaling invariance with respect to the transformations (8.36), (8.37) occurs only due to a symmetry of perturbed Hamiltonian (8.35). The analysis of this property will be also given below.

8.3.3 Separatrix Maps of the System

To construct the separatrix maps for this system we introduce sections Σ_{mn} ($m, n = 0, \pm 1, \pm 2, \ldots$) centered at the hyperbolic fixed points (x_m, y_n) shown in Fig. 8.9. Each of sections Σ_{mn} consists of two segments perpendicularly crossing each other at the hyperbolic point with $45°$ to the x-axis. There are four branches of the each section Σ_{mn}, denoted as $p = 1, 2, 3, 4$. Define maps as

$$z_{k+1}^{(m'n')} = \hat{M} z_k^{(mn)},$$

where $z_k^{(mn)} = (t_k, h_k)$ is the crossing point of orbit with the section Σ_{mn} at the k-th step of the map. We denote them as $\hat{X}_{m,m\pm1}^{(n)}, \hat{Y}_{n,n\pm1}^{(m)}$. The map $\hat{X}_{m,m\pm1}^{(n)}$ transforms the point z_k at the section Σ_{mn} to z_{k+1} at $\Sigma_{m\pm1,n}$ along the horizontal axis x at fixed $y = y_n$. Similarly $\hat{Y}_{n,n\pm1}^{(m)}$ connects points at the sections Σ_{mn} and $\Sigma_{m,n\pm1}$ along the vertical axis y at fixed $x = x_m$. Because of the periodicity of the system in the (x, y) space with the period 1 there are following symmetry properties of the maps

$$\hat{X}_{m,m\pm1}^{(n+2)} = \hat{X}_{m,m\pm1}^{(n)}, \qquad \hat{Y}_{n,n\pm1}^{(m+2)} = \hat{Y}_{n,n\pm1}^{(m)} ,$$
$$\hat{X}_{m+2,m+2\pm1}^{(n)} = \hat{X}_{m,m\pm1}^{(n)} , \qquad \hat{Y}_{n+2,n+2\pm1}^{(m)} = \hat{Y}_{n,n\pm1}^{(m)} . \qquad (8.38)$$

Since there are only eight independent saddle–saddle connections, there are only eight independent maps $\hat{M}$ which fully determine the dynamics of system.

For small perturbations $\epsilon \ll 1$ when the stochastic layer near the unperturbed separatrices is sufficiently thin the maps (8.38) can be approximated by the separatrix maps of type (5.46) along the each saddle–saddle connections. They describe the evolution of energy (h) and time (t) variables at the sections Σ_{mn}, i.e., $(h_{k+1}, t_{k+1}) = \hat{M}(h_k, t_k)$, $(\hat{M} = \hat{X}_{m,m\pm1}^{(n)}, \hat{Y}_{n,n\pm1}^{(m)})$. Using the orbits at the separatrices (8.31) and the perturbed Hamiltonian (8.34) one can show that the integrals K_n and L_n (5.34) along the saddle saddle connections are determined by

$$K_{m,m\pm1}^{(n)} = (-1)^n a_x K^{(\pm)}, \qquad \text{for } m+n = 2k ,$$
$$K_{n,n\pm1}^{(m)} = (-1)^m a_y K^{(\pm)} , \qquad \text{for } m+n = 2k+1 , \qquad (8.39)$$

where $k = 0, \pm1, \pm2, \dots$ and

$$K^{(\pm)} = \frac{\Omega}{2\pi} \frac{\exp(\pm\Omega/4)}{\sinh(\Omega/2)} .$$

The coefficients L_n identically vanish. Then the separatrix maps along the saddle saddle connections take the form

$$h_{k+1} = h_k + \epsilon K_{\alpha,\alpha\pm1}^{(\beta)} \cos\left(\varphi_k + \frac{\Omega}{4\pi} \ln \frac{2}{\pi|h_k|} + \chi \right) ,$$
$$\varphi_{k+1} = \varphi_k + \frac{\Omega}{4\pi} \left[\ln \frac{2}{\pi|h_k|} + \ln \frac{2}{\pi|h_{k+1}|} \right] . \qquad (8.40)$$

Because of the symmetry of the perturbed Hamiltonian (8.35) with respect to translation along the x- and y-axes we have the following properties of the coefficients $K_{\alpha,\alpha\pm1}^{(\beta)}$:

$$K_{m,m\pm1}^{(n)} = -K_{m\mp1,m}^{(n+1)} , \qquad K_{n,n\pm1}^{(m)} = -K_{n\mp1,n}^{(m+1)} . \qquad (8.41)$$

Existence of rescaling invariance of motion near saddle points (8.36), (8.37) established by numerical integration of the equations of motion can be proven using the separatrix maps (8.40). It is given in Abdullaev (2000) by constructing maps $\hat{F}_{mn}^{(p)}$, $(p = 1, 2, 3, 4)$ for fixed points $(\varphi_{q,s}, H_{q,s})$ near the saddle points (x_m, y_n) at the each branches p of the section Σ_{mn}, i.e.,

$$(\varphi_{q,s} + 2\pi s, H_{q,s}) = \left(\hat{F}_{mn}^{(p)} \right)^q (\varphi_{q,s}, H_{q,s}) , \qquad (8.42)$$

as similar to those maps $\hat{F}_p$ introduced for determination of the fixed points (8.24) in the model studied in Sect. 8.2.1. The rescaling transformation (8.36) is equivalent to the following transformation applied to the mappings (8.42):

$$\epsilon \to \lambda^{1/2}\epsilon\,, \qquad\qquad \chi \to \chi - \pi/2\,,$$

$$H \to -\lambda^{1/2}H, \qquad t \to -t\,. \tag{8.43}$$

The last expression in the second line in (8.43) corresponds to $x \to -x, y \to -y$.

It was shown by Abdullaev (2000) that for even $m+n$ the rescaling of the mapping parameter (8.43) transform the mappings $\hat{F}_{mn}^{(p)}$ in (8.42) according to the following way:

$$\begin{aligned}
\hat{F}_{m,n}^{(1)} &\to \hat{F}_{m,n+1}^{(3)}\,, & \hat{F}_{m,n}^{(2)} &\to \hat{F}_{m,n+1}^{(4)}\,, \\
\hat{F}_{m,n}^{(3)} &\to \hat{F}_{m,n+1}^{(1)}\,, & \hat{F}_{m,n}^{(4)} &\to \hat{F}_{m,n+1}^{(2)}\,.
\end{aligned} \tag{8.44}$$

Similarly, for odd $m+n$ we have

$$\begin{aligned}
\hat{F}_{m,n}^{(1)} &\to \hat{F}_{m+1,n}^{(3)}\,, & \hat{F}_{m,n}^{(2)} &\to \hat{F}_{m+1,n}^{(4)}\,, \\
\hat{F}_{m,n}^{(3)} &\to \hat{F}_{m+1,n}^{(1)}\,, & \hat{F}_{m,n}^{(4)} &\to \hat{F}_{m+1,n}^{(2)}\,.
\end{aligned} \tag{8.45}$$

These transformation properties (8.44), (8.45) of the phase space near the four saddle points with respect to the rescaling transformations (8.43) are fully equivalent to the rescaling properties (8.36), (8.37) found by the numerical integration of Hamiltonian system (8.29), (8.34).

Poincaré sections of orbits at the Σ_{00} and Σ_{01} obtained using the separatrix maps are shown in Fig. 8.10a for the same parameters as in Fig. 8.8a: $-\epsilon_a = 0.0208$ and $\chi_a = 0$. Similar plots for the rescaled parameters $\epsilon_b = \lambda^{-1/2}\epsilon_a = 0.0052$ and $\chi_b = \chi_a + \pi/2$ are presented in Fig. 8.10b at the sections Σ_{01} and Σ_{11}. The rescaling parameter $\lambda = \exp(4\pi^2/\Omega) = 16$. Corresponding plots at the sections Σ_{11} and Σ_{10} may be obtained from the Poincaré plots at Σ_{00} and Σ_{10} by shifting the phase φ by π, respectively. Note that the axes φ and H in Fig. 8.10b are inverted. These plots confirm the rescaling invariance of motion with regard of transformations (8.36), (8.37).

8.3.4 On the Validity Conditions of the Rescaling Invariance Property

The rescaling invariance of phase space of Hamiltonian system near the saddle points with respect to transformations given by (8.6), (8.20), (8.21) is exact property of the separatrix mapping (8.19). Therefore the validity of the rescaling law coincides with the validity of the separatrix mapping itself. As was shown in Chap. 5 (see Sects. 5.3.1, 5.3.3) the separatrix mapping (8.19) is derived at the following conditions:

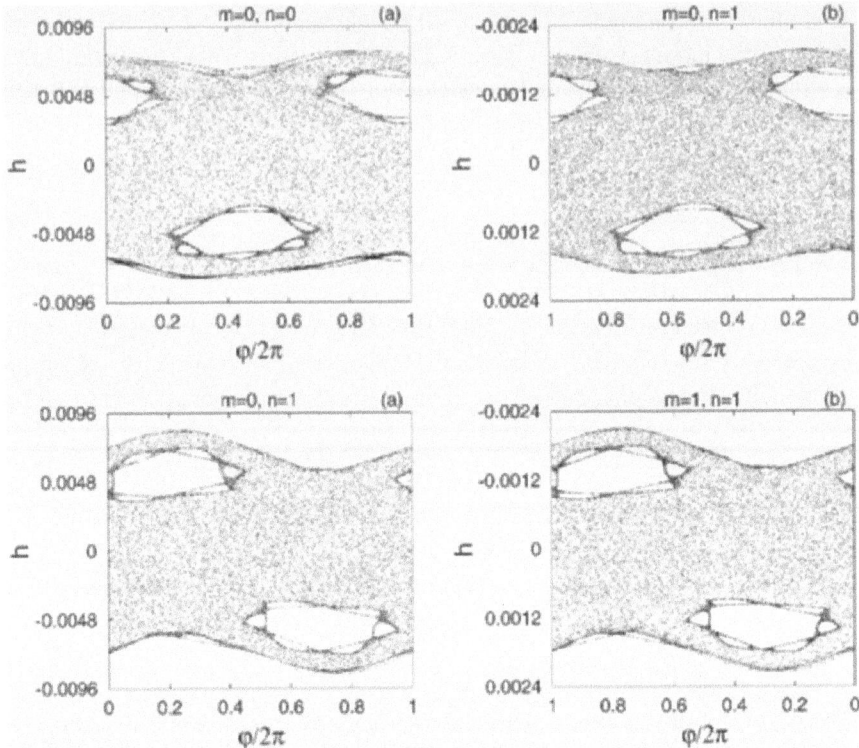

Fig. 8.10. Poincaré sections of orbits of the periodic vortical flow (8.29), (8.34) obtained by the separatrix mapping: (**a**) at the sections Σ_{00} and Σ_{10} for the parameters: $\epsilon_a = 0.0208$ and $\chi_a = 0$; (**b**) at the sections Σ_{01} and Σ_{00} for $\epsilon_b = \lambda^{-1/2}\epsilon_a = 0.0052$ and $\chi_b = \chi_a + \pi/2$. The rescaling parameter $\lambda = \exp(4\pi^2/\Omega) = 16$. Other parameters are $a_x = 1$ and $a_y = 0.5$

a. in the first order of perturbation parameter ϵ by neglecting all terms of order of ϵ^n $(n \leq 2)$;

b. the frequency of motion $\omega(h)$ is replaced by its logarithmic asymptotics (8.5);

The condition (a) means that the rescaling invariance is an effect in the first order in a small perturbation parameter ϵ. From the second condition (b) follows that the effect is valid in the regions of phase space sufficiently close to the separatrix where the frequency of motion $\omega(h)$ can be replaced by its asymptotics (8.5).

8.4 Summary

In this chapter we have studied the structure of the stochastic layer in one degree of freedom Hamiltonian systems subjected to a time-periodic perturbation. It was found the structure of phase space of these systems in the neighborhood of saddle points is invariant with respect to the rescaled transformation of the perturbation amplitude with the shift its phase and the rescaling the coordinates (8.6). Beside this universal property of Hamiltonian systems we have established additional rescaling invariance properties of motion due to symmetries of system in phase space (8.22). These properties gives rise to the periodic change of the topology of phase–space near the saddle point with varying the perturbation amplitude ϵ. As we will see in the next chapter it leads to the $\log \epsilon$-periodic dependence of statistical characteristics of chaotic motion in the stochastic layer.

9 Chaotic Transport in Stochastic Layers

In this chapter we study the statistical properties of chaotic motion in a stochastic layer in the context of their relation with the structure of phase space near saddle points. Before discussing this problem we briefly recall the statistical methods of description of chaotic transport in a stochastic layer of dynamical systems.

9.1 Statistical Description of Chaotic Dynamical Systems

In dynamically chaotic systems the motion of particles becomes practically unpredictable: any small error in initial conditions leads to the enormous error in the final conditions. In this situation it does not have a sense to follow each individual orbit, and the statistical description of the dynamics of system becomes appropriate. This approach is based on the concepts and methods of statistical mechanics. Below we recall the main notions of statistical description to study dynamically chaotic Hamiltonian systems.

9.1.1 Ergodicity and Mixing

Let $(q(t), p(t))$ be the position of system in phase space (q, p) at the time instant t. The time-evolution of system over time period T is given by the map

$$(q_{k+1}, p_{k+1}) = \hat{M}(q_k, p_k) \,,$$

where $(q_k, p_k) \equiv (q(t_k), p(t_k))$, $(t_k = kT)$. Consider the evolution of the arbitrary integrable function $f(q, p)$ of (q, p). Suppose the system's orbits lie in a finite domain W of phase space. Then the motion is called *ergodic* if the time averaging of the function f defined as

$$\bar{f} = \lim_{N \to \infty} \frac{1}{N} \sum_{k=1}^{N} f(q_k, p_k) \,, \tag{9.1}$$

is equal to its value, $\langle f \rangle$ obtained by averaging over phase space,

S.S. Abdullaev: *Construction of Mappings for Hamiltonian Systems and Their Applications*,
Lect. Notes Phys. **691**, 197–218 (2006)
www.springerlink.com

$$\langle f \rangle = \int_W f(q,p)\,dq\,dp \,, \tag{9.2}$$

i.e.,

$$\bar{f} = \langle f \rangle \,. \tag{9.3}$$

In order to introduce the mixing property of dynamical systems we define the *correlation function*. Let f and g be two arbitrary integrable functions of (q,p): $f = f(q,p)$ and $g = g(q,p)$. The correlation function $C_{fg}(\tau)$ is defined as

$$C_{fg}(\tau) = \overline{M^s f \cdot g} - \bar{f}\bar{g}, \qquad \tau = sT \,, \tag{9.4}$$

where

$$\overline{M^s f \cdot g} \equiv \overline{f(q_{k+s}, p_{k+s}) \cdot g(q_k, p_k)}$$
$$= \lim_{N \to \infty} \frac{1}{N} \sum_{k=1}^{N} f(q_{k+s}, p_{k+s}) g(q_k, p_k) \,. \tag{9.5}$$

The correlation function $C_{fg}(\tau)$ gives a quantitative measure of correlation of functions f and g separated by time delay τ.

The dynamical system is called a *mixing* if the correlation function (9.4) vanishes at the limit $\tau \to \infty$:

$$C_{fg}(\tau) = 0, \qquad \text{for } \tau \to \infty \,. \tag{9.6}$$

Mixing represents a more subtle property of system than the ergodicity and it is related with the exponential divergence of neighboring orbits (7.14). In systems with the mixing property the initial small finite domain of phase space acquires the complicated shape with the same area (due to Liouville theorem) but with the stretched and bended regions as illustrated in Fig. 9.1. Ergodicity of system follows from the mixing property. However, the opposite statement that mixing follows from the ergodicity is not generally true.

The law according to the correlation function tends to zero significantly depends on the structure of phase space. In the case of systems with a highly developed chaotic zone with no KAM stability islands the correlation function decays exponentially with τ,

$$C_{fg}(\tau) \sim \exp(-\tau/\tau_c) \,, \tag{9.7}$$

where τ_c is the *correlation time*. Usually, it is inverse proportional to the Lyapunov exponent σ (7.21) averaged over the phase space area W:

$$\tau_c \sim \frac{1}{\langle \sigma \rangle} \,.$$

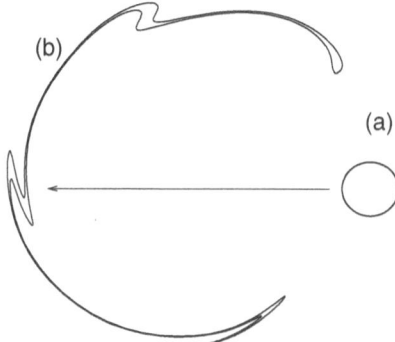

Fig. 9.1. Example of mixing property of system: initial phase space area (**a**) of circular shape turns into the stretched and bended area (**b**) with complicated shape

However, in typical partially chaotic Hamiltonian systems with the embedded KAM stability islands in the chaotic zone the correlation function decays slowly according to power-like law,

$$C_{fg}(\tau) \sim \tau^{-p}, \tag{9.8}$$

with the constant exponent p. As we will see later it significantly depends on the structure of chaotic zone.

In analogy with the notion of entropy in the statistical mechanics Kolmogorov (1958) introduced a *dynamical entropy*, h_K, for dynamically chaotic Hamiltonian systems to characterize a degree of stochasticity (see, e.g. Zaslavsky (1985)). The dynamical entropy also known as the *Kolmogorov entropy* has an order of the averaged Lyapunov exponent

$$h_K \sim \langle \sigma \rangle \sim \tau_c^{-1} . \tag{9.9}$$

For dynamically chaotic systems the Kolmogorov entropy is positive, $h_K > 0$.

9.1.2 Kinetic Description

In statistical mechanics of many body problems kinetic equations are obtained from the dynamical equations of motion with assumptions like random phase approximation. Dynamically chaotic systems with a few degrees of freedom can be also described in terms of the kinetic equations describing the *irreversible behavior of system*. One of the features of dynamically chaotic systems is concluded in that the kinetic equation for them can be obtained without above mentioned assumptions of random phases. In typically highly developed chaotic systems the kinetic equation can be obtained thanks to exponential decay (9.7) of the correlation functions (9.4).

Specifically consider a dynamically chaotic system with the Hamiltonian

$$H = H_0(I) + \epsilon H_1(I, \vartheta, t) ,$$
$$\epsilon H_1(I, \vartheta, t) = \epsilon \sum_{mn} H_{mn}(I) \cos(m\vartheta - n\Omega t + \chi_{mn}) . \tag{9.10}$$

Let $F(I, \vartheta, t)$ be a probability distribution function (PDF) of particles in phase space at the moment of time t. Then $dN = F(I, \vartheta, t) dI d\vartheta$ is a number of particles in the element of area $dV = dI d\vartheta$. According to (1.3), (1.4), (9.10) the time-evolution of the function $F(I, \vartheta, t)$ is described by the equation

$$\frac{\partial F}{\partial t} + \omega \frac{\partial F}{\partial \vartheta} = \epsilon \left(\frac{\partial F}{\partial I} \frac{\partial H_1}{\partial \vartheta} - \frac{\partial F}{\partial \vartheta} \frac{\partial H_1}{\partial I} \right) . \tag{9.11}$$

We introduce the distribution function, $f(I, t)$, averaged over phases ϑ,

$$f(I, t) = \frac{1}{2\pi} \int_0^{2\pi} F(I, \vartheta, t) d\vartheta .$$

Supposing that in dynamically chaotic systems the correlation functions $C_{mm'}(\tau)$ of phases ϑ decay exponentially,

$$C_{mm'}(\tau) = \lim_{T \to \infty} \frac{1}{T} \int_0^T e^{im\vartheta(t+\tau) - im'\vartheta(t)} dt \sim \exp(-\tau/\tau_c) ,$$

from the equation (9.11) one can obtain the following Fokker–Planck equation for the function $f(I, t)$ (see Lichtenberg and Lieberman (1992); Zaslavsky (1985); Sagdeev et al. (1988)):

$$\frac{\partial f}{\partial t} = \frac{1}{2} \frac{\partial}{\partial I} \left(D(I) \frac{\partial f}{\partial I} \right) , \tag{9.12}$$

where

$$D(I) = \pi \epsilon^2 \sum_{mn} m^2 |H_{mn}(I)|^2 \delta(m\omega(I) - n\Omega) \tag{9.13}$$

is the diffusion coefficient in the space of action variable I. This statistical approach to study irreversible behavior of chaotic dynamics is known also as the *quasilinear theory* because of its analogy with the corresponding theory of plasma oscillations[1] (see, e.g., Sagdeev and Galeev (1969)). The diffusion coefficient (9.13) is called the *quasilinear diffusion coefficient.* The quasilinear

---

[1] Recently Elskens and Escande (2002a,b) have rigorously proven that diffusion coefficient $D(I)$ takes on the quasilinear value (9.13) for the chaotic motion of an electron in plasmas with a set of strongly overlapping Langmuir waves with random phases. As a result, the macroscopic irreversible evolution of a plasma is described by its microscopic chaotic dynamics.

approximation usually is valid for the statistical description of Hamiltonian systems with highly developed chaos.

Qualitatively, this diffusion process can be considered a random walk along the action variable I with a step size ΔI and with time step $\Delta t = 2\pi/\Omega$. The diffusion coefficient of such a process defined as

$$D(I) = \frac{\langle (\Delta I)^2 \rangle}{2\Delta t} ,$$

is given by (9.13).

9.1.3 Anomalous Diffusion

In typical Hamiltonian systems the zone of chaotic motion is not uniform, especially the stochastic layer near the separatrix. It consists of KAM-stability islands embedded in a so-called stochastic sea. The structure of the stochastic layer is determined by the mutual positions and sizes of KAM-islands. Existence of these islands leads to the deviation of chaotic motion from the normal diffusion processes described by the Gaussian random walk approximation. This is because of long-time range correlations of type (9.8) due to particles trapped near the islands. In general, such a chaotic motion in the stochastic layer is not described by the normal random transport process with the Fokker–Planck equation (9.12). It is one of the important features of typical deterministic chaotic systems. Departure of the statistics of chaotic motion from the Gaussian one is called *anomalous diffusion*. It has been the subject of extensive studies for more two decades starting from pioneering works by Karney (1983); Chirikov and Shepelyansky (1984) (see, also reviews by Bouchaud and Georges (1990); Chirikov (1991); Shlesinger et al. (1993); Klafter et al. (1996); Metzler and Klafter (2000); Zaslavsky (2002)). In one dimension the anomalous diffusion along space, x, (or momentum p) coordinate is characterized by a nonlinear time dependence of a second moment displacement,

$$\sigma^2(t) = \langle (x(t) - \langle x \rangle)^2 \rangle = 2Dt^\gamma , \qquad (\gamma \neq 1) \qquad (9.14)$$

of random coordinate x. For the normal diffusion process the exponent $\gamma = 1$ and D determines a diffusion coefficient. The case $\gamma > 1$ is known as enhanced (*superdiffusive*) transport, while the case $\gamma < 1$ describes a reduced (*subdiffusive*) transport. At the present time it is well established that the anomalous transport occurs in partially chaotic Hamiltonian systems where the regions of phase space with the regular motion [so-called the Kolmogorov-Arnold-Mozer (KAM) stability islands] are embedded in a stochastic zone. Due to stickiness of orbits to the KAM stability islands chaotic motion alternates with a time intervals with a regular behavior. Then the type and rate of anomalous transport depend on the structure of the stochastic layer determined by the mutual positions of the KAM islands and their size.

Presently several theoretical approaches have been developed to describe anomalous diffusion in partially chaotic Hamiltonian systems. Among them one should mention the continuous time random walk (CTRW) method (see, e.g., Metzler and Klafter (2000)), the fractional kinetic equations (see Zaslavsky (2002)), the lobe dynamics (see, Wiggins (1990b)). Particularly, the kinetic theory of the standard map in different diffusion regimes has been studied by Balescu (2000a,b).

9.2 Non-Gaussian Statistics in Stochastic Layers

In the case of the stochastic layer formed near the separatrices the chaotic transport is mainly determined by its structure near the saddle points where particles spend more time than in other parts of the phase space. As was shown in the previous Chap. 8 the perturbed motion near saddle points is invariant with respect to the rescaled transformation of the perturbation amplitude ϵ and the shift of phase χ and the phase space coordinates (8.6), i.e., the topology of the stochastic layer in the neighborhood of saddle points is a periodic function of $\log \epsilon$ with the period $\log \lambda$. Therefore, one can expect that by varying ϵ one can periodically change the statistical and transport properties of chaotic motion in a stochastic layer. Below we demonstrate these properties for the different models of Hamiltonian systems considered in the previous chapters. All calculations are carried out using the separatrix mappings obtained in Chaps. 6 and 8.

9.2.1 Mean Residence Time

We first consider a statistics of residence time of particle in one of wells in a double–well potential field affected by time-periodic perturbation. The system is described by Hamiltonian (6.1). Suppose that in the absence of perturbation a particle is trapped in one of the wells. A motion of particle may be described in the (x, p) plane by the closed curve 1 in Fig. 6.1b, and it is separated from the other well by the separatrix (curve 2). The time-periodic perturbation destroys the separatrix replacing it by a stochastic layer which is illustrated in Fig. 8.4. If the initial position of particle is inside the stochastic layer it leaves the potential well by crossing the unperturbed separatrix during a certain residence time τ which is known as a *residence time*. Its value is very sensitive to the small changes of initial coordinate of particle. The statistics of residence time depends on the structure of the stochastic layer. Below we study a dependence of mean residence time on perturbation amplitude ϵ.

In the quasilinear approximation the mean residence time $\langle \tau \rangle$ on ϵ can be estimated using the following arguments. If one does not take into account trapped particles one could expect that $\langle \tau \rangle$ is proportional to the characteristic period of particle's orbit $T(h_w)$ inside the stochastic layer with

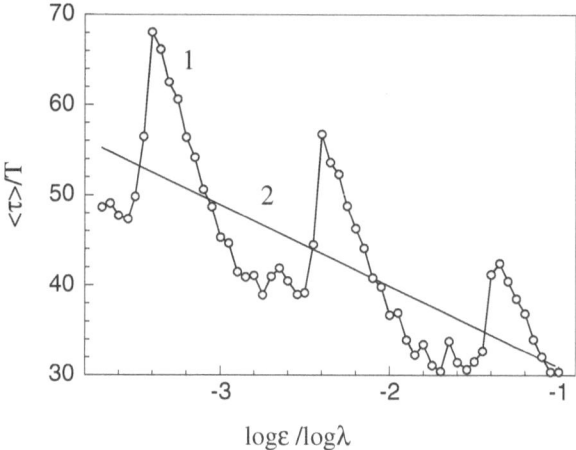

$\log\varepsilon / \log\lambda$

Fig. 9.2. A mean residence time $\langle\tau\rangle$ in the potential well normalized to the perturbation period T_0 versus a perturbation parameter ϵ (curve 1). Curve 2 describes the fitting of $\langle\tau\rangle$ vs ϵ by a linear function $a - b\log\epsilon$ ($a = 21.87\pm2.64, b = 3.92\pm0.462$). The rescaling parameter $\lambda = 10$

effective energy h_w, i.e., $\langle\tau\rangle \sim T(h_w)$. The latter, h_w, counted with respect to the separatrix energy, is of order of the width of the stochastic layer $w_s(\epsilon) \sim \epsilon$, and taking into account the logarithmic asymptotics of the period $T(h_w) \sim \ln(A/h_w)$ (6.8), we have $\langle\tau\rangle \sim a - b\log\epsilon$, where a and b are constants independent on ϵ. Therefore, neglecting the stickness of particles to the KAM stability islands one expects that the mean residence time $\langle\tau\rangle$ linearly decreases with $\log\epsilon$.

The calculations of $\langle\tau\rangle$ were performed using the separatrix map (6.21) and it is presented in Fig. 9.2. The value of the perturbation frequency Ω is chosen to have the rescaling parameter $\lambda = \exp(2\pi/\Omega) = 10$. Averaging is made over $N = 10^6$ orbits. Curve 1 describes a dependence of $\langle\tau\rangle$ on ϵ, curve 2 corresponds to it's fitting with the linear-log law $\langle\tau\rangle \sim a - b\log\epsilon$. From Figure one can clear see that the mean residence time does not monotonically depend on ϵ. There are strong periodic oscillations around linear–log dependence. These oscillations are due to a periodical variation of the topology of phase–space near saddle point with the change of perturbation amplitude ϵ. The period of oscillations are determined by the rescaling parameter λ, i.e., equals to $\log\lambda$.

9.2.2 Statistics of Poincaré Recurrences

According to Poincaré recurrence theorem any orbit of bounded Hamiltonian system returns to the small neighborhood of its initial position (see, e.g., Arnold (1989)). The distribution function $P_{rec}(\tau)$ of recurrence time τ is the important statistical characteristics of dynamically chaotic systems. It is defined as

$$P_{rec}(\tau) = \frac{N(\tau)}{N} ,$$

where $N(\tau)$ is the number of recurrences with $t > \tau$ and N is the full number of recurrences. The function $P_{rec}(\tau)$ is related to the correlation function of dynamical variables, $C(\tau) = \langle \dot{x}(t+\tau)\dot{x}(t) \rangle$ (see, e.g., Chirikov and Shepelyansky (1984)):

$$C(\tau) \sim \tau P_{rec}(\tau)/\langle \tau \rangle ,$$

where $\langle \tau \rangle$ is the mean recurrence time. The diffusion coefficient (rate) D is directly related to the correlations,

$$D \sim \int_0^\infty C(\tau)d\tau .$$

In a fully developed chaotic system $P_{rec}(\tau)$ decays exponentially with τ, $P_{rec}(\tau) \sim \exp(-a\tau)$ (Lichtenberg and Lieberman (1992)). Numerous studies show that in partially chaotic systems with the KAM islands the recurrence distribution has a power-law $P_{rec}(\tau) \sim \tau^{-p}$ at a large time. First calculations of the exponent p performed by Chirikov and Shepelyansky (1984); Karney (1983) for the separatrix map and others different maps gave $p \approx 1.5$. During last decade the values of $p \approx 1-2.5$ have been found for different Hamiltonian systems (Chirikov and Shepelyansky (1984); Chirikov (1991); Geisel et.al (1987); Shlesinger et al. (1993); Klafter et al. (1996); Artuso (1999); Zaslavsky (2002)). However, Chirikov (1983) presented some arguments that the value of p should be equal to 3, which is strongly different from $p \approx 1.5$. Murray (1991) maintained that in order to achieve the exponent $p = 3$ one requires larger times. Recently in Chirikov and Shepelyansky (1999) the power−law decay $P_{rec}(\tau) \sim \tau^{-p}$ with $p = 3$ was numerically observed at very large times for the dynamical chaos in the standard map with the critical golden KAM-invariant curve, i.e., $m : n = \Omega/\omega = (\sqrt{5} - 1)/2$.

We have studied a dependence of the Poincaré recurrence statistics on the structure of the stochastic layer by varying perturbation parameter ϵ. For this purpose we have calculated statistics of first return times to the 4-th branch of the section Σ_s in Fig. 6.3 in the double–well potential affected by the time-periodic perturbation. All calculations are performed using the separatrix map (6.21) for the same parameters as in Sect. 9.2.1. The mean recurrence time $\langle \tau \rangle$ as a function of the perturbation parameter ϵ is shown in Fig. 9.3. Similar to the mean residence time (see Fig. 9.2) it is also a quasiperiodic function of $\log \epsilon$ with the period $\log \lambda$. However maxima of the mean residence time correspond to minima of the mean recurrence time $\langle \tau \rangle$ and vice versa.

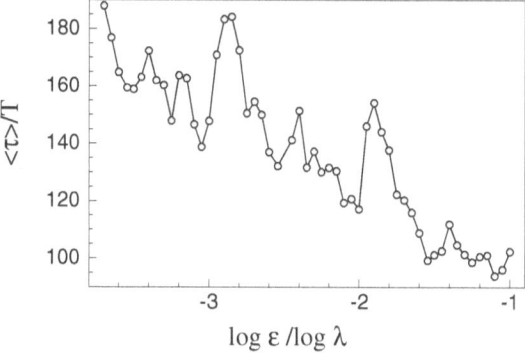

Fig. 9.3. Average recurrence time $\langle \tau \rangle$ as a function of perturbation amplitude ϵ

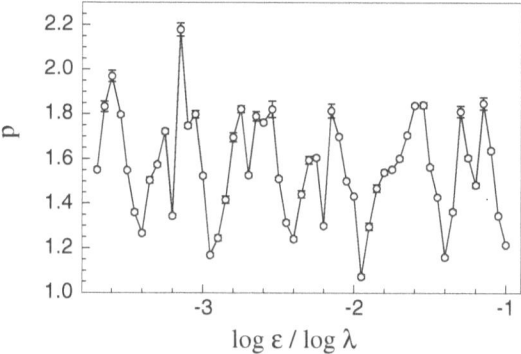

Fig. 9.4. Exponents p of asymptotics of $P_{rec}(\tau) \sim \tau^{-p}$ as a function of the perturbation parameter ϵ

Probability of recurrences $P(\tau)$ computed up to the moderate times $\tau \leq 10^5 2\pi/\Omega$ shows that it decays oscillating near to the power-law τ^{-p} which were also observed in Chirikov and Shepelyansky (1984). The amplitude of these oscillations varies with perturbation parameter ϵ. The estimations of the exponent p for different ϵ were performed by a fitting of the probability of recurrences $P(\tau)$ with the power-law $C\tau^{-p}$ in the time interval $10^2 \leq \tau/T \leq 10^5$. The latter is of order of oscillation period of $P(\tau)$ around the power-law τ^{-p}. The dependence of the exponent p on ϵ is shown in Fig. 9.4 from which one can recognize a periodic dependence of p on $\log \epsilon$ with the period $\log \lambda = \log 10$. The values of p vary between 1 and 2, and it is oscillating near the average value $p \approx 1.5$ which was observed in most previous calculations mentioned above.

9.3 Chaotic Transport in Stochastic Layers. Three-Wave Field Model

In this section we consider a chaotic transport of particles in a stochastic layer for three-wave field model (6.23)–(6.25) (see Sect. 6.2). We will study statistical properties of transport along the (infinite) x - axis, in particularly, an advection and diffusion by calculating the first, $\langle x \rangle$, and the second moments of the spatial displacement, $\sigma^2(t)$ (9.14), respectively, as well as the probability density function (PDF) $P(x, t)$ for a particle with position x at time instant t as a function of perturbation amplitude ϵ.

Calculations of the statistical moments were performed using the separatrix map (6.49) as well as direct numerical integrations of motion equations for the Hamiltonian (6.23)–(6.25) using the symplectic integrator (1.58), (1.59). The rescaling parameter is chosen equal to $\lambda = 4$ which corresponds the perturbation frequency $\Omega = 4.53236\omega_0$. A set of initial data at $t = 0$ consisting of $N = 5 \times 10^3$ trajectories were taken in a square region centered at the hyperbolic fixed point $(x = \pi, p = 0)$.

9.3.1 Advection

An advection in the stochastic layer takes place in the direction of the perturbation wave with the larger amplitude $a_{max} = \max(A, B)$. The maximum advection occurs if only one perturbation wave is present. We consider this case putting $A = 1$ and $B = 0$. Calculations show that at least up to $t \leq 2 \times 10^4 T$ the mean coordinate $\langle x(t) \rangle$ is linear function of time t, i.e., $\langle x(t) \rangle = vt$ with an advection speed v. It has been found that the advection speed v is not a monotonic function of perturbation parameter ϵ. Similar to the mean residence time $\langle \tau \rangle$ (see Fig. 9.2) it varies quasi-periodically with $\log \epsilon$ with the period $\log \lambda$ as shown in Fig. 9.5.

9.3.2 Anomalous Diffusion

To study an anomalous diffusion we have calculated the second moments $\sigma^2(t)$ (9.14) for $A = B = 1$. In this case the perturbation in the Hamiltonian (6.23)–(6.25) acts symmetrically on particles traveling in both positive and negative directions along the x-axis without advection, i.e., the mean value $\langle x \rangle = 0$.

Figure 9.6 shows the dependence of $\sigma^2(t)$ on the perturbation amplitude ϵ at two different time instants: curve 1 corresponds to $t = (2\pi/\Omega) \times 10^4$ and curve 2 to $t = (4\pi/\Omega) \times 10^4$. The thick curves correspond to the numerical integration of the equations of motion, while the thin curves corresponds to the separatrix map calculations (6.49) [with an average over $N = 10^4$ orbits]. One can see that the separatrix map (6.49) correctly reproduces the results of direct numerical integrations with a good accuracy up to the relatively

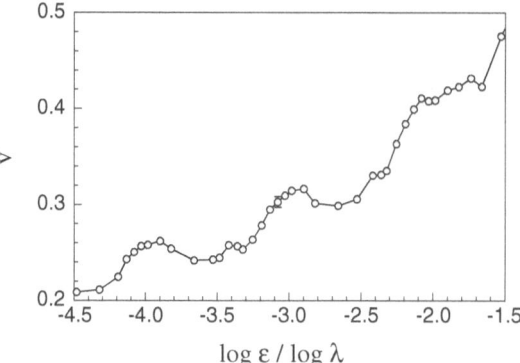

Fig. 9.5. Advection velocity v versus perturbation amplitude ϵ as obtained by direct numerical integration. Rescaling parameter $\lambda = \exp(2\pi\omega_0/\nu) = 4$

large value of $\epsilon = 0.1$. Fig. 9.6 clearly shows the strong quasi-periodical dependence of the second moment $\sigma^2(t)$ on perturbation parameter ϵ. There are local maxima of $\sigma^2(t)$ at the values $\epsilon_{max}^{(j)} = \lambda^{-j}\epsilon_{max}$, $\epsilon_{max} \approx 0.192$, and local minima at $\epsilon_{min}^{(j)} = \lambda^{-j}\epsilon_{min}$, $\epsilon_{min} \approx 0.08$, $(j = 1, 2, \cdots)$. For large perturbation amplitudes $\epsilon > 0.1$ the quasi-periodical behavior of $\sigma^2(t)$ is less pronounced since the rescaling property of Hamiltonian system starts to violate for large perturbations. With increasing time, the periodic dependence of $\sigma^2(t)$ on ϵ becomes even more pronounced.

A quasiperiodic dependence of $\sigma^2(t)$ on ϵ is a consequence of the periodic change of the structure of the stochastic layer near saddle points. Minima of $\sigma^2(t)$ correspond to the situations when a majority of particles are stuck to

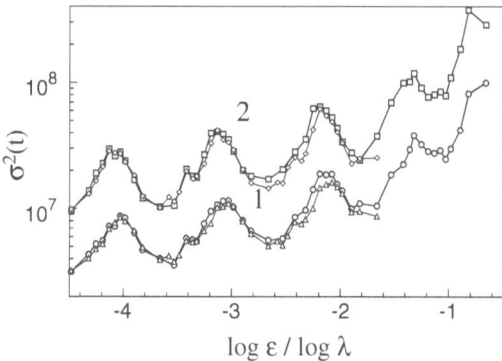

Fig. 9.6. Second moment $\sigma^2(t)$ versus perturbation amplitude ϵ as obtained by direct numerical integration. Solid curve 1 corresponds to $t = (2\pi/\Omega) \times 10^4$ and solid curve 2 to $t = (4\pi/\Omega) \times 10^4$ (thick lines). The corresponding thin line curves describe the results obtained by the separatrix map (6.49)

the islands of type 2, 3, 5, 6 (see Fig. 8.2) responsible for trapping particles by the main wave. Similarly, maxima of $\sigma^2(t)$ describes the case when a majority of particles are stuck to the islands of type 1, 4 (see Figure 8.2) in which particles are running along x-axis for a long time without changing the direction. The variation of ϵ periodically alternates the domination of these different types of islands. More detailed discussion of this phenomenon can be found in Abdullaev (2000).

For large times t we have the following asymptotics of $\sigma^2(t) \sim t^\gamma$. The exponent γ is also a strong quasi-periodic function of $\log \epsilon$ with the period $\log \lambda$. The dependence γ on ϵ obtained using the separatrix map (6.49) is shown in Fig. 9.7. The chaotic transport along the x-axis is superdiffusive ($\gamma > 1$) for all perturbation amplitudes. The exponent γ takes maxima and minima values at the same ϵ values as $\sigma^2(t)$ does. The regions with $\gamma > 2$ correspond to the acceleration regimes.

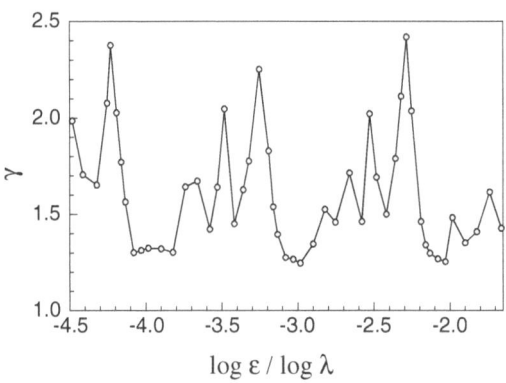

Fig. 9.7. Exponent γ versus perturbation amplitude ϵ. It is obtained by fitting $\sigma^2(t)$ with $2Dt^\gamma$ in the large time interval $10^4 T_0 \leq t \leq 10^5 T_0$

9.3.3 Probability Density Function

The separatrix map (6.49) is also applied to calculate PDF $P(x,t)$. It was calculated at the time instant $t = (2\pi/\Omega) \times 10^4$ for perturbation parameters ϵ in the interval $[0.002, 0.1]$. The statistics is taken averaging over $N = 10^5$ orbits. The PDF is almost symmetrically localized near $x = 0$. The width $2\Delta\sigma$ of $P(x,t)$ is defined as an area $-\Delta\sigma < x < \Delta\sigma$ where a half of orbits is localized, i.e.,

$$\int_{-\Delta\sigma}^{\Delta\sigma} P(x,t)dx = 0.5 \ .$$

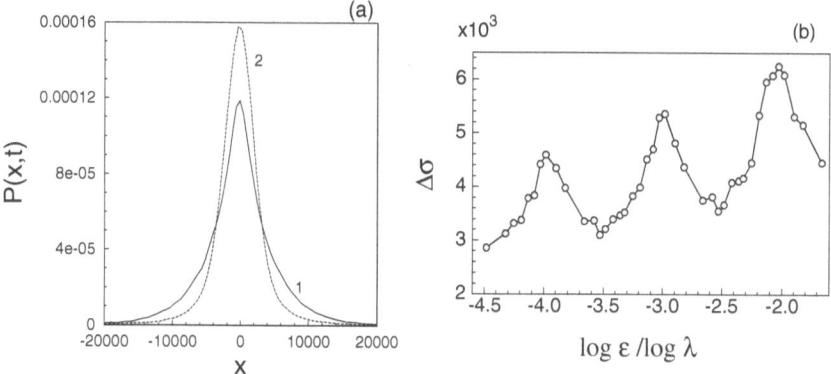

Fig. 9.8. (a) PDF $P(x,t)$ for the two values of ϵ: curve $1 - \epsilon = 0.016$, curve $2 - \epsilon = 0.03$; (b) Width of the PDF $\Delta\sigma$ versus perturbation amplitude ϵ

The PDF $P(x,t)$ for two values of ϵ and the dependence of the width $\Delta\sigma$ on the perturbation parameter ϵ are plotted in Fig. 9.8a, b, respectively. Curve 1 in Fig. 9.8a corresponds to $\epsilon = 0.016$, and curve 2 – to $\epsilon = 0.03$. These values of ϵ correspond to the local maxima and minima of $\Delta\sigma$. Similar to the second moment $\sigma^2(t)$ it has also a strong periodical dependence on $\log \epsilon$ with the period $\log \lambda$.

One should note that the square root of the second moment $\sigma(t)$ and the width $\Delta\sigma$ describe the width of the probability density functions. However, in the case of anomalous (non-Gaussian) transport they describe different physical situations of transport process. The width $\Delta\sigma$ describe the PDF near its central part where half of particles are located. The main contribution to $\Delta\sigma$ comes from random particles and particles trapped by islands due to stickiness. On the other hand, contributions to $\sigma(t)$ mainly comes from particles with long distance flights. Therefore, $\sigma(t)$ always exceeds the width $\Delta\sigma$: $\sigma(t) > \Delta\sigma$. In the case of normal Gaussian transport both $\sigma(t)$ and $\Delta\sigma$ would have the same physical nature.

The main feature of $P(x,t)$ is its long tail asymptotics for $|x| \gg \Delta\sigma$. The latter significantly depends on the perturbation parameter ϵ. The comparison, for instance, of the two PDF at $\epsilon = 0.048$ and $\epsilon = 0.08$ for which the second moments $\sigma^2(t)$ have maximum and minimum values, respectively, shows that while the PDF for $\epsilon = 0.048$ has a slowly decaying tail, the PDF for $\epsilon = 0.08$ decays much faster. We have approximated $P(x,t)$ asymptotically by power–exponential law,

$$P(x,t) \sim |x|^{-\alpha} e^{-\beta|x|} ,$$

with the fitting exponents α and β. They are found at the time instant $t = (4\pi/\Omega) \times 10^4 T_0$ and presented in Fig. 9.9: (a) describes α versus ϵ, while (b) describes β versus ϵ. It shows the strong quasi-periodic dependence of

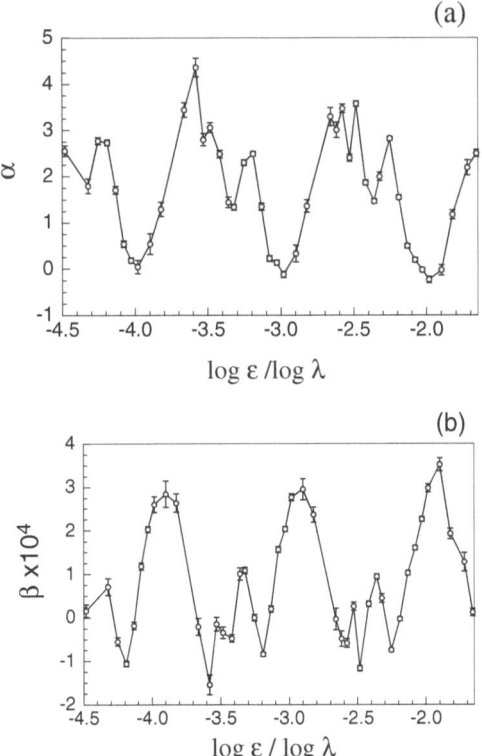

Fig. 9.9. Fitting parameters α and β for a power–exponential law $P(x,t) \sim |x|^{-\alpha} \exp(-\beta|x|)$: **(a)** α versus ϵ; **(b)** β versus ϵ. A time instant $t = (4\pi/\Omega) \times 10^4$

these parameters on $\log \epsilon$ appears with the period $\log \lambda$ similar to that for the exponent γ in the time-asymptotics of $\sigma^2(t)$.

Overall the results show that the asymptotics of PDF $P(x,t)$ for $|x| \gg \Delta\sigma$ significantly depends the structure of the stochastic layer, and it is mainly determined by the outermost KAM-stability islands at the chaos border.

9.4 Chaotic Transport in 2D-periodic Vortical Flow

In this section we consider a chaotic transport in a stochastic web of the two dimensional time-dependent periodic vortical flow. This example shows that depending on the perturbation parameter ϵ the chaotic transport may have the superdiffusive, normal or subdiffusive characters.

Suppose the periodic vortical flow (8.29) is perturbed by the time-dependent perturbation

$$H_1(x,y,t) = \frac{\epsilon}{2\pi}[\sin(2\pi y) - \sin(2\pi x)]\cos(\Omega t + \chi) \,, \qquad (9.15)$$

which may be considered as combination wave perturbations $\sin(2\pi x - \Omega t - \chi) + \sin(2\pi x + \Omega t + \chi)$, $\sin(2\pi y - \Omega t - \chi) + \sin(2\pi y + \Omega t + \chi)$ propagating in opposite directions of x- and y-axes, respectively. Thanks to the symmetry property (8.35) of the perturbation (9.15) there exists an additional rescaling invariance property (8.36), (8.37) of the system. Therefore one expects that all statistical characteristics of chaotic transport in such a system are to be quasi-periodical functions of the perturbation amplitude $\log \epsilon$ with the period $(\log \lambda)/2$.

For the perturbation (9.15) one can expect that the chaotic transport along x- and y-directions are equivalent, in particularly, mean spatial displacements $\langle x \rangle = \langle y \rangle = 0$, and mean squared displacements $\langle x^2 \rangle = \langle y^2 \rangle$.

The chaotic transport in a stochastic web was studied using the separatrix map constructed in Sect. 8.3.3. The separatrix map for the perturbation (9.15) has the form (8.40) with the perturbation coefficients

$$K^{(\beta)}_{\alpha,\alpha\pm1} = (-1)^\beta \frac{\Omega}{4\pi} \frac{1}{\sinh(\Omega/4)} \ . \tag{9.16}$$

9.4.1 Variation of Diffusion Regimes

The rescaling parameter $\lambda = \exp(4\pi^2/\Omega)$ has been chosen equal to 16. The statistics of transport is studied averaging over $N = 10^4$ orbits. The second moment of the radial displacement $\sigma_r^2(t) = \langle x^2(t) + y^2(t) \rangle$ is displayed in Fig. 9.10 as a function of ϵ at the time instant $t = (2\pi/\Omega) \times 10^4$. As seen $\sigma_r^2(t)$ is a quasiperiodic function of $\log \epsilon$. Its period, $0.5 \log \lambda$ is twice smaller than in the case of transport along the 1D stochastic layer (Section 9.3). There are sharp periodic peaks of $\sigma^2(t)$ at certain values of ϵ. As we will see below they appear due to flights in the stochastic web.

The exponent γ of asymptotics $\sigma_r^2(t) \sim t^\gamma$ shown in Fig. 9.11 also periodically varying with $\log \epsilon$. There are large periodic intervals of ϵ where γ is less but close to 1. Minima values of the exponent γ are about 0.9

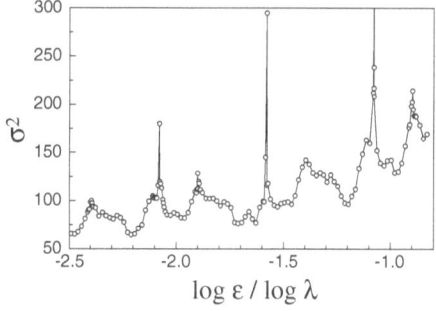

Fig. 9.10. Second moment of radial displacement $\sigma^2 = \langle x^2 + y^2 \rangle$ versus ϵ. A time instant $t = (2\pi/\Omega) \times 10^4$. Rescaling parameter $\lambda = \exp(4\pi^2/\nu) = 16$

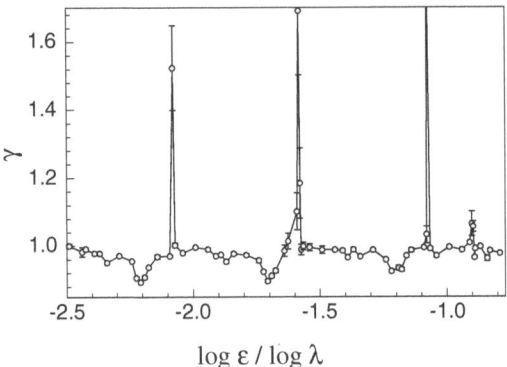

$$\log \varepsilon / \log \lambda$$

Fig. 9.11. Exponents γ of the asymptotics $\sigma^2 \sim t^{\gamma}$. Parameters of the system are the same as in Fig. 9.10

in small intervals of ϵ, where the transport may be considered as a weakly subdiffusive.

For the most values of ϵ the exponent γ is close 1 and the chaotic transport may be well approximated by the normal diffusion (Gaussian) process introducing a diffusion coefficient D. We have calculated the diffusion coefficients using its two definitions: (*i*) through the squared radial displacement:

$$D_{\sigma} = \sigma_r^2(t)/2t , \qquad t \to \infty ,$$

and (*ii*) through the Gaussian PDF

$$P_G(r,t) = \frac{r}{D_G t} \exp\left(-\frac{r^2}{2D_G t}\right) ,$$

which describes PDF to find a particle with radial position $r = \sqrt{x^2 + y^2}$ at a time instant t corresponding to the initial distribution $P_G(r,0) = 2\pi r\delta(r)$ of particles at the time instant $t = 0$. For the normal diffusion process the diffusion coefficients D_{σ} and D_G should coincide.

Diffusion coefficients D_{σ} and D_G are presented in Fig. 9.12 as functions of ϵ. Curve 1 describes the diffusion coefficient D_{σ} determined from the mean squared radial displacement $\sigma_r^2(t)$ in the time interval $10^4 \leq t/T \leq 10^5$, and curve 2 corresponds to D_G obtained by fitting the numerically determined PDF $P(r,t)$ at the time instant $t = 10^5 T$ with the radial Gaussian PDF $P_G(r,t)$. Figure 9.12 also shows a quasi-periodical dependence of both values of D on $\log \epsilon$ similar to those ones in Figs. 9.10, 9.11. For most values of ϵ the diffusion coefficients D_{σ} and D_G are close, but D_{σ} systematically exceeds D_G. The reason of such a behavior was discussed in Sect. 9.3.3, and consists of fact that there exists a difference between the second moment $\sigma_r^2(t)$ and the width of the PDF $P(r,t)$ in a case of anomalous diffusion. The coefficient D_G is mainly determined by the central part of the PDF while rare events

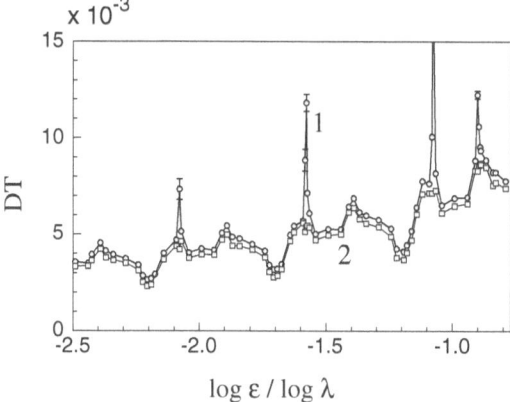

Fig. 9.12. Diffusion coefficients D determined by the asymptotics $D_\sigma = \sigma^2/2t$ (curve 1) and by fitting with the radial Gaussian distribution $P_G(r,t) = r(D_G t)^{-1} \exp[-r^2/2D_G t]$ (curve 2). Parameters of the system are the same as in Fig. 9.10

with long flights may contribute to the D_σ. Results shown in Fig. 9.12 suggest that in large periodic intervals of ϵ where differences between D_σ and D_G are small the transport process may be considered as a normal diffusion process.

9.4.2 Superdiffusive Regime. Levý Flights

As seen from Fig. 9.12 there are large difference between D_σ and D_G in the narrow periodic intervals of ϵ located near values $\epsilon = \lambda^{k/2}\epsilon_0$, $\epsilon_0 = 0.0031$ ($k = 0, \pm1, \pm2, \ldots$). They corresponds to the peaks in $\sigma_r^2(t)$ (see Fig. 9.10) and to the large exponents γ in Fig. 9.11. For these values of ϵ the Gaussian approximation fails and the chaotic transport becomes superdiffusive. Enhanced transport is connected with long distance flights, known as the *Levý flight* at these values of ϵ. A single flight event is shown in Fig. 9.13a in the (x,y) plane and in Fig. 9.13b in the (t,H) plane for the specific value of $\epsilon = 0.0124$.

These flights are connected with stickiness of orbits to the specific KAM–stability islands. They are shown in Fig. 9.14 on the Poincaré sections of orbits in the (x,y) plane near the saddle point: $(x_s = 0.25, y_s = 0.25)$. Four types of islands continuously labeled by 1–4 are seen as dark sticks. A close up view of the region near the 1-st island is shown in Fig. 9.14b. The islands 1–4 have different flight directions: 1-st island flies in the direction of $45°$ with respect to positive direction of the x-axis, the 2-nd island in $315°$, the 3-rd island in $225°$ and the 4-th island in $135°$.

Figure 9.15a shows the structure of these KAM-stability islands on the section Σ_{11} obtained by the separatrix map: (b) shows close up view of the region near the tiny-size KAM-stability island shown in (a). The island with

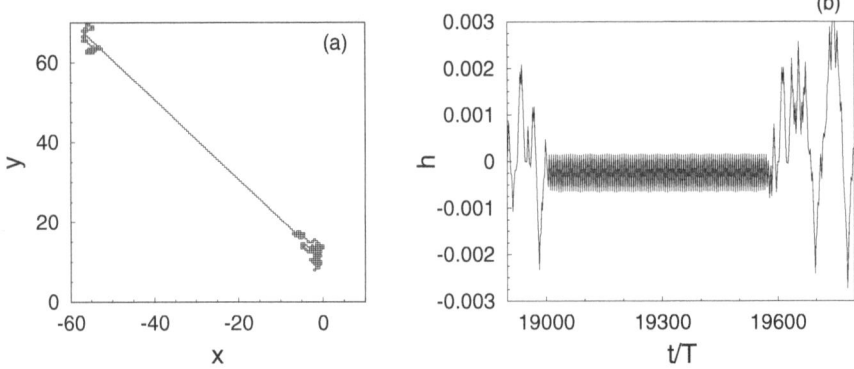

Fig. 9.13. Long distance flight event: (a) orbit in the (x, y) plane, (b) orbit in the (H, t) plane. Parameters are $\epsilon = 0.0124, \chi = 0$

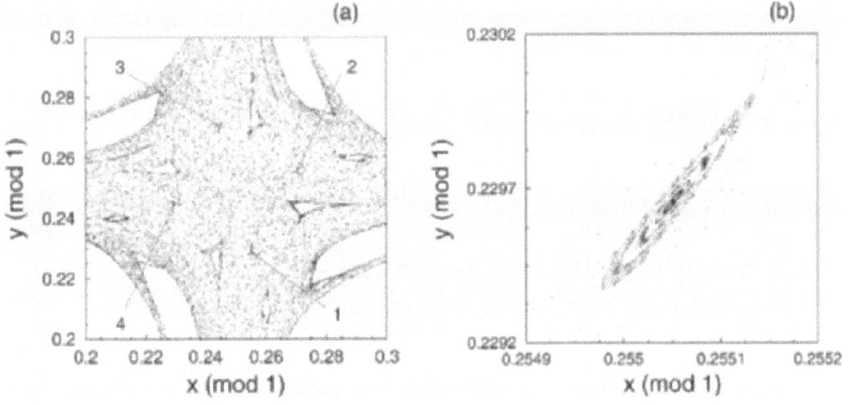

Fig. 9.14. (a) Poincaré section near the saddle points $- (x_s = 0.25, y_s = 0.25)$; Four tiny-size islands of regular motion responsible for long distance flights are shown by arrows 1–4; (b) Close up view of the region near the 1-st island shown in (a). Parameters are the same as in Fig. 9.13

$H > 0$ ($H < 0$) in Fig. 9.15a corresponds 1-st (2-nd) and 3-rd (4-th) islands in Fig. 9.14a, respectively.

The elliptic fixed points ($\varphi_e = 0, \pi, H_e \approx \pm 1.495 \times 10^{-4}$ in the (φ, h)-plane correspond to the fixed of the 1- and 3-islands (x_e, y_e), and the elliptic fixed points ($\varphi_e = \pi/2, 3\pi/2, H_e \approx 6.4565 \times 10^{-4} -$ correspond to the 2- and 4-islands in Fig. 9.14a.

One should note that although the applied numerical integration scheme sufficiently well determines the positions of these island on the phase-space (x, y) but it cannot resolve fine details of their structure because of loosing accuracy. The fine structure of these tiny-size KAM-stability islands may be

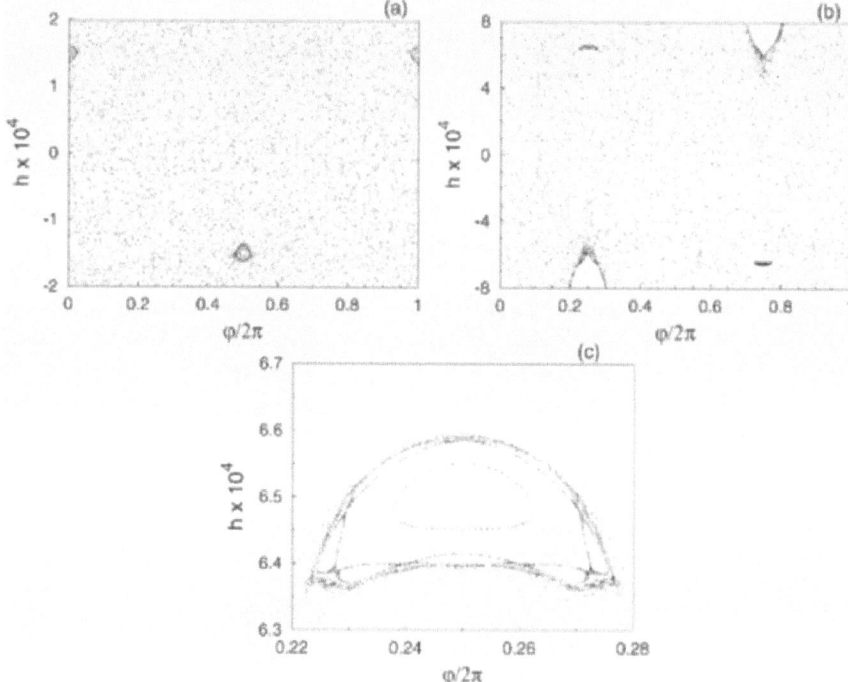

Fig. 9.15. Poincaré plots obtained by the separatrix map in the (H, t) plane at the sections: (**a**) Σ_{10} (**b**) Σ_{11}. The small-size islands (with dark edges) are responsible long distance flights. (**c**) − close up view of one of theses islands shown in (a). Parameters are the same as in Fig. 9.13

seen on Poincaré plots in the (t, H) plane obtained by the separatrix map (see Fig. 9.15c).

9.4.3 Fixed Points of Flight Islands

Islands responsible for flights have a specific feature. As seen from Figs. 9.13b, 9.14 and 9.15 the energy h for trapped orbits takes successively positive and negative values near the saddle points, for instant, if $h_k > 0$ then $h_{k+1} < 0$ and vice versa. The elliptic fixed points $(\varphi^{(e)}, H^{(e)})$ of flight islands at any section Σ_{mn} (or at the equivalent sections $\Sigma_{m\pm2q, n\pm2p}$, $q, p = 1, 2, \ldots$) can be determined by the fixed points of the map $\hat{F}_{mn}^{(p)}$ (8.42) imposing the requirement that if $h_k^{(e)} > 0$ then $h_{k+1}^{(e)} < 0$ and vice versa. The maps $\hat{F}_{mn}^{(p)}$ are constructed consecutive mappings $\hat{X}_{m,m'}^{(n)}$, $\hat{Y}_{n,n'}^{(m)}$. For example, the map

$$\hat{F}_{m,n}^{(2)} = \hat{Y}_{n+1,n+2}^{(m+2)} \hat{X}_{m+1,m+2}^{(n+1)} \hat{Y}_{n,n+1}^{(m+1)} \hat{X}_{m,m+1}^{(n)} \tag{9.17}$$

transforms the fixed point of the 1-st island at the section Σ_{mn} with the even $m + n$ to the equivalent one at the $\Sigma_{m+2,n+2}$. It is illustrated in Fig. 9.16.

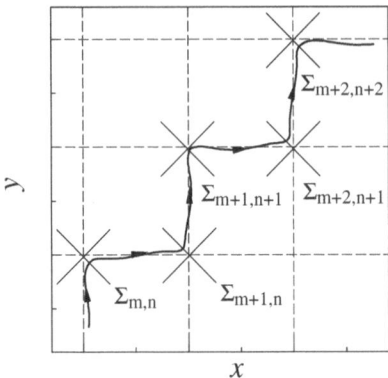

Fig. 9.16. Construction of the map $F_{mn}^{(2)}$ for determination of fixed points of the 1-st island responsible for flights

To be specific we determine the principal fixed points of the 1-st island corresponding to $q = 1$. Using (8.40), (9.16) the equations for the fixed points $(\varphi^{(e)}, h^{(e)})$ may be written as

$$
\begin{aligned}
h_2 &= h_1 + \epsilon K \cos w_1 , & w_2 &= w_1 + g(h_2) , \\
h_3 &= h_2 - \epsilon K \cos w_2 , & w_3 &= w_2 + g(h_3) , \\
h_4 &= h_3 - \epsilon K \cos w_3 , & w_4 &= w_3 + g(h_4) , \\
h_1 &= h_4 + \epsilon K \cos w_4 , & w_1 + 2\pi s &= w_4 + g(h_1) ,
\end{aligned}
\tag{9.18}
$$

where $g(h) = (\Omega/2\pi) \ln(2/\pi|h|)$, $w = \varphi + g(h)/2 + \chi$, and $s = 1, 2, \dots$ In (9.18)

$$
\begin{aligned}
(w_1, h_1) &\in \Sigma_{m,n}, \Sigma_{m+2,n+2} , & (w_2, h_2) &\in \Sigma_{m+1,n} , \\
(w_3, h_3) &\in \Sigma_{m+1,n+1} , & (w_4, h_4) &\in \Sigma_{m+2,n+1} .
\end{aligned}
$$

Fixed points of the 1-st flight island should also satisfy the condition: $h_1 < 0$, $h_2 > 0$, $h_3 < 0$, and $h_4 > 0$. From the equations for the angular variables $w_i, (i = 1, 2, 3, 4)$ in (9.18) follow that

$$
\sum_{i=1}^{4} g(h_i) = 2\pi s, \quad \text{or} \quad \prod_{i=1}^{4} |h_i| = 16/(\pi^4 \lambda^s) ,
\tag{9.19}
$$

where λ is the rescaling parameter. First we consider the case when $h_1 = h_3$ and $h_2 = h_4$. In this case (9.18) may be reduced to

$$
\begin{aligned}
h_2 &= h_1 + \epsilon K \cos w_1 , \\
h_1 &= h_2 - \epsilon K \cos(w_1 + \Omega(h_2)) , \\
|h_1 h_2| &= 4/(\pi^2 \lambda^{s/2}) .
\end{aligned}
$$

Taking into account that $h_1 < 0$, $h_2 > 0$, we obtain the transcendental algebraic equation for w_1:

$$\cos w_1 - \cos \left[w_1 + \Omega\big(h_2(w_1)\big) \right] = 0 , \qquad (9.20)$$

where $h_2(w_1)$ is a positive solution of the quadratic equation

$$h_2^2 - \epsilon K \cos w_1 h_2 + 4/(\pi^2 \lambda^{s/2}) = 0 . \qquad (9.21)$$

We determined the fixed points for the values ϵ near its specific value $\epsilon = 0.0124$. The corresponding number s is equal to 11. Numerical study of (9.20), (9.21) shows that the fixed points, $(\varphi^{(e)}, h^{(e)})$, exist for $0.0121 \leq \epsilon \leq 0.031$. At the section Σ_{00} the phase $\varphi^{(e)} = \pi$ (mod 2π), and $h^{(e)}$ changes in the interval $[-1.542623 \times 10^{-4}, -7.518702 \times 10^{-5}]$, and at Σ_{01} the phase $\varphi^{(e)} = 3\pi/2$ (mod 2π) and $6.263842 \times 10^{-4} \leq h^{(e)} \leq 1.285161 \times 10^{-3}$, respectively. However, the fixed points are elliptic only for $0.0121 \leq \epsilon \leq 0.0131$, and they are hyperbolic for $\epsilon > 0.0131$.

Determination of fixed points in the cases $h_1 \neq h_3$ or $h_2 \neq h_4$ is more difficult. We have studied them by direct plotting Poincaré sections in the (φ, h) plane. Such islands appear for $\epsilon \geq 0.0132$ when the fixed point with $h_1 = h_3$, $h_2 = h_4$ becomes a hyperbolic and generating two elliptic fixed points. With increasing the perturbation ϵ the stochastic layer near the separatrix grows. The islands dissappear for $\epsilon > 0.0133$.

Therefore the flights may occur for the perturbation parameters $0.0121 \leq \epsilon \leq 0.0132$ due to stickiness to the KAM islands. One can obtain similar results for the principal fixed points with $q = 1$ of 2-nd, 3-rd and 4-th islands.

One can conclude that the chaotic transport in the stochastic web of two dimensional time-dependent periodic vortical flow may exhibit three types of stochastic processes: subdiffusive, normal Gaussian and superdiffusive. Varying the perturbation parameter ϵ one can control the types of chaotic transport.

9.5 Conclusions

Obtained in this chapter results reveal the important relationship between the structure of the stochastic layer and statistical properties of chaotic transport in it. It was shown that systems with topologically similar stochastic layers have similar statistical properties of transport. This property of dynamical systems was extensively studied in one- and half-degree of freedom Hamiltonian systems. Established in the previous chapter the rescaling invariant property of systems near saddle points gives rise to the periodic change of the topology of phase–space near saddle points with varying the perturbation amplitude. It leads in turn to the quasi-periodical oscillations of statistical characteristics of transport with the change of perturbation amplitude. The period of these oscillations is determined by the universal rescaling parameter

$\lambda = \exp(2\pi\gamma/\Omega)$ which depends only on the expansion coefficient γ of the unperturbed Hamiltonian near saddle point and the frequency of external perturbation Ω.

This effect is universal for one-degree-of-freedom Hamiltonian systems subjected to small time-periodic perturbations regardless on the specific features of the system. One can expect the effect occurs in chaotic transport problems in structured flows, for instance, in chaotic mass transport in a chain of vortices in a shear layer (see, for instance in del Castillo-Negreto (1998)). Similarly, it may also be observed in models of physical systems which are described by a stochastic web (see e.g., Zaslavsky et al. (1991); Shlesinger et al. (1993); Klafter et al. (1996)).

Our study shows that the chaotic transport rate is not monotonic function of the perturbation amplitude ϵ, in spite of that the width of the stochastic layer linearly increases with ϵ. This suggests, first, that the width of the stochastic layer, determination of which was a primary goal of many works (see, e.g., Treschev (1998) and references therein) does not completely characterize a chaotic motion. The existence of KAM-stability islands embedded in a stochastic layer is one of its essential features, and particularly the outermost islands play a crucial role in chaotic transport. This situation is not taken into account by qualitative transport theories, for instance, by the quasilinear theory (see Sect. 9.1.2), which predicts the monotonic dependence of the diffusion coefficient D (9.13) on perturbation amplitude ϵ.

The established effect also shows the possible range of controlling Hamiltonian chaos Lai et al. (1993), in particularly, a chaotic transport by varying perturbation amplitude. It may be useful to control a transport of heat and particles in magnetic fusion devices with stochastic magnetic field lines Ghendrih et al. (1996), a transport of passive scalars in a chain of vortices del Castillo-Negreto (1998), or a mixing of fluids Ottino (1989).

One should note that the oscillations of normal diffusion coefficient D as a function of stochasticity parameter K ($K > 1$) with the period 2π have been observed in the standard map (3.24) (see Chirikov (1979); Rechester and White (1980); Rechester et al. (1981)). However, this quasi-oscillatory behavior related with the existence of accelerator modes (Karney et al. (1982)) is the exclusive property of the standard mapping, so is unlike the universal quasi-oscillations of chaotic transport in a stochastic layer on $\log \epsilon$ considered in this work.

10 Magnetic Field Lines in Fusion Plasmas

Study of mappings as a part of Hamiltonian dynamics of magnetic field lines in plasmas were initiated by the research in magnetically confined plasma devices in a quest for the controlled fusion[1]. Actually, a fusion research in early sixties gave a huge impact on the development of Hamiltonian dynamics, particularly, on mapping methods. Since, particles predominantly follow magnetic field lines the determination of their structure is important to confine particles. In fusion devices, like tokamaks and stellarators the confinement of charged particles in a bounded area is achieved by specially created (by external coils and plasma current itself) magnetic fields whose field lines lie on nested (magnetic) surfaces (Wesson (2004)). In a study of field lines the most important was the fact that a divergence free magnetic field is equivalent to Hamiltonian system with $1 + 1/2$ degrees of freedom (see, e.g., Cary and Littlejohn (1983); Boozer (1983); Morrison (2000)). The Hamiltonian formulation of magnetic field lines has been instrumental in early studies of the problems of stability and destruction of magnetic surfaces in tokamaks and stellarators due to the presence of magnetic perturbations (Kerst (1962); Rosenbluth et al. (1966); Filonenko et al. (1967); Freis et al. (1973); Hamzeh (1974); Finn (1975); Matsuda and Yoshikawa (1975)).

The use of mappings to describe magnetic field lines in tokamaks and stellarators is intended to simplify the study of different problems in plasma physics ranging from a stability of magnetic surfaces, particle motion in electromagnetic fields to the transport of heat and particles in plasmas. In this chapter we discuss the mapping methods to describe magnetic field lines in magnetically confined plasmas.

10.1 Magnetic Field Lines as Hamiltonian System

Below we recall some necessary elements of plasma physics and Hamiltonian formulation of the equations of magnetic field lines. A description of field line equations in action-angle variables will be also given.

---

[1] There is a number of books devoted to the physics of magnetic confinement of plasmas. The most complete among of them is, probably, *Tokamaks* by Wesson (2004). The reviews accessible for more wider audience are given by Boozer (1992, 2004)

S.S. Abdullaev: *Construction of Mappings for Hamiltonian Systems and Their Applications*, Lect. Notes Phys. **691**, 219–254 (2006)
www.springerlink.com   © Springer-Verlag Berlin Heidelberg 2006

10.1.1 Equilibrium Magnetic Field

Let $\mathbf{B}(x, y, z)$ be a magnetic field vector at the spatial point with coordinates (x, y, z). In plasma the magnetic field is determined by the current density $\mathbf{j}(x, y, z)$ and satisfied a divergent-free condition:

$$\nabla \times \mathbf{B} = \mu_o \mathbf{j} \,, \qquad \nabla \cdot \mathbf{B} = 0 \,, \qquad (10.1)$$

where the constant μ_o is the magnetic permeability of free space. The magnetic field lines, $\mathbf{r}(\tau) = (x(\tau), y(\tau), z(\tau))$, are three-dimensional curves tangent to the magnetic vector field $\mathbf{B}$. They are defined as

$$\frac{d\mathbf{r}}{d\tau} = \mathbf{B} \,, \qquad (10.2)$$

where τ is independent parameter related to the length element of curve $ds = (dx^2 + dy^2 + dz^2)^{1/2}$: $d\tau = |\mathbf{B}|^{-1} ds$.

The high temperature plasma can be considered as an ideal gas of fully ionized ions and electrons with the equation of state $p = 2nT$, where p is a pressure, $n \equiv n_e = n_i$ is the density of electrons and ions, and T is a temperature of the plasma. In magnetic fusion devices the plasma is confined by the magnetic field $\mathbf{B}$ in order to isolate it from the walls of the vessel and to hold it in a bounded area. The equilibrium of this system can be maintained when the Ampere force $\mu_o \mathbf{j} \times \mathbf{B}$ acting on the plasma column is balanced with the pressure gradient ∇p: $\mu_o \mathbf{j} \times \mathbf{B} = \nabla p$. From this equation it follows the equations for the *equilibrium magnetic field*

$$\mathbf{B} \cdot \nabla p = 0 \,, \qquad \mathbf{j} \cdot \nabla p = 0 \,. \qquad (10.3)$$

It means that the pressure $p(x, y, z)$ is constant along the magnetic field $\mathbf{B}$ and the current $\mathbf{j}$. Since ∇p is perpendicular to the surface $p(x, y, z) = $ const the field lines lie on the surface of constant pressure $p(x, y, z) = $ const. Surfaces of different constant pressure should not cross each other and they should be bounded in a finite domain. These conditions are satisfied if the surfaces are nested toroidal surfaces as illustrated in Fig. 10.1. Therefore, the field lines of the equilibrium magnetic field cover nested toroidal surfaces, known also *magnetic surfaces*.

10.1.2 Hamiltonian Field Line Equations

The equations of magnetic field lines in toroidal devices can be conveniently represented in the terms of variables, a so-called *toroidal flux*, ψ, and a *poloidal flux*, H, an *intrinsic poloidal angle*, ϑ, and a *toroidal angle*, φ. In terms of these variables a divergence – free magnetic field $\mathbf{B}$ can be always presented in the following form (see, Hinton and Hazeltine (1976); Boozer (1983); Balescu (1988); Boozer (1992, 2004))

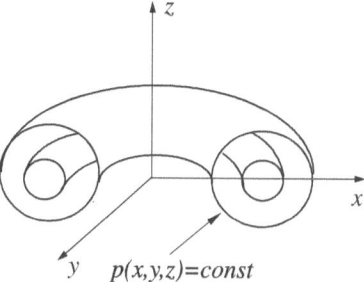

p(x,y,z)=const

Fig. 10.1. Nested magnetic surfaces in a toroidal system

$$\mathbf{B} = \nabla\psi \times \nabla\vartheta + \nabla\varphi \times \nabla H \,, \tag{10.4}$$

known as the *Clebsch representation*. The magnetic field lines, $(\psi(\varphi), \vartheta(\varphi))$, in this representation satisfy the Hamiltonian equation

$$\frac{d\psi}{d\varphi} = -\frac{\partial H}{\partial \vartheta} \,, \qquad \frac{d\vartheta}{d\varphi} = \frac{\partial H}{\partial \psi} \,, \tag{10.5}$$

with (ϑ, ψ) as canonical variables, φ as independent time-like variable. The function $H = H(\vartheta, \psi, \varphi)$ plays the role of a Hamiltonian. It is a 2π-periodic function of ϑ, φ.

The unperturbed case corresponds to the equilibrium magnetic field configuration with nested magnetic surfaces. Physically, the quantity $2\pi\psi$ is the amount of toroidal magnetic flux enclosed by the magnetic surface of constant ψ, while $2\pi H$ is the poloidal magnetic flux outside a constant H surface. It is the function of toroidal flux ψ, $H = H_0(\psi)$. In this case the equations of field lines (10.5) are integrable

$$\psi = \text{const} \,, \qquad \vartheta = \omega(\psi)(\varphi - \varphi_0) \,, \tag{10.6}$$

where

$$\omega(\psi) = \frac{\partial H_0(\psi)}{\partial \psi} = \frac{1}{q(\psi)} \,, \tag{10.7}$$

is the frequency of "motion" known as a *winding number*. The quantity $q(\psi)$ inverse to $\omega(\psi)$, i.e., $q(\psi) = 1/\omega(\psi)$ is known as the *safety factor*. The latter has a meaning of the number of turns along the toroidal angle ϑ per one turn along the poloidal angle φ.

10.1.3 Hamiltonian Formulation of Field Line Equations in a Toroidal System

The plasma column with toroidal magnetic surfaces is called a toroidal plasma system. We consider the toroidal plasma column in the cylindrical coordinate

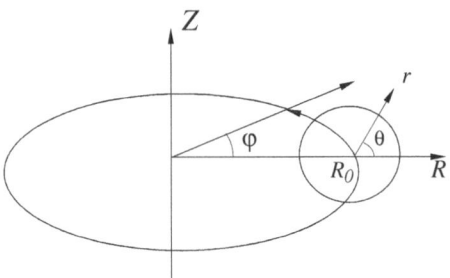

Fig. 10.2. Geometry of a toroidal system

system (R, φ, Z) shown in Fig. 10.2. Usually the radial distance R_0 called a major radius of plasma coincides with the center of plasma, and the angle φ is directed along the plasma column. The equations of magnetic field lines in this system are

$$\frac{1}{R}\frac{dZ}{d\varphi} = \frac{B_Z}{B_\varphi}, \qquad \frac{1}{R}\frac{dR}{d\varphi} = \frac{B_R}{B_\varphi}. \qquad (10.8)$$

The magnetic field **B** can be presented through the vector potential $\mathbf{A}(R, Z, \varphi) = (A_R, A_\varphi, A_Z)$: $\mathbf{B} = \nabla \times \mathbf{A}$. Because of the gauge invariance of the vector potential one can always choose $A_R = 0$. Then one can express the magnetic field through the components of the vector potential,

$$B_R = \frac{1}{R}\frac{\partial A_Z}{\partial \varphi} - \frac{\partial A_\varphi}{\partial Z}, \qquad B_\varphi = -\frac{\partial A_Z}{\partial R}, \qquad B_Z = \frac{1}{R}\frac{\partial R A_\varphi}{\partial R}. \quad (10.9)$$

The A_Z component of the vector potential determines the main toroidal component of the magnetic field B_φ which typically decays inverse proportional to the radial coordinate R: $B_\varphi \propto R^{-1}$. In typical plasmas the deviation of the toroidal field B_φ from this law due to, for instance, a diamagnetic current, is small. For this reason one can neglect a dependence of A_Z on the Z coordinate and the toroidal angle φ. We introduce canonical variables (z, p_z) of the Hamiltonian system related with the geometrical coordinate (R, Z) and the magnetic field, **B**, according to

$$z = \frac{Z}{R_0}, \qquad p_z = \frac{1}{B_0 R_0}\int_{R_0}^{R} B_\varphi dR = -\frac{A_z(R) - A_z(R_0)}{B_0 R_0}. \qquad (10.10)$$

Then the equations for field lines (10.8) can be transformed to the Hamiltonian form

$$\frac{dz}{d\varphi} = \frac{\partial H}{\partial p_z}, \qquad \frac{dp_z}{d\varphi} = -\frac{\partial H}{\partial z}. \qquad (10.11)$$

The variables (z, p_z) are canonical coordinate and momentum, the toroidal angle φ plays the role of a time-like independent variable, and the Hamiltonian function H is determined by the normalized φ- component of the vector potential

$$H \equiv H(z, p_z, \varphi) = -\frac{RA_\varphi}{B_0 R_0^2} \; . \tag{10.12}$$

In the axisymmetric case the magnetic field does not depend on the toroidal angle φ: $A_\varphi = A_\varphi(R, Z)$, and thus $H = H(z, p_z)$. In this case the Hamiltonian system (10.11) is completely integrable. The field lines lie on the nested toroidal surfaces, determined by the surface function $H(z, p_z) = f(Z, R) = $ const. The section of a toroidal surface with the plane $\varphi = $ const is shown in Fig. 10.3.

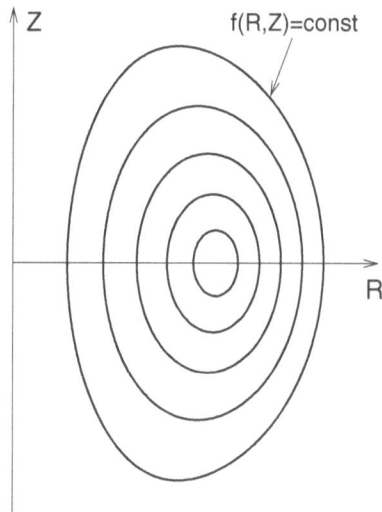

Fig. 10.3. Magnetic flux surfaces $H(z, p_z) = f(Z, R) = $ const

One can introduce the action-angle variables (I, ϑ):

$$I = \frac{1}{2\pi} \oint_C p_z dz \; , \qquad \vartheta = \frac{\partial}{\partial I} \int^z p_z(z', I) dz' \; , \tag{10.13}$$

where the integration is taken along the closed contour C consisting of cross–section of the surface function $f(R, Z) =$const with the poloidal plane $\varphi = $ const (see Fig. 10.3). Actually the action variable I coincides with the normalized toroidal flux ψ. Indeed, according to the definition of variables p_z, and z we have

$$I = \frac{1}{2\pi} \oint_C p_z dz = \frac{1}{2\pi} \int_S dp_z dz$$

$$= \frac{1}{2\pi R_0^2 B_0} \int_S B_\varphi(R, Z) dR dZ = \psi \; , \tag{10.14}$$

which has a meaning of the normalized flux of the toroidal field B_φ through the area S enclosed by the closed contour C on the poloidal plane $\varphi = \text{const}$ (see Fig. 10.2). The angle variable ϑ is no more than the intrinsic poloidal angle. In the action-angle variables (ψ, ϑ) the Hamiltonian $H = H(\psi)$ and the field lines are determined by (10.6), (10.7). The frequency of motion $\omega(\psi) = dH(\psi)/d\psi$ determines the safety factor $q(\psi) = 1/\omega(\psi)$. It can be also found from the equation of field lines (10.11). According to the definition of q it is equal to the number of toroidal turns per one poloidal turn, i.e., $q = \Delta\varphi/2\pi$, where $\Delta\varphi$ is the increment of the toroidal angle φ when field line make one full poloidal turn. Then from the first equation (10.11) it follows that

$$q(\psi) = \frac{\Delta\varphi}{2\pi} = \int_C \frac{dz}{\partial H/\partial p_z} \, , \tag{10.15}$$

where the integral is taken along the closed contour C of $H = H_0(z, p_z) = \text{const}$.

The geometrical coordinates (R, Z) of field lines are periodic functions of the angle variable ϑ: $R(\vartheta, \psi) = R(\vartheta + 2\pi, \psi)$, $Z(\vartheta, \psi) = Z(\vartheta + 2\pi, \psi)$, and they can be presented by Fourier series:

$$Z(\vartheta, \psi) = \sum_m Z_m(\psi)e^{im\vartheta} \, , \qquad R(\vartheta, \psi) = \sum_m R_m(\psi)e^{im\vartheta} \, , \tag{10.16}$$

with coefficients $R_m(\psi), Z_m(\psi)$ depending on a toroidal flux ψ.

In the following subsections we consider some simple equilibrium magnetic configurations and formulate the Hamiltonian equations of field lines.

10.1.4 The Standard Magnetic Field

Consider first the following model of magnetic field

$$\mathbf{B}_0(r, \theta) = B_\varphi \mathbf{e}_\varphi + B_\theta \mathbf{e}_\theta = \frac{B_0}{1 + \varepsilon \cos\theta} \left(\mathbf{e}_\varphi + \mathbf{e}_\theta \frac{\varepsilon}{q(r)} \right) \, , \tag{10.17}$$

known as the *standard magnetic field* (Balescu (1988)). In (10.17) B_0 is the strength of the toroidal magnetic field at magnetic axis $R = R_0$, and $q(r)$ is the safety factor, $\varepsilon = r/R_0$ is the inverse aspect ratio (the ratio R/r is called *aspect ratio*), (r, θ) is polar coordinates in the minor cross section. The radial coordinate r is called a *minor radius*. The relation between the cylindrical coordinates (R, Z) and the toroidal coordinates (r, θ) is $R = R_0 + r\cos\theta, Z = r\sin\theta$ (see Fig. 10.2).

The component of the magnetic field along the toroidal angle φ, B_φ is called a *toroidal field*, and its component along the poloidal angle θ, B_θ is a *poloidal field*. This model describes the main features of equilibrium magnetic field of the toroidal plasma, i.e., the radial decay of the toroidal field $B_\varphi \propto R^{-1}$, and nested, circular magnetic surfaces.

The standard magnetic field (10.17) can be presented through the vector potential $\mathbf{A}$:

$$\mathbf{A}(r,\theta) = (0, A_\varphi(R,Z), A_z(R,Z)) \ ,$$

$$A_\varphi(R,Z) = \frac{B_0}{R} \int \frac{d\phi}{q\,(r(\phi))}, \qquad A_z(R) = -B_0 R_0 \ln R \ , \qquad (10.18)$$

where $\phi = r^2/2 = [(R - R_0)^2 + Z^2]/2$.

The magnetic surfaces of the field (10.17) represent of circular surfaces $r = $ const. According to (10.10) and (10.17) the canonical momentum p_z is

$$p_z = \ln(R/R_0) = \ln(1 + \sqrt{r^2/R_0^2 - z^2}) \ ,$$

and the normalized toroidal flux ψ is given by

$$\psi = \frac{1}{2\pi R_0^2 B_0} \int_{C_r} B_\varphi(r,\theta) r dr d\theta = 1 - \left(1 - r^2/R_0^2\right)^{1/2} \ . \qquad (10.19)$$

According to (10.13) the relation between the intrinsic poloidal angle ϑ and the geometrical poloidal angle θ is determined by the integral

$$\vartheta = \frac{\partial}{\partial \psi} \int_0^z \ln(1 + \sqrt{r^2/R_0^2 - z'^2}) dz' = \frac{1}{2R_0^2} \frac{\partial r^2}{\partial \psi} \int_0^\theta \frac{d\theta}{1 + r \cos\theta/R_0} \ ,$$

the integration of which gives

$$\vartheta = 2 \arctan \frac{\sqrt{1 - r^2/R_0^2} \tan(\theta/2)}{1 + r/R_0} \ . \qquad (10.20)$$

The Hamiltonian $H(\psi)$ is determined by $H(\psi) = \int d\psi/q(r(\psi))$, where $r(\psi) = R_0[1 - (1 - \psi)^2]^{1/2}$.

For the large aspect ratio tokamaks, $\varepsilon = r/R_0 \ll 1$, one can approximate

$$\psi \approx \frac{r^2}{2R_0^2} \ , \qquad \vartheta = \theta - \varepsilon \sin\theta + O(\varepsilon^2) \ . \qquad (10.21)$$

10.1.5 Equilibrium Magnetic Field with the Shafranov Shift

In the standard magnetic field configuration all circular magnetic surfaces $r = $ const have a common center $r = 0$. However, in a real toroidal plasma due to the plasma pressure and the electric current the magnetic surfaces are shifted outwardly along the radial coordinate R and slightly deformed. Schematically it is shown in Fig. 10.4. Here $R_0(r)$ is the position of the center of the magnetic surface of radius r, and a is a radius of the the last

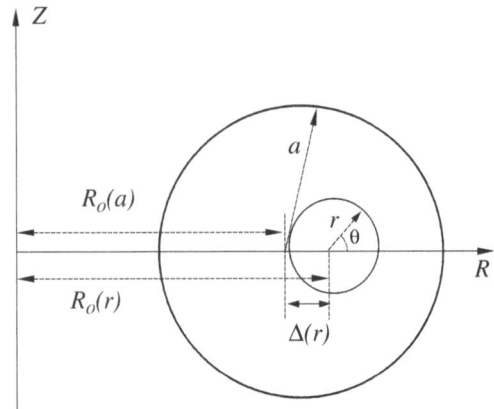

Fig. 10.4. Shifted magnetic flux surfaces in a toroidal plasma

magnetic surface. The shift of $R_0(r)$ with respect $R_0(a)$ is denoted $\Delta(r)$, i.e., $\Delta(r) = R_0(r) - R_0(a)$. This quantity is known as the *Shafranov shift*.

Below we consider the model of equilibrium plasma magnetic field which quite well describes a real plasma with the Shafranov shift. It consists of nested, circular magnetic surfaces. The Shafranov shift $\Delta(r)$ of the magnetic surface of radius r is given by

$$\Delta(r) = \left[R_0^2(a) + (\Lambda + 1)\left(a^2 - r^2\right)\right]^{1/2} - R_0(a) , \qquad (10.22)$$

where $\Lambda = \beta_{pol} + l_i/2 - 1$. Here, β_{pol} is the ratio of the plasma pressure, $\langle p \rangle$ to the magnetic pressure, $B_\theta^2/8\pi$ of the poloidal field B_θ:

$$\beta_{pol} = \frac{8\pi \langle p \rangle}{B_\theta^2} ,$$

l_i is the internal inductance. The toroidal, B_φ, and poloidal, B_θ, fields on each magnetic surface of radius r are given by (see, Kadomtsev (1988))

$$B_\varphi(R) = \frac{\mu_0 I_\varphi}{2\pi R} = \frac{R_0(a)}{R_0(r)} \frac{B_0}{1 + \frac{r}{R_0(r)} \cos\theta} , \qquad B_0 = \frac{\mu_0 I_\varphi}{2\pi R_0(a)} ,$$

$$B_\theta(R, Z) = \frac{\mu_0 I_p(r)}{2\pi r} \left(1 + \Lambda \frac{r}{R_0} \cos\theta\right) , \qquad (10.23)$$

where I_φ is the current of toroidal field, and $I_p(r)$ is the plasma current flowing through the section enclosed by the magnetic surface of radius r, B_0 is the strength of toroidal field at the center of the last magnetic surface, and $R = R_0(r) + r\cos\theta$, $Z = r\sin\theta$.

The equations of field lines are

$$\frac{dr}{R d\varphi} = 0 , \qquad \frac{d\theta}{d\varphi} = \frac{r}{R} \frac{B_\theta(r, \theta)}{B_\varphi(r, \theta)} . \qquad (10.24)$$

From (10.24) follows that the magnetic surfaces are circular $r = $ const. Integrating the second equation in (10.24) over the poloidal angle θ and using that $\vartheta = \varphi/q(r)$, where $q(r)$ is the safety factor, we obtain

$$\varphi - \varphi_0 = \int_0^\theta \frac{r B_\varphi(r, \theta')}{(R_0(r) + r \cos\theta') B_\theta(r, \theta')} d\theta' = q(r)\vartheta \ . \qquad (10.25)$$

Since ϑ is a periodic function of θ with the property $\Delta\vartheta = \vartheta(\theta + 2\pi) - \vartheta(\theta) = 2\pi$, we find the safety factor

$$q(r) = \frac{1}{2\pi} \int_0^{2\pi} \frac{r B_\varphi(r, \theta)}{(R_0(r) + r \cos\theta) B_\theta(r, \theta)} d\theta \ . \qquad (10.26)$$

According to (10.14) one can obtain the following relationships between the normalized toroidal flux ψ and the radius r of a magnetic surface:

$$\psi = \frac{R_0(r)}{R_0(a)} \left[1 - \left(1 - \frac{r^2}{R_0^2(r)} \right)^{1/2} \right] \ . \qquad (10.27)$$

For the small Shafranov shift, $\Delta(r) \ll R_0(a)$, these formulas may be simplified to

$$\psi \approx 1 - \left(1 - \frac{r^2}{R_0^2(a)} \right)^{1/2} , \qquad r \approx R_0(a) \left[1 - (1 - \psi)^2 \right]^{1/2} . \qquad (10.28)$$

The relation between the intrinsic poloidal angle ϑ and the geometrical one θ given by the integral (10.25) is rather complicated (Nguyen et al. (1995)) and it is hard to find the inverse relationship $\theta = \theta(\vartheta, r)$. Below we give a new relation between ϑ and θ as a series in powers of the inverse aspect ratio $\varepsilon = r/R$. The details of calculations is given in Appendix D [see also Abdullaev et al. (1999)].

Expanding the integral in (10.25) in a series of powers of ε the relation $\vartheta = \vartheta(\theta, r)$ may be presented in the form

$$\vartheta(\theta, r) = \theta + \sum_{m=1}^M \alpha_m \sin m\theta + O(\varepsilon^{M+1}) \ , \qquad (10.29)$$

where the expansion coefficients α_m are series in powers of ε:

$$\alpha_m = \sum_{k=0}^M \alpha_m^{(k)} \varepsilon^{m+k} + O(\varepsilon^{M+1}) \ .$$

The coefficients α_m for the case $M = 4$ are given by

$$\alpha_1 = a_1 \varepsilon + \left(\frac{3a_3}{4} - \frac{a_1 a_2}{2} \right) \varepsilon^3 + O(\varepsilon^5) \ ,$$

$$\alpha_2 = \frac{\varepsilon^2}{4}\left[a_2 + \left(a_4 - \frac{a_2^2}{2}\right)\varepsilon^2\right] + O(\varepsilon^5),$$

$$\alpha_3 = \frac{1}{12}a_3\varepsilon^3 + O(\varepsilon^5), \qquad \alpha_4 = \frac{1}{32}a_4\varepsilon^4 + O(\varepsilon^5). \qquad (10.30)$$

The coefficients a_m are polynomial functions of the plasma parameter Λ:

$$a_m = (-1)^m \sum_{k=0}^{m}(m - k + 1)\Lambda^k .$$

The safety factor $q(r)$ may be also presented as a series of powers ε:

$$q(r) = \frac{r^2}{R_0^2(r)}\frac{I_\varphi}{I_p}\left(1 + \frac{1}{2}a_2\varepsilon^2 + \frac{3}{8}a_4\varepsilon^4\right) + O(\varepsilon^8) . \qquad (10.31)$$

The relation (10.29), expressing the intrinsic poloidal angle ϑ via the geometrical poloidal angle θ, can be inverted to find θ in terms of ϑ,

$$\theta(\vartheta, r) = \vartheta + \sum_{m=1}^{M} \alpha_m^* \sin m\vartheta + O(\varepsilon^{M+1}) . \qquad (10.32)$$

The expansion coefficients α_m^* can be expressed in terms of coefficients α_m. They are calculated in Appendix D for the case $M = 4$.

A typical dependence ϑ on θ is shown in Fig. 10.5a at the magnetic surface $r = 0.43$ m, and the radial profile of the safety factor $q(r)$ is plotted in Fig. 10.5b. The plasma parameters are chosen: $\beta_{pol} = 1$, $l_i = 1.2$ and $R_0(a) = 1.75$ m. The radial profile of the plasma current density is chosen as $j(r) = (2I_p/\pi a^2)(1 - r^2/a^2)$ where I_p is the full plasma current. The toroidal field $B_\varphi = 2.25$ T.

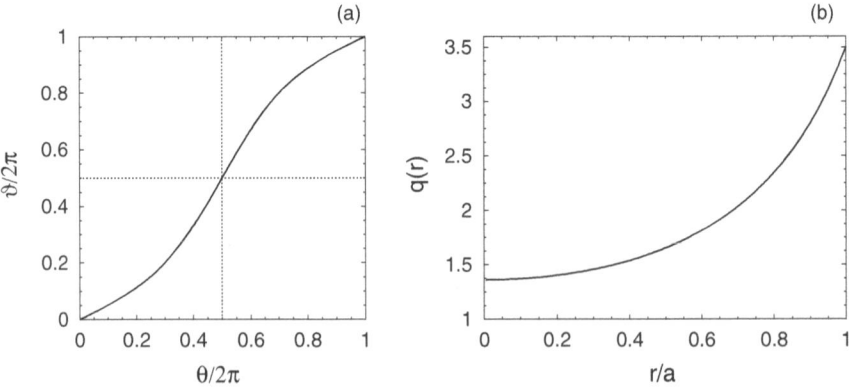

Fig. 10.5. (a) Dependence of the intrinsic angle ϑ on the poloidal angle θ; (b) Radial profile of the safety factor $q(r)$

The accuracy of the relations (10.29) and (10.32) is sufficiently high. At $M = 4$ they deviate from the exact formulas by less than 1%. However, the higher order derivatives, $d^k \vartheta / d\theta^k$ $(k > 2)$, calculated from these formulas are less accurate because of divergence of series.

10.2 Hamiltonian Equations in the Presence of Magnetic Perturbations

In real tokamak plasmas magnetic field deviates from the ideal equilibrium field. Typically the deviation of magnetic field from the background equilibrium field called the *magnetic field perturbation* is small. The physical nature of magnetic perturbations is diverse, and it ranges from the error fields produced by imperfect technical installation of poloidal and toroidal field coils, helical magnetic fields generated by the *magneto hydrodynamic (MHD)* instabilities of plasmas, up to the artificially created magnetic fields by external coils as in so-called ergodic divertors (see Chap. 11).

The magnetic perturbations, in general, are not uniform along the toroidal, φ, and poloidal, θ, axes and they break the symmetry of the equilibrium field along the toroidal axis φ. In the presence of these *non-axisymmetric magnetic perturbation* the poloidal flux H can be presented as a sum of the unperturbed flux $H_0(\psi)$ and the perturbed flux $\epsilon H_1 = \epsilon H_1(\psi, \vartheta, \varphi)$ depending on the poloidal and toroidal angles:

$$H = H_0(\psi) + \epsilon H_1(\psi, \vartheta, \varphi) \, , \qquad H_0(\psi) = \int \frac{d\psi}{q(\psi)} \, . \qquad (10.33)$$

The dimensionless perturbation parameter ϵ introduced in (10.33) stands for the relative strength of magnetic perturbations. Since perturbed Hamiltonian H_1 is a 2π periodic function of ϑ, φ, it can be always presented as a Fourier series:

$$H_1(\psi, \vartheta, \varphi) = \sum_{m,n} H_{mn}(\psi) \cos(m\vartheta - n\varphi + \chi_{mn}) \, . \qquad (10.34)$$

The integer numbers m and n are called the *poloidal* and *toroidal mode numbers*, respectively.

In typical situations when the safety factor $q(\psi)$ (or the winding number $\omega(\psi)$) satisfies the twist condition (7.3), $dq/d\psi \neq 0$, the behavior of toroidal magnetic surfaces in the presence of magnetic perturbations is described by the KAM theory (see Sect. 7.1). The resonant magnetic surfaces, $\psi_{m,n}$, defined by

$$q(\psi_{m,n}) = \frac{m}{n} \, , \qquad (10.35)$$

are destroyed by any small perturbation forming chain of magnetic islands. The width of magnetic islands, $W_{m,n}$, are determined by (7.12). The majority of sufficiently irrational magnetic surfaces ψ, satisfying the condition

$|q^{-1}(\psi) - n/m| > C\sqrt{\epsilon}m^{-\nu}$, are survived, but they are slightly deformed. With increasing the magnitude of magnetic perturbations, ϵ, the neighboring resonance magnetic surfaces start to interact destroying invariant magnetic surfaces between them. It leads to the global chaos and unrestricted diffusion of field lines which guides the particles out of the bounded area to the vessel wall. This phenomenon known also as a *magnetic stochasticity* is one of the main undesirable physical processes responsible for breaking the plasma confinement.

On the other hand the phenomenon of magnetic stochasticity can be applied in magnetically confined plasma devices to control a particle and energy transport at the plasma edge. One of these applications is the *ergodic divertor* concept introduced to divert particles and a heat releasing from the plasma to special plates in a controlled way. We study this artificially created magnetic stochasticity in Chap. 11.

In the pioneering studies of destruction of magnetic surfaces in magnetic fusion devices by Kerst (1962); Rosenbluth et al. (1966); Filonenko et al. (1967); Freis et al. (1973); Hamzeh (1974); Finn (1975); Matsuda and Yoshikawa (1975) the onset of global chaos of field lines has been investigated using the Chirikov criteria of overlapping of resonances (7.29). The most convenient and computationally efficient method to study the dynamics of magnetic field lines is symplectic mappings. In Sect. 10.4 we will discuss the most prominent mappings used to model a magnetic stochasticity in tokamak plasmas.

The study of magnetic stochasticity can be significantly simplified if the equations of field lines in the presence of magnetic perturbations are reduced to the Hamiltonian form (10.33), (10.34). It would allow one to apply directly the qualitative and quantitative methods of Hamiltonian dynamics and chaos to the problem. Particularly, the Chirikov criteria can be used to obtain qualitative estimations for the conditions of formation of magnetic stochasticity. The computationally effective symplectic mappings can be directly applied to integrate the Hamiltonian system (10.33), (10.34).

One of the main problems in the Hamiltonian formulation of field line equations is to find the spectrum of the perturbed Hamiltonian, $H_{mn}(\psi)$, through magnetic perturbation, $\mathbf{B}^{per}$. Below we consider this problem for the cylindrical and toroidal plasma models. Particularly, we will study a generic asymptotic behavior of the spectrum of perturbations, $H_{mn}(\psi)$, for the large poloidal mode numbers m in toroidal plasmas.

10.2.1 Cylindrical Model of Plasmas

The tokamak plasma with the large aspect ratio $R/r \gg 1$ can be modeled by the cylindrical model. In this model the plasma column is straight along the z coordinate and confined along the radial coordinate r. The equilibrium magnetic field of the cylindrical plasma is given by

$$\mathbf{B}_0(r) = \mathbf{e}_\theta B_\theta(r) + \mathbf{e}_z B_0 , \tag{10.36}$$

where $\mathbf{e}_\theta$, $\mathbf{e}_z$ are unit vectors in the cylindrical coordinate system (r, θ, z). Here B_0 is a main component of the magnetic field corresponding to the toroidal field, $B_\theta(r)$ is a poloidal field.

Let $\mathbf{B}_1(r, \theta, z)$ be a non-axisymmetric magnetic perturbation. We impose a periodic boundary conditions on this field: $\mathbf{B}_1(r, \theta, z) = \mathbf{B}_1(r, \theta, z + 2\pi R)$ with the period $L = 2\pi R$ along the z-axis. The full magnetic field then can be presented in the form

$$\mathbf{B} = \mathbf{B}_0(r) + \mathbf{B}_1(r, \theta, z) .$$

In tokamaks the z-component of the unperturbed field B_0 is much larger larger than the poloidal field, $B_\theta(r)$ and the perturbation field, $\mathbf{B}_1$: $B_z \gg B_\theta(r)$, $B_z \gg |B_1|$. In typical cases the perturbation field can be presented through the $z-$ component of the vector potential $A_z(r, \theta, z)$ $(\mathbf{B} = \nabla \times \mathbf{A})$:

$$\mathbf{B}_1(r, \theta, z) = -\mathbf{e}_r \frac{1}{r} \frac{\partial A_z}{\partial \theta} + \mathbf{e}_\theta \frac{\partial A_z}{\partial r} . \tag{10.37}$$

Because of the periodicity of magnetic perturbations along the z and θ axes one can present the vector potential A_z in a Fourier series:

$$A_z(r, \theta, z) = \sum_{m,n} A_{m,n}(r) \cos(m\theta - n\varphi + \chi_{mn}) , \tag{10.38}$$

where $\varphi = z/R$, and χ_{mn} are phases.

The equations of field lines, $(r(z), \theta(z))$, in the cylindrical coordinate system, given by

$$\frac{dr}{dz} = \frac{B_r}{B_z} , \qquad\qquad \frac{d\theta}{dz} = \frac{B_\theta}{B_z} , \tag{10.39}$$

can be transformed into the Hamiltonian form (10.5) with the Hamiltonian (10.33), (10.34) by introducing the toroidal flux $\psi = r^2/2R^2$, the intrinsic poloidal angle $\vartheta = \theta$, the toroidal angle $\varphi = z/R$. The safety factor q and the perturbed Hamiltonian are determined by

$$q(\psi) = \frac{r(\psi)B_0}{RB_\theta(r(\psi))} , \qquad r(\psi) = R\sqrt{2\psi},$$

$$\epsilon H_1(\psi, \vartheta, \varphi) = -\frac{1}{RB_0} A_z(r(\psi), \vartheta, R\varphi) , \tag{10.40}$$

According to (10.38) and (10.40) the spectrum of perturbations, $H_{mn}(\psi)$, are determined by the Fourier coefficients $A_{mn}(r)$ of the vector potential A_z:

$$\epsilon H_{mn}(\psi) = -\frac{1}{RB_0} A_{mn}(r(\psi)) . \tag{10.41}$$

10.2.2 Magnetic Perturbations in Toroidal Plasmas

In the presence of non-axisymmetric magnetic perturbations the toroidal component of the vector potential $A_\varphi^{(per)}(R, Z, \varphi)$ can be presented as a sum:

$$A_\varphi = A_\varphi^{(0)}(R, Z) + A_\varphi^{(per)}(R, Z, \varphi) \,. \qquad (10.42)$$

In typical situations the perturbed part of the component A_Z is small and it can be neglected in comparison with the unperturbed part $A_Z^{(0)}(R)$ which determines the toroidal magnetic field B_φ. Then the magnetic perturbation $\mathbf{B}^{(per)}(R, Z, \varphi)$ has only two nonzero components B_R and B_Z which, according to (10.9), are expressed via the perturbed part of the vector potential $A_\varphi^{(per)}(R, Z, \varphi)$:

$$\mathbf{B}^{(per)}(R, Z, \varphi) = \left(-\mathbf{e}_R \frac{1}{R} \frac{\partial}{\partial Z} + \mathbf{e}_Z \frac{1}{R} \frac{\partial}{\partial R} \right) R A_\varphi^{(per)}(R, Z, \varphi) \,. \quad (10.43)$$

Introducing the toroidal flux ψ and the intrinsic poloidal angle ϑ (10.13) (or the action-angle variables) for the equilibrium magnetic field, the Hamiltonian equations for perturbed field lines (10.11), (10.12) can be presented as

$$\frac{d\vartheta}{d\varphi} = \frac{1}{q(\psi)} + \epsilon \frac{\partial H_1}{\partial \psi}, \qquad \frac{d\psi}{d\varphi} = -\epsilon \frac{\partial H_1}{\partial \vartheta} \,, \qquad (10.44)$$

with the perturbed Hamiltonian $\epsilon H_1 \equiv H_1(\psi, \vartheta, \varphi)$ given by

$$H_1(\psi, \vartheta, \varphi) = -\frac{R(\psi, \vartheta)}{R_0^2 B_0} A_\varphi^{(per)}(R(\psi, \vartheta), Z(\psi, \vartheta), \varphi) \,, \qquad (10.45)$$

where $R(\psi, \vartheta) = R_0 + r(\psi, \vartheta) \cos\theta(\psi, \vartheta)$.

The Fourier components $H_{mn}(\psi)$ in (10.34) of the perturbation (10.45) are found by the Fourier integrals:

$$\epsilon H_{mn}(\psi) = -\mathrm{Re} \int_0^{2\pi} \int_0^{2\pi} \frac{R(\psi, \vartheta)}{(2\pi)^2 R_0^2 B_0}$$
$$\times A_\varphi^{(per)}(R(\psi, \vartheta), Z(\psi, \vartheta), \varphi) e^{-im\vartheta + in\varphi} d\vartheta d\varphi \,. \qquad (10.46)$$

In a toroidal system any magnetic perturbations are periodic along the toroidal, φ, and the poloidal, θ, angles, and thus the perturbed vector potential, $A_\varphi^{(per)}$ can be expanded in a Fourier series

$$A_\varphi^{(per)}(R, Z, \varphi) = \sum_{m,n} A_{mn}(r) \cos(m\theta - n\varphi + \chi_{mn}) \,, \qquad (10.47)$$

where $A_{mn}(r)$ are radial dependent Fourier coefficients. In the case of the tokamak plasma with circular magnetic surfaces the radial coordinate r coincides with the radius of magnetic surfaces. In general plasma configurations the coordinate r is a radius of the magnetic surface averaged over the poloidal angle θ. It is a function of the poloidal flux ψ, $r = r(\psi)$, while the poloidal angle $\theta = \theta(\vartheta, \psi)$.

Introducing the inverse aspect ratio $\varepsilon = r(\psi)/R_0$ and the dimensionless perturbation parameter $\epsilon = \max |A_\varphi^{(per)}|/(R_0 B_0)$ one can present R as $R = R_0(1 + \varepsilon(\psi)\cos\theta(\vartheta, \psi))$ and $A_{mn}(r) = \epsilon R_0 B_0 a_{mn}(\psi)$, where a_{mn} are coefficients of order of 1. Then using the expansion (10.47) one can transform the coefficients (10.46) into

$$H_{mn}(\psi) = \sum_{m'} S_{mm'}(\psi) a_{m'n}(\psi) , \qquad (10.48)$$

where $S_{mm'}(\psi)$ is a transformation matrix defined by

$$S_{mm'}(\psi) = -\mathrm{Re}\frac{1}{2\pi} \int_0^{2\pi} [1 + \varepsilon\cos\theta(\vartheta, \psi)] e^{im'\theta(\vartheta, \psi) - im\vartheta} d\vartheta . \quad (10.49)$$

These Fourier integrals depend on the relation between the poloidal angle θ and the intrinsic poloidal angle ϑ. For the large aspect ratio tokamaks, $\varepsilon \ll 1$, with circular magnetic surfaces $r = \mathrm{const}$, this relation can be approximated by (10.21): $\theta \approx \vartheta - \varepsilon\sin\vartheta$. Then the diagonal terms $S_{mm}(\psi)$ have an order 1, and the non-diagonal terms $S_{mm\pm k}(\psi)$, $(k = 1, 2, \cdots)$ have an order of ε^k. Therefore, the main contribution to $H_{mn}(\psi)$ comes from the harmonics A_{mn} of the magnetic field with the same poloidal mode m, while the contribution to this poloidal mode m from the sideband modes $A_{m\pm k,n}$ decreases as ε^k.

However, this commonly accepted feature of mode transformation matrix, $S_{mm'}(\psi)$, is valid only for small poloidal mode numbers m, m'. As was established in Abdullaev et al. (1999) for large mode numbers m, m' the matrix $S_{mm'}(\psi)$ may grow with m' at the fixed m. It means that the main contributions to $H_{mn}(\psi)$ may come not from the poloidal mode m and neighboring modes $m' = m \pm 1$ of magnetic field perturbations, but from those modes m' located far from m. Below we study these new features of the mode transformation matrix $S_{mm'}(\psi)$ using the method of asymptotic estimations of the integral (10.49). Details of calculations are given in Appendix E.

10.2.3 Asymptotics of the Transformation Matrix Elements $S_{mm'}(\psi)$

First of all we recall a typical feature of the dependence $\theta = \theta(\vartheta) \equiv \theta(\vartheta, \psi)$. According to a definition we have $\theta(\vartheta = 0) = 0$ (the high field side) and $\theta(\vartheta = \pi) = \pi$ (the low field side) (see Fig. 10.5a). These points are the critical points of the function $\theta = \theta(\vartheta)$. At these points the second derivatives $d^2\theta/d\vartheta^2$ vanish. As we will see below that the asymptotics of the integral (10.49) at large m, m' is mainly determined by the behavior of the function $\theta = \theta(\vartheta)$ near these points, namely by the first and third derivatives at the points $\vartheta = 0$ and $\vartheta = \pi$:

$$\gamma_1 = \frac{d\theta}{d\vartheta}\Big|_{\theta=0} , \qquad \gamma_3 = \frac{d^3\theta}{d\vartheta^3}\Big|_{\theta=0} ,$$

$$\beta_1 = \frac{d\theta}{d\vartheta}\Big|_{\theta=\pi} , \qquad \beta_3 = \frac{d^3\theta}{d\vartheta^3}\Big|_{\theta=\pi} . \qquad (10.50)$$

One should note that $\gamma_1 > 1$, $\gamma_3 < 0$, $0 < \beta_1 < 1$, and $\beta_3 > 0$. The coefficients γ_1 and β_1 determine the tangents of the curves θ versus ϑ at the low field side and at the high field side, respectively (see Fig. 10.5a).

It is shown in Appendix E that for fixed values of m the main contribution to the integral (10.49) comes from two intervals of m'. For small values of m'

$$m' \leq m/\gamma_1 + x_c(m'|\gamma_3|/2)^{1/3} , \qquad (x_c \approx 2)$$

the integral $S_{mm'}(\psi)$ may be expressed as a product

$$S_{mm'}(\psi) = f(0)A_{mm'}(\psi), \qquad f(0) = 1 + \varepsilon , \qquad (10.51)$$

where $f(\vartheta) = 1 + \varepsilon\cos\theta(\vartheta)$ and $A_{mm'}(\psi)$ is the the Fourier integral having the following asymptotics at large m', m:

$$A_{mm'}(\psi) = \frac{1}{2\pi}\int_0^{2\pi} e^{i(m'\theta(\vartheta,\psi)-m\vartheta)}d\vartheta$$
$$\approx \left(\frac{2}{|\gamma_3|m'}\right)^{1/3}\mathrm{Ai}\left(-\frac{\gamma_1 m' - m}{(|\gamma_3|m'/2)^{1/3}}\right) , \qquad (10.52)$$

expressed via the Airy function $\mathrm{Ai}(z)$. The Airy function $\mathrm{Ai}(x)$ oscillates for $x < 0$ and exponentially decays for $x > 0$. It has a local maximum at $x_c \approx -1$. Similarly, for the values of m',

$$m' > m/\beta_1 - x_c(m'\beta_3/2)^{1/3} , \qquad (x_c \approx 3)$$

the quantity $S_{mm'}(\psi)$ may be approximated by the product

$$S_{mm'}(\psi) = f(\pi)A_{mm'}(\psi) , \qquad f(\pi) = 1 - \varepsilon , \qquad (10.53)$$

with the following asymptotics

$$A_{mm'}(\psi) \approx (-1)^{m+m'}\left(\frac{2}{\beta_3 m'}\right)^{1/3}\mathrm{Ai}\left(\frac{\beta_1 m' - m}{(\beta_3 m'/2)^{1/3}}\right) . \qquad (10.54)$$

For the intermediate values of m': $m/\gamma_1 + x_c(m'|\gamma_3|/2)^{1/3} < m' < m/\beta_1 + x_c(m'\beta_3/2)^{1/3}$ unlike (10.52) and (10.54) the integral (10.49) is proportional to $1/\sqrt{m'}$. Contributions from these terms may be neglected due to the rapid oscillations of $A_{mm'}$ with m'.

An example of the dependence of the transformation matrix $S_{mm'}(\psi)$ on m' at the fixed value $m = 12$ is shown in Fig. 10.6 for the plasma model considered in Sect. 10.1.5 with the parameter $\Lambda = 1$ and the minor radius

$r = 0.43\,\mathrm{m}$. The other plasma parameters are the same as in Fig. 10.5. The derivatives in (10.50) are $\beta_1 = 0.560678$, $\beta_3 = 0.142254$, $\gamma_1 = 2.035916$, and $\gamma_3 = -4.198625$. Figure 10.6a shows $S_{mm'}(\psi)$ itself, and Fig. 10.6b shows it after multiplication by $(-1)^{m'}$. The solid curves correspond to the exact numerical evaluation the integral (10.49), and the dashed curves describe their asymptotics by the Airy functions. As seen from Figs. 10.6a,b the asymptotic formulas well describe the transformation matrix $S_{mm'}(\psi)$ near the values $m' \approx m/\gamma_1$ and $m' \approx m/\beta_1$.

As seen from Figures 10.6a, b the main contribution to $H_{mn}(\psi)$ with the fixed m comes from a few magnetic perturbation modes $a_{m'n}$ located near $m' \approx m/\gamma_1$ or $m' \approx m/\beta_1$. The contributions from other modes are negligible because of rapid oscillations of the transformation matrix $S_{mm'}(\psi)$. Below we analyze this phenomenon.

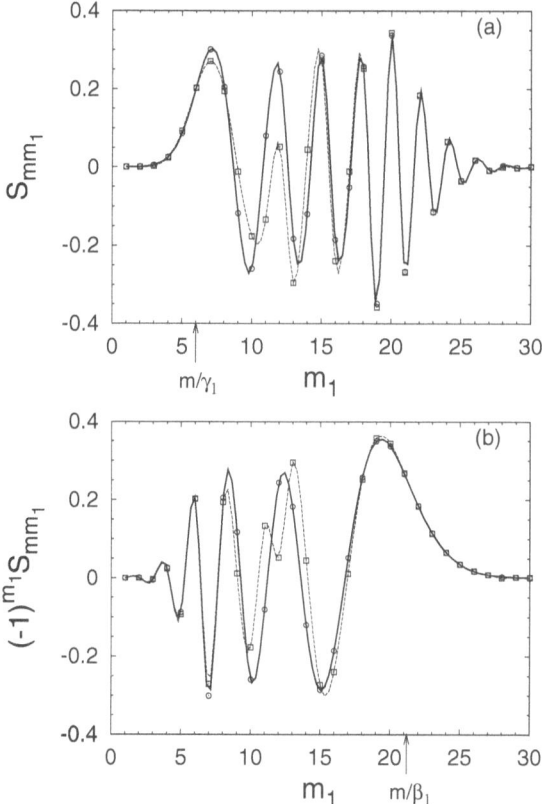

Fig. 10.6. Transformation integral (10.49) S_{mm_1} as a function of m_1 for the fixed $m = 12$: (a) S_{mm_1}; (b) $(-1)^{m_1} S_{mm_1}$. *Solid curve* describes the exact numerical integration, *dashed curve* corresponds to the asymptotics given by (10.52), (10.54). The plasma parameter are the same as in Fig. 10.5

10.2.4 Asymptotic Behavior of $H_{mn}(\psi)$

Using the asymptotic formulae (10.52) and (10.54), the sum (10.48) can be presented as a sum of two terms,

$$H_{mn}(\psi) = f(0)h_{mn}^{(l)}(\psi) + f(\pi)h_{mn}^{(h)}(\psi) , \qquad (10.55)$$

where

$$h_{mn}^{(l)}(\psi) \approx \sum_{m_{min}}^{m} a_{m'n}(\psi)\left(\frac{2}{|\gamma_3|m'}\right)^{1/3} \mathrm{Ai}\left(-\frac{\gamma_1 m' - m}{(|\gamma_3|m'/2)^{1/3}}\right)dm' , \qquad (10.56)$$

$$h_{mn}^{(h)}(\psi) \approx (-1)^m \sum_{m}^{m_{max}} (-1)^{m'} a_{m'n}(\psi)\left(\frac{2}{\beta_3 m'}\right)^{1/3} \mathrm{Ai}\left(\frac{\beta_1 m' - m}{(\beta_3 m'/2)^{1/3}}\right)dm' . \qquad (10.57)$$

The sums (10.56), (10.57) for $H_{mn}(\psi)$ significantly depend on how the sign of the amplitude, $a_{m'n}$ changes with m'.

When $a_{m'n}$ changes with m' smoothly, the term $h_{mn}^{(h)}(\psi)$ (10.57) can be neglected because of rapid oscillations of terms under the sum. Taking into account that the parameter $|\gamma_3| \ll 1$ we replace the summation in (10.56)) over m' by integration by introducing a new variable

$$x = \frac{\gamma_1 m' - m}{|\gamma_3|m'/2} \approx \frac{\gamma_1 m' - m}{(|\gamma_3|m/2)^{1/3}} .$$

Equation (10.55) can be written as the integral

$$H_{mn}(\psi) \approx \frac{f(0)}{\gamma_1} \int_{-\infty}^{\infty} \mathrm{Ai}(x)\, a\left(\frac{m - x(|\gamma_3|m/2)^{1/3}}{\gamma_1}\right) dx , \qquad (10.58)$$

where $a(m) \equiv a_{mn}$. Since the function $a(m)$ in (10.58) is a smooth function of m, the integral can be estimated by the Laplace method (see, e.g., Fedoryuk (1989)). The integration yields

$$H_{mn}(\psi) \approx \frac{f(0)}{\gamma_1} B a\left(\frac{m - x_c(|\gamma_3|m/2)^{1/3}}{\gamma_1}\right) , \qquad (10.59)$$

where $B = \sqrt{2\pi/|x_c|}\,\mathrm{Ai}(x_c)$, and $x_c \approx -1$ is the local maximum of the Airy function.

Consider the case when the magnetic perturbation amplitude, $a_{m'n}$, changes rapidly with m' as $a_{m'n} = (-1)^{m'}\tilde{a}(m')$, where $\tilde{a}(m')$ is a smooth function of m'. In this case the first term in (10.55) can be neglected because of rapid oscillations in the sum (10.56). Then main contribution to $H_{mn}(\psi)$

comes the second term (10.57) which can be estimated similarly to the above case. The corresponding asymptotic formula for $H_{mn}(\psi)$ is

$$H_{mn}(\psi) \approx (-1)^m \frac{f(\pi)}{\beta_1} B \, \tilde{a} \left(\frac{m + x_c(\beta_3 m/2)^{1/3}}{\beta_1} \right) . \qquad (10.60)$$

Asymptotic formulas (10.59) and (10.60) qualitatively describe the main features of conversations of magnetic perturbations in toroidal plasmas. They can be also used for quantitative estimations of $H_{mn}(\psi)$. To improve their accuracy the two parameters B and x_c in formulas (10.59) and (10.60) can be considered as fitting parameters and chosen to have a better agreement with the numerically calculated values of $H_{mn}(\psi)$.

To illustrate the mode transformation consider the following example when the magnetic perturbation modes, a_{mn}, are given by

$$a_{mn} = C_m \frac{\sin(m - m_0)\theta_c}{\pi m(m - m_0)} . \qquad (10.61)$$

We consider two case type of coefficients C_m: (i) $C_m = (-1)^m$ and (ii) $C_m = 1$. The modes are localized near the central mode with a width $\Delta m = 2\pi/\theta_c$. Suppose that the equilibrium magnetic field is described by the plasma model considered in Section (10.1.5). The plasma parameters are the same as in Figures 10.5 and 10.6.

As seen from Fig. 10.7a in the first case (i) the center of the mode distribution m_0 of magnetic mode perturbations, a_{mn}, (solid curve 1) is shifted to the lower mode $m_0^* \approx m_0\beta_1$ (solid curve 2 and dashed curve 3). The width of the distribution $\Delta m = 2\pi/\theta_c$ of magnetic perturbations became narrow: $\Delta m^* \approx \beta_1 2\pi/\theta_c$. In the second case (ii) the central mode number m_0 of magnetic perturbations is shifted to higher mode $m_0^* \approx m_0\gamma_1$ and the distribution became wider: $\Delta m^* \approx \gamma_1 2\pi/\theta_c$. The asymptotical formulae (10.59), (10.60) qualitatively correct describe the transformation of magnetic perturbations into Hamiltonian perturbations.

From the asymptotic formulae (10.59), (10.60) it follows that a particular mode m of Hamiltonian perturbations is determined by the mode number $m' \approx m/\gamma_1$ (or $m' \approx m/\beta_1$) of magnetic perturbations. They reflect the specific features of mode conversations between the spectra of Hamiltonian perturbation $H_{mn}(\psi)$ and of magnetic perturbations is toroidal plasmas. We will use these features in Chap. 11 to analyze the spectrum of poloidal modes in ergodic divertor tokamaks.

10.3 Mapping of Field Lines

A symplectic integration of Hamiltonian equations (10.5) with the Hamiltonian (10.33) can be performed using the mapping method presented in Sect. 4.1. The geometry of the mapping of field lines in toroidal plasmas

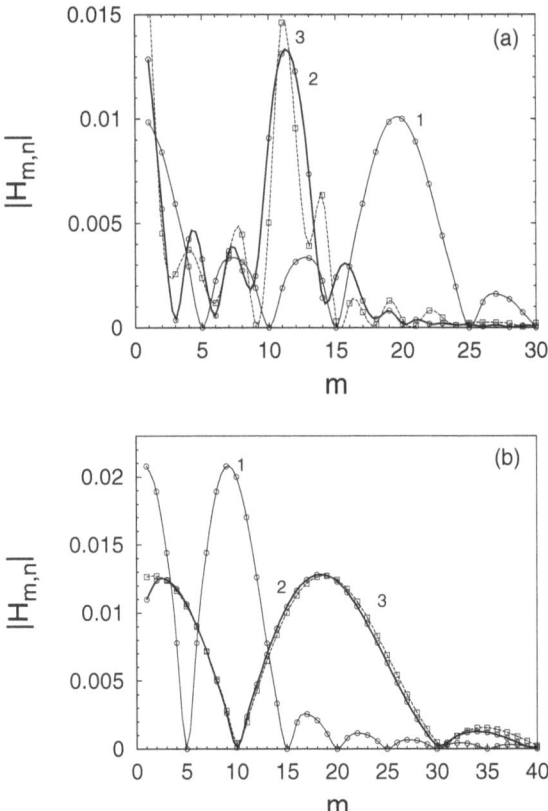

Fig. 10.7. Spectrum of magnetic a_{mn} (curve 1) and Hamiltonian perturbations $H_{mn}(\psi)$ obtained from (10.48) by numerical integration (10.49) (*solid curve* 2) and using the asymptotics (10.60) or (10.59) (*dashed curve* 3): (**a**) describes the case $C_m = (-1)^m$, $m_0 = 20$, $\theta_c = \pi/5$, fitting parameters: $B = 1.3$, $x_c = -1.$; (**b**) the case (*ii*) $C_m = 1$, $m_0 = 10$, $\theta_c = \pi/5$. Fitting parameters: $B = 1$, $x_c = 0$

is illustrated in Fig. 10.8. The mapping of field lines in tokamaks is constructed by the following way. Introduce poloidal sections $\varphi = \varphi_k = (2\pi/s)k$, ($k = 0, \pm 1, \pm 2, \ldots$), where s ($s \geq 1$) is an integer number which stands for a number of map steps per one toroidal rotation along the torus.

The mapping

$$(\vartheta_{k+1}, \psi_{k+1}) = \hat{M}(\vartheta_k, \psi_k) \,, \qquad (10.62)$$

relates the points, (ϑ_k, ψ_k), of the field line $(\vartheta(\varphi), \psi(\varphi))$ at the poloidal section $\varphi = \varphi_k$, with the ones $(\vartheta_{k+1}, \psi_{k+1})$ at $\varphi = \varphi_{k+1}$. The Poincaré return map to the same poloidal section φ_k (mod 2π) can be obtained by applying the map (10.62) s times or setting $s = 1$ (see Fig. 10.8b).

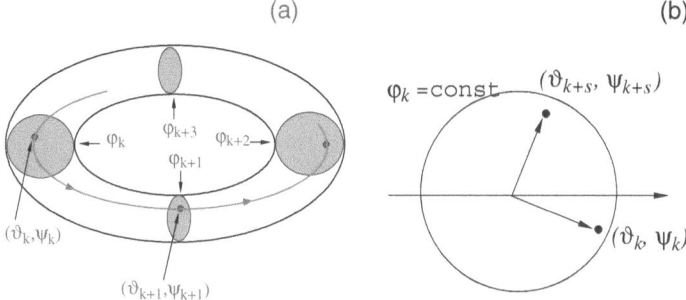

Fig. 10.8. Geometry of mapping in a toroidal system: (**a**) Scheme of the mapping along the toroidal angle for the case $s = 4$; (**b**) Poincaré return map of field lines in the poloidal plane $\varphi_k = \text{const}$

According to (4.6)–(4.8) the general form of the mapping (10.62) for the Hamiltonian system (10.33), (10.34) has the following symmetric flux-preserving form

$$\Psi_k = \psi_k - \epsilon \frac{\partial S_k}{\partial \vartheta_k} \, , \qquad \Theta_k = \vartheta_k + \epsilon \frac{\partial S_k}{\partial \Psi_k} \, ,$$

$$\Psi_{k+1} = \Psi_k \, , \qquad \Theta_{k+1} = \Theta_k + \frac{\varphi_{k+1} - \varphi_k}{q(\Psi_k)} \, , \qquad (10.63)$$

$$\psi_{k+1} = \Psi_{k+1} + \epsilon \frac{\partial S_{k+1}}{\partial \vartheta_{k+1}}, \quad \vartheta_{k+1} = \Theta_k - \epsilon \frac{\partial S_{k+1}}{\partial \Psi_{k+1}} \, ,$$

with the generating function $S_k = S(\vartheta, \Psi, \varphi; \epsilon)|_{\varphi = \varphi_k}$. In the first order of perturbation parameter ϵ the generating function, $S(\vartheta, \Psi, \varphi; \epsilon)$, according to (2.10), and (2.35) is given by

$$S(\vartheta, \Psi, \varphi) = -(\varphi - \varphi_0) \sum_m H_{mn}(\Psi) \big[a(x_{mn}) \sin{(m\vartheta - n\varphi + \chi_{mn})}$$
$$+ b(x_{mn}) \cos{(m\vartheta - n\varphi + \chi_{mn})} \big] \, , \qquad (10.64)$$

where the toroidal angle φ is located in the interval $\varphi_{k+1} < \varphi < \varphi_k$, and

$$a(x) = \frac{1 - \cos x}{x} \, , \qquad b(x) = \frac{\sin x}{x} \, ,$$

$$x_{mn} = \left(\frac{m}{q(\Psi)} - n \right)(\varphi - \varphi_0) \, .$$

We recall that the free parameter φ_0 lies in the interval $\varphi_k \leq \varphi_0 \leq \varphi_{k+1}$, and the mapping can be applied to systems with moderately large perturbation ϵ by taking the map step $\Delta\varphi = \varphi_{k+1} - \varphi_k$ sufficiently small.

As was shown in Sect. 4.1.1 by appropriately choosing the parameter φ_0 one can obtain the nonsymmetric forms of the mapping. Particularly, taking

$\varphi_0 = \varphi_{k+1}$ and $\Delta\varphi = \vartheta_{k+1} - \vartheta_k = 2\pi$ we obtain the Poincaré return map in the form

$$\psi_{k+1} = \psi_k - \epsilon \frac{\partial S(\vartheta_k, \psi_{k+1})}{\partial \vartheta_k} \ ,$$

$$\vartheta_{k+1} = \vartheta_k + \frac{2\pi}{q(\psi_{k+1})} + \epsilon \frac{\partial S(\vartheta_k, \psi_{k+1})}{\partial \psi_{k+1}} \ , \tag{10.65}$$

where $S(\vartheta_k, \psi_{k+1}) \equiv S(\vartheta_k, \psi_{k+1}, \varphi_k, \varphi_k + 2\pi, \epsilon)$. According to (10.64), in the first order of ϵ the generating function $S(\vartheta_k, \psi_{k+1})$ is determined by

$$S(\vartheta, \psi) = 2\pi \sum_m H_{mn}(\psi) \left[a(x_{mn}) \sin\left(m\vartheta + \chi'_{mn}\right) + b(x_{mn}) \cos\left(m\vartheta + \chi'_{mn}\right) \right] \ ,$$
$$\tag{10.66}$$

where $x_{mn} = 2\pi \left(m/q(\psi) - n \right)$, and $\chi'_{mn} = \chi_{mn} - n\varphi_k$.

The mapping of the form (10.65) is widely used in the literature referring to it as the *perturbed twist map*, when the safety factor $q(\psi)$ is a monotonic function of ψ.

10.4 Mappings as Models for Magnetic Field Lines

Since early 80's iterative maps to study magnetic field line in plasmas have been introduced in order to avoid the time consuming integration of field line equations. Although, in most cases the maps were not rigorously derived from the equations of field lines they are constructed in the symplectic form (10.65) which conserves the flux-preserving property of magnetic field. Another requirement was that the map should have a simple form as possible. Below we shortly discuss some of these important maps.

10.4.1 The Standard Map and its Generalizations

This class of mappings corresponding to the symplectic mapping (10.65) with the generating function S depending only on the poloidal angle ϑ (Mendonça (1991)). The general form of corresponding generating function is given by

$$\epsilon S(\vartheta) = \frac{K}{2\pi} \sum_m g_m \cos(m\vartheta + \chi_m) \ , \tag{10.67}$$

where g_m and χ_m are the constant amplitudes and phases of perturbation modes. The constant parameter K describes the relative strength of magnetic perturbation. The corresponding map has the following form

$$\psi_{k+1} = \psi_k - \frac{K}{2\pi} \sum_m g_m \sin(m\vartheta_k + \chi_m) \ ,$$

$$\vartheta_{k+1} = \vartheta_k + \frac{2\pi}{q(\psi_{k+1})} \ . \tag{10.68}$$

The *standard map* (Chirikov (1979)) can be obtained from (10.68) by keeping in the generating function S (10.67) only one mode $m = 1$ with $g_m = 1$ and $\chi_m = 0$ and choosing the safety factor $q(\psi) = 1/\psi$:

$$\psi_{k+1} = \psi_k - \frac{K}{2\pi} \sin \vartheta_k \ ,$$
$$\vartheta_{k+1} = \vartheta_k + 2\pi \psi_{k+1} \ . \tag{10.69}$$

This map has been widely used to model a field line stochasticity in tokamaks by many authors (see, e.g., Rechester et al. (1979); Rechester and White (1980); Rechester et al. (1981); Ichikawa et al. (1987); Rax and White (1992). Thanks to its simple form it allowed one to efficiently calculate the diffusion coefficients of field lines, the transport of test particles in a magnetic field with destroyed magnetic surfaces. Recently, the standard map has been used to model the test particle transport in a tokamak caused by the drift-wave turbulence (Horton et al. (1998); Kwon et al. (2000)). The more general maps (10.68) proposed by Mendonça (1991) have been used by Tabet et al. (1998, 2000); Miskane et al. (2001) to study stochastic magnetic field lines in tokamaks.

10.4.2 The Wobig–Mendonça Map

In toroidal systems typical magnetic perturbations are radially dependent, i.e., $H_1(\psi, \vartheta, \varphi)$ depends on the torodial flux ψ. In order to take into account this feature of the toroidal magnetic field, the generating function S should be chosen as a function that depends also on the toroidal flux ψ. The simplest form of such a map proposed by Wobig (1987) and later generalized by Mendonça (1991)) has the following form

$$\psi_{k+1} = \psi_k - \frac{K}{2\pi} f(\psi_{k+1}) \sum_m g_m \sin m\vartheta_k,$$
$$\vartheta_{k+1} = \vartheta_k + \frac{2\pi}{q(\psi_{k+1})} - \frac{K}{2\pi} f'(\psi_{k+1}) \sum_m g_m \cos m\vartheta_k \ , \tag{10.70}$$

which corresponds to the map (10.65) with the generating function

$$\epsilon S(\vartheta, \psi) = -\frac{K}{2\pi} f(\psi) \sum_m g_m \cos m\vartheta, \qquad f(\psi) = 1 - \exp(-\psi) \ , \tag{10.71}$$

where $f'(\psi) \equiv df(\psi)/d\psi$. The case when $f(\psi) = \psi$ and $q(\psi) = 1/\psi$ was considered by Wobig (1987). The corresponding map can be obtained from the map (10.70) at a limiting case. Indeed, near the magnetic axis ($\psi = 0$), when $\psi \ll 1$, the function $f(\psi)$ can be approximated by a linear function ψ and the map (10.70) is reduced to (Wobig (1987)):

$$\psi_{k+1} = \psi_k \left(1 + \frac{K}{2\pi} \sum_m g_m \sin m\vartheta_k \right)^{-1},$$

$$\vartheta_{k+1} = \vartheta_k + \frac{2\pi}{q(\psi_{k+1})} - \frac{K}{2\pi} \sum_m g_m \cos m\vartheta_k . \tag{10.72}$$

At regions far from the magnetic axis, when $\psi \gg 1$, the map (10.70) approaches the mappings (10.68).

As was noted by Balescu et al. (1998) that the toroidal flux ψ in the standard map (10.69) may take negative values after some iterations of the maps for arbitrary value of K. A similar situation occurs in the Wobig map (10.72) for $K > 2\pi$. This fact is not compatible with the magnetic field line behavior in a toroidal system where the toroidal flux $\psi \sim r^2$ is always positive. On the other hand in the both maps the profile of the safety factor $q \sim r^{-2}$ does not represent any realistic case in tokamaks.

10.4.3 The Tokamap

The specific form of the map called a *tokamap*, which is compatible with the toroidal geometry has been proposed by Balescu et al. (1998); Balescu (1998). It describes the global behavior of magnetic field lines in tokamaks. The tokamap has been constructed as an iterative symplectic map, (10.62) representing a global picture of a tokamak cross section $\varphi_k = 2\pi k$ (mod 2π). The generating function S is chosen to be compatible with the toroidal geometry, i.e., satisfying the following constraints:

1. The map should be Hamiltonian (or symplectic).
2. It should be *compatible with toroidal geometry* that the canonical momentum ψ (toroidal flux) be always positive number. For instance, if $\psi_0 > 0$ at the section $k = 0$ then $\psi_k > 0$ for all k ; and if $\psi_0 = 0$, then $\psi_k = 0$ for all k.

The toroidal flux ψ is normalized to its value at the plasma edge, and takes values in the interval, $0 \leq \psi < 1$. Balescu et al. (1998) has shown that the following map satisfies these constraints

$$\psi_{k+1} = \psi_k - \frac{K}{2\pi} \frac{\psi_{k+1}}{1 + \psi_{k+1}} \sin \vartheta_k ,$$

$$\vartheta_{k+1} = \vartheta_k + \frac{2\pi}{q(\psi_{k+1})} - \frac{K}{2\pi} \frac{1}{(1 + \psi_{k+1})^2} \cos \vartheta_k , \tag{10.73}$$

where K is a *stochasticity parameter*, similar to the parameter in the standard map (3.26). It takes values in the interval: $0 < K < 2\pi$. (We have used in the safety factor $q(\psi)$ instead of the winding number $W(\psi) = 1/q(\psi)$ in the original work by Balescu et al. (1998)).

The generating function $S(\vartheta, \psi)$ associated with the tokamap is

$$\epsilon S(\vartheta, \psi) = -\frac{K}{2\pi} \frac{\psi}{1 + \psi} \cos \vartheta , \qquad (10.74)$$

and the safety factor $q(\psi)$ can be chosen arbitrary. In Balescu et al. (1998) the following analytical form has been used,

$$q(\psi) = \frac{4q(0)}{(2 - \psi)(2 - 2\psi + \psi^2)} , \qquad (10.75)$$

where $q(0)$ is the value of the safety factor on the magnetic axis $\psi = 0$. At the plasma edge $\psi = 1$ it takes four times the central value, $q(1) = 4q(0)$. This profile of $q(\psi)$ shown in Fig. 10.9a has been derived by Misguich and Weyssow (see Appendix A in Misguich et al. (2002b)) for the following density, $n(r) = n(0)[1 - r^2]$, and the electron temperature, $T(r) = T(0)[1 - r^2]^2$, profiles in tokamaks.

If the profile of $q(\psi)$ is a non-monotonic function of ψ then the tokamap becomes the *non-twist map* (see Sect. 7.3). The corresponding tokamap called *revtokamap* has been studied by Balescu (1988) for the following $q-$ profile

$$q(\psi) = \frac{q_m}{1 - a(\psi - \psi_m)^2} , \qquad (10.76)$$

where q_m is a minimum value of q at $\psi = \psi_m$. The profile of the safety factor (10.76) is shown in Fig. 10.9b. Parameters a and ψ_m can be written in terms of the values of $q(\psi)$ at the axis $\psi = 0$, $q_0 = q(0)$, and at the plasma edge $\psi = 1$, $q_1 = q(1)$, respectively,

$$\psi_m = \left[1 + \left(\frac{1 - q_m/q_1}{1 - q_m/q_0} \right)^{1/2} \right]^{-1} , \qquad a = \frac{1 - q_m/q_0}{\psi_m^2} .$$

The revtokamap describes the dynamics of field lines in a tokamak magnetic configuration with a so-called *reversed magnetic shear*. The latter corresponds to the profile of the safety factor $q(\psi)$ which has a minimum not at the magnetic axis $\psi = 0$, but at a normalized radius of 0.3-0.4 and regularly increases towards the center ($\psi = 0$) and the edge of the plasma ($\psi = 1$). Recent tokamak experiments (see reviews by Litaudon (1998); Wolf (2003)) have clearly shown the appearance of improved confinement regimes, *internal transport barriers* (ITB's), in the presence of a reversed magnetic shear.

The first equation in (10.73) can be explicitly solved with respect to variable ψ_{k+1}:

$$\psi_{k+1} = \frac{1}{2} \left[\sqrt{P^2(\psi_k, \vartheta_k) + 4\psi_k} - P(\psi_k, \vartheta_k) \right] , \qquad (10.77)$$

where

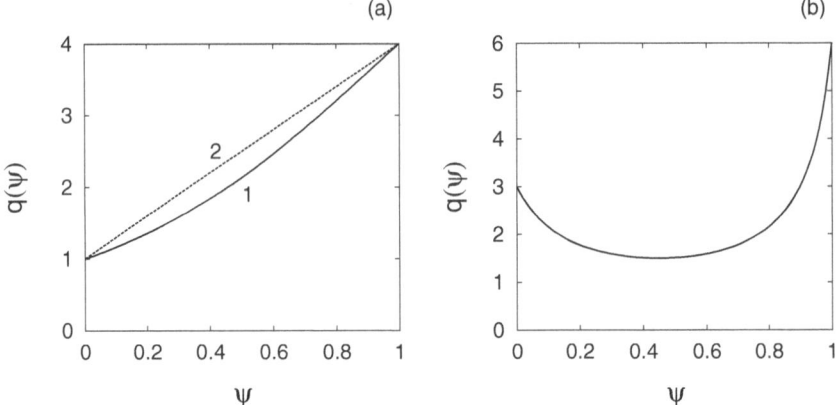

Fig. 10.9. Profiles of the safety factor $q(\psi)$: (**a**) a monotonic profile (10.75) (curve 1) and a linear profile $q(\psi) = q_0 + (q_1 - q_0)\psi$ (curve 2) at values $q_0 = 1$, $q_1 = 4$. (**b**) The non-monotonic profile (10.76) at the values: $q_0 = 3$, $q_m = 1.5$, and $q_1 = 6$

$$P(\psi_k, \vartheta_k) = 1 - \psi_k + \frac{K}{2\pi} \sin \vartheta_k .$$

The tokamap has been used to model the formation of transport barriers, and anomalous subdiffusion of field lines in tokamaks Misguich (2001); Misguich et al. (2002a,b).

Although the tokamap has not been directly derived from the equations of field lines it has recently been recovered by Weyssow and Misguich (1999) from the particle map of the guiding center motion in a toroidal system described by the standard magnetic field (10.17) in the presence of the magnetic field perturbation. As was stated by Weyssow and Misguich (1999) it is deduced as a particular case of the particle map in the limit of zero magnetic moment and when only a simple $m = 0$ non-resonant magnetic perturbation is applied. The relation between the tokamap and the continuous field line equations has been also discussed by Eberhard (1999). Below we study this problem in detail.

10.5 Continuous Hamiltonian System and Tokamap

One should note that the derivation of the tokamap from a continuous Hamiltonian system is not well-defined problem unless we make some additional assumptions. The main difficulty lies in determining to which kind of a continuous Hamiltonian function corresponds the tokamap. In order to make this problem unambiguous we suppose that the perturbation Hamiltonian $H_1(\psi, \vartheta, \varphi)$ as well as the generating function $S(\vartheta, \psi)$ of the tokamap do not depend on the safety factor.

As was shown in Sect. 4.3 general mappings of type (3.5), (10.65) with the simple generating function (10.74) cannot be constructed for Hamiltonian systems of type (10.33) with radial- dependent perturbation functions $H_1(\psi, \vartheta, \varphi)$. It means that the tokamap cannot be rigorously derived from Hamiltonian field line equations under the constraint on the Hamiltonian and the generating function imposed above in order to have a simple mapping model. We study this problem by constructing a mapping for the following continuous Hamiltonian system corresponding to the tokamap:

$$H = \int \frac{d\psi}{q(\psi)} + \epsilon H_1(\psi) \sum_{n=-M}^{M} \cos(\vartheta - n\varphi) \, ,$$

$$\epsilon H_1(\psi) = -\frac{K}{(2\pi)^2} \frac{\psi}{1 + \psi} \, , \tag{10.78}$$

with a large number, $2M + 1 \gg 1$, of toroidal modes n. The magnetic perturbation $\mathbf{B}_1$ corresponding to this system can be found using the relation (10.45). From the last equation we obtain the vector potential A_φ of perturbed field

$$A_\varphi^{(per)}(r, \theta, \varphi) = -\epsilon \frac{R_0^2 B_0 H_1(\psi(r))}{R(r, \theta)} \sum_{n=-M}^{M} \cos(\vartheta(r, \theta) - n\varphi) \, . \tag{10.79}$$

Particularly, for the cylindrical model of magnetic field, $R = R_0$, $\psi = r^2/2R_0^2$, and $\vartheta = \theta$, we have

$$A_\varphi^{(per)}(r, \theta, \varphi) = \frac{K B_0}{2(2\pi)^2 R_0} \frac{r^2}{1 + r^2/2R_0^2} \sum_{n=-M}^{M} \cos(\theta - n\varphi) \, . \tag{10.80}$$

In the limit $M \to \infty$ the Hamiltonian (10.78) is reduced to

$$H = \int \frac{d\psi}{q(\psi)} + \epsilon H_1(\psi) \cos\vartheta \sum_{k=-\infty}^{\infty} \delta(\varphi - 2\pi k) \, . \tag{10.81}$$

This Hamiltonian has been used by Eberhard (1999) to derive the tokamap–like map. The Hamiltonian equations of field lines are

$$\frac{d\psi}{d\varphi} = -\frac{\partial H}{\partial \vartheta} = -\epsilon H_1(\psi) \sin\vartheta \sum_{k=-\infty}^{\infty} \delta(\varphi - 2\pi k) \, ,$$

$$\frac{d\vartheta}{d\varphi} = \frac{\partial H}{\partial \psi} = \frac{1}{q(\psi)} - \epsilon \frac{\partial H_1}{\partial \psi} \cos\vartheta \sum_{k=-\infty}^{\infty} \delta(\varphi - 2\pi k) \, . \tag{10.82}$$

As was discussed in Sect. 3.3 the integration of these equations along the δ-functions is not well-defined procedure. We would obtain the tokamap (10.73) if the integrals of type,

$$\int_{2\pi k}^{2\pi (k+1)} f(\psi(\varphi),\vartheta(\varphi))\delta(\varphi - 2\pi k) \,,$$

where $f(\psi(\varphi),\vartheta(\varphi))$ is a function depending on the orbit $(\psi(\varphi),\vartheta(\varphi))$, are replaced by $f(\psi_{k+1},\vartheta_k)$. Eberhard (1999) integrated the system (10.82) from φ_k to φ_{k+1} using the symmetric definition of the δ-function,

$$\int_{2\pi k}^{2\pi k \pm \epsilon} f(\psi,\vartheta)\delta(\varphi - 2\pi k) = \pm\frac{1}{2}f(\psi_k \pm 0, \vartheta_k \pm 0) \,, \quad (\epsilon > 0) \,,$$

and obtained the symmetric map corresponding the tokamap

$$\psi_k^+ = \psi_k - \frac{1}{2}\epsilon H_1(\psi_k)\sin\vartheta_k \,,$$

$$\vartheta_k^+ = \vartheta_k - \frac{1}{2}\epsilon \frac{\partial H_1(\psi_k)}{\partial\psi_k}\cos\vartheta_k \,,$$

$$\vartheta_{k+1}^- = \vartheta_k^+ + \frac{2\pi}{q(\psi_k^+)} \,, \quad \psi_{k+1}^- = \psi_k^+ \,, \tag{10.83}$$

$$\psi_{k+1} = \psi_{k+1}^- - \frac{1}{2}\epsilon H_1(\psi_{k+1})\sin\vartheta_{k+1} \,,$$

$$\vartheta_{k+1} = \vartheta_{k+1}^- - \frac{1}{2}\epsilon \frac{\partial H_1(\psi_{k+1})}{\partial\psi_{k+1}}\cos\vartheta_{k+1} \,.$$

The numerical integration of the continuous system (10.82) by replacing the δ-function by its continuous representation $\delta(t) = \exp(-t^2/a^2))/a\sqrt{\pi}$, $(a \ll 1)$ gives a good agreement with the symmetric form of the map (10.83) rather than with the tokamap (10.73). The difference between the numerical integration and the tokamap is relatively large (see Sect. 3.3). In this sense the tokamap is not good discrete replacement of the continuous Hamiltonian system (10.81).

However, the constructed symmetric map (10.83) has a serious shortcoming: it is not symplectic, i.e., $|\partial(\psi_{k+1},\vartheta_{k+1})/\partial(\psi_k,\vartheta_k)| \neq 1$. This is a consequence of uncertain procedure of integration along $\delta-$ functions. In order to obtain the symplectic map from the continuous system (10.81) we can use the rigorous procedure presented in Sect. 4.3.

10.5.1 The Symmetric Tokamap

The continuous Hamiltonian (10.81) is a particular case of the Hamiltonian (4.24). According to results of Sect. 4.3 only the symmetric form (4.31) - (4.33) of the map (10.62) corresponding to the system (4.24) or (10.81) has the simple generating function (4.34). The generating function of the non-symmetric mappings (4.38) or (10.65) has a rather complicated form (4.37). It depends not only on the harmonics $H_m(\psi)$ of the perturbation Hamiltonian but also on the safety factor $q(\psi)$.

According to (4.31)–(4.33) the symmetric map corresponding the tokamap Hamiltonian (10.81) can be written as (see Abdullaev (2004a))

$$\Psi_k = \psi_k - \frac{K}{4\pi} \frac{\Psi_k}{1+\Psi_k} \sin\vartheta_k \ ,$$

$$\Theta_k = \vartheta_k - \frac{K}{4\pi} \frac{1}{(1+\Psi_k)^2} \cos\vartheta_k \ ,$$

$$\Theta_{k+1} = \Theta_k + \frac{2\pi}{q(\Psi_k)} \ , \qquad \Psi_{k+1} = \Psi_k \ ,$$

$$\psi_{k+1} = \Psi_{k+1} - \frac{K}{4\pi} \frac{\Psi_{k+1}}{1+\Psi_{k+1}} \sin\vartheta_{k+1} \ ,$$

$$\vartheta_{k+1} = \Theta_{k+1} - \frac{K}{4\pi} \frac{1}{(1+\Psi_{k+1})^2} \cos\vartheta_{k+1} \ , \tag{10.84}$$

by replacing the action variable I by the toroidal flux ψ, the frequency of motion, $\omega(J)$, by the inverse safety factor $q^{-1}(\Psi)$, and putting the perturbation frequency $\Omega = 1$. We call the mapping (10.84) a *symmetric tokamap*. The generating function $S(\vartheta, \psi)$ associated with this map is

$$\epsilon S(\vartheta, \Psi) = -\frac{K}{4\pi} \frac{\Psi}{1+\Psi} \cos\vartheta \ . \tag{10.85}$$

In general, the symmetric tokamap is a implicit map unlike the tokamap. The first equation in (10.84) can be explicitly resolved with respect to Ψ_k, similarly to the tokamap,

$$\psi_{k+1} = \frac{1}{2}\left[\sqrt{P^2(\psi_k, \vartheta_k) + 4\psi_k} - P(\psi_k, \vartheta_k) \right] \ ,$$

$$P(\psi_k, \vartheta_k) = 1 - \psi_k + \frac{K}{4\pi} \sin\vartheta_k \ .$$

However, the last equation in (10.84) cannot be explicitly resolved with respect to the variable ϑ_{k+1}. The latter should be found numerically using, for instance, the Newton method. The zero approximation to ϑ_{k+1} can be chosen

$$\bar{\vartheta}_{k+1} = \Theta_{k+1} - \frac{K}{4\pi} \frac{1}{(1+\Psi_{k+1})^2} \cos\Theta_{k+1} \ .$$

The symmetric tokamap is invariant with respect to the translation $\vartheta \leftrightarrow \pi - \vartheta$. It reflects the invariance of the symmetric map with respect to transformation $k \leftrightarrow k+1$ with the simultaneous change $K \to -K$ and $q \to -q$. The property corresponds to the symmetry of the continuous Hamiltonian system with respect to the formal transformation $t \to -t$, $H \to -H$.

10.5.2 Comparison of the Tokamap and the Symmetric Tokamap

Fixed points (ϑ_s, ψ_s) of the tokamap and the symmetric tokamap, defined by $\psi_{k+1} = \psi_k$, $\vartheta_{k+1} = \vartheta_k + 2\pi n$ where n is an integer number, are the same. They either lie in the *polar axis* $\psi_s = 0$, or in the *equatorial plane* $\vartheta_s = 0, \pi$, $\sin\vartheta_s = 0$. For the last case ψ_s are roots of the equation (Balescu et al. (1998))

$$\frac{1}{q(\psi)} \mp \frac{K}{(2\pi)^2} \frac{1}{(1+\psi)^2} - n = 0 \, ,$$

where the signs ($\mp$) correspond to $\vartheta_s = 0$ and $\vartheta_s = \pi$, respectively.

However, the periodic fixed points, $(\vartheta, \psi) = \hat{M}^q(\vartheta, \psi)$, $(q > 1)$, of the tokamap and the symmetric tokamap are different. For the tokamap they have been studied in Misguich et al. (2002b). Since the tokamap cannot be exactly obtained from the symmetric map the the fixed points of the symmetric tokamap cannot be recovered from those of the tokamap. The deviation of these maps significantly depends on the perturbation parameter K. It is of order of $\epsilon^2 = K^2/4\pi^2$. For the small values $K \ll 2\pi$ it is small, however, for $K \sim 1$ the difference becomes relatively large. Below we will study the difference between the symmetric tokamap and the tokamap by comparing their phase portraits. More detailed study of fixed points of the symmetric map requires a separate investigation and we will not discuss this problem here.

Figure 10.10 shows phase portraits of the tokamap (10.73) and the symmetric tokamap (10.84) with the profile of the safety factor $q(\psi)$ (10.75) at the perturbation parameter $K = 2.55$: (a), (b) correspond the tokamap, and (c), (d) − the symmetric tokamap. Figures 10.10a,c,e describe the phase portrait in the (ϑ, ψ) -plane, while 10.10b, d, f − in the polar plane $(\psi \cos \vartheta, \psi \sin \vartheta)$.

For comparison we have also integrated the continuous Hamiltonian (10.78) with a finite number of toroidal modes $M = 5$ using the mapping (10.63). They are plotted in Fig. 10.10e, f. The map step along the toroidal angle φ is taken $\Delta\varphi = 2\pi$.

As seen from Figures the symmetric tokamap very closely describes the continuous Hamiltonian system (10.78) with a finite number of toroidal modes M. However, the phase curves of the tokamap are more distorted than in the symmetric tokamap. The positions of the periodic fixed points are shifted not only radially, but also along the poloidal angle ϑ. The difference between the tokamap and the symmetric tokamap becomes more pronounced for large perturbation parameter K.

To illustrate this we compare the perturbation thresholds K_g of destruction of the so-called *golden KAM curve* corresponding to the tokamap and the symmetric tokamap, respectively. For the tokamap this problem was discussed by Balescu et al. (1998).

The golden KAM curve is the most robust KAM barrier in the standard map, and it corresponds to the winding number W equal to $g_* = G_*^{-1} = 0.6180339\ldots$, where G_* is the golden section defined by the equation: $G_*^2 = G_* + 1$. For the safety factor $q(\psi) = 1/W(\psi)$ (10.75) the golden KAM curve is located at $\psi_g = 0.31599$. It lies between two periodic orbits (m, n) $(q(\psi_{mn}) = m/n)$ corresponding to $q(\psi_{2,1}) = 2 : 1$ located at $\psi_{2,1} \approx 0.45631$ and $q(\psi_{4,3}) = 4 : 3$ located at $\psi_{4,3} \approx 0.18946$. For the finite perturbation parameter $K > 0$ the KAM curves are deformed or broken. The periodic orbits are replaced by islands. With increasing K the width of islands grows

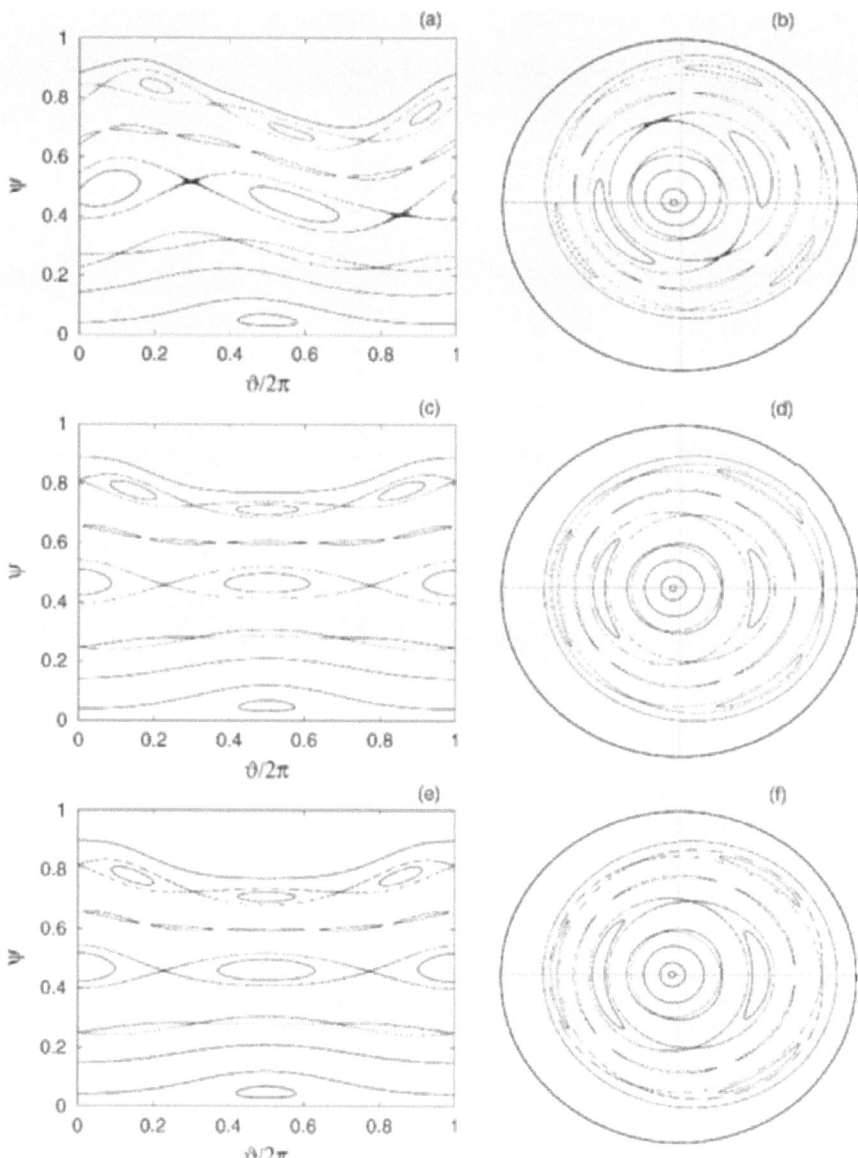

Fig. 10.10. Phase portraits: (**a**) and (**b**) correspond to the tokamap (10.73); (**c**) and (**d**) correspond the symmetric tokamap (10.84); (**e**) and (**f**) correspond to the symmetric map (10.63) with the Hamiltonian (10.78) with $M = 5$. The q-profile is given by (10.75). Parameters $K = 2.55$, $q(0) = 1$

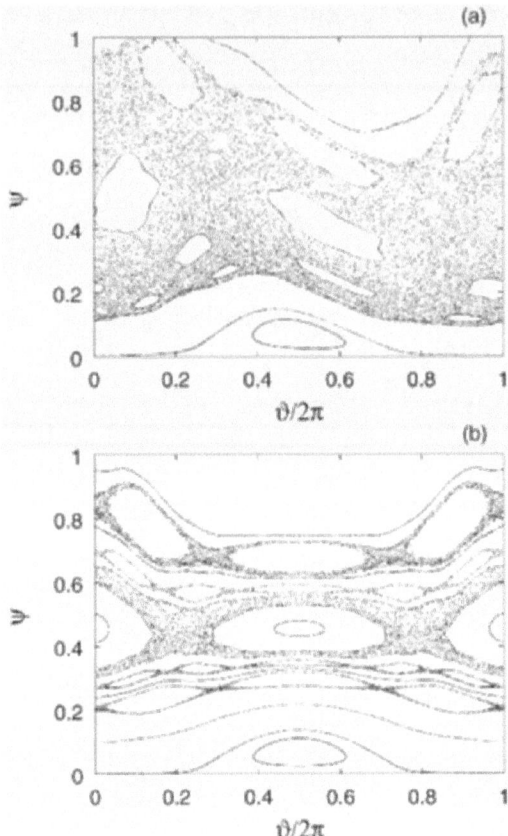

Fig. 10.11. Stochastic belt of the tokamap (10.73) (**a**) and the symmetric tokamap (**b**) with the q-profile (10.75) for $K = 4.5$, $q(0) = 1$

destroying the KAM curves located between two neighboring islands. According to Balescu et al. (1998) the golden KAM curve ψ_g corresponding to the tokamap is destroyed at $K_g = 4.3$. However, at this value of K *the golden KAM curve of the symmetric tokamap is still survived*. This is shown in Fig. 10.11 for the perturbation parameter $K = 4.5$: a) the tokamap; b) the symmetric tokamap.

In the case of the tokamap several islands including 2 : 1 and 4 : 3 are already overlapped forming a single stochastic belt. But in the case of the symmetric tokamap many of these islands are still isolated. From Fig. 10.11b one can see a gap between the islands 2 : 1 and 4 : 3. The golden KAM curve of the symmetric tokamap is destroyed at larger value K equal to $K_g \approx 4.8$. The Phase portrait of the symmetric tokamap at $K = 5$ is shown in Fig. 10.12.

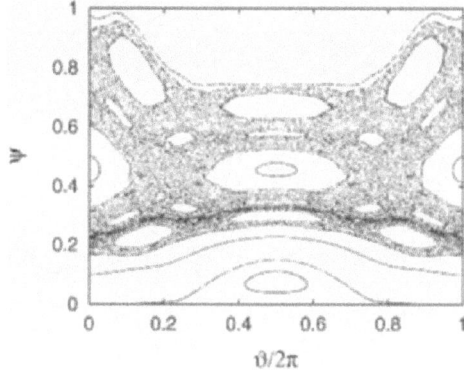

Fig. 10.12. The same as in Fig. 10.11b but for $K = 5.0$, $q(0) = 1$

10.5.3 The Revtokamap and the Symmetric Revtokamap

Consider another example illustrating a difference between the tokamap (or revtokamap) and the symmetric tokamap (revtokamap) in the case of reversed magnetic shear configuration described by the safety factor (10.76) (see also Fig. 10.9b). The mappings in this case are non-twist maps which were studied in Sect. 7.3. The dynamics of the revtokamap in such a reversed magnetic shear has been studied by Balescu (1998) with a great detail. Here, we will only compare the symmetric revtokamap and the revtokamap at large values of the perturbation parameter K when the difference between these maps is expected to be large.

Phase portraits of the maps are plotted in Fig. 10.13 for the perturbation parameter $K = 6.3$: a) the revtokamap; b) the symmetric tokamap. The values of the safety factor $q(\psi)$ are chosen equal to $q_0 = 3$ at the magnetic axis, $q_m = 1.5$ at the minimum, and $q_1 = 6$ at the edge, as in Balescu (1998). Then the position of the shearless curve is located at $\psi_m = 0.44948974$ where $q'(\psi_m) = 0$.

The most robust invariant curves are located near the shearless curve $\psi = \psi_m$. According to the revtokamap the upper region $\psi > \psi_m$ and the lower region $\psi < \psi_m$ of field lines are developed into the chaotic belts: the upper region $\psi > \psi_m$ is open, while the lower region is localized. These regions are separated from each other by the transport barrier located near the perturbed shearless curve (see Fig. 10.13a). However, the symmetric revtokamap with the same perturbation parameter K gives rather a different picture. Although, the upper region $\psi > \psi_m$ of open field lines is completely chaotic, the lower region $\psi < \psi_m$ consists of almost regular field lines with a few isolated islands (see Fig. 10.13b). This result can be also confirmed by direct integration of the continuous Hamiltonian system (10.78). In general, the tokamap systematically gives lower critical values of the perturbation parameter K_g of destruction of the KAM curves than the symmetric tokamap.

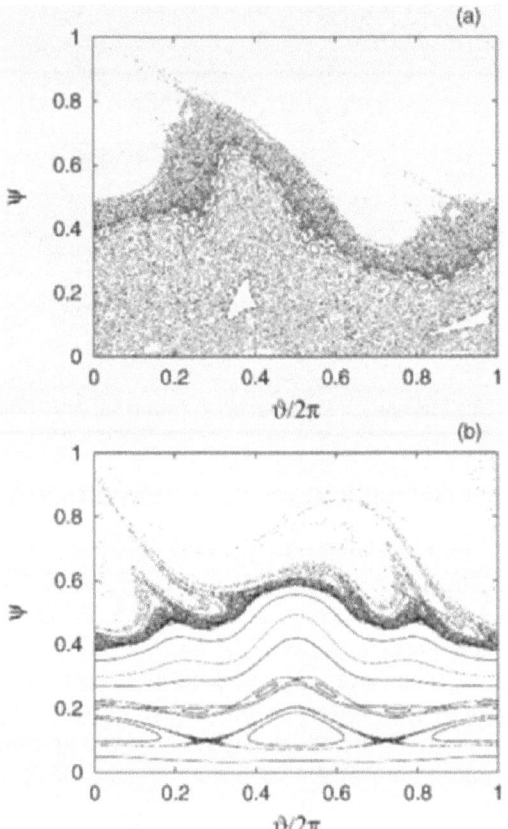

Fig. 10.13. Phase portraits of the tokamap (**a**) and the symmetric tokamap (**b**) with the reversed q-profile (10.76). Parameters $K = 6.3$, $q_0 = 3$, $q_m = 1.5$ and $q_1 = 6$

10.6 Other Mapping Models of Field Lines

10.6.1 Analytical Models

We will also mention some other areas of fusion plasma physics where mapping methods have been a useful tool to study the transport of energy and particles. Particularly, in Albert and Boozer (1988); White (1998); Wobig and Pfirsch (2001) (and references therein) a chaotic motion of particles in a tokamak due to magnetic perturbations have been studied by replacing the continuous Hamiltonian equations for the guiding center motion with the discrete mappings. The standard map and its generalizations have been used in numerous works (see, e.g., Rechester et al. (1979); Rechester and White (1980); Rechester et al. (1981); Rax and White (1992); Wobig (1987); Wobig and Fowler (1988); Mendonça (1991); Tabet et al. (1998, 2000); Miskane et al. (2001); Wobig and Pfirsch (2001)) to simulate a transport of the field lines and test particles in a magnetic field with destroyed magnetic surfaces. It has been also used by Horton et al. (1998); Kwon et al. (2000) to model a test particle transport in a tokamak due to a drift-wave turbulence.

Beside of these mappings describing the global dynamics of field lines in tokamaks and stellarators there were proposed specific mapping models of field lines at the plasma periphery of poloidal and ergodic divertor tokamaks by Martin and Taylor (1984); Regianni and Sakanaka (1994); Abdullaev et al. (1998); Fischer and Cooper (1998); Punjabi et al. (1992); Abdullaev and Zaslavsky (1996); Abdullaev and Finken (1998) and others. Particularly, we shall discuss the application of separatrix mappings to study field lines in poloidal divertor tokamaks in next chapter. Heat and test particle transport in a tokamak perturbed by externally created magnetic field have been studied by McCool et al. (1990); Wootton et al. (1991); Feron and Ghendrih (1997) using different mapping models of magnetic field lines.

The Reversed Field Pinch (RFP) is another fusion device of a magnetically confined plasma. It is known as the equilibrium state of plasma with the minimal magnetic energy. The magnetic field of this state has intrinsic turbulent components due to a turbulent transient to the minimal energy state. The transport and diffusion processes in the RFP are mostly due to the magnetic field line stochasticity (D'Angelo and Paccagnella (1996, 1999)). In Bazzani et al. (1989, 1998) the perturbed twist maps have been derived to study a field line diffusivity and transport in the RFP.

10.6.2 Numerical Mapping Models

Beside analytical mapping models numerical mapping methods have been also developed to study field lines and modeling heat and particle transport in tokamak plasmas. The numerical mapping technique has been proposed for integrating Hamiltonian systems, particularly, the equations of magnetic field lines and particles in a realistic toroidal geometry. Below we briefly describe the main idea of the method known as *Interpolated Cell Mapping* (Tongue (1987); Montvai and Düchs (1993); de Rover et al. (1996)).

The method uses real space coordinates $\mathbf{x} = (x, y, \varphi)$, which may be cylindrical coordinates (R, Z, φ), or toroidal coordinates (r, θ, φ). One introduces N_φ cross sections $\varphi_k = $ const equally spaced along the toroidal angle φ. We define a mapping,

$$(x_{k+1}, y_{k+1}) = \hat{M}(x_k, y_k) = (X_k(x_k, y_k), Y_k(x_k, y_k)) , \qquad (10.86)$$

relating field line coordinates (x_{k+1}, y_{k+1}) at the section φ_{k+1} with the ones (x_k, y_k) at φ_k. The mapping (10.86) is determined by the set of functions $X_k(x, y), Y_k(x, y), k = 1, 2, \ldots, N_\varphi$. In order to create these functions the field line equations (10.2) are integrated numerically from grid points $(x_{k,ij}, y_{k,ij})$ $(i = 1, \ldots, N_x, j = 1, \ldots, N_y)$ at the section φ_k to the intersection point $(X_k(x_{k,ij}, y_{k,ij}), Y_k(x_{k,ij}, y_{k,ij}))$ at the section φ_{k+1}. The functions $X_k(x, y), Y_k(x, y)$ are interpolated by the cubic splines from their computed values at the grid points $(x_{k,ij}, y_{k,ij})$. Once the set of functions $X_k(x, y), Y_k(x, y)$, $k = 1, 2, \ldots, N_\varphi$, have been created one can follow field lines using the mapping (10.86). The main advantage of this numerical mapping technique is that it runs much faster than direct numerical integration of field line equations.

This numerical mapping and its modifications have been employed by Kasilov et al. (1997); Runov et al. (2001); Feng et al. (2002) for Monte–Carlo simulations of transport processes in tokamak plasmas with stochastic field lines.

However, the numerical mapping (10.86) has a fundamental shortcoming because it is not symplectic, i.e., flux-preserving. Since the mapping variables (x, y) are not canonical variables it is hard to impose conditions of the functions $X_k(x, y), Y_k(x, y)$ that would make mappings symplectic.

10.7 Conclusions

In conclusion of this chapter we would like to make general remarks on mapping models of magnetic field lines in toroidal plasmas in the presence of magnetic perturbations. Usually these perturbations are strongly radial dependent which is manifested in the dependence of perturbed Hamiltonian H_1 (10.34) on the toroidal flux ψ. In general studies of field lines one wishes to replace the continuous equations of magnetic field lines in toroidal plasmas with the simple discrete model, mapping, of field lines. These model mappings should satisfy several requirements or constraints which would make them compatible with the properties of Hamiltonian equations of field lines and the toroidal geometry of plasmas. The example of such requirements was given in Sect. 10.4.3. Beside of being symplectic and compatible with the toroidal geometry one wishes to construct a mapping whose generating function would have a simple dependence on the radial–dependent perturbation field. As was shown in Sect. 4.3 that only the symmetric symplectic map of type (4.31)–(4.33) derived from the Hamiltonian equations satisfies this constraint. The non-symmetric form of mapping (4.38) which can be also derived from a Hamiltonian system has a complicated dependence on perturbation field H_1 and the frequency of motion $\omega(J)$ (or the safety factor $q(\psi)$) (see the generating function 4.37).

Therefore, when one wishes to construct a model map for magnetic field lines in a toroidal plasmas, one should impose an *additional constraint* along with two ones mentioned in Sect. 10.4.3, namely the *map should be constructed in a symmetric form* of type (4.31)–(4.33) with the generating function $S(\vartheta, \Psi)$ in the form (4.34). The form of perturbation functions $H_m(\Psi)$ can be established by two other constraints listed in Sect. 10.4.3. Such mappings would be more compatible with Hamiltonian equations of field lines.

11 Mapping of Field Lines
in Ergodic Divertor Tokamaks

Mappings have been very useful tool to study magnetic field lines in divertor tokamaks. In this and the next chapters we study the structure of magnetic field lines in so-called ergodic and poloidal divertor tokamaks using mappings. We shall study the general structure of magnetic field, chaotic and statistical properties of field lines in ergodic divertor tokamaks. Mappings constitute an important and computationally efficient tool to study magnetic field lines in both types of tokamaks. In ergodic divertor tokamaks we shall apply the mapping procedure of magnetic field lines presented in Sect. 10.3.

11.1 Ergodic Divertor Concept

In magnetically fusion devices particles and energy cannot be hold in a confinement area for a long time. Due to different turbulent processes particles and heat are transported to the plasma edge and deposited to the device wall. A large energy and particle out-flux may damage wall components and release the impurities into the plasma core. In order to control these processes at the plasma edge the concept of the *ergodic divertor* has been introduced (see Engelhard and Feneberg (1978); Feneberg and Wolf (1981); Samain et al. (1982)). The idea of the ergodic divertor is based on the creation of a perturbed magnetic field at the plasma edge by a special coil system installed outside the plasma vessel. The perturbation field creates the zone of chaotic field lines at the plasma edge which are open to the plasma wall. These field lines guide ionized particles to the special divertor plates. The schematic view of the poloidal section of ergodic divertor tokamak is shown in Fig. 11.1. In the literature on plasma physics this zone is more known as an *ergodic zone*.

During last two decades the ergodic divertor concepts were implemented in the Texas Experimental Tokamak (TEXT) in Austin, Texas, USA (more known as a *ergodic magnetic limiter*) (Gentle (1981); McCool et al. (1990)), Tore-Supra in Cadarache, France (see a review Ghendrih et al. (1996) and references therein) and others fusion devices Kawamura et al. (1982). Recently, the *dynamic ergodic divertor* (DED) (see Finken (1997); Abdullaev et al. (2003)) has been installed for the Tokamak Experiment for the Technology Oriented Research (TEXTOR) in Jülich, Germany. Beside the conventional concept of the ergodic divertor the DED also permits the operation with a

S.S. Abdullaev: *Construction of Mappings for Hamiltonian Systems and Their Applications*, Lect. Notes Phys. **691**, 255–273 (2006)
www.springerlink.com

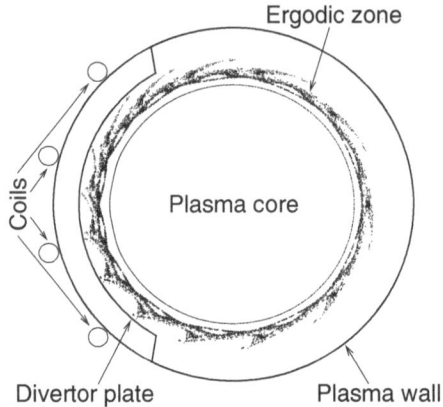

Fig. 11.1. Schematic view of the section of an ergodic divertor tokamak

rotating magnetic field which allows, in particular, to spread the heat and particles over the large area on the divertor plates.

11.1.1 Mappings to Study Ergodic Divertors

In the ergodic divertor the perturbed magnetic field created by external coils destroys rational magnetic surfaces at the plasma edge forming magnetic islands. At the certain level of perturbation field the interaction of magnetic islands creates the stochastic layer (ergodic zone) of magnetic field lines. The main problem in divertor physics is to study particle and heat transport through the ergodic zone and to determine heat and particle deposition patterns on the divertor plates. Since charged particles, ions and electrons, predominantly follow magnetic field lines a determination of the structure of ergodic zone of field lines at the plasma edge and the patterns of striking points of field lines, *magnetic footprints*, on divertor plates are first important problems. The latter problems can be studied by solving the equations of magnetic field lines in the presence of magnetic perturbations created by external divertor coils. Usually, the direct integration of the equations of magnetic field lines requires long computational times and has a several disadvantages. For instance, standard numerical integration methods, like Runge–Kutta, does not conserve the flux–preserving property of magnetic field.

The most convenient and natural approach to this problem is to use the Hamiltonian equations of field lines and to study them using symplectic methods of integration, particularly, symplectic mappings. They always preserve a magnetic flux, and run much faster. Several mapping models have been proposed to study the formation of ergodic zone at the plasma edge. The first mapping model has been proposed by Martin and Taylor (1984) for the rectangular model of a tokamak. This model has been used in Regianni and

Sakanaka (1994) to calculate the Lyapunov exponents and diffusion coefficients. The generalization of the Martin and Taylor mapping model to the toroidal plasma has been proposed by Viana and Caldas (1992); Caldas et al. (1996); Viana and Vasconcelos (1997); Ullmann and Caldas (2000); Portela et al. (2003) to study different statistical properties of magnetic field lines.

More realistic and generic mapping models for magnetic field lines in the ergodic divertor, namely, the perturbed twist mappings have been proposed by Abdullaev et al. (1998); Fischer and Cooper (1998); da Silva et al. (2001a,b, 2002a,b). The numerical study and Hamiltonian analyses of magnetic field lines have been also studied in Viana (2000). Rigorous mapping methods developed by Abdullaev et al. (1999); Finken et al. (1999); Abdullaev et al. (2001) were applied to study a magnetic structure of the DED of the TEXTOR.

Below we describe the main features of the perturbed magnetic field, the formation of ergodic zone and a specific features of field lines in typical ergodic divertor tokamaks. Specifically, we consider the DED of the TEXTOR tokamak. The detailed description of these problems can be found in Abdullaev et al. (1999); Finken et al. (1999); Abdullaev et al. (2001, 2003).

11.2 Magnetic Structure of the DED

The DED for the TEXTOR tokamak has been proposed (see Finken (1997)) as a new tool to control the plasma edge by the external magnetic perturbations. The new feature of the DED in comparison with the conventional concept of the ergodic divertor is that it permits to create a rotating magnetic field. In this section we consider the coil configuration of the DED and the magnetic field created by this coil system.

11.2.1 Set of Divertor Coils and Magnetic Perturbations

The set of external coils designed to create the resonant magnetic perturbations at the plasma edge consists of 16 helical coils covering the finite poloidal section $\pi - \theta_c < \theta < \pi + \theta_c$ of toroidal surface of radius $r = r_c$. They are located on the *high field side (HFS)* , $\theta = \pi$, of the torus where the toroidal field $B_\varphi \sim (R_0 + r\cos\theta)^{-1}$ is higher than at the center of torus $r = 0$. Similarly the side of the torus $\theta = 0$ is called *low field side (LFS)* , where the magnitude of the toroidal field is lower than at $r = 0$. The poloidal extension of the of coils is $\Delta\theta = 2\theta_c \approx 72°$. Each coil is winding once around the torus starting at a toroidal, $\varphi_j = j\pi/8$, and a poloidal angle $\theta_j = \pi - \theta_c$ and ending after one toroidal turn at $\theta_j = \pi + \theta_c$, where j $(j = 1, \ldots, 16)$ stands for a coil number. A schematic view of the coil positions in the (φ, θ)-plane is shown in Fig. 11.2a.

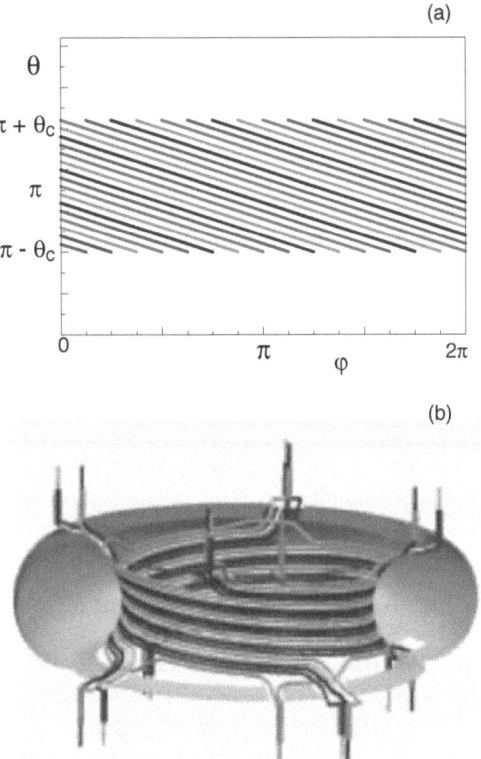

Fig. 11.2. (**a**) Ideal DED coil configuration in the (φ, θ)-plane; (**b**) Sketch of the technically implemented DED coil configuration

Described configuration of coils is called ideal. Because of difficulties of technical implementation of such an ideal coil configuration, the real coil configuration has been installed in the tokamak device. The coils are bundled in four quadruples and two additional so called compensations coils are added to compensate a net poloidal magnetic field resulted from such a bundling (Kaleck et al. (1997)). The sketch of the technically implemented divertor coils is shown in Fig. 11.2b.

There are three possible operational regimes of the DED which are intended to create the resonant $m : n$ magnetic perturbations (10.34). In the first (or *standard*) regime the toroidal mode $n = 4$ and a several poloidal modes m ($10 < m < 14$) create the ergodic zone near the resonant magnetic surface $q = 3$. In the second regime the ergodic zone is created by the toroidal mode $n = 2$ and the poloidal modes $m = 5 - 7$. And finally $n = 1$ $m = 3$ in the third regime. For the *standard DED operation* ($n = 4$) the current distribution in the coils is given by

$$I_j = I_c \sin(\pi j/2 + \Omega t), \qquad (j = 1, \ldots, 16), \qquad (11.1)$$

where I_c is a current amplitude ($I_c \leq 15$ kA), and Ω is the rotation frequency of perturbed magnetic field.

The coil system (11.1) creates the magnetic field perturbations at the plasma edge localized on the HFS of the torus and radially decaying toward inside the plasma (Abdullaev et al. (1999); Finken et al. (1999)). It is mainly determined by the toroidal component of vector potential

$$A_\varphi(r, \theta, \varphi) = \epsilon B_0 R_0 \sum_m a_{mn}(r) \cos(m\theta - n\varphi - \Omega t), \qquad (11.2)$$

where the dimensionless perturbation parameter ϵ is defined as

$$\epsilon = \frac{B_c}{B_\varphi} \frac{r_c}{R_0}, \qquad B_c = \frac{2\mu_0 I_c n}{\Delta \theta_c r_c}, \qquad (11.3)$$

where B_c is the characteristic strength of magnetic field perturbation. For the large aspect ratio tokamaks, $R_0/a \gg 1$, the Fourier coefficients $a_m(r)$ can be presented in the asymptotical form (see Abdullaev et al. (1999); Finken et al. (1999)):

$$a_{mn}(r) \approx \frac{g_m}{\sqrt{1 - r/R_0}} \left(\frac{r - \Delta(r)}{r_c} \right)^m, \qquad (11.4)$$

where

$$g_m = (-1)^{m+m_c} \frac{\sin[(m - m_c)\theta_c]}{\pi m (m - m_c)},$$

describes the poloidal spectrum of magnetic perturbations formed due to the finite poloidal extension of the coil set. The central mode number m_c is determined by $\Delta \theta_c$ and the toroidal mode number n: $m_c = 2\pi n/\Delta \theta_c$. The poloidal, m, spectrum of perturbation A_m is localized near the central mode m_c. The perturbed field has the toroidal mode $n = 4$ and possesses a strong radial decay $A_\varphi \propto r^{m_c}$ ($m_c \approx 20$). The factor $1/\sqrt{1 - r/R_0}$ in (11.4) corresponds to the first order toroidal corrections.

The radial component of perturbed magnetic field $B_r(\theta, \varphi) = r^{-1} \partial A_\varphi / \partial \theta$ in the (θ, φ) plane is shown in Fig. 11.3 for the standard DED operation ($n=4$) and the ideal coil configuration. The TEXTOR-DED parameters are chosen as: $R_0 = 175$ cm, $r_c = 53.25$ cm, the current $I_c = 15$ kA, $\Delta \theta_c = 2\pi/5$. The characteristic strength of perturbed magnetic field $B_c = 2.251 \times 10^3$ G. The perturbed field profiles for the technically implemented coil configuration are calculated by Kaleck et al. (1997).

11.3 Spectrum of Magnetic Perturbations

Since the perturbation field is localized in the finite interval of poloidal angles: $\pi - \theta_c < \theta < \pi + \theta_c$ at the HFS the spectrum a_{mn} of magnetic perturbation (11.2), (11.4) contains the factor $(-1)^m$. As have been shown in

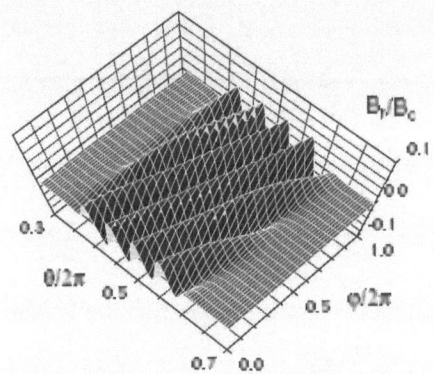

Fig. 11.3. Radial component of the perturbed magnetic field B_r in the (θ, φ) plane at the radial distance $r = 46$ cm

Sections 10.2.2, 10.2.4 the spectra of Hamiltonian perturbations $H_{mn}(\psi)$ in this case originate from the spectra of magnetic perturbation $a_{m'n}(\psi)$ with m' located near m/β_1, and it is approximately described by (10.60). Using (11.4) the spectra $H_{mn}(\psi)$ for large mode numbers m can be presented by Abdullaev et al. (1999):

$$H_{mn}(\psi) \approx (-1)^m C \sqrt{1 - \frac{r}{R_0}} \left(\frac{r - \Delta(r)}{r_c} \right)^{m^*} \frac{\sin(m^* - m_c)\theta_c}{\pi \beta_1 m^* (m^* - m_c)} \, , \quad (11.5)$$

where

$$C = \sqrt{\frac{2\pi}{|x_c|}} \mathrm{Ai}(x_c) \, , \qquad m^* = \frac{m + [m\beta_3/2\beta_1]^{1/3}}{\beta_1} \, .$$

The radial coordinate r is considered as a function of the toroidal flux ψ. For the plasma model with nested circular magnetic surfaces the relation $r = r(\psi)$ is given by (10.27). The poloidal spectrum $H_{mn}(\psi)$ (11.5) is located near the central mode $m_c^* \approx m_c \beta_1$ with the width $\Delta m \approx \pi \beta_1 / \Delta \theta_c$. These quantities are mainly determined by the parameter β_1, which depends on the plasma parameter β_{pol} as well as the flux coordinate ψ. The spectra $H_{mn}(\psi)$ can be controlled by simple varying β_{pol}.

Although the asymptotic formula for (11.5) qualitatively correct describes the spectrum of Hamiltonian perturbation its accuracy is not enough to obtain the quantitatively correct the critical perturbations for the onset of global chaos of field lines. Since it is hard to obtain the exact analytical formulas for the perturbation spectrum $H_{mn}(\psi)$ we have numerically calculated them using the integral (10.46) and unperturbed field lines $R(\psi, \vartheta)$, $Z(\psi, \vartheta)$ obtained from the numerical integration of the equations of field lines (10.11) over one poloidal turn in the (R, Z) -plane. The values of $H_{mn}(\psi)$ were calculated for the linear grid of ψ): $\psi_i = \psi_0 + i\Delta\psi$, $(i = 1, \ldots, N)$. The safety factor $q(\psi)$

has been also found numerically at this grid. The values of $q(\psi)$ and $H_{mn}(\psi)$ for the arbitrary values of ψ are interpolated using the cubic splines.

In order to display the field lines in the geometrical space of the cylindrical coordinate system (R, Z, φ) (or the toroidal coordinate system (r, θ, φ)) the relationship with these coordinates and the flux, ψ, and intrinsic angle, ϑ, variable are found by integrating of the unperturbed equations of field lines. This relationship is sought in the following general form

$$r = r_0 + \sum_{m=1}^{M} \left(r_m^{(s)}(\psi) \sin m\vartheta + r_m^{(c)}(\psi) \cos m\vartheta \right) ,$$

$$\theta = \vartheta + \sum_{m=1}^{M} \alpha_m(\psi) \sin m\vartheta .$$

The Fourier coefficients $r_m^{(s)}(\psi)$, $r_m^{(s)}(\psi)$, and $\alpha_m(\psi)$ are calculated numerically by integrating the equations (10.11) over one poloidal rotation in the (R, Z)-plane for the same grid coordinates of ψ. Their values for arbitrary ψ are interpolated by the cubic splines.

The safety factor $q(\psi)$ and typical properties of the perturbation spectra $H_{mn}(\psi)$ obtained in such a way are illustrated in Figs. 11.4 and 11.5 for the plasma model with circular magnetic surfaces considered in Sect. 10.1.5.

Figure 11.4 shows profiles of $q(\psi)$ for three different values of the plasma parameter β_{pol}: curve 1 corresponds to $\beta_{pol} = 0.2$, curve 2 – to $\beta_{pol} = 1$, and curve 3 – to $\beta_{pol} = 1.8$. The dependence of $H_{mn}(\psi)$ on the poloidal mode number m at the resonant magnetic surface ψ_{mn}, $m : n = 12 : 4$ is plotted in Fig. 11.5a for the same values of the plasma β_{pol}. The $\psi-$ (radial) dependencies of $H_{mn}(\psi)$ of the four main poloidal modes ($m = 11, 12, 13, 14$) are shown in Fig. 11.5b for the value $\beta_{pol} = 1$.

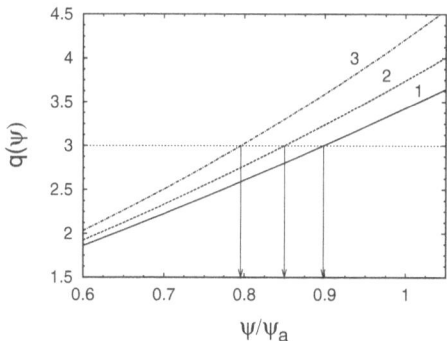

Fig. 11.4. Safety factor profile $q(\psi)$ for three different values of the plasma parameter: curve 1 – $\beta_{pol} = 0.2$, curve 2 – to $\beta_{pol} = 1$, curve 3 – to $\beta_{pol} = 1.8$. The crossing points of these curves with the horizontal line $q = 3$ determine the resonant magnetic flux ψ_{mn}: $q(\psi_{mn}) = 3$

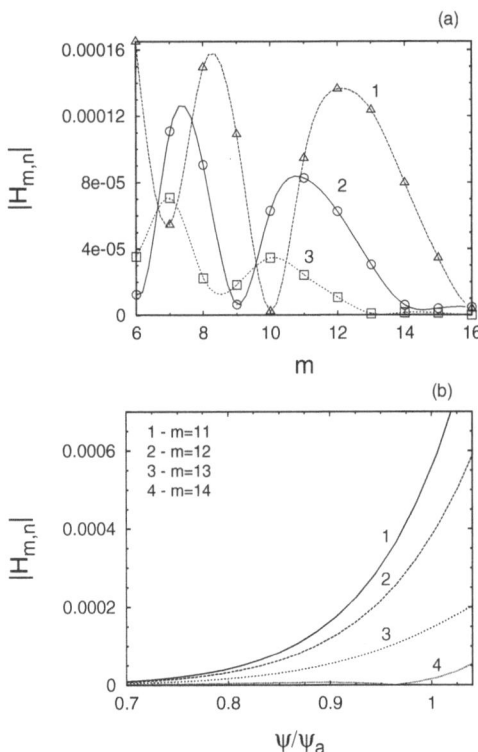

Fig. 11.5. (a) Hamiltonian perturbation spectrum $|H_{mn}(\psi)|$ $(n = 4)$ for different values of β_{pol} at the resonant magnetic surface ψ_{mn} $(m : n = 12 : 4)$: curve 1 corresponds to $\beta_{pol} = 0.2$, curve 2 $-$ to $\beta_{pol} = 1$, curve 3 $-$ to $\beta_{pol} = 1.8$. (b) Dependence of $|H_{mn}(\psi)|$ on the toroidal flux ψ normalized to one ψ_a corresponding the last magnetic surface with the radius $r = a$. Plasma current $I_p = 0.42$ MA, toroidal field $B_\varphi = 1.875$ T. The plasma radius $a = 46.7$ cm, $R_0 = 175$ cm, $R_a = 174$ cm

As seen from Fig. 11.4 the q-profiles are significantly modified with a variation of β_{pol}. The locations, ψ_{mn}, of the resonant magnetic surface, $q(\psi_{mn}) = m/n$ are shifted toward inward the plasma with increasing the plasma β_{pol}. Particularly, the resonant $q = 3$ magnetic surface is shifted by $\Delta\psi \approx 0.05\psi_a$ when β_{pol} is changed from 1.8 to 1. Since $\psi/\psi_a \approx r^2/a^2$, one can obtain that $\Delta r \approx (a/2)\sqrt{\psi_a/\psi}(\Delta\psi/\psi_a)$.

For the plasma of radius $a = 46$ cm the radial shift of the resonant surface is $\Delta r \approx 1.2$ cm. It shows that the threshold of the formation of stochastic layer changes drastically with the plasma β_{pol}. This is because of the rapid variation of the perturbation field modes $H_{mn}(\psi)$ with ψ (or r). According to (11.5), $H_{mn}(\psi) \propto \psi^{m/2\beta_1}$ (see also Fig. 11.5b), and it changes a several times when ψ varies to $\Delta\psi \approx 0.05\psi_a$.

From Fig. 11.5a also follows that only a group of few poloidal modes contributes to the destruction of resonant magnetic surfaces located near the magnetic surface with $q = 3$. The central mode m_c^* of this group is shifted to small modes m with increasing β_{pol}.

11.4 Formation of the Ergodic Zone

Below we study the formation of the stochastic layer of field lines at the plasma edge in the presence of magnetic perturbations. The Hamiltonian equations of field lines (10.44) in this case are integrated using the mappings (10.63). The perturbation modes $H_{mn}(\psi)$ and the safety factor $q(\psi)$ are found using the cubic spline interpolations of their precomputed values.

The stochastic layer at the plasma edge is formed due to the interactions of several magnetic islands created on the resonant magnetic surfaces ψ_{mn}, $q(\psi_{mn}) = m/n$, $(m_c^* - \Delta m/2 < m < m_c^* + \Delta m/2)$. The typical picture of the formation of the ergodic zone is illustrated in Fig. 11.6 by plotting the Poincaré sections of field lines in the (ϑ, ψ) plane for the different levels of magnetic perturbations. At the sufficiently small magnetic perturbation the resonant magnetic surfaces $(m : n)$, $(m = 9, \dots 14)$, are destroyed and the chains of isolated islands are formed as shown in Fig. 11.6a. The increase of the current I_c on the divertor coils up to $I_c = 7.5$ kA results in overlapping of three magnetic island with the poloidal mode numbers $m = 11, \dots 13$, and in the formation of the stochastic layer of field lines bounded in a finite area along the radial coordinate ψ. The magnetic island with $m = 14$ is isolated. Chaotic field lines is this stochastic layer are closed and they cannot reach the divertor plates. Fig. 11.6b shows this case. Further increase of the perturbation current to $I_c \approx 10.5$ kA destroys the barrier located between the resonant magnetic surfaces $m = 14$ and the stochastic layer. However, the field lines are still confined and they do not reach the divertor plate. At the maximal current $I_c = 15$ kA one obtains the well–developed stochastic layer with field lines open to the divertor plates which is shown in Fig. 11.6c, d. Figure 11.6d displays the structure of ergodic zone in the geometrical coordinate system (θ, r). A small rectangular area on the top of the plot corresponds to the divertor plate.

In general the formation of the ergodic zone depends on the many parameters of the plasma. It can be varied by changing the profile of the safety factor $q(\psi)$, the plasma parameter β_{pol}, the perturbation current I_c. Different aspects of this problem including the statistical properties of field lines are discussed by Abdullaev et al. (1999); Finken et al. (1999); Abdullaev et al. (2001, 2003).

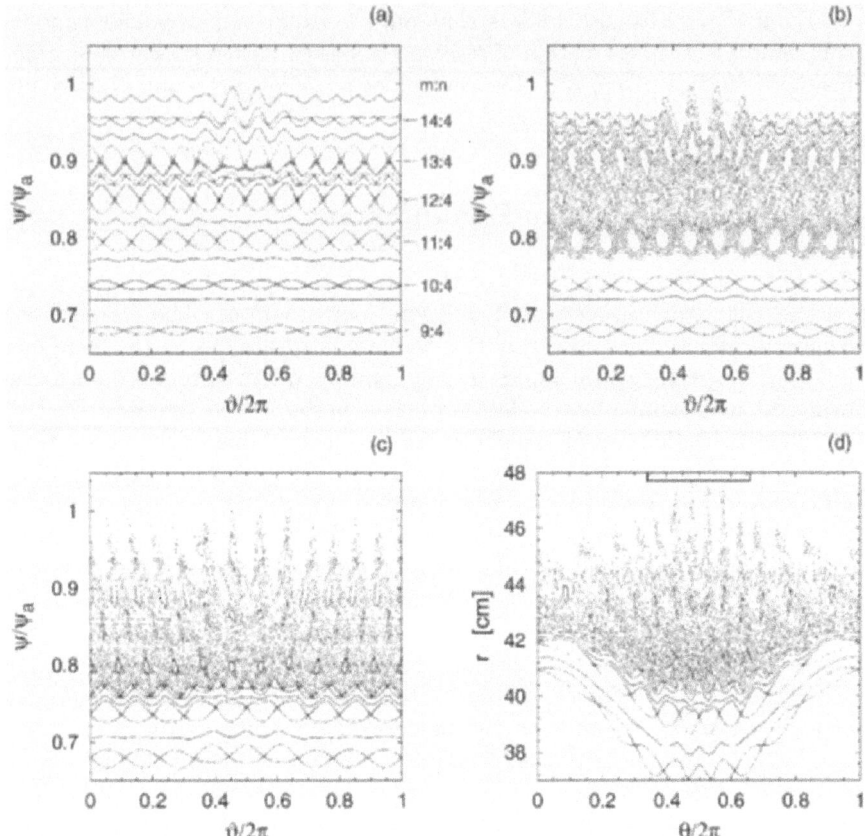

Fig. 11.6. Formation of stochastic layer of magnetic field lines with increasing the perturbation current I_c: (**a**) perturbation current $I_c = 3$ kA; (**b**) $I_c = 7.5$ kA; (**c**) $I_c = 15$ kA; (**d**) the same as in (**c**) but on the geometrical space (θ, r). The plasma parameter $\beta_{pol} = 1$, the plasma current $I_p = 420$ kA

11.5 Statistical Properties of Field Lines

The chaotic field lines in the ergodic layer contribute to the radial energy and particle transport at the plasma edge in addition to the perpendicular energy transport caused by turbulent processes in the plasma. To illustrate this consider for simplicity a collisionless plasma. Since electrons predominantly follow magnetic field lines the radial transport of electrons is determined by the radial deviation of field lines from the magnetic surfaces. Indeed, the radial diffusion coefficient of electrons is defined as $D_e = \langle (\Delta r)^2 \rangle / 2 \Delta t$ where Δr is a random radial advance of particle during time period Δt, and $\langle (\cdots)^2 \rangle$ stands for averaging over magnetic surface. Suppose that v_e is the thermal velocity of electrons. They make a full toroidal turn in time $\Delta t = \Delta l / v_e$

where Δl is the length of field line. Therefore

$$D_e = \frac{\langle (\Delta r)^2 \rangle}{2\Delta t} = \frac{\Delta l}{\Delta t} \frac{\langle (\Delta r)^2 \rangle}{2\Delta l} = D_{FL} v_e \ , \tag{11.6}$$

where

$$D_{FL} = \frac{\langle (\Delta r)^2 \rangle}{2\Delta l} \ , \tag{11.7}$$

defines the diffusion coefficient of field lines. It means that in the collisionless plasma the diffusion of charged particles is mainly determined by the radial diffusion of field lines[1]. Below we discuss the problem of determination of field line diffusion coefficients.

11.5.1 Global and Local Diffusion Coefficients

Let $\sigma^2(l)$ is the second order displacement moment,

$$\sigma(l) = \langle (r - \langle r(l) \rangle)^2 \rangle = \frac{1}{N} \sum_{i=1}^{N} (r_i(l) - \langle r \rangle)^2 \ , \tag{11.8}$$

where $\langle (\ldots) \rangle$ means over a set of initial conditions at $l = 0$ taken on a certain magnetic surface.

First consider the definition of diffusion coefficients in an unlimited domain stochastic layer[2]. In this case the asymptotics of $\sigma^2(l)$ at the large distance l has the following general behavior,

$$\sigma^2(l) = 2D_{FL} l^\gamma \ , \qquad l \to \infty \ , \tag{11.9}$$

where D_{FL} is a constant coefficient (see also Sect. 9.1.3 and (9.14)). For a normal Gaussian process the exponent $\gamma = 1$, and the coefficient D_{FL} is defined as a field line diffusion coefficient, $D_{FL} = \sigma^2(l)/2l$, $l \to \infty$. Such a diffusion coefficient can be called as *global* since it characterizes a global diffusive behavior of a system.

However, in the stochastic system with a finite domain, as the ergodic zone of field lines, the asymptotics (11.9) is not valid at long distances l, since in this case $\sigma^2(l) \to$ constant at $l \to \infty$, i.e., one cannot introduce the global diffusion coefficient D_{FL} defined as a ratio $D_{FL} = \sigma^2(l)/2l$, $l \to \infty$.

Nevertheless, in order to describe a transport in the stochastic layer one can introduce a *local diffusion coefficient* $D_{FL}(r)$. It gives a quantitative measure of field line diffusion near the given magnetic surface of radius r. Below we give analytical and numerical determinations of diffusion coefficients.

---

[1] Actually the radial transport of charged particles in the plasma is more complicated. It is determined not only due to collisions of particles but also the different kind of electromagnetic instabilities developing in the plasma (see Wesson (2004))

[2] As an example one can refer to the astrophysical and space plasmas where the magnetic field line diffusion represents the main sources of particle transport (see the paper by Pommois et.al (1999) and references therein).

11.5.2 Quasilinear Diffusion Coefficients

For the highly developed ergodic zone the diffusion coefficient D_{FL} can be obtained from the Hamiltonian equations of field lines with the Hamiltonian (10.33), (10.34). According to the quasilinear approximation (see Sect. 9.1.2) the diffusion coefficient D for the Hamiltonian system (9.10) is given by (9.13). The quasilinear diffusion coefficient, D_M, of magnetic field lines defined as a random walk process along the toroidal flux ψ, i.e., $D_M = \langle (\Delta\psi)^2 \rangle / 2\Delta\varphi$, can be obtained from (9.13) by replacing the action I by the toroidal flux ψ and the frequency of motion $\omega(I)$ by the inverse safety factor $q^{-1}(\psi)$ and the perturbation frequency Ω by n:

$$D_M = \pi\epsilon^2 \sum_{m,n} m^2 \, |H_{mn}(\psi)|^2 \, \delta\left(\frac{m}{q(\psi)} - n\right) .$$

The relation between the diffusion coefficient D_{FL} and D_M in the magnetic flux, ψ, and the radial r coordinates can be found using the relation $\psi = r^2/2R^2$:

$$D_{FL}^{(Q)} = \frac{R_0^3}{r^2} D_M = \pi\epsilon^2 \frac{R_0^3}{r^2} \sum_{m,n} m^2 \, |H_{mn}(\psi)|^2 \, \delta\left(\frac{m}{q(\psi)} - n\right) . \quad (11.10)$$

It coincides with the traditionally used quasilinear formula for D_{FL} (see Ghendrih et al. (1996)) if the term $\epsilon m H_{mn}$ is replaced by $r B_{mn}/B_0 R_0$ where B_{mn} is the Fourier expansion coefficient of the perturbed field B_r obtained using the relations (10.37), (10.38) and (10.41), i.e.,

$$D_{FL}^{(Q)} = \pi R_0 \sum_{m,n} \left(\frac{B_{mn}(r)}{B_0}\right)^2 \delta\left(\frac{m}{q(\psi)} - n\right) . \quad (11.11)$$

11.5.3 Numerical Calculation of Field Line Diffusion Coefficients

To determine a local diffusion coefficient, $D_{FL}(r)$, numerically we calculate the second order radial displacement moments according to (11.8) performing the averaging over a set of initial field lines with initial angle θ being uniformly distributed on the magnetic surface $r = r_0(\theta)$.

A typical dependence $\sigma_{r_0}^2(l)$ on l in the ergodic zone is shown in Fig. 11.7a for the two cases of open and closed field lines: curve 1 describes the case of closed ergodic zone; curve 2 corresponds the case of open field lines. In the first case field lines are confined in the ergodic zone and they don't reach the divertor plate, while in the second case the field lines are open and they reach the divertor plate after a certain number of poloidal turns. It has following features: $\sigma_{r_0}^2(l)$ grows with l up to a certain distance, when field lines reach the boundaries of the ergodic zone, and then it tends to a constant value as

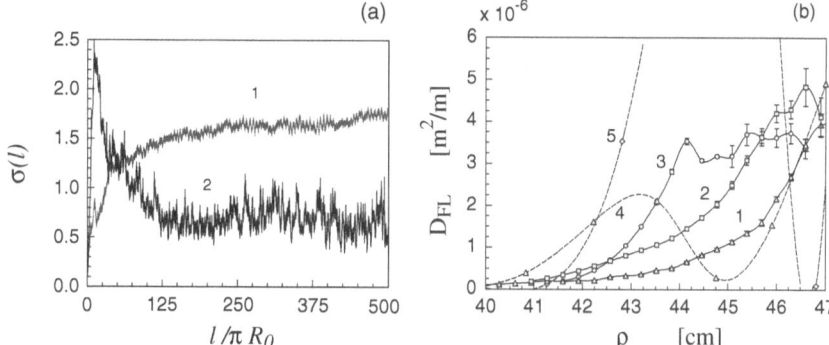

Fig. 11.7. (a) Typical behavior of the mean square radial displacement $\sigma^2(l)$ vs the length l along the toroidal angle φ for the two cases: curve 1 describes the case of the closed ergodic zone; curve 2 corresponds to the case with open field lines. (b) Radial profiles of local field line diffusion coefficients D_{FL} for different plasma currents: curve 1 corresponds to $I_p = 460$ kA, curve 2 − to $I_p = 520$ kA, curve 3 − to $I_p = 580$ kA. Quasilinear diffusion coefficients D_Q are plotted by dashed curves: 4 − for $I_p = 460$ kA, and 5 − for $I_p = 520$ kA. The plasma $\beta_{pol} = 1$

shown by curve 1 in Fig. 11.7a or it decreases when the field lines leave the ergodic zone hitting the divertor plates (curve 2 Fig. 11.7a).

As was mentioned above in this situation one cannot introduce a global diffusion coefficient $D = \sigma_r^2(l)/2l$, $(l \to \infty)$ as in the case of unlimited stochastic domain. However, one can introduce a local diffusion coefficient $D_{FL}(r)$ as $\sigma_r^2(l)/2l$ which is valid for the initial linear growth regime of $\sigma_r^2(l)$ with l (Abdullaev et al. (1999)). Typical profiles of $D_{FL}(r)$ as a function of magnetic surface radius r are presented in Fig. 11.7b for the different plasma currents. The corresponding quasilinear diffusion coefficients D_{FL} (11.10) are also plotted in this figure. One can see that D_{FL} grows with r monotonically up to the certain radius ρ_l then it decays in the zone $r > r_l$ where the field lines hit the divertor plates in very short lengths. The radius r_l characterizes the inner boundary of the zone of almost regular (non-chaotic) field lines with very short wall to wall connection lengths. This zone is called as the *laminar zone* because of resemblance with the laminar flow in hydrodynamics. The width of the laminar zone grows with the plasma current I_p.

Usually the quasilinear diffusion coefficients overestimates the diffusion transport rate in the DED ergodic zone. The quasilinear theory is not valid in this case since the ergodic zone is formed by overlapping only a few neighboring magnetic islands.

One should note that the local and quasilinear diffusion coefficients do not completely describe the transport processes in the ergodic zone. They are valid only for the highly developed ergodic zone. The typical ergodic zone at the plasma edge is not well-developed, and it consists of areas with the remnants of magnetic islands as well as areas with almost regular field lines with short connection lengths (the laminar zone). The transport processes in such an ergodic zone cannot be simply described by the quasilinear theory.

11.6 Ergodic Divertor as a Chaotic Scattering System

Field lines at the plasma edge with initial coordinates located in the stochastic layer eventually leave the plasma region hitting the divertor plate (expect for those field lines being trapped inside the magnetic islands). In this sense the ergodic zone with open field lines can be viewed as chaotic scattering system[3], whereby field lines enter into the plasma edge from the divertor plate and leave when hitting it again after a certain number of poloidal turns, N_p, as illustrated in Fig. 11.8.

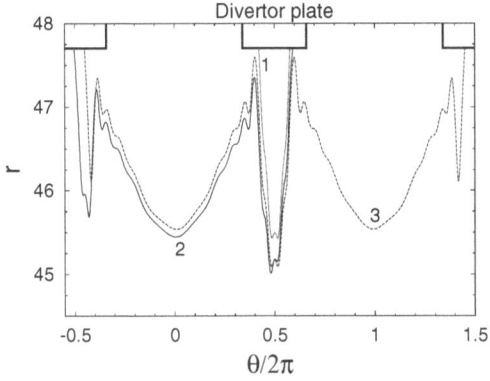

Fig. 11.8. Field lines connecting the divertor plate with itself after a several turns, N_p, along the poloidal angle: curve 1 describes $N_p = 0$, curve 2 corresponds to $N_p = 1$, and curve 3 – to $N_p = 2$

In chaotic scattering systems, a trajectory may leave a system in one of several different ways. The space of initial coordinates corresponding to the various exit ways are separated by a boundary, which may be a fractal Bleher et al. (1990). The set of initial conditions for which trajectories leave the system in a particular way is called a *basin* of a particular mode. In the case of the ergodic divertor it is convenient to classify field lines by the number of poloidal turns N_p. Indeed, the perturbation field created by divertor coils are localized on the HFS, and field lines enter into the plasma and leave it on this side making almost full poloidal turns (see Fig. 11.8). Therefore, the set of initial conditions for which field lines cross the section $\theta = 0$ (mod 2π) with the same number of times may be referred as the basin of a particular number

---

[3] A definition of chaotic scattering systems can be found in Eckhard (1988); Bleher et al. (1989); Tel and Ott (1993) and references therein.

of poloidal turns N_p. Field lines with $N_p = 0$ that do not cross the section $\theta = 0$ belong to a special set called a *private flux zone*. Spatial structures of boundaries of basins belonging to different N_p reveal fine details of the structure of field lines which cannot be seen in Poincaré sections. Below we study the structure of these basins by plotting the contours of N_p within the plasma edge and on the divertor plate.

11.6.1 Basin Boundary Structure at the Plasma Edge

First we consider the structure of basin boundaries referring to the basin as the set of points (ϑ, ψ) at the given poloidal section $\varphi = $ constant which are crossed by field lines with a particular number of poloidal turns N_p. The procedure to obtain these plots is the following.

At the poloidal plane $\varphi = 0$ the field line with a given initial coordinate (ϑ, ψ) is traced by iterating the map (10.63) along the positive and negative directions of the toroidal angle φ until a field line reaches the divertor plate. Then we determine a fractional number of poloidal turns N_{pol} as the ratio of the total change of the poloidal angle $\Delta\theta$ to the full circle 2π, i.e., $N_{pol} = \Delta\theta/2\pi$. The values of N_{pol} computed in this way are close to integer numbers although they are not exactly integer. Let N_p be the integer number closest to N_{pol}. Areas in the (ϑ, ψ)-plane with different poloidal turns N_p are topologically different. The dependence of N_{pol} on the initial coordinates (ϑ, ψ) is displayed by a contour plot with contour lines separating the basins of different poloidal turns N_p. The example such a plot is presented in Fig. 11.9.

The basin with one poloidal turn has a non-fractal boundary with the private flux zone. But it may have fractal boundaries with the basins corresponding to two and more poloidal turns. For $N_p \geq 2$ there are several topologically different basins related to the same N_p. As seen from Fig. 11.9a the relatively large basins of a few poloidal turns $N_p \leq 3$ at the plasma edge are clearly separated by non-fractal boundaries. But they are alternating with the long dark elongated areas (or stripes) containing the basins for a few poloidal turns up to very large N_p. In Eich et al. (2000) these stripes were called *"fingers"*. At the HFS some of these stripes are radially extended toward the divertor plate.

The structure of stripes has a complicated fractal nature. In order to study the fine structure of stripes we have magnified the area of the stripe with the fine resolution of basins of higher poloidal turns. In Fig. 11.9b shows a blow up of the rectangular area in Fig. 11.9a. It shows that the basins at the stripe are highly elongated and the boundaries between them have fractal structure, i.e., the stripes consists of layered basins of different poloidal turns with a self–similar behavior at different spatial scales. As seen from Fig. 11.9b the basins of field lines with a few poloidal turns N_p are "sandwiched" between basins for field lines with large numbers of poloidal turns $N_p \gg 1$.

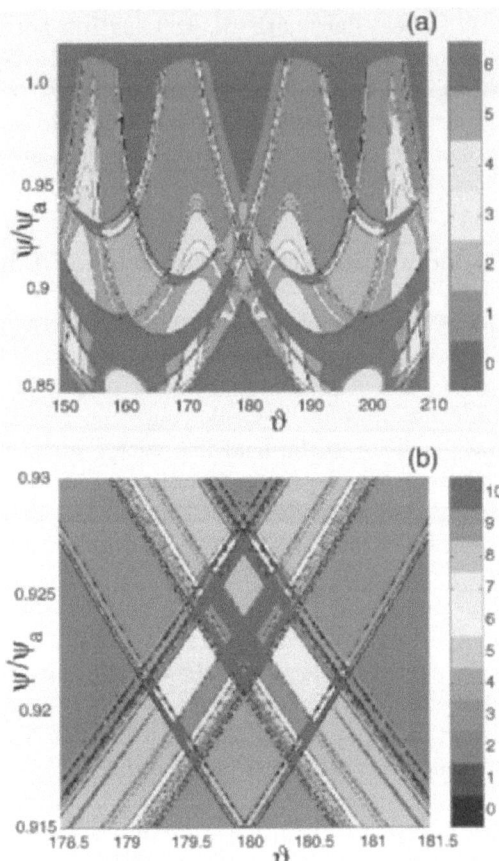

Fig. 11.9. Contour plots of N_p in the (ϑ, ψ)-plane for the plasma currents $I_p =$ 580 kA. (**a**) shows the sector $150° < \vartheta < 210°$, (**b**) Expanded view of the rectangular area in (**a**)

11.6.2 Magnetic Footprints

In order to study the basin boundary structure on the divertor plate we will use the following procedure. We follow a field line which enters into the plasma starting from the divertor plate with a given initial coordinate (φ, θ) and returns back to the plate after a certain number of poloidal turns N_p. The set of initial conditions (φ, θ) with a particular number N_p determines a basin. The whole picture of basin boundaries with $N_p \geq 1$ on the plasma wall determines a structure known as *magnetic footprints*. Similar to the stripe at the plasma edge the magnetic footprints have a fractal structure as well. A typical structure of magnetic footprints is displayed in Fig. 11.10: a) in the finite poloidal section; b) shows the expanded view of the rectangular region in Fig. 11.10a.

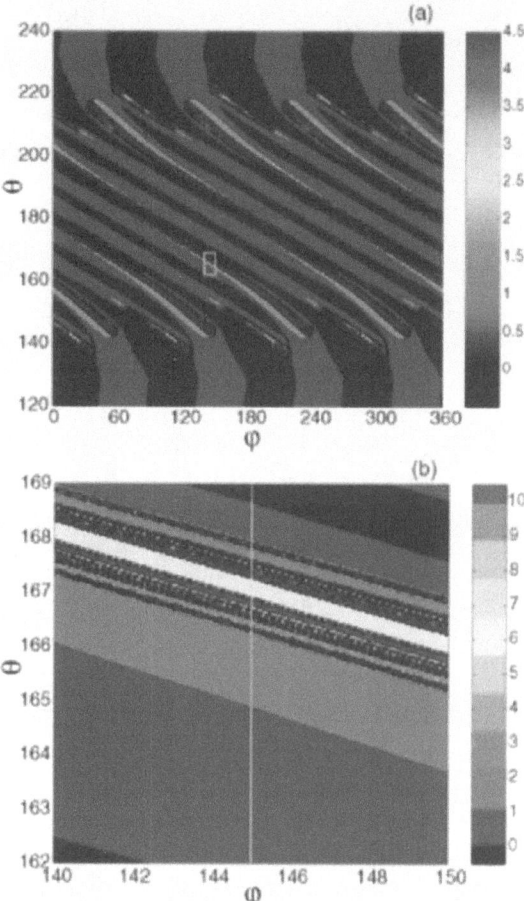

Fig. 11.10. Basin boundary structure (magnetic footprints) on the (φ, θ) –plane at the divertor plane: (**a**) On the entire plane; (**b**) expanded view of the rectangular region on the stripe shown in (a). The plasma parameters are the same as for the case shown Fig. 11.9

One can see from Figs. 11.10 that the field lines can enter into the plasma (or hit the divertor plate from the plasma side) only along the four pairs of narrow helical stripes. (Dark blue areas in Figures correspond to the field lines in a private flux zone). The distance between stripes of each pairs depends on the plasma current I_p Abdullaev et al. (2001). Each helical stripe has a fractal structure and it consists of layered basins of different poloidal turns (see Fig. 11.10a, b). The width of layers is changing along the toroidal direction φ. The area of the basin with one poloidal turn $N_p = 1$ is the largest. For the higher $N_p > 1$ corresponding areas of basins are drastically decreased in size. The boundaries between these basins are fractal.

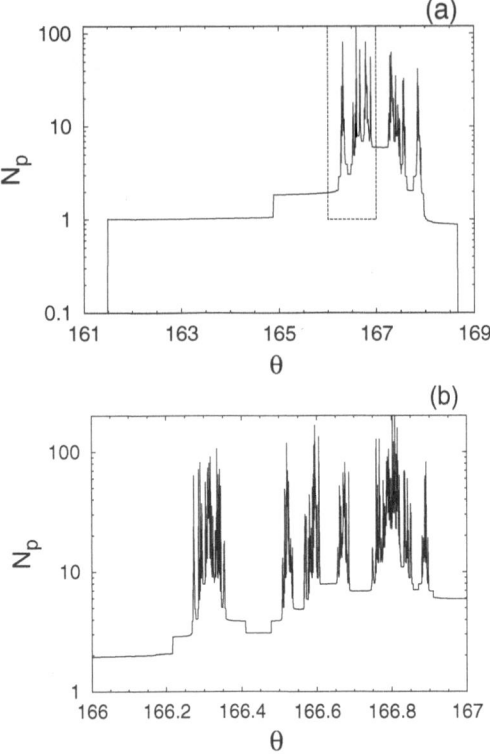

Fig. 11.11. (a): Fractal dependence of N_{pol} on θ along the line with the fixed toroidal angle $\varphi = 145°$ (the yellow line in Fig. 11.10b); (b): Expanded view of the dashed rectangular area shown in (a)

The fine structure of helical stripes can be revealed by studying the dependence of N_p on the poloidal angle θ at fixed toroidal angle φ. Such a dependence of N_p on θ is described by Cantor-like, fractal curves. It is presented in Fig. 11.11 at the fixed toroidal angle $\varphi = 145°$. The curve in Fig. 11.11a describes the poloidal dependence of N_p along the (yellow) straight line shown in Fig. 11.10b, while Fig. 11.11b shows the expanded view of the dashed rectangular area in Fig. 11.11a. They clearly show areas of field lines connecting plate to plate in one, two, three and more poloidal turns N_{pol}. These areas are described by almost horizontal steps in the fractal curve. The width of layers becomes smaller with increasing N_{pol}.

The structure of the helical stripes plays an important role for heat and particle deposition on the divertor plates. Indeed, the basins with $N_p \gg 1$ correspond to the field lines coming from deep within ergodic zone. These field lines may bring high energetic particles to the wall because the particles predominantly move along field lines. Therefore one can expect that the spatial distribution of power deposition within the helical stripes will depend on the spatial structure of basins with $N_p > 1$. The cross–field diffusion of particles broadens the spatial distribution of power deposition around the maxima located at the basins corresponding to large number of poloidal turns. This conclusion is confirmed with the Monte–Carlo simulations of heat and energy transport in the ergodic zone (Runov et al. (2001)). Recent experimental measurements of heat deposition patterns on the divertor plates performed by Jakubowski et al. (2004) in the TEXTOR-DED tokamak the confirm the above theoretical predictions on helical stripe structures.

11.7 Conclusion

In this section we have considered the application of the mapping method to study magnetic lines in divertor tokamaks. In ergodic divertor tokamaks the mapping approach is based on the fact that the equations of field lines in the presence of external nonaxisymmetric magnetic perturbations can be always formulated in the generic Hamiltonian form, given by (10.5), (10.33), (10.34) in terms of the toroidal flux, ψ, and the intrinsic poloidal angle, ϑ. Then the general mapping method developed in the Chap. 4 can be directly applied to study a magnetic system. It runs much faster than the the direct integration of field lines equations and it can be applied the case with the moderately large perturbations.

12 Mappings of Magnetic Field Lines in Poloidal Divertor Tokamaks

Another important concept to control the plasma edge in tokamaks is the so-called *poloidal divertor tokamaks* (see Wesson (2004)). The magnetic configuration of these tokamaks contains a magnetic surface (a *magnetic separatrix*) sharply separating closed field lines on nested magnetic surfaces from open field lines hitting the walls of a fusion device. It has one (or two) singular points, *X-points*, on the poloidal section where the poloidal components of the magnetic field are zeros. These configurations are schematically shown in Fig. 12.1): a) a so-called single–null poloidal divertor; b) a double–null poloidal divertor. Such configurations of the magnetic field are created by one or two external current coils parallel to the plasma current, respectively. Magnetic fusion devices with a poloidal divertor provide an improved energy confinement of the plasma and diverts particles and heat efficiently into divertor plates in a special volume, from where they are pumped away. The future International Thermonuclear Experimental Reactor, *ITER*, is designed as a poloidal divertor tokamak.

Magnetic field lines in such a magnetic configuration is described by the Hamiltonian system with hyperbolic fixed points. The magnetic separatrix and the X-points correspond to the separatrices and the hyperbolic saddle points, respectively. Typically any small non-axisymmetric magnetic perturbations destroy the magnetic separatrix replacing it by the stochastic layer of field lines.

The nature of these magnetic perturbations may range from magnetic fluctuations produced by plasma instabilities, field errors etc. Typically, magnitudes of magnetic perturbation fields are small. For instance, the amplitude of error field are $\delta B/B_0 \approx 10^{-4}$ (Pomphrey and Reiman (1992)). In present day tokamaks one uses specially created external magnetic fields to null the magnetic perturbation caused by magnetohydrodynamic instabilities. Recent experiments in the Doublet III D (DIII-D) tokamak by Evans et al. (2004), which has a magnetic configuration with the separatrix, showed that the external magnetic perturbation created at the plasma edge suppresses the so-called large edge–localized modes (ELMs) in a high confinement regime known as H-mode. The ELMs are known as a magnetohydrodymical turbulent processes at the plasma edge which cause large, repetitive heat and particle loading to the divertor plates. In general, they are considered as

S.S. Abdullaev: *Construction of Mappings for Hamiltonian Systems and Their Applications*, Lect. Notes Phys. **691**, 275–298 (2006)
www.springerlink.com

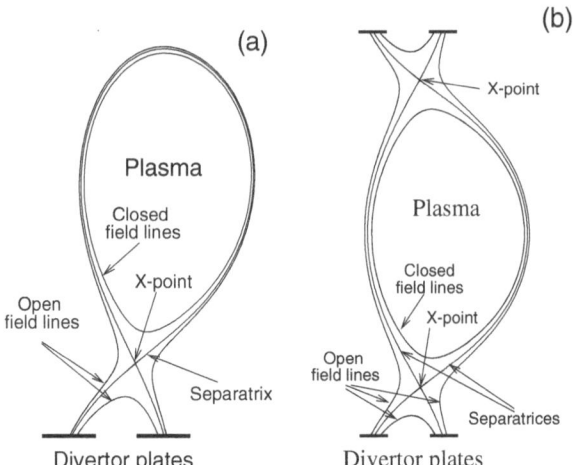

Fig. 12.1. Schematic view of magnetic configurations in poloidal divertor toka-maks: (a) Single-null divertor; (b) Double-null divertor

undesirable for burning plasma experiments since they deteriorate the plasma confinement and performance, reduce the lifetime of divertor plates due to increased erosion from impulsive heat and particle fluxes. A stochastic layer at the plasma boundary might reduce ELMs. Therefore, a study of stochastic field lines near the separatrix and their structure, the pattern of magnetic footprints on the divertor plates are important issues to understand the particle and energy transport at the edge of plasmas in tokamak fusion devices (Boozer and Rechester (1978); Pomphrey and Reiman (1992); Reiman (1996); Evans et al. (2002)).

Below we shortly recall the main theoretical studies of magnetic field lines in poloidal divertor tokamaks. In the early theoretical works by Tomita et al. (1977, 1978); Boozer and Rechester (1978) the effect of magnetic perturbations on the divertor separatrix has been studied by employing the Chirikov overlapping criteria to estimate the width of the stochastic layer in the absence of divertor plates. The direct numerical integration of the equations of field lines in the DIII-D tokamak in the presence of field irregularities has been performed by LaHaye (1991). The effect of magnetic perturbations created by special coils in this tokamak has been analyzed by Evans et al. (2002) (see also references therein) using a numerical code which takes into account a real magnetic geometry of the system. 'Wire' models have been used by Pomphrey and Reiman (1992); Reiman (1996) to study the effect of magnetic field errors on the formation of the stochastic layer near the magnetic separatrix and magnetic footprints on the divertor plates.

Mapping approaches to study magnetic field lines near the separatrix in divertor tokamaks have been developed by Punjabi et al. (1992, 1994, 1996,

1997) and Yamagishi (1995); Abdullaev and Zaslavsky (1995, 1996); Abdullaev and Finken (1998). The mappings are computationally efficient and they run much faster than the numerical integration of field line equations. Punjabi et al. (1992, 1994, 1996) proposed simple algebraic mapping models, called *tokamak divertor maps* to describe the field lines in poloidal divertor tokamaks. Later Punjabi et al. (1997, 2003); Ali et al. (2004) obtained the symmetric form of this map and its more sophisticated generalization. These area – preserving maps are simple algebraic difference equations. They allow to study generic features of the structure of field lines near X-points and the magnetic footprints affected by asymmetric magnetic perturbations. However, since these simple maps are not deduced from the field line equations it is not clear how parameters of maps are related to magnetic field configuration. It makes difficult to apply these maps to analyze the magnetic structure of real poloidal divertor tokamaks.

A separatrix mapping approach to estimate the width of the stochastic layer formed near the magnetic separatrix has been considered by Yamagishi (1995). Following Chirikov (1979) he derived the separatrix map for field lines near the magnetic separatrix in a single null poloidal divertor tokamak. This separatrix map allowed to plot the structure of the stochastic layer and estimate its width. In Abdullaev and Zaslavsky (1995, 1996); Abdullaev and Finken (1998) the separatrix mapping method has been generalized to describe field line near the X-points and on the divertor plates. The method of construction of the separatrix mapping from the equations of field lines in poloidal divertor tokamaks with an arbitrary magnetic configuration has been proposed. It allowed to obtain not only the structure of the stochastic layer, and also the magnetic footprint patterns on the divertor plates.

Below we present the separatrix mapping method to describe the magnetic field line near the magnetic separatrix for arbitrary divertor tokamaks. We will use the general separatrix mappings derived in Sect. 5.

12.1 Field Lines in Equilibrium Plasmas Near the Separatrix

According to Hamiltonian equations (10.11), (10.12) field lines are determined by the poloidal flux, H, related to the vector potential $A_\varphi(R, Z, \varphi)$: $H(z, p_z, \varphi) = -R(p_z)A_\varphi(R(p_z), R_0 z, \varphi)/R_0^2 B_0$. Consider the unperturbed case when the magnetic field is homogeneous along the toroidal angle φ: $A_\varphi = A_\varphi(R, Z)$. The magnetic surfaces then are determined by contour line of $RA_\varphi(R, Z) = $ const. The field line equations are determined by the unperturbed Hamiltonian $H_0(z, p_z) = -R(p_z)A_\varphi(R(p_z), R_0 z)/R_0^2 B_0$. The normalized toroidal flux, ψ, and the safety factor $q(H)$ on the magnetic surface $H = H_0(z, p_z) = $ const determined from the field line equations are given by (10.14) and (10.15), respectively. On the magnetic surface one has

$H = H(\psi)$, and according to (10.7), the safety factor, $q(H)$, is determined by $q(H) = d\psi/dH$.

According to (10.9) the X-points (R_s, Z_s), i.e., the nulls of the poloidal field (B_R, B_Z), on the poloidal section correspond to the hyperbolic fixed points (z_s, p_s) of Hamiltonian system (10.11) where

$$\frac{dz}{d\varphi} = \frac{\partial H_0}{\partial p_z} = 0, \qquad \frac{dp_z}{d\varphi} = -\frac{\partial H_0}{\partial z} = 0 . \qquad (12.1)$$

In the single null divertor tokamaks there is only one X-point connecting with itself by the homoclinic orbit on the magnetic separatrix. In the double null divertor tokamaks there are two X-points and they are connected with two heteroclinic orbits on the two separatrices, respectively (see Fig. 12.1).

Consider field lines near the X-point (see Fig. 12.2). The unperturbed Hamiltonian $H_0(z, p_z)$ can be expanded in a series of powers of $(z - z_s)$, $(p_z - p_s)$ near the X-points:

$$H_0(z, p_z) = H_0(z_s, p_s) + \frac{1}{2}H_{zz}(z - z_s)^2 + H_{zp}(z - z_s)(p_z - p_s)$$
$$+ \frac{1}{2}H_{pp}(p_z - p_s)^2 + O[(z - z_s)^3, (p_z - p_s)^3] , \qquad (12.2)$$

where H_{zz}, H_{zp}, and H_{pp} are second derivatives of H_0 with respect to (z, p_z) taken at the hyperbolic fixed point. By the linear transformation of variable (see Fig. 12.2):

$$\xi = (z - z_s)\cos\alpha - (p_z - p_s)\sin\alpha ,$$
$$\eta = (z - z_s)\sin\alpha + (p_z - p_s)\cos\alpha ,$$

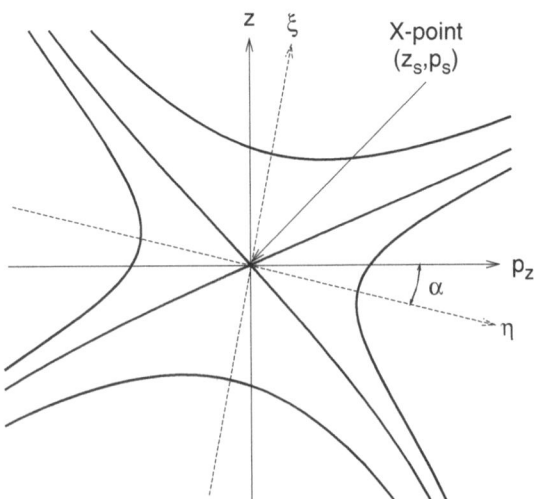

Fig. 12.2. Field lines near the X-point

the Hamiltonian (12.2) can be diagonalized:

$$H_0(z, p_z) = H_0(z_s, p_s) - \frac{|\lambda_1|}{2}\xi^2 + \frac{\lambda_2}{2}\eta^2 + O(\xi^3, \eta^3) , \qquad (12.3)$$

where λ_1, λ_2, ($\lambda_1 < 0, \lambda_2 > 0$) are the eigenvalues of the matrix

$$\begin{pmatrix} H_{zz} & H_{zp} \\ H_{zp} & H_{pp} \end{pmatrix} ,$$

and they are given by

$$(\lambda_1, \lambda_2) = \frac{H_{zz} + H_{pp}}{2} \pm \sqrt{\frac{(H_{zz} - H_{pp})^2}{4} + H_{zp}^2} .$$

The angle α is determined by $\tan 2\alpha = H_{zp}/(H_{zz} - H_{pp})$.

We introduce the relative poloidal magnetic flux h:

$$h = H_0(z, p_z) - H_0(z_s, p_s) . \qquad (12.4)$$

For the closed magnetic field lines the relative flux is negative, $h < 0$, and for the open field lines it is positive, $h > 0$. At the separatrices, $h = 0$ (see Fig. 12.1). A behavior of field lines near the separatrix and the X-points is generic (see Sect. 8.1.1). Particularly, according to (12.3), (8.2)–(8.5) the safety factor $q(h)$ (or inverse frequency of oscillations) has the following asymptotics:

$$q(h) = \frac{1}{2\pi\gamma} \ln \frac{Q}{|h|} + O(h) , \qquad |h| \to 0 , \qquad (12.5)$$

where γ is a parameter determined the expansion parameter λ_1, and λ_2 in (12.3): $\gamma = \sqrt{|\lambda_1 \lambda_2|}$, and Q is a positive constant. Unperturbed field line trajectories lie on the magnetic surfaces of constant h and they can be presented in the form $(R, Z) = R(\varphi; h), Z(\varphi; h)$.

The increment of toroidal angle $\Delta\varphi$ along the field lines near the X-point has also universal behavior. We calculate, for instance, $\Delta\varphi$ along field lines from the axis ξ at $\eta = 0$ (or from the axis η at $\xi = 0$) to the line $\eta = \text{const}$ ($\xi = \text{const}$). Using the field line equations near the X-point

$$\frac{d\xi}{d\varphi} = \lambda_2\eta , \qquad \frac{d\eta}{d\varphi} = \lambda_1\xi ,$$

and the relation $\eta = \sqrt{2h + |\lambda_1|\xi^2}/\sqrt{\lambda_2}$, we have

$$\Delta\varphi = \frac{1}{\sqrt{\lambda_2}} \int_{\xi_0}^{\xi} \frac{dx}{\sqrt{2h + |\lambda_1|x^2}} = \frac{1}{\gamma}\ln\left(\frac{\xi}{b} + \sqrt{\frac{\xi^2}{b^2} + p}\right) . \qquad (12.6)$$

Here $\xi_0 = 0$, $p = 1$ for $h > 0$ and $\xi_0 = b$, $p = -1$ for $h < 0$, where $b = \sqrt{2|h|/|\lambda_1|}$. At the limit $|h| \to 0$ this integral has the following asymptotics

$$\Delta\varphi \approx \frac{1}{\gamma}\ln\frac{2\xi}{\xi_0} = \frac{1}{2\gamma}\ln\frac{2\lambda_1\xi^2}{|h|} .$$

12.1.1 Magnetic Perturbations

Suppose that the magnetic perturbations are described by the toroidal component of the vector potential $A_\varphi^{(per)}(R, Z, \varphi)$ which represented as a Fourier series:

$$A_\varphi^{(per)}(R, Z, \varphi) = \epsilon B_0 R_0^2 R^{-1} \sum_n A_n(R, Z) \cos(n\varphi + \chi_n) , \qquad (12.7)$$

where n is the toroidal mode number, $\epsilon = \max|A_\varphi^{(per)}|/B_0 R_0$ is the dimensionless perturbation parameter. Using the expansion (12.7) we present the perturbed Hamiltonian $H_1(z, p_z, \varphi)$ in the form

$$H_1(z, p_z, \varphi) = \sum_n H_n(z, p_z) \cos(n\varphi + \chi_n) , \qquad (12.8)$$

where $H_n(z, p_z) = -A_n(R(p_z), R_0 z)$. In the presence of non-axisymmetric perturbations (12.7) the magnetic separatrices are destroyed. The field lines with the initial coordinates located within a certain distance from the separatrix become chaotic. These field lines are not confined in the plasma region, and they leave the plasma region hitting the divertor plates after a certain number of poloidal turns.

Below we describe the method of the separatrix mapping to study open chaotic field lines near the destroyed separatrix. This method is computationally efficient to study the structure of the stochastic layer and to obtain magnetic footprints on divertor plates.

12.1.2 Separatrix Map

We will use the mappings near the separatrix constructed in the Sect. 5, and particularly in Sect. 5.3.3. It is convenient to define these sections (Σ_s) near the X-points where field lines stay longer. In order to determine magnetic footprints we should construct also mappings to sections coinciding with the divertor plates (Σ_d). The locations of these sections in the poloidal plane and the magnetic configuration of the systems in the absence of the divertor plates are shown in Fig. 12.3: a) corresponds to the single null divertor tokamak; b) corresponds to the double null divertor tokamak. In the case of the single null divertor there are one X-point and two homoclinic saddle–saddle connections, C_p and C_c, in the plasma and coil regions, respectively (see Fig. 12.3a). In the double null divertor tokamak there are two X-points and two heteroclinic saddle saddle connections, C_{p1} and C_{p2}, in the plasma region, two homoclinic saddle saddle connections, C_{c1} and C_{c2}, in the lower and upper coil regions, respectively.

According to definitions given in Sect. 5, the cross section Σ_s consists of two stripes (segments in the poloidal (R, Z)-plane) along the ξ and η axes transversely crossing each other along the X-lines (at the X-point on the

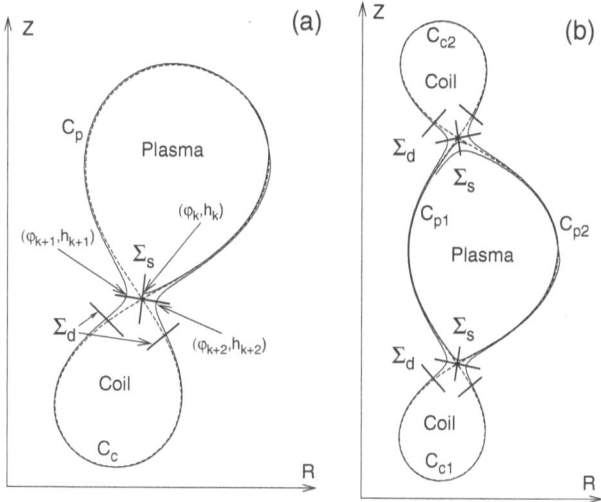

Fig. 12.3. Geometry of the separatrix map in a single null poloidal divertor toka-mak: **(a)** for single null divertor; **(b)** for the double null divertor. *Solid curves* describe perturbed field lines, *dashed curves* — the unperturbed separatrix

(R, Z)-plane). The unperturbed field lines cross these stripes transversely (see Fig. 12.2).

We study perturbed field lines near the separatrix. Let (φ_k, h_k) be the toroidal angle, φ and the poloidal flux, h, at the k-th crossing point of field lines with the cross section Σ_s. Our aim is to construct the mapping

$$(\varphi_{k+1}, h_{k+1}) = \hat{M}(\varphi_k, h_k) , \qquad (12.9)$$

connecting two consecutive crossing points of field line at the section(s) Σ_s.

In the case of the single null divertor there are two mappings $\hat{M}^{(j)}$, $j = (p, c)$ (12.9) along the separatrices C_p and C_c in the plasma and coil regions, respectively, which completely determine dynamics of perturbed field lines near the separatrix (see 12.3a).

For the double null divertor (Fig. 12.3b) there are four mappings corresponding to the four separatrices: C_{p1} and C_{p2} on the plasma region, C_{c1} and C_{c2} in the coil regions, respectively.

According to results obtained in Sect. 5 for the Hamiltonian system $H_0(z, p_z)$ in the presence of perturbation (12.8) the mapping along each sad-dle –saddle connection is given by (5.45) with the perturbation functions F and G given by (5.44). Replacing the frequency $\omega(h)$ by the inverse effective safety factor $q^{(j)}(h)$, the perturbation frequency Ω_n by the toroidal mode number n, and the time variable t by the toroidal angle φ we obtain the fol-lowing form of the mapping (12.9) along the j-th saddle–saddle connection:

$$h_{k+1} = h_k - \epsilon F^{(+)}(\varphi_k, h_{k+1}, h_k) ,$$

$$\varphi_{k+1} = \varphi_k + \pi \left[q^{(j)}(h_k) + q^{(j)}(h_{k+1}) \right] + \epsilon G^{(+)}(\varphi_k, h_{k+1}, h_k) \ ,$$

$$(12.10)$$

where

$$F^{(+)}(\varphi_k, h_{k+1}, h_k) = \sum_n n \Bigg(K_n(h_{k+1}) \sin \Phi_n^{(+)}(\varphi_k, h_k)$$

$$+ L_n(h_k) \cos \Phi_n^{(+)}(\varphi_k, h_k) \Bigg) \ ,$$

$$G^{(+)}(\varphi_k, h_{k+1}, h_k) = - \sum_n \Bigg(\frac{dK_n(h_{k+1})}{dh_{k+1}} \cos \Phi_n^{(+)}(\varphi_k, h_k)$$

$$- \frac{dL_n(h_{k+1})}{dh_{k+1}} \sin \Phi_n^{(+)}(\varphi_k, h_k) \Bigg) \ , \quad (12.11)$$

$$\Phi_n^{(+)}(\varphi_k, h_k) = n \left(\varphi_k + \pi q^{(j)}(h_k) \right) + \chi_n \ . \quad (12.12)$$

The effective safety factor $q^{(j)}(h)$ near the j-th saddle–saddle connection is given by

$$q^{(j)}(h) = \frac{1}{2\pi} \int\limits_{C^{(j)}} \frac{dz}{\partial H_0/\partial p_z} \ , \quad (12.13)$$

where the integral is taken along the contour $C^{(j)}$ of $H = H_0(z, p_z) = $ const connecting the sections Σ_s. In the single null divertor and in the coils regions in the double null divertor it coincides with the safety factor (10.15). Near the separatrix the safety factors $q^{(j)}(h)$ have the universal asymptotics of type (12.5) with the same parameter γ but different constant parameters $Q = Q_j$ corresponding to the different contours $C^{(j)}$.

The mappings (12.10)–(12.12) trace field lines along the positive direction of the toroidal angle φ. We call them the *forward maps*. For the complete description we need to determine the *backward map* to trace field lines along the negative direction of the toroidal angle φ. This map can be obtained from the forward maps (12.10) by the transformation $k \to k+1$:

$$h_{k+1} = h_k + \epsilon F^{(-)}(\varphi_k, h_{k+1}, h_k) \ ,$$

$$\varphi_{k+1} = \varphi_k - \pi \left[q^{(j)}(h_k) + q^{(j)}(h_{k+1}) \right] - \epsilon G^{(-)}(\varphi_k, h_{k+1}, h_k) \ , \quad (12.14)$$

where

$$F^{(-)}(\varphi_k, h_{k+1}, h_k) = \sum_n n \left(K_n(h_{k+1}) \sin \Phi_n^{(-)}(\varphi_k, h_k) \right.$$

$$\left. + L_n(h_k) \cos \Phi_n^{(-)}(\varphi_k, h_k) \right) ,$$

$$G^{(-)}(\varphi_k, h_{k+1}, h_k) = -\sum_n \left(\frac{dK_n(h_{k+1})}{dh_{k+1}} \cos \Phi_n^{(-)}(\varphi_k, h_k) \right.$$

$$\left. - \frac{dL_n(h_{k+1})}{dh_{k+1}} \sin \Phi_n^{(-)}(\varphi_k, h_k) \right) , \qquad (12.15)$$

$$\Phi_n^{(-)}(\varphi_k, h_k) = n \left(\varphi_k - \pi q^{(j)}(h_k) \right) + \chi_n . \qquad (12.16)$$

According to (5.27) the Melnikov type integrals $K_n(h)$ and $L_n(h)$ in (12.11) are given

$$K_n(h) + iL_n(h) = R_n(h) = \int\limits_{-\pi q(h)}^{\pi q(h)} V_n(h, \tau) e^{in\tau} d\tau , \qquad (12.17)$$

taken over the functions $V_n^{(j)}(h, \tau) \equiv H_n(z^{(j)}(h, \varphi - \varphi_0), p_z^{(j)}(h, \varphi - \varphi_0))$ along the unperturbed field lines near the j-th separatrix. At $h = 0$ the integral is taken along the unperturbed separatrix.

As was shown in Sect. 5.2.5 the integrals $R_n(h)$ can be presented as a sum of regular, $R_n^{(reg)}(h)$, and oscillatory, $R_n^{(osc)}(h)$, parts, i.e.,

$$R_n(h) = R_n^{(reg)}(h) + R_n^{(osc)}(h) . \qquad (12.18)$$

The regular part, $R_n^{(reg)}(h)$, is a smooth function of the relative poloidal flux h. The oscillatory part, $R_n^{(osc)}(h)$, is a rapidly oscillating function of h. The zeros of $R_n^{(reg)}(h)$, i.e., $R_n^{(reg)}(h_{mn}) = 0$, coincide with the resonant poloidal fluxes of primary resonances where $q(h_{mn}) = m/n$. As was shown in Sect. 5.3.1 that since field lines are mostly affected near the primary resonances where oscillatory terms of the integrals $R_n(h)$ vanish, we can retain only the smooth regular parts $R_n^{(reg)}(h)$.

Near the separatrix the mapping (12.10) can be simplified by replacing the integrals $K_n(h)$, $L_n(h)$ by their values $K_n(0)$, $L_n(0)$ at the separatrix $h = 0$ similar to the mapping (5.46).

$$h_{k+1} = h_k \mp \epsilon \sum_n n \left(K_n^{(j)}(0) \sin \left[n \left(\varphi_k \pm \pi q^{(j)}(h_k) \right) + \chi_n \right] \right.$$

$$\left. + L_n^{(j)}(0) \cos \left[n \left(\varphi_k \pm \pi q^{(j)}(h_k) \right) + \chi_n \right] \right) ,$$

$$\varphi_{k+1} = \varphi_k \pm \pi [q^{(j)}(h_k) + q^{(j)}(h_{k+1})] , \tag{12.19}$$

where the signs ($\pm$) correspond to the forward map, and the lower sign ($\pm$) correspond to the backward map.

The set of mappings $\hat{M}^{(j)}$ given by (12.19) completely describe perturbed field lines near the separatrices of magnetic system. They are determined only by the safety factors $q^{(j)}(h)$ (12.13) and the Melnikov type integrals $K_n^{(j)}(h)$, $L_n^{(j)}(h)$ (12.17).

12.1.3 Mappings to the Divertor Plates

To determine magnetic footprints on the divertor plates we need to construct the mapping of field lines from the sections Σ_s to the divertor sections Σ_d. Since these sections are located close to the X-point it takes for field lines less than half of poloidal turns to reach the divertor plate from the section Σ_s. These field lines can be traced along unperturbed field lines starting from its coordinates (φ_k, h_k) at Σ_s until they reach the divertor plate Σ_d. Let (φ_d, h_d) be the toroidal angle and the relative poloidal flux on the divertor plate. Then the map (φ_k, h_k) → (φ_d, h_d) is given by

$$h_d = h_k , \qquad \varphi_d = \varphi_k + \Delta\varphi(h_k) , \tag{12.20}$$

where $\Delta\varphi(h)$ is the increment of the toroidal angle φ necessary to reach the plate Σ_d along unperturbed field lines on the magnetic surface of constant $h = h(z, p_z)$. If the divertor plate Σ_d is sufficiently close to the X-point the quantity $\Delta\varphi(h)$ can be estimated by (12.6).

Magnetic footprints on the divertor plates are obtained by tracing field lines by the mappings (12.19) and (12.20) using the following procedure. We choose a set of field lines inside the plasma with coordinates (φ_0, h_0) at the section Σ_s. The relative poloidal flux h inside the plasma in negative, $h < 0$. We apply the forward and backward mappings given by (12.19) in the plasma region until field lines cross the unperturbed separatrix, when $h_k > 0$. Then using the map (12.20) we find the coordinates (φ_d, h_d) on the divertor plate.

12.2 Two-Wire Model of the Plasma

To demonstrate the described method of the separatrix mapping we consider the simple model of a tokamak plasma. The model consists of two current loops of radius R_0 located at the planes $Z_p = a$ and $Z_c = -a$ as shown in

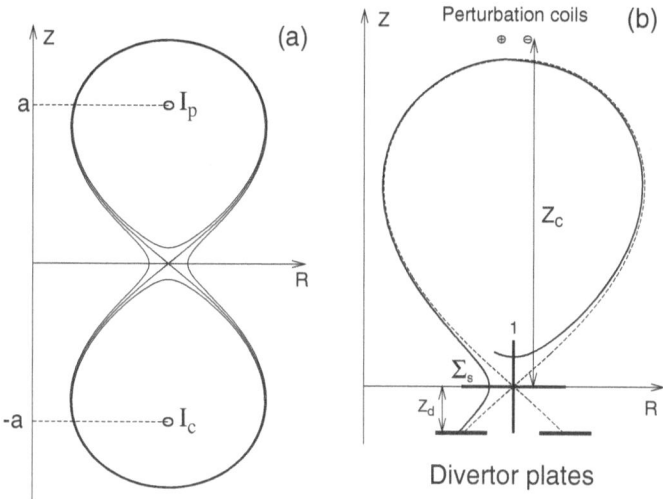

Fig. 12.4. (a) Magnetic configuration of the two wire model. (b) Magnetic field lines in the presence of the magnetic perturbation and the divertor plates. *Solid line* corresponds to *open field lines* hitting divertor plates, the *dashed curve* is the unperturbed separatrix

Fig. 12.4a. The first loop carries the plasma current I_p, and the second one describes the divertor current I_c. Such a model of the plasma well describes the single null divertor tokamak plasma configuration near the separatrix which does not depend on the radial profiles of plasma current I_p. However, this model does not include a plasma response to a perturbation field.

The vector potential $A_\varphi(R, Z)$ of the each current loop is given by (see Morozov and Solov'ev (1966))

$$A_\varphi(R, Z) = \sum_{j=p,c} \frac{\mu_o I_j}{\pi k_j} \sqrt{\frac{R_0}{R}} \left[\left(1 - \frac{k_j^2}{2} \right) K(k_j) - E(k_j) \right] , \qquad (12.21)$$

where $K(k)$ and $E(k)$ are the complete elliptic integrals with the module

$$k_j^2 = \frac{4R_0 R}{(R + R_0)^2 + (Z - Z_j)^2} .$$

Here $j = (p, c)$ stands for the plasma and the current loops, respectively.

For the large aspect ratio tokamaks $R_0/a \gg 1$, $|1 - k^2| \ll 1$ and for the toroidal field $B_\varphi(R) = B_0 R_0/R$ the unperturbed Hamiltonian $H_0(z, p_z) = -R A_\varphi(R, Z) / B_0 R_0^2$ up to constant terms can be approximated by

$$H_0(z, p_z) = L_p R_0^{-1} \ln r_p^2 r_c^{2\delta} , \qquad (12.22)$$

where $r_j^2 = x^2(p_z) + (z - z_j)^2$, x and z are the normalized radial coordinates $x = (R - R_0)/R_0$ and $z = Z/R_0$, and $z_j = Z_j/R_0$. According to (10.10) the momentum p_z is related with x: $p_z = \ln(1 + x)$ or $x = \exp(p_z) - 1$. The quantity $L_p = \mu_o I_i / 4\pi B_0$ is the length scale, and $\delta = I_c/I_p$ is the ratio of the coil current I_c to the plasma current I_p.

The Hamiltonian equations of field lines are

$$\frac{dz}{d\varphi} = \frac{\partial H_0}{\partial p_z} = 2L_p R_0^{-1} x \left(r_p^{-2} + \delta r_c^{-2} \right) \exp(p_z) \,,$$

$$\frac{dp_z}{d\varphi} = -\frac{\partial H_0}{\partial z} = -2L_p R_0^{-1} \left(\frac{z - a/R_0}{r_p^2} + \delta \frac{z + a/R_0}{r_c^2} \right) \,. \tag{12.23}$$

The coordinates of the X-point is located at

$$(z_s, p_s) = \left(\frac{a}{R_0} \frac{1 - \delta}{1 + \delta}, 0 \right) \,.$$

One can show that the second derivatives $H_{zz} = -H_{pp}$ and the mixed derivative H_{zp} at this point vanishes, and the coefficients λ_1 and λ_2 are

$$- \lambda_1 = \lambda_2 = \gamma = \frac{L_p R_0}{2a^2} \frac{(1 + \delta)^3}{\delta} \,. \tag{12.24}$$

On the separatrix we have

$$H_0(z_s, p_s) = (1 + \delta)L_p R_0^{-1} \ln \frac{2a^2 \delta}{(1 + \delta)R_0^2} - L_p R_0^{-1} \ln \delta \,. \tag{12.25}$$

Magnetic flux surfaces, $f(x, z) = H(z, p_z(x)) = \text{const}$, are given by

$$r_p^2 r_c^{2\delta} = \frac{1}{\delta} \left(\frac{2a^2 \delta}{(1 + \delta)R_0^2} \right)^{1+\delta} \Lambda \,, \qquad \Lambda = \exp(hR_0/L_p) \,, \tag{12.26}$$

where h is the relative poloidal flux (12.4). Magnetic flux surfaces for $\delta = 1$ are shown in Fig. 12.4a. The width of the separatrix along R and Z axes are $\Delta R = a$ and $\Delta Z = \sqrt{2}a$, respectively. The width of the separatrix is maximum at the planes $Z = \pm\sqrt{3/2}a$. Elongation $\kappa = \Delta Z/\Delta R$, i.e., the ratio of the widths of the separatrix ΔZ and ΔR is equal to $\kappa = \sqrt{2}$.

Analytical integration of field line equations (12.23) in the toroidal system is rather difficult task. Further to simplify the problem we consider the case of large aspect ratio tokamak plasmas, $|x| \ll 1$, putting $p_z = \ln(1 + x) \approx x$. Moreover, we suppose that the plasma, I_p, and coil, I_c, currents are equal, i.e., $\delta = 1$. Then, according to (10.15), (12.23) and (12.22), we have

$$q(h) = \frac{R_0}{4\pi L_p} \oint_{C_p} \frac{dz}{x(z, h)(r_p^{-2} + r_c^{-2})}$$

$$= \frac{a^2 \Lambda}{4\pi L_p R_0} \begin{cases} K\left(\Lambda^{1/2}\right), & \text{for } h < 0, \ (\Lambda < 1) \\ \Lambda^{-1/2} K\left(\Lambda^{-1/2}\right), & \text{for } h > 0, \ (\Lambda > 1) \,, \end{cases} \tag{12.27}$$

where integration is taken over closed magnetic surface C_p. Here $K(k)$ is the complete elliptic integral of the first kind with modulus k. Using the asymptotics of $K(k)$ at $k \to 1$ ($|h| \to 0$), $K(k) \approx \ln(4/\sqrt{1-k^2})$ we obtain the asymptotic formula for the safety factor (12.5) with the constants

$$\gamma = \frac{4L_p R_0}{a^2}, \qquad Q = \frac{16L_p}{R_0}.$$

Using the relation, $q(h) = d\psi/dh$, for the safety factor $q(h)$ one finds the toroidal flux ψ as a function of the relative poloidal flux h:

$$\psi = \int q(h)dh = \frac{a^2}{4\pi L_p R_0} \begin{cases} \int K\left(\Lambda^{1/2}\right)\Lambda dh, & \text{for } h < 0, \\ \int \Lambda^{1/2} K\left(\Lambda^{-1/2}\right) dh, & \text{for } h > 0, \end{cases}$$

$$= \frac{a^2}{2\pi R_0^2} \begin{cases} E\left(\Lambda^{1/2}\right) - (1 - \Lambda)K\left(\Lambda^{1/2}\right), & \text{for } h < 0, \\ \Lambda^{1/2} E\left(\Lambda^{-1/2}\right), & \text{for } h > 0. \end{cases} \tag{12.28}$$

At the separatrix, $h = 0$ ($\Lambda = 1$), the toroidal flux ψ is equal to $\psi_a = a^2/(2\pi R_0^2)$.

The safety factor profile q (12.27) as a function of the toroidal flux ψ is plotted in Fig. 12.5. We have chosen the value of the length scale L_p equal to 0.0472 in order to fix the value $q_{95} \approx 3.395$, where q_{95} stands for the value of the safety factor at the magnetic surface which has 95 % of the total magnetic flux ψ_a, i.e., $\psi/\psi_a = 0.95$. The major radius is $R_0 = 6.2$ m, and $a = 2$ m.

According to (12.23) the unperturbed field lines can be found by the integral

$$\varphi - \varphi_0 = \frac{R_0}{2L_p} \int_{z_0}^{z} \frac{dz}{x(z,h)(r_p^{-2} + r_c^{-2})}, \tag{12.29}$$

where z_0 is the highest z-coordinate of field line on the flux surface $f(x,z) \equiv H_0(z, p_z(x)) = $ const. The integration gives

$$\varphi - \varphi_0 = \frac{a^2 \Lambda}{4L_p R_0} \begin{cases} F\left(\nu, \Lambda^{1/2}\right), & \text{for } h < 0, (\Lambda < 1) \\ \Lambda^{-1/2} F\left(\nu, \Lambda^{-1/2}\right), & \text{for } h > 0, (\Lambda > 1), \end{cases} \tag{12.30}$$

where $F(\nu, k)$ is the incomplete elliptic integral of the first kind. The argument ν is related to the coordinate $z = Z/R_0$:

$$\sin^2 \nu = \frac{2 + \Lambda^{1/2} - (\Lambda + 4Z^2/a^2)^{1/2}}{\Lambda^{1/2} + (\Lambda + 4Z^2/a^2)^{1/2}} \begin{cases} \Lambda^{-1/2}, & \text{for } h < 0, (\Lambda < 1) \\ \Lambda^{1/2}, & \text{for } h > 0, (\Lambda > 1). \end{cases} \tag{12.31}$$

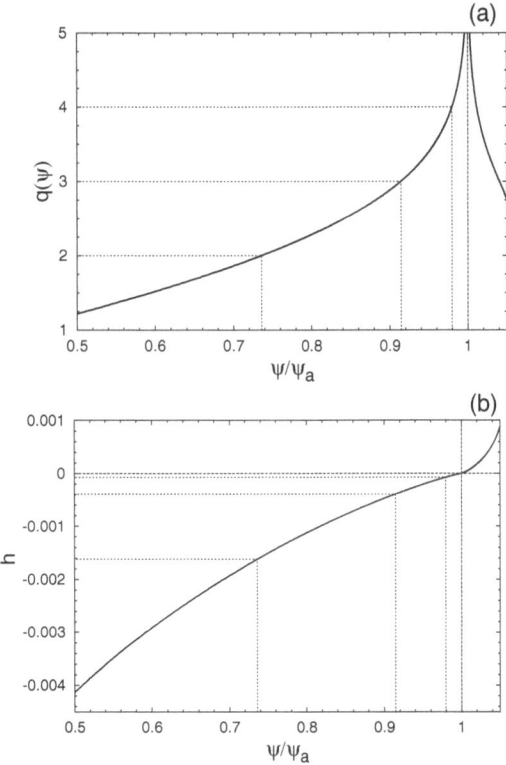

Fig. 12.5. (a) Safety factor q as a function of the toroidal flux ψ; (b) Relation between h and ψ. *Dotted lines* in both Figures show the values of ψ and h corresponding to the rational magnetic surfaces $q = 2$, $q = 3$, and $q = 4$. The length scale L_p is chosen equal to 0.0472 to fix $q_{95} = 3.394585$ at $\psi/\psi_a = 0.95$. Other parameter are $R_0 = 6.2\,\mathrm{m}$, $a = 2\,\mathrm{m}$

The field line on the separatrix, $z_s(\varphi), p_s(\varphi)$, is given by

$$z_s(\tau) = \pm \frac{\sqrt{2}a}{R_0 \cosh(\gamma\tau)[1 + \tanh^2(\gamma\tau)]} \ ,$$

$$p_s(\tau) \approx x_s(\tau) = \mp \frac{\sqrt{2}a \tanh(\gamma\tau)}{R_0 \cosh(\gamma\tau)[1 + \tanh^2(\gamma\tau)]} \ , \qquad (12.32)$$

where $\tau = \varphi - \varphi_0$, upper signs stand for the plasma region, and the lower ones for the coil region. Note that when $\varphi = \varphi_0$ the field line coordinates are located at the farthest point from the X-point $z_s(\varphi_0) = \pm\sqrt{2}a/R_0$.

12.2.1 Magnetic Field Perturbations

Suppose that non-axisymmetric magnetic perturbations are created by the pair of perturbation coils with opposite flowing currents $\pm I_d$. The coils are located at $(R_0 \pm d/2, Z_c)$, respectively as shown in Fig. 12.4b. The distance between perturbation coils d slowly varies along the toroidal angle φ, $d = d(\varphi)$ and it is much smaller than the minor radius a, $d \ll a$. Then the perturbation magnetic field created by these coils is mainly determined by the toroidal component $A_\varphi^{(per)}$. It is given by equation of type (12.21), where $I_j = \pm I_d$, $Z_j = Z_c$, $R_0 \rightarrow R_0 \pm d/2$. For the large aspect ratio tokamak the perturbed Hamiltonian $H_1(z, p_z, \varphi)$ takes the form similar to (12.22):

$$\epsilon H_1(z, p_z, \varphi) = \epsilon L_p R_0^{-1} \left(\ln r_{c+}^2 - \ln r_{c-}^2 \right), \tag{12.33}$$
$$r_{c\pm}^2 = (x \mp d(\varphi)/2R_0)^2 + (z - z_c)^2 ,$$

where the perturbation parameter ϵ is introduced as the ratio of the perturbation current I_d to the plasma current I_p: $\epsilon = I_d/I_p$. At the distances $|x| \gg \max|d|/R_0$ the Hamiltonian (12.33) can be approximated by

$$\epsilon H_1(z, p_z, \varphi) \approx -\epsilon \frac{L_p}{R_0^2} \frac{2xd(\varphi)}{r_c^2} , \qquad r_c^2 = x^2 + (z - z_c)^2 . \tag{12.34}$$

Suppose that $d(\varphi)$ is the periodic function of the toroidal angle φ. Furthermore we specify it as

$$d(\varphi) = d_0 \cos(n\varphi + \chi)$$

with the single toroidal mode number n. Therefore, according to the presentation (12.8) we have

$$H_n(z, p_z) = -\frac{L_p}{R_0^2} \frac{2xd_0}{x^2 + (z - z_c)^2} , \qquad x = x(p_z) = e^{p_z} - 1 \approx p_z . \tag{12.35}$$

The Melnikov Integrals $K_n(h)$ and $L_n(h)$.

They are given by

$$R_n(h) = K_n(h) + iL_n(h) = \int\limits_{-\pi q(h)}^{\pi q(h)} V_n(h, \varphi) e^{in\varphi} d\varphi , \tag{12.36}$$

$$V_n(h, \varphi) = H_n \left(z(h, \varphi), p_z(h, \varphi) \right) ,$$

where the functions $z(h, \varphi)$, $p_z(h, \varphi)$ are the unperturbed orbits of field lines given (12.30), (12.31).

Since the function $V_n(h, \varphi)$ is antisymmetric with respect to the change of the sign of its argument, i.e., $V_n(h, -\varphi) = -V_n(h, \varphi)$, the integral K_n vanishes, $K_n \equiv 0$. The integral $L_n(h)$ can be found numerically by integrations along the unperturbed field lines given by (12.30), (12.31). Then its

regular part, $L_n^{(reg)}(h)$, can be obtained by substacting from $L_n(h)$ its oscillatory part, $L_n^{(osc)}(h)$, i.e., $L_n^{(reg)}(h) = L_n(h) - L_n^{(osc)}(h)$. For $L_n^{(osc)}(h)$ the
asymptotical formulas near the separatrix were constructed in Appendix B.

Field lines near the X-point of the model shown in Fig. 12.4 correspond
to the phase space of Hamiltonian system near the saddle point plotted in
Fig. B.1: the axes (R, Z) (or (p, z)) corresponds to the axes (ξ, η), respectively; the plasma region corresponds to the region of the phase space $\eta > 0$
and located inside the separatrix. It is obtained by the connection of the
second branch II with the first branch I, i.e, the plasma region $h < 0$ of the
model corresponds to the region $h > 0$, $\eta > 0$ in Fig. B.1. In this case the
asymptotical formulae for $L_n^{(osc)}(h)$ is given by (B.34) where one should make
appropriate changes: the frequency $\omega(h)$ should be replaced by the inverse
safety factor $1/q(h)$, the perturbation frequency Ω − by the toroidal mode
number n, the case, $h < 0$, − by $h > 0$ and vise versa. Then we obtain the
following

$$L_n^{(osc)}(h) = 2\sqrt{|h|}A_\xi \left\{ \begin{array}{ll} \gamma \sin[\pi n q(h)], & \text{for } h < 0 , \\ n \cos[\pi n q(h)], & \text{for } h > 0 , \end{array} \right\} + 2C|h| \sin[\pi n q(h)] , \tag{12.37}$$

where

$$A_\xi = \frac{\sqrt{2}a_\xi}{\beta(\gamma^2 + n^2)} , \qquad C = \frac{2b_{\xi\eta}}{4\gamma^2 + n^2} . \tag{12.38}$$

In (12.38) the coefficients a_ξ and C are defined as

$$a_\xi = \frac{\partial H_n}{\partial x}\bigg|_{x=z=0} = -\frac{2L_p d_0}{R_0^2 z_c^2}, \qquad b_{\xi\eta} = \frac{\partial^2 H_n}{\partial x \partial z}\bigg|_{x=z=0} = -\frac{4L_p d_0}{R_0^2 z_c^3} .$$

The numerical integration of the integral $L_n(h)$ are performed for the
plasma parameters: the major radius $R_0 = 6.2$ m, $a = 2$ m, the lenght scale
$L_p = 0.0472$, which are close the ITER plasma parameters. The vertical
position of the perturbation coils is $Z_c = 3.2$ m, and the distance d_0 between
them is taken equal to 0.2 m. The dependences of the integral $L_n(h)$ and its
regular part $L_n^{(reg)}(h)$ on h for the toroidal mode number $n = 4$ is shown
in Fig. 12.6a. The value of $L_n(0)$ on the separatrix, $h = 0$, of the plasma
region versus the toroidal mode number n are presented in Fig. 12.6b. The
corresponding values of $L_n(0)$ in the coil region are small and have an order
of 10^{-5}.

In a certain region near the separatrix which includes the rational magnetic surfaces $q = 3$ the function $L_n^{(reg)}(h)$ can be well fitted by the quadratic
function of h:

$$L_n^{(reg)}(h) = L_n(0) + a_n h + b_n h^2 , \tag{12.39}$$

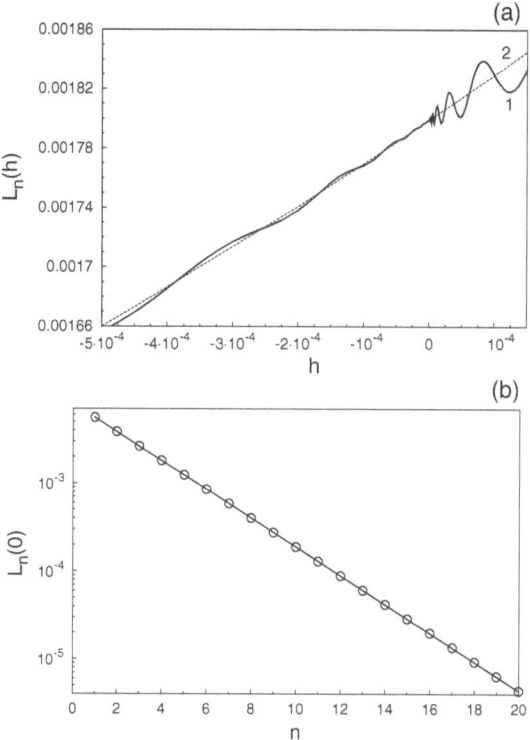

Fig. 12.6. (a) Melnikov type integral $L_n(h)$ (12.36) for the toroidal mode number $n = 4$: curve 1 corresponds to $L_n(h)$ itself, and curve 2 – its regular parts $L_n^{(reg)}$. (b) Dependence of $L_n(0)$ on the toroidal mode number n. The plasma parameters are: a major radius $R_0 = 6.2$ m, $a = 2$ m, the lenght scale $L_p = 0.0472$. The vertical position of the perturbation coils $Z_c = 3.2$ m, and the distance between them is $d_0 = 0.2$ m

with the constant coefficients $L_n(0)$, a_n, b_n. In the interval $-1.5 \times 10^{-3} < h < 1.5 \times 10^{-4}$ we obtained the following values $L_n(0) = 1.81 \times 10^{-3}$, $a_n = 0.31277$, and $b_n = 40.794$ for the toroidal mode $n = 4$.

As seen from Fig. 12.6b the integral $L_n(0)$ decays exponentially with increasing the toroidal mode number n. It is well described by the functions

$$L_n(0) = L_0 \exp(-An) , \tag{12.40}$$

with $L_0 = 8.13457 \times 10^{-3}$ and $A = 0.375677$.

Mappings

Below we write down the mappings (12.10), (12.14) for the wire model. Since $K_n(h) = 0$, these mappings for the plasma region are reduced to

$$h_{k+1} = h_k \mp \epsilon L_n^{(reg)}(h_{k+1}) \cos \Phi_n^{(\pm)}(\varphi_k, h_k) \, ,$$

$$\varphi_{k+1} = \varphi_k \pm \pi \left[q(h_k) + q(h_{k+1}) \right] \mp \epsilon \frac{d L_n^{(reg)}(h_{k+1})}{dh_{k+1}} \sin \Phi_n^{(\pm)}(\varphi_k, h_k) \, ,$$

$$(12.41)$$

where

$$\Phi_n^{(\pm)}(\varphi_k, h_k) = n \left(\varphi_k \pm \pi q(h_k) \right) + \chi_n \, , \tag{12.42}$$

where the upper signs corresponds to the forward map, and the lower signs corresponds to the backward map.

Using the asymptotics of the safety factor $q(h)$ (12.5) in the region close to the separatrix the mapping (12.41) can be replaced by the separatrix map of type (12.19):

$$h_{k+1} = h_k \mp \epsilon n L_n^{(reg)}(0) \cos \left(n\varphi_k \pm \frac{n}{2\gamma} \ln \frac{16 L_p R_0^{-1}}{|h_k|} + \chi_n \right) \, ,$$

$$\varphi_{k+1} = \varphi_k \pm \frac{1}{\gamma} \ln \frac{16 L_p R_0^{-1}}{\sqrt{|h_k h_{k+1}|}} \, . \tag{12.43}$$

Upper signs in (12.43) corresponds to the forward map and the lower signs − to the backward map.

12.2.2 The Structure of the Stochastic Layer

In this section we study the structure of the stochastic layer near the magnetic separatrix and X-point using the mappings (12.41), (12.43). Specifically, we consider the Poincaré section of field lines in the toroidal plane (φ, Z), $(x = 0, Z > 0)$ coinciding with the 1-st branch of the section Σ_s (see Fig. 12.4b). The Z-coordinate of field lines in this plane is related to the relative poloidal flux h:

$$Z = a\sqrt{1 - \exp(hR_0/2L_p)} \, , \qquad h < 0 \, . \tag{12.44}$$

First we consider only field lines in the plasma region, terminating them when they hit the divertor plates.

Figure 12.7a shows the Poincaré section obtained by the mapping (12.41) and Fig. 12.7b shows one obtained by the numerical integration of the Hamiltonian equations of field lines (10.11) with the Hamiltonian (12.22), (12.34) using the Runge–Kutta scheme. The plasma parameters R_0, B_0, L_p, and the perturbation coil positions, Z_c, d are chosen as in Fig. 12.6. The toroidal mode is taken $n = 4$ and the perturbation parameter $\epsilon = I_c/I_p = 0.002$. As seen from Fig. 12.7 the separatrix map quantitatively well reproduces results of the direct numerical integration: the positions and the widths of magnetic

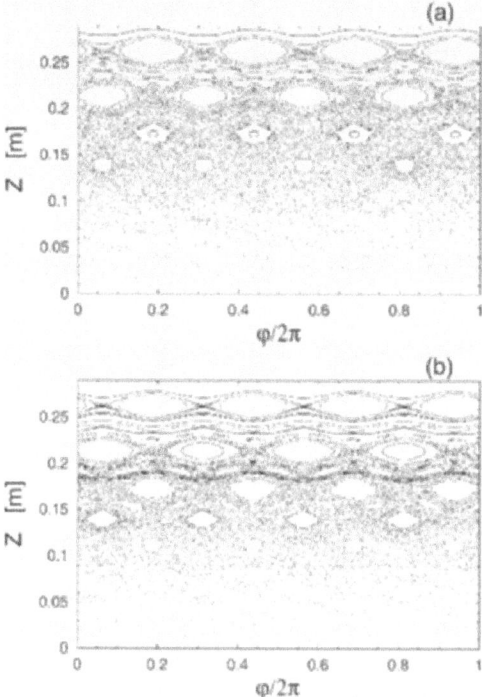

Fig. 12.7. Poincaré section of field lines on the branch 1 of the section Σ_s in the presence of the divertor plates: (**a**) obtained by the separatrix map; (**b**) by integration of field line equations. Toroidal mode number $n = 4$ and the perturbation $\epsilon = 0.002$. The parameters are the same as in Fig. 12.6

islands. The main advantage of the mapping is its computational speed: it runs almost three order faster than the numerical integration.

As seen from Fig. 12.7 the stochastic layer of field lines is not uniformly chaotic. The chaos is highly developed near the separatrix, and the "level" of chaos decreases towards the plasma, i.e., with the increase the distance Z: the sizes of KAM stability islands increase. The edge of the stochastic layer is usually formed by the largest KAM islands. For the case shown in Fig. 12.7 there is a chain of magnetic islands at the edge of the stochastic layer corresponding to the primary resonance $q(h_{mn}) = m : n = 14 : 4$ at the $Z_{mn} \approx 0.214$ m, $h_{mn} \approx -1.75 \times 10^{-4}$ or $\psi/\psi_a \approx 0.957$.

In the presence of divertor plates, where chaotic field lines are terminated, the structure of the stochastic layer near the separatrix consists of long stripes along which field lines leave the plasma region. It is clear seen from the stochastic layer plotted in Fig. 12.8 for the higher perturbation current with $\epsilon = 0.01$ and the same toroidal mode $n = 4$. Field lines leave the plasma region after a certain number of poloidal turns.

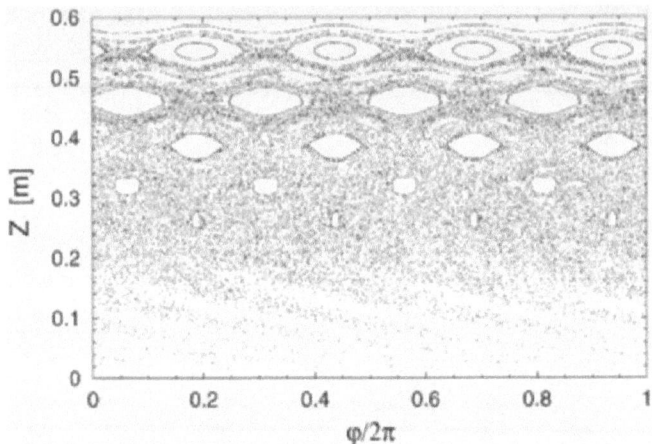

Fig. 12.8. The same as in Fig. 12.7 but for the perturbation $\epsilon = 0.01$

Similar to chaotic field lines in ergodic divertor tokamaks field lines in a poloidal divertor tokamak can be considered as a chaotic scattering system (see Sect. 11.6). Field lines can be classified by a number of poloidal turns, N_p, necessary for field lines to connect the left and right divertor plates (see Fig. 12.4). The fine structure of the stochastic layer in the presence of the divertor plates can be seen by plotting the boundaries of basins with the different wall to wall connection turns N_p. The basin boundary structure of N_p in the (φ, Z)-plane are constructed using the same procedure as in Sect. 11.6: a field line with a particular initial coordinate (φ_i, Z_i) in the (φ, Z)-plane is traced by the forward and backward maps (12.41) along the positive and the negative directions of the toroidal angle φ, respectively, until they reach the divertor plates. The complete number of map iterations gives the number poloidal turns, N_p, since a field line makes a full poloidal turn after each map iteration. The countour plot of N_p corresponding to the Poincare section in Fig. 12.8 is shown in Fig. 12.9. The lowest colorbar corresponds to the basins with the smallest number of $N_p = 2$, and the highest one corresponds to the basins with $N_p \geq 6$. The latter basins are not resolved. Fine details of these basin boundaries are seen in Fig. 12.9b where the toroidal dependence of N_p along the straight line $Z = 0.02$ m (a blue line shown in Fig. 12.9a) is plotted.

The structure of basin boundaries is similar to the corresponding structure in ergodic divertor tokamaks shown in Fig. 11.9. The regions with small connection numbers, $N_p = 2$ cover relatively large area. As seen from Fig. 12.9 field lines with the large number connection turns $N_p \geq 3$ are connected along the long helical stripes. Each helical stripe has a layered fractal structure as shown in Fig. 12.9c.

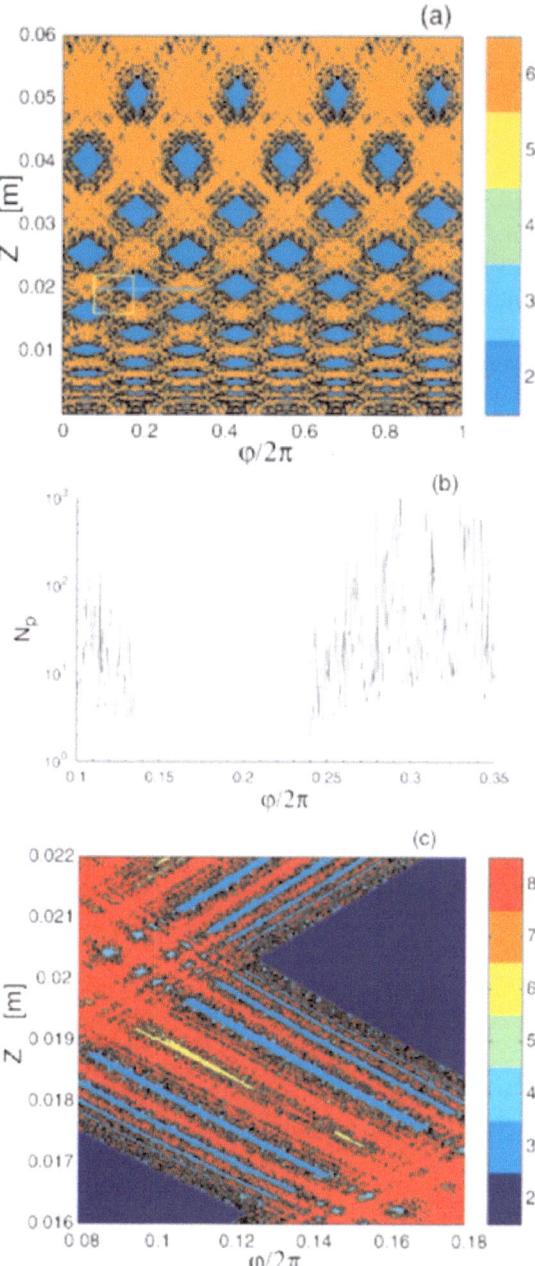

Fig. 12.9. (a) Basin boundary structure of N_p on the (φ, Z) plane $(Z > 0)$ of the section Σ_s. (b) Toroidal dependence of N_p along line $Z = 0.02$ m (a blue line in (a)) in the region $0.1 < \varphi/2\pi < 0.35$; (c) Expanded view of the rectangular area in (a)

12.2.3 Structure of Magnetic Footprints

The magnetic footprints on divertor plates give a pattern of heat deposition. They have been obtained using the following procedure. For simplicity we have supposed the simple geometry of divertor plates shown in Fig. 12.4b. They consists of two segments parallel to the R-axis located at $Z = Z_d$ and crossing the two branches of the separatrix in the coil region. We have taken a set of field lines with initial coordinates (φ_0, h_0), $(h_0 < 0)$, in the (φ, Z)-plane located in the stochastic layer of the plasma region. In order to obtain magnetic footprints on the left divertor plate field lines were traced using the forward separatrix map (12.43) until $h_k > 0$, i.e., the field lines intersect the left branch of the section Σ_s. Then the field line with (φ_k, h_k) is mapped to the footprint (φ_d, h_d) on the divertor plane using the mapping (12.20). The increment of the toroidal angle, $\Delta\varphi$, for the geometry of the divertor plates shown in Fig. 12.4b is determined by (12.6), where $\xi = Z_d$. The radial coordinate R of the footprint can be found using the relation (12.4) between the relative poloidal flux, h_d, and the unperturbed Hamiltonian, i.e., $f(R_d) \equiv H_0(z_d, p_z(R_d)) - H_0(z_s, p_s) = h_d$, where $p_z(R) = \ln(R/R_0) \approx (R - R_0)/R_0$. If the divertor plates are close to the X-point then according to (12.3) we have $R_d = R_0 - \sqrt{2h_d R_0^2/\gamma + Z_d^2}$.

The magnetic footprint corresponding to the stochastic layer in Fig. 12.8 is shown in Fig. 12.10. The distance Z_d is taken equal to 0.2 m. It consists of four clusters corresponding the toroidal mode number $n = 4$. A fine structure of the cluster is shown in Fig. 12.11: a) represents the structure of boundaries of basins corresponding to the field lines with the different poloidal numbers N_p; b) Toroidal dependence of N_p along the straight line $R = R_0 - 0.225$ m.

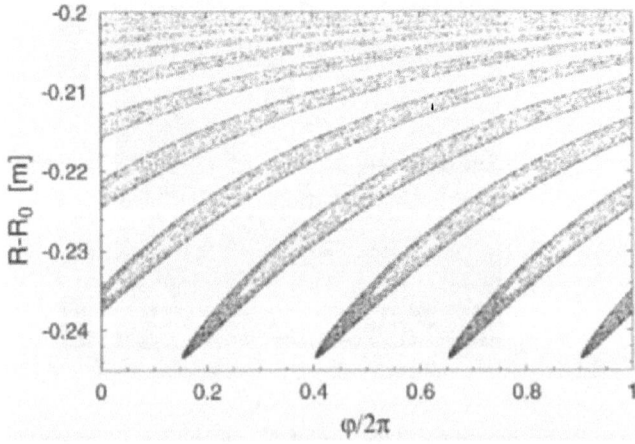

Fig. 12.10. Magnetic footprints on the left divertor plate. The plasma and perturbation parameters are the same as in Fig. 12.8 and $Z_d = 0.2$ m

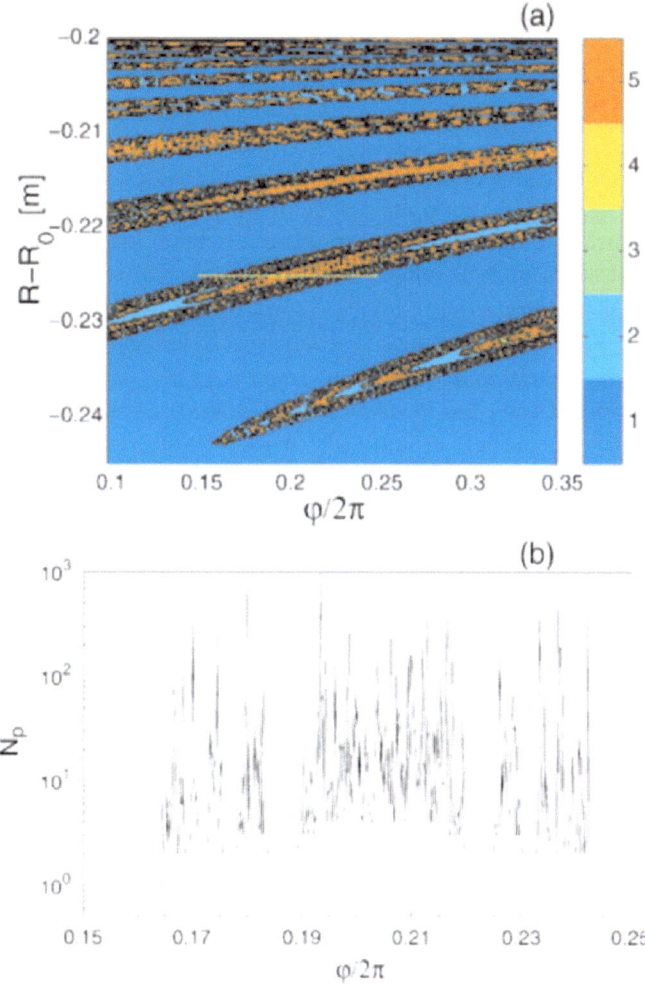

Fig. 12.11. (a) Basin boundary structure of N_p on the divertor plate in the interval of toroidal angle: $0.1 < \varphi/2\pi < 0.35$; (b) Dependence N_p on the toroidal angle φ along the straight line $R = R_0 - 0.225$ m (a yellow line in (a)) covering the toroidal angle region, $0.15 < \varphi/2\pi < 0.25$

As seen from the Fig. 12.11 each cluster has a generic spiral-like structure with spirals corresponding to the different N_p.

The white area on the divertor plate in Fig. 12.10 and the area with $N_p = 1$ in Fig. 12.11 correspond to field lines connecting the divertor plates without entering into the plasma region. The areas with $N_p \geq 2$ correspond to field lines entering the plasma region. Therefore particles, electrons and ions, followed these field lines are predominantly deposited along the spirals.

12.3 Conclusion

The mapping method of the described in this chapter is a very effecient tool to describe magnetic field lines in magnetically confienement devices with magnetic separatrices. It allows to straightforwardly obtain the structure of the stochastic field lines near the separatrix and the magnetic footprints on the divertor plates. The mapping method can be applied for arbitrary magnetic configuration and the mappings have a generic form given by (12.10), (12.14) determined only by a few functions, namely, the safety factor, $q^{(j)}(h)$, and the regular parts of the Melnikov type integrals, $R_n^{(j)}(h)$ (12.17) along each saddle–saddle connections. In the case of the separatrix map (12.19) one needs only values of $R_n^{(j)}(h)$ at the separatrix $h = 0$. In the example of a two-wire model of the plasma we were able to find the safety factor, $q^{(j)}(h)$, as well as, the coefficients γ and $Q^{(j)}$ in its asymptotical form (12.5) analytically. For the more general magnetic configurations of the plasma these quantities can be found by the numerical integration of field line equations.

13 Miscellaneous

In this conclusive chapter we shall briefly discuss some other areas of physics and the astronomy where methods of Hamiltonian mappings play an important role. These areas are related to the problems of ray propagation in waveguide media, particle acceleration and planetary motion. In a good approximation these problems are described by Hamiltonian systems, and therefore they can be studied by Hamiltonian mappings.

13.1 Ray Dynamics in Waveguide Media

A description of short-wavelength waves by the concept of rays is one of the powerful methods to study the propagation of waves in inhomogeneous media. A ray approximation, also known as the method of *geometrical optics*, has been extensively used in the problems of long distance wave propagation in waveguide media, particularly to low frequency sound propagation in the ocean and atmosphere, optical radiation in fibers, radiowave propagation in the ionosphere (see, e.g., Brekhovskih and Lysanov (2003); Marcuse (1982); Gurevich and Tsedilina (1985)). Importantly that the rays can be described as a Hamiltonian system which allows to apply the methods of Hamiltonian dynamics including symplectic mappings to study wave propagation problems. It also reveals the possibility of a qualitatively new behavior of rays, namely, the chaos of rays in wave propagation problems in inhomogeneous media (see reviews by Abdullaev and Zaslavsky (1991); Brown et al. (1991b); Smirnov et al. (2001); Brown et al. (2003) and the book by Abdullaev (1993)). In this section we briefly describe the mapping description of ray dynamics which demonstrates the main phenomena which occur in wave propagation problems.

13.1.1 Rays as a Hamiltonian System

Ray optics is the oldest area of physics in which Hamilton first developed his method, known as Hamilton method that was later applied to the classical (Newtonian) mechanics (Hamilton (1828, 1834)). The Hamilton description of rays is originated from *Fermat's principle* (the principle of least time) according to which a ray trajectory in a medium takes a path with the minimal

S.S. Abdullaev: *Construction of Mappings for Hamiltonian Systems and Their Applications*,
Lect. Notes Phys. **691**, 299–316 (2006)
www.springerlink.com

propagation time. Formally, it is written as the functional integral of the path $\mathbf{r}(\tau) = (x(\tau), y(\tau), z(\tau))$ connecting its initial, P_0, and final, P_1, positions at space (see Born and Wolf (1986)):

$$t[\mathbf{r}(\tau)] = \int_{P_0}^{P_1} \frac{ds}{c(x, y, z)} = \min , \qquad (13.1)$$

where $c(x, y, z)$ is the speed of light in a medium, $ds = (dx^2 + dy^2 + dz^2)$ is the element of length along the path.

In waveguide media there is a preferential direction of wave propagation. For instance, in the case of sound propagation in the ocean it coincides with the horizontal coordinates x (or y) where z serves as a vertical coordinate. The equations for rays $\mathbf{r}(\tau)$ can be obtained from Fermat's principle (13.1) by choosing the spatial coordinate x along the preferential wave propagation direction as an independent time-like variable τ, i.e., $\tau \equiv x$. Then the ray equations can be presented in the following Hamiltonian form (see, e.g., Marcuse (1982); Abdullaev (1993))

$$\frac{dy}{dx} = \frac{\partial H}{\partial p_y}, \qquad \frac{dp_y}{dx} = \frac{\partial H}{\partial y} ,$$
$$\frac{dz}{dx} = \frac{\partial H}{\partial p_z}, \qquad \frac{dp_z}{dx} = \frac{\partial H}{\partial z} . \qquad (13.2)$$

with the Hamiltonian function

$$H(y, z, p_y, p_z, x) = -n(x, y, z)\sqrt{1 - p_y^2 - p_z^2} . \qquad (13.3)$$

The transversal coordinates (y, z) are canonical coordinates, and (p_y, p_z) are canonical momenta, defined as

$$p_y = \frac{n(x, y, z)\dot{y}}{\sqrt{1 + \dot{y}^2 + \dot{z}^2}} , \qquad p_z = \frac{n(x, y, z)\dot{z}}{\sqrt{1 + \dot{y}^2 + \dot{z}^2}} , \qquad (13.4)$$

where $\dot{y} \equiv dy/dx, \dot{z} \equiv dz/dx$.

One should note that a ray concept constitutes the approximate description of wave propagation in inhomogeneous media. Therefore the ray equations should be obtained from wave equations as their approximate solutions in the limit of short wavelengths. Let $u(x, y, z, t) = u(x, y, z) \exp(-i\nu t)$ be a monochromatic wave field of frequency ν (for instance, a pressure variations during a sound propagation in the ocean). The propagation of this wave is governed by the wave equation

$$\nabla^2 u - k^2 n^2(x, y, z)u = 0 , \qquad (13.5)$$

where $k = 2\pi/\lambda = \nu/c_0$ is the wavenumber, and λ is the wavelength.

When the wavelength λ is much smaller than the characteristic scale-length, l, of variations of the speed of wave $c(x, y, z)$, $\lambda/l \ll 1$, the solution of (13.5) can be presented in the form

$$u(x, y, z) = A(x, y, z) \exp[ik\Phi(x, y, z)] \,, \qquad kl \gg 1 \,, \qquad (13.6)$$

where the slowly varying phase function $\Phi(x, y, z)$ and the amplitude $A(x, y, z)$ satisfy the equations

$$(\nabla\Phi)^2 = n^2(x, y, z) \,, \qquad \nabla(A^2\nabla\Phi) = 0 \,. \qquad (13.7)$$

The first equation in (13.7) known as the *eikonal equation* is similar to the Hamilton–Jacobi equation (1.19). Introducing the canonical momenta $p_y = \partial\Phi/\partial y$, $p_z = \partial\Phi/\partial z$ and the Hamiltonian $H = -\partial\Phi/\partial x$, we obtain the Hamilton function (13.3) from the eikonal equation (13.7).

A geometrical meaning of the momenta, p_y, p_z, and the Hamiltonian H is the following. Let θ be the angle between the normal vector $\mathbf{p} = (p_y, p_z, p_z) = \nabla\Phi$ to the surfaces of constant phase $\Phi(x, y, z) = \text{const}$ and the x-coordinate, and φ be the angle between the transversal vector $\mathbf{p}_\perp = (p_y, p_z)$ and the coordinate y. Then, according to (13.7) we have the relations

$$\begin{aligned}
p_y &= n(x, y, z) \sin\theta \cos\varphi \,, \\
p_z &= n(x, y, z) \sin\theta \sin\varphi \,, \\
H &= -p_x = -n(x, y, z) \cos\theta \,.
\end{aligned} \qquad (13.8)$$

In weakly inhomogeneous media the variations of a refractive index n is small in respect to the reference index n_0, i.e., $n(x, y, z) = n_0 + \Delta n(x, y, z)$, $|\Delta n| \ll n_0$. In such media a wave propagates under small angles θ with respect to the axis x, i.e., $\theta \ll 1$. Then The Hamiltonian of rays (13.3) can be simplified by neglecting the small terms of order of $|\Delta n|^2/n_0^2 \ll 1$, $p_y^4 \ll 1$, $p_z^4 \ll 1$ and higher,

$$H(y, z, p_y, p_z, x) = -n_0 + \frac{1}{2n_0}(p_y^2 + p_z^2) - \Delta n(x, y, z) \,. \qquad (13.9)$$

This approximation known as a *paraxial* is equivalent to the Hamiltonian describing the two-dimensional motion of non-relativistic particle in a time-dependent potential field $U = -\Delta n(x, y, z)$.

In waveguide media the speed of wave $c(x, y, z)$ (or the refractive index $n(x, y, z)$) non-monotonically depends on the transversal coordinates (y, z) and it may slowly vary along the longitudinal coordinate x. For instance, the low-frequency sound speed in the ocean c has a minimum at the certain level $z = z_m$ of the vertical coordinate z (see Fig. 13.1). Thanks to refraction of rays in this medium they oscillate along the vertical coordinate z about the level $z = z_0$ whereas they propagate along the horizontal coordinate x without restrictions.

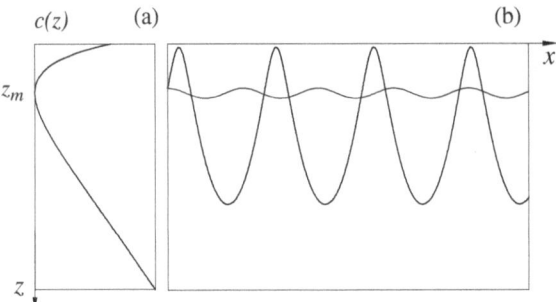

Fig. 13.1. (a) Vertical profile of the sound speed $c(z)$ in a deep ocean. (b) Ray trajectories in an ocean waveguide

When the speed of wave c (or the refractive index n) does not change along the preferential direction of wave propagation coinciding with the x-coordinate, i.e., $c = c(y, z)$, the ray system (13.2) is described by the autonomous Hamiltonian $H = H(y, z, p_y, p_z)$. Variations of the speed c along x-coordinate corresponds to the time-dependent Hamiltonian system. In some practically interesting cases variations of c along x-axis are small and periodic:

$$c(x, y, z) = c_0 + \Delta c_0(y, z) + \delta c(x, y, z) \ ,$$
$$\delta c(x, y, z) = \delta c(x + L, y, z) \ ,$$
$$|\delta c(x + L, y, z)| \ll c_0 \ , \tag{13.10}$$

where L is the spatial period of variations. A problem of ray propagation in this medium becomes equivalent to the dynamics of Hamiltonian system subjected to time-periodic perturbation, and it is described by the Hamiltonian system

$$H = H_0(y, z, p_y, p_z) + \epsilon H(y, z, p_y, p_z, x) \ ,$$
$$H_0(y, z, p_y, p_z, x) = -1 + \frac{1}{2}(p_y^2 + p_z^2) - \frac{\Delta c_0(x, y)}{c_0} \ ,$$
$$\epsilon H_1(y, z, p_y, p_z, x) = -\frac{\delta c(x, y, z)}{c_0} \ . \tag{13.11}$$

This problem has been studied using the methods of Hamiltonian dynamics. The review of these studies are given by Abdullaev and Zaslavsky (1991); Abdullaev (1993); Smirnov et al. (2001); Brown et al. (2003). Here we discuss the mapping models of ray propagation in waveguide media.

13.1.2 Mapping Models of Ray Propagation in Waveguide Media

Consider a model of ray dynamics in a waveguide with periodically corrugated wall. This model proposed by Abdullaev and Zaslavsky (1988) describes the main features of chaotic and regular dynamics of rays.

The model waveguide consists of two stiff walls along the x-coordinate and the medium between them is homogeneous with the refractive index $n = n_0$. The upper wall $z = 0$ is flat, and the bottom wall $z = h(x)$ is periodically corrugated along the x-coordinate, i.e. $h(x) = a + bf(x)$, where a is the unperturbed width of the waveguide, b is the amplitude of deviations of the corrugated wall from the level $z = a$, $f(x)$ is a periodic function along the x-coordinate with a period L. Suppose that $a \gg b$.

The ray trajectory in the waveguide can be easily found, since they are straight line which successively reflecting the walls. The example of orbit is drawn in Fig. 13.2. Here x_k $(k = 0, 1, \ldots)$ are longitudinal coordinates of the ray at the reflecting points from the upper unperturbed wall, θ_k is the corresponding angle between the wall and the ray. Ray dynamics is fully determined by the mapping

$$(x_{k+1}, \theta_{k+1}) = \hat{M}(x_k, \theta_k) , \qquad (13.12)$$

connecting the coordinates of rays at the wall $z = 0$ after reflections from the corrugated wall.

The map (13.12) corresponds the Poincaré mapping (4.54) of time, t, and energy, H, variables to the section $z = 0$ in one-degree-of-freedom Hamiltonian system. However, it is written in non-canonical variables (x, θ): the variable conjugated to the coordinate x is the Hamiltonian H. Using the relation (13.8) between H angle the geometrical angle θ, $H = -n_0 \cos \theta$, and one can write the condition of symplecticness for the mapping (13.12)

$$\left| \frac{\partial(x_{k+1}, \cos \theta_{k+1})}{\partial(x_k, \cos \theta_k)} \right| = 1 . \qquad (13.13)$$

The map (13.12) can be easily constructed using the geometry of rays in the waveguide shown in Fig. 13.2. First, suppose that the ray is reflected only once from the bottom corrugated wall before returning to the upper wall $z = 0$ (see Fig. 13.2). Let ψ_k be a x-coordinate of the reflection point of the ray from the corrugated wall. Then it is not difficult to find the form of the mapping (13.12):

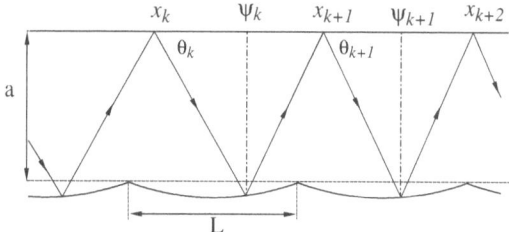

Fig. 13.2. Ray in a waveguide with a corrugated wall

$$\theta_{k+1} = \theta_k - 2\arctan\left[bf'(\psi_k)\right] ,$$

$$x_{k+1} = \psi_k + [a + bf(\psi_k)]\cot\theta_{k+1} , \qquad (13.14)$$

where the intermediate coordinate ψ_k is determined by the equation

$$\psi_k = x_k + [a + bf(\psi_k)]\cot\theta_k , \qquad f'(x) \equiv \frac{df(x)}{dx} . \qquad (13.15)$$

The equation (13.15) is implicit with respect to the coordinate ψ_k. For the small values of b it can be find by the Newton method with the initial input $\psi_{k0} = x_k + a\cot\theta_k$. One can also easily check that the mapping (13.14) satisfies the condition (13.13).

For the parabolic profile of the wall corrugation

$$f(x) = 4\xi(1-\xi) , \qquad \xi = \left\{\frac{x}{L}\right\} , \qquad (13.16)$$

the equation (13.14) can be explicitly solved with respect to ψ_k. Here $\{x\}$ stands for the fractional part of number x, i.e., $\{x\} = x - [x]$ ($[x]$ is the integer part of x). The corresponding value of ψ_k is

$$\psi_k = x_k + [a + 4b\xi_k(1-\xi_k)]\cot\theta_k ,$$

$$\xi_k = -\frac{1}{2}(A-1) + \frac{1}{2}\sqrt{(A-1)^2 + 4A\xi_k^0} , \qquad (13.17)$$

where

$$A = \frac{L\tan\theta_k}{4b} , \qquad \xi_k^0 = \left\{\frac{x_k + a\cot\theta_k}{L}\right\} .$$

The mappings (13.14) and (13.17) describe the case when a ray is reflected from the corrugated wall only once before reaching the upper wall. However, the rays propagating at small angles θ_k to the z axis may reflect from the corrugated wall a several times over one period. An example of such a case is shown in Fig. 13.3. The condition of such reflections is

$$4b\xi_k(1-\xi_k)\cot\theta_{k+1} > L(1-\xi_k) \quad \text{or} \quad \frac{4b\xi_k}{L} - \tan\theta_{k+1} > 0 . \qquad (13.18)$$

The coordinates $(\bar{\psi}_k, \bar{\theta}_k)$ of the second reflection point of the ray from the corrugated wall can be found from the geometrical consideration. They are related to the first reflection point coordinates as

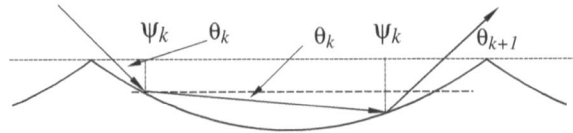

Fig. 13.3. Two consecutive reflections of the ray from corrugated wall in one period of the corrugation

$$\bar{\theta}_k = \theta_k - 2\arctan[bf'(\psi_k)] \ ,$$

$$\bar{\xi}_k = 1 - \xi_k + L\tan\bar{\theta}_k/4b \ ,$$

$$\bar{\psi}_k = \psi_k + L(\bar{\xi}_k - \xi_k) \ . \tag{13.19}$$

Then the coordinates (ψ_k, θ_k) in the map (13.14) should be replaced by $(\bar{\psi}_k, \bar{\theta}_k)$.

Similar to the Fermi accelerator mapping (3.30) the mapping of rays is exact. The both mappings have similar properties. The perturbation function $f(x)$ (13.16) is a continuous function of x but it has discontinuous first derivatives at points $x_s = sL$, $(s = 0, \pm 1, \pm 2, \cdots)$. The mapping (13.14) belongs to the class of non-smooth maps discussed in Sect. 7.4 and therefore the KAM theory is not applicable to this problem. There is no invariant curve for any small corrugations and one expects that rays with special initial conditions may diffuse far from their initial state.

13.1.3 Ray Dynamics in the Waveguide Model

Consider first the phase space of the mapping (13.14) in the plane $(\theta, \{x/L\})$. It is plotted in Fig. 13.4 for the following values of waveguide parameters: $a = 1$, $L = 5$, and $b = 0.01$. One can see that rays propagating under small angles θ, $\theta < \theta_c \approx 6°$ are chaotic with some remnants of KAM-stability islands. Because of absence invariant curves the rays from the chaotic zone may diffuse rays along the angle θ through the chaotic web structure, formed along the destroyed separatrices.

However, the majority of rays are trapped in the islands formed near the resonant angles θ_{mn} determined by the commusurability of the spatial frequency of ray oscillations, $\omega(\theta) = \pi\tan\theta/a$, along the x-coordinate with the spatial frequency of corrugation, $2\pi/L$:

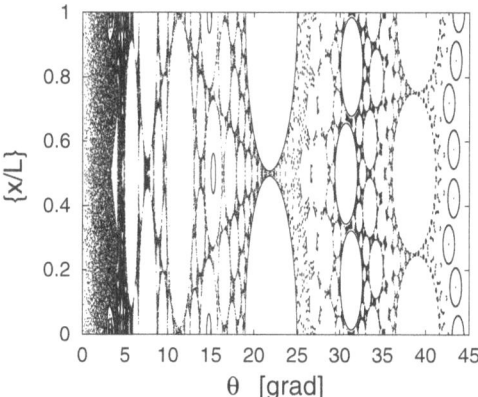

Fig. 13.4. Mapping on the $(\theta, \{x/L\})$ plane. Parameters are $a = 1$, $L = 5$, and $b = 0.01$

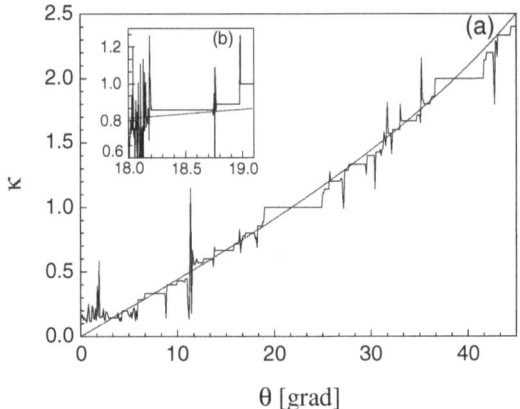

θ [grad]

Fig. 13.5. Normalized spatial frequency $\kappa = LN/x_N$, $(N \gg 1)$, of ray oscillations along the x-coordinate as a function of initial ray angle θ: (a) In the interval $0 < \theta < 45°$; (b) Expanded view of the region $18° < \theta < 19.1°$. Parameters are the same as in Fig. 13.4

$$m\omega(\theta_{mn}) - 2n\pi/L = 0 , \quad \text{or} \quad \tan\theta_{mn} = \frac{n}{m}\frac{2a}{L} , \qquad (13.20)$$

where m, n are integer numbers.

Consider some important properties of rays. First we study the spatial oscillation frequency of rays $w(\theta) = N/x_N$ $(N \to \infty)$ along the axis x in the corrugated waveguide. In Fig. 13.5 we have plotted the normalized frequency $\kappa = Lw(\theta)$ against the initial angle θ: solid curve corresponds to the corrugated waveguide, and dashed curve describes the unperturbed case $b = 0$ when $\kappa = L\omega(\theta)/2\pi = (L/2a)\tan\theta$. Figure 13.5a shows the κ vs θ dependence in the interval $[0, 45°]$, and Fig. 13.5b shows the expanded view of this dependence in the interval $[18°, 19.1°]$. In the waveguide with unperturbed walls the dependence of the frequency κ on the initial angle θ is described by the smooth function. Any small corrugation of the wall dramatically changes this dependence and it becomes the fractal known as *"devil's staircase"*. As seen from Fig. 13.5 there are intervals of angle θ located near the resonant angles θ_{mn} where the frequency κ remains constant and takes the rational values $\kappa_{mn} = m/n$.

The devil's staircase is a special case of fractal objects (see Mandelbrot (1982); Bak (1986))[1]. It has been also found in other Hamiltonian problems, for instance, in the one-dimensional Ising model by Bak and Bruinsma (1982) and the Frenkel–Kontorova model by Aubry (1978).

One of the practically interesting characteristics of waves is the signal propagation time along the rays. Let $t(\theta, x)$ be a transmission time of a

---

[1] The role of fractals in wave processes, particularly, in acoustics is discussed in a review papaer by Zosimov and Lyamshev (1995).

signal along a ray with the initial angle θ from the plane $x = 0$ to the plane $x =$ const. An optical path length along this ray $S(\theta, z)$ is defined as $S(\theta, x) = c_0 t(\theta, x)$, where c_0 is a reference signal speed. In the waveguide with a homogeneous refractive index the optical path length is equal to the geometrical length of ray. From the geometry of rays in Fig. 13.2 one can easily obtain the optical length

$$S(\theta_0, x) = \frac{\psi_0}{\cos\theta_0} + \sum_{k=1}^{N} \frac{\psi_{k+1} - \psi_k}{\cos\theta_k} + \frac{x - \psi_N}{\cos\theta_{N+1}}, \qquad (13.21)$$

where N is integer number satisfying the condition $\psi_N < x < \psi_{N+1}$.

The normalized relative optical length $(S(\theta, x) - x)/L$ versus the initial angle θ is plotted in Fig. 13.6 for the same waveguide parameters as in Fig. 13.4: solid curve corresponds to the corrugated waveguide, and dashed curve − to the unperturbed one. As seen from Fig. 13.4 the optical path length S, similar to the frequency of ray oscillations, has a "devil's staircase" − like dependence on the initial launching angle θ. As seen from Fig. 13.6b the optical lengths of rays with the small launch angles $\theta < 3°$ are distributed randomly. They correspond to chaotic rays. For the rays with $\theta > 3°$ there exist intervals of angle θ located near the resonant values θ_{mn} where the optical length S practically remains constant, while it varies significantly in the unperturbed waveguide. The widths $\Delta\theta_{mn}$ these intervals are equal to the width of resonant islands shown in Fig. 13.4. The obtained result means that when a signal propagated along the resonant rays does not broaden significantly than in the case of unperturbed rays. The effect may be used to suppress broadening signals caused by intermode dispersion in waveguides. This phenomenon has been first found numerically by Abdullaev and Zaslavsky (1988), and it has been

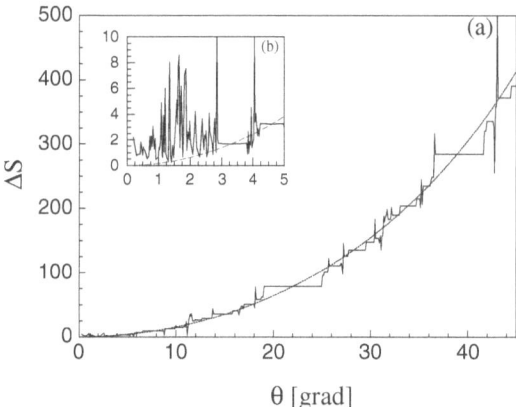

Fig. 13.6. Relative optical length of ray, $\Delta S = (S - x)/L$, propagated from $x = 0$ to x, as a function of initial angle θ: (**a**) In the interval $0 < \theta < 45°$; (**b**) Expanded view of the small angle region $0° < \theta < 5°$. Parameters are the same as in Fig. 13.4

explained later in Abdullaev (1991) (see, also a review by Abdullaev and Zaslavsky (1991) and a book by Abdullaev (1993)) as a consequence of ray trapping by nonlinear resonance islands.

One should mention that the numerical studies of sound ray propagation in a realistic range–dependent deep ocean environment by Tappert and Tang (1996) revealed the existence of a group of eigenrays (rays connecting the source of acoustic signal with a receiver) with equal signal transmission times. This result can be explained by the above phenomenon of nonlinear resonance. All rays belonging to the group of eigenrays with equal transmission times are trapped by the same resonant island and therefore they have similar properties.

13.1.4 Other Mapping Models of Rays

There were proposed also several mapping models of ray propagation in inhomogeneous media. Ray mapping models in waveguide media with a refractive index $n(z, x)$ depending on the vertical coordinate z have been derived by Abdullaev (1991, 1994a). The chaos of rays in electromagnetic waveguides with strongly corrugated walls has been studied by Vatrunin et al. (1997). Area-preserving mappings in sound ray propagation problems have been obtained by Brown et al. (1991a); Tappert et al. (1991) (see also Brown et al. (2003)). Particularly, in Brown et al. (1991a) the explicit map was derived for the bilinear model of the deep ocean with a constant sound speed gradient above and below the sound waveguide axis. The mapping of rays of type (3.23) with a discontinuously derivative of frequency $\omega(I)$ was obtained by Tappert et al. (1991) for the model of sound channel with a linear sound speed profile and a periodic bottom. These mappings are used to study the onset of chaotic ray motion in waveguides with a range–dependent environment.

13.2 Mapping Methods in Accelerator Physics

Particle accelerators are devices used to accelerate charged elementary particles and ions to high energies (Edwards and Syphers (1993)). One of the aspects of accelerator physics is the study of single particle motion in accelerators, the determination of *dynamical aperture*, the stable region in transverse phase space or the region of long-time stable orbits. In this section we briefly discuss the main contemporary mapping methods to study this problem.

The schematic view of a circular accelerator (a synchrotron) is shown in Fig. 13.7a. A particle accelerated during passing through one (or several) accelerating device(s) (stations) returns from turn to turn by bending magnetic fields directed perpendicular to the orbit of particle. The accelerator is designed to have a closed (reference) orbit. The investigation of stability of particle orbits with respect to their deviations from the reference orbit is one

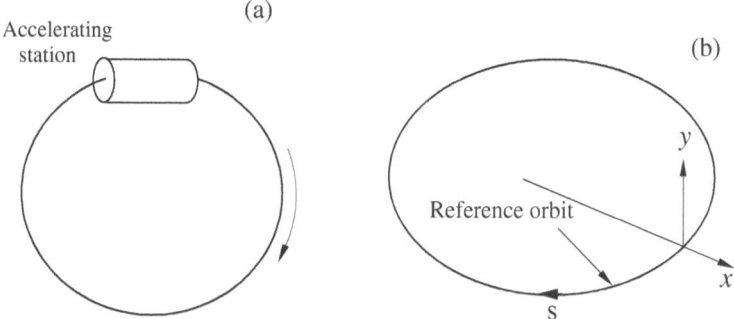

Fig. 13.7. (a) Schematic view of a circular accelerator (a synchrotron); (b) Coordinate system (x, y, s) with respect to the reference orbit

of the main tasks of accelerator physics. The Poincaré return map is a powerful tool to study this stability problem. In the accelerator theory this map is also known as *full-turn transfer map*. In a good approximation when dissipative effects due to synchrotron radiation are neglected a particle motion in accelerators is described by a Hamiltonian system. Therefore, the problem can be reduced to the study of stability of Hamiltonian systems (Ruth (1987); Edwards and Syphers (1993)).

Let s be the coordinate representing arclength along this orbit, and x, y be coordinates describing transverse displacements with respect to the reference orbit. The coordinate system is plotted in Fig. 13.7b. The corresponding canonical momenta are p_x, p_y. Since particles move along the coordinate s it is convenient to use the formulation of Hamilton equations by choosing s as an independent time-like variable (see Sect. 1.1.3). Let $t_0(s)$ and E_0 be a flight time along the reference orbit and a designed energy, respectively. Then the deviation of time t from $t_0(s)$, $\tau = t - t_0(s)$, and the deviation of particle energy E from E_0, taken with opposite sign, $p_0 = -(E - E_0)$, are the canonical conjugated variables. Then the motion of particle is described by a Hamiltonian

$$H = H(x, y, \tau, p_x, p_y, p_0, s) . \tag{13.22}$$

The Hamiltonian H is a periodic function of s with a period equal to the circumference L of the reference orbit. This is a system with three degrees of freedom and periodic dependence on the independent variable s (Ruth (1987); Berg et al. (1994)).

The Hamiltonian of the system can be also written in the action-angle variables $(\vartheta, I) = (\vartheta_x, \vartheta_y, \vartheta_\tau, I_x, I_y, I_\tau)$ introduced for motion transverse to the reference orbit, in the phase space (x, p_x, y, p_y) and for the energy - time variables (τ, p_0) (Berg et al. (1994); Wan and Cary (1998)). The Hamiltonian H can be presented in the form

$$H = H_0(I) + \epsilon H_1(\vartheta, I, s) \ ,$$

$$\epsilon H_1(\vartheta, I, s) = \sum_{m,n} H_{mn}(I) \cos(m \cdot \vartheta - 2\pi n s/L + \chi_{mn}) \ , \qquad (13.23)$$

as the sum of unperturbed Hamiltonian $H_0(I)$ and perturbed one ϵH_1 depending on the angle variables ϑ and the longitudinal variable s. The frequencies of motion are determined by the unperturbed Hamiltonian $H_0(I)$:

$$\omega_i = \frac{\partial H_0}{\partial I_i} \ , \qquad i = x, y, \tau \ .$$

The oscillation frequencies in the transversal plane (x, y), ω_x, ω_y, are known as *betatron oscillations* , while the frequency ω_τ represents the oscillations of a flight time and it is called *synchrotron oscillations*. They have a nonlinear dependence on oscillation amplitude I_x, I_y. The unperturbed Hamiltonian $H_0(I)$ can be obtained by averaging (Cary (1982, 1984)).

The main aim of dynamics of single particle motion in a circular accelerator is to construct the full turn transfer map for variables $\mathbf{z} = (x, y, \tau, p_x, p_y, p_0)$:

$$\mathbf{z}_{k+1} = \hat{M}\mathbf{z}_k, \qquad s = 0, 1, 2, \cdots \ , \qquad (13.24)$$

where

$$\mathbf{z}_k \equiv \mathbf{z}(s_0 + kL) \qquad (13.25)$$

are variables at the certain section $s = s_0 = \text{const}$.

Contemporary methods to construct the mappings (13.24) have been overviewed by Dragt (1996) (see also the introduction of a paper by Berg et al. (1994)). A direct way to obtain the mapping (13.24) is to integrate the equations of motion numerically with the small steps of s along the accelerator system using the symplectic integrators. However, this procedure known as *tracking* requires long computational times, especially, in large accelerators when one needs to evaluate the stability of orbits for more than 10^6 turns. Because of that one wishes to have a full-turn map expressed in a single formula one iteration of which is equivalent to one full turn. This would enormously simplify the study of a stability problem by running orbits with many different initial conditions.

There are several requirements on such a kind of mappings (see Berg et al. (1994); Dragt (1996)): a) they should reproduce the full turn map obtained from the Hamiltonian of the system with high accuracy; b) they should be symplectic; c) they should run faster as possible with reasonable computer storage requirements. Below we discuss the main methods to construct such mappings.

Taylor maps. A widely used approach to construct the full turn map has been based on *Taylor expansions* of the right hand side of the mapping (13.24) in the powers of $\mathbf{z}$:

$$z_{k+1}^{(i)} = A^{(i)} + \sum_j B^{(ij)} z_k^{(j)} + \sum_{jl} C^{(ijl)} z_k^{(j)} z_k^{(l)} + \cdots , \qquad (13.26)$$

where $z^{(j)} = (z^{(1)}, \ldots, z^{(6)}) = (x_k, y, \tau, p_x, p_y, p_0)$. Equation (13.26) represents an explicit map known as a *Taylor map*. The coefficients $A^{(i)}$, $B^{(ij)}$, $C^{(ijl)}, \ldots$ in the Taylor map (13.26) are to be found from the tracking code. The zero order coefficients $A^{(i)}$ describe the translations in phase space caused by imperfections in devices, the first order coefficients $B^{(ij)}$ describe the linear properties of the system, and the higher order coefficients $C^{(ijl)}$, etc. describe its nonlinear properties. Because the map (13.26) must be symplectic, these coefficients are not independent. The coefficients of different orders are interrelated by a set of complicated conditions. Since in practical calculations one uses the truncated Taylor series the corresponding mappings become not exactly symplectic. Beside of this there is another shortcoming of the Taylor map: expansion series (13.26) are not always suitable for the representation of orbits which are not too close to the reference orbit.

A powerful method to construct the full turn map by a series of explicit symplectic nonlinear maps has been proposed by Dragt (1979) (see also Dragt (1996); Dragt and Abel (1996); Dragt (2000)) using *Lie algebraic methods*. It presents the map as the factored product of symplectic maps describing correspondingly translations in, linear and nonlinear effects. Usually in computer codes one uses symplectic nonlinear mappings expanded in truncated Taylor series with the limited number of Taylor coefficients. This is one of main disadvantages of the method since the truncated Taylor maps are not exactly symplectic. These maps cannot be used to study a long time evolution of particles. For this reason methods to symplectify these nonlinear maps have been proposed (see for details in Dragt and Abel (1996) are references therein).

Symplectic maps with a mixed variable generating function. These mappings are suitable to study the long term evolution of particle motion (Berg et al. (1994); Warnock and Berg (1996)). The construction of these maps is based on the method which we called in Sect. 3.2 the method of a priori assumption. The corresponding map in terms of action-angle variables (ϑ, I) has the symplectic form of type (3.13), (3.14):

$$I_{k+1} = I_k + \frac{\partial S(\vartheta_{k+1}, I_k)}{\partial \vartheta_{k+1}} , \qquad \vartheta_{k+1} = \vartheta_k - \frac{\partial S(\vartheta_{k+1}, I_k)}{\partial I_k} , \qquad (13.27)$$

determined by a generating function $S(\vartheta_{k+1}, I_k)$ of mixed variables (ϑ_{k+1}, I_k). The generating functions should be found from the relation

$$I_{k+1} = I_k + R(\vartheta_k, I_k) , \qquad \vartheta_{k+1} = \vartheta_k + \Phi(\vartheta_k, I_k) , \qquad (13.28)$$

between the variables (ϑ_k, I_k) and $(\vartheta_{k+1}, I_{k+1})$ after the full turn. The functions $R(\vartheta, I), \Phi(\vartheta, I)$ are usually computed by tracking particles with different initial coordinates (ϑ, I). Since the generating function $S(\vartheta, I)$ is a periodic function of angle variables ϑ with period 2π, it can be presented as a Fourier series:

$$S(\vartheta, I) = S_0(I) + \sum_{m, m \neq 0} S_m(I) \exp(im \cdot \vartheta) .$$

From (13.27) and (13.28) it follows that $\partial S(\vartheta_{k+1}, I_k)/\partial \vartheta_{k+1} = R(\vartheta_k, I_k)$, and thus the Fourier coefficients $S_m(I)$ are calculated through the function $R(\vartheta, I)$. For arbitrary values of I these coefficients are interpolated by B-spline functions of I (see for details in Berg et al. (1994)). However, a spline expansion in action variables I has a disadvantage because of coordinate singularities in systems with more than one degree of freedom. Such singularities does not occur in Cartesian coordinates. For this reason Warnock and Ellison (1999) have presented the generating function of maps by a spline expansion in these coordinates. The effectiveness of mappings with a mixed variable generating function to study long term dynamics of particles in circular accelerators has been discussed by Berg et al. (1994); Warnock and Berg (1996, 1997)).

One should note that the mapping (13.27) is not only one and the best form of symplectic mappings corresponding to the Hamiltonian (13.22). This is the main shortcoming of the presented method. As we have discussed in Chaps. 4 and 10 the symmetric maps (4.6)–(4.8) (or (4.31)–(4.33)) more accurately describe Hamiltonian systems than their non-symmetric forms (4.9), (4.10) (or (4.38)) with the same generating functions. Therefore, the construction of the full turn map in the symmetric form would improve the accuracy of tracking particles in accelerators without significant increase of a computational time.

Four-dimensional maps. In many cases one can neglect the influence of the longitudinal (synchrotron) motion in (τ, p_0) on the transversal (betatron) motion of particles in the phase space (x, y, p_x, p_y). Then the system can be described by $2 + 1/2$ degrees of freedom Hamiltonian system with $H = H(x, y, p_x, p_y, s)$. Correspondingly, the transversal motion in this case can be also studied by reducing the six-dimensional map (13.24) to the four-dimensional one

$$(x_{k+1}, y_{k+1}, p_{k+1}^x, p_{k+1}^y) = \hat{M}(x_k, y_k, p_k^x, p_k^y) , \qquad (13.29)$$

for transversal variables (x, y, p_x, p_y). This simplifies the study of a stability of betatron motion by reducing a degree of freedom of system by one.

During the last decade the 4D symplectic mapping (13.29), have been extensively exploited to study the transverse betatron motion (see Todesco (1999) and references therein). Particularly, 4D Hénon map Hénon (1969) and its modifications have been used by Bazzani et al. (1994); Giovannozzi et al. (1998) to study the different aspects of particle motion in accelerators, long term stability, the dynamical aperture and others. On the other hand there is another aspect of the problem related with increasing a dynamical aperture. This problem has been discussed by Wan and Cary (1998, 2001) (and references therein) and the method to find 4D symplectic mappings (13.29) with reduced chaos has been proposed.

13.3 Mappings in Dynamical Astronomy

The motion of planets and planetary objects in the Solar system has been one of the oldest problems of classical mechanics (Poincaré (1892–99); Arnold et al. (1988)). In a good approximation it constitutes a non-relativistic Hamiltonian problem of gravitationally interacting bodies. A present interest to this problem is mostly related with its new aspects concerning the chaotic evolution of the Solar system (Sussman and Wisdom (1988, 1992); Lissauer (1999) and references therein). The study of this problem requires the long term integration (over hundreds million years) of the equations of planetary motion. Direct symplectic integrations of the whole planetary system are computationally expensive even on the present–day computers. An important breakthrough in computational efforts has been made by Wisdom (1982) who developed a mapping method to integrate the equations of motion. Later there were proposed other mapping methods to study these problems. In this section we briefly discuss the main mapping methods used in dynamical astronomy during the last two decades.

Method of Delta Functions. The method developed by Wisdom (1982) was inspired by Chirikov's derivation of the standard map by introducing a series of delta functions into perturbed Hamiltonian (Chirikov (1979)). It belongs to the type of methods to construct symplectic mappings which we called the method of delta functions (Sect. 3.3). The idea of the method was based on the averaging principle: if removing fast oscillating terms in perturbed Hamiltonian does not significantly change the dynamics of system, then adding these terms should not also significantly change it. In the early version of the method Wisdom (1982) obtained explicit mappings by replacing the resonant terms in Hamiltonian by the series of delta functions with a period of order of the orbital period. Later Wisdom and Holman (1991, 1992) generalized this method to study the problem of n gravitationally interacting bodies, and it is known now as *the n body mapping method*. In the new version they presented the Hamiltonian of n gravitationally interaction bodies of masses m_i

$$H = \sum_{i=0}^{n-1} \left(\frac{p_i^2}{2m_i} - \sum_{i<j}^{n-1} \frac{Gm_i m_j}{r_{ij}} \right) , \qquad (13.30)$$

where $r_{ij} = |\mathbf{x}_i - \mathbf{x}_j|$ is a distance between bodies, as a sum

$$H = H_{Kepler} + H_{int} , \qquad (13.31)$$

of the Kepler Hamiltonian, H_{Kepler}, describing the interaction of a dominant mass m_0 (Sun) with other $n - 1$ small bodies, and interaction Hamiltonian H_{int} describing the interaction between $n - 1$ small bodies. In an appropriate coordinate system related with the center of mass of the whole system they are reduced to

$$H_{Kepler} = \sum_{i=1}^{n-1} \left(\frac{p_i'^2}{2m_i'} - \frac{Gm_i m_0}{r_i'} \right) , \tag{13.32}$$

and

$$H_{int} = \sum_{i=1}^{n-1} Gm_i m_0 \left(\frac{1}{r_i'} - \frac{1}{r_{i0}} \right) - \sum_{0<i<j} \frac{Gm_i m_j}{r_{ij}} , \tag{13.33}$$

where m_i' are reduced masses, r_i' and p_i' are relative distances and momenta (see Wisdom and Holman (1991) for the exact definitions).

The Kepler motion described by Hamiltonian H_{Kepler} is completely integrable. In order to integrate the system in the presence of interaction H_{int} it is replaced by $H_{int} T \delta_T(t)$, where

$$\delta_T(t) = \sum_{k=-\infty}^{\infty} \delta(t - kT)$$

is the series of delta functions with a period T. Between "kicks" at the time instants $t_k = kT$ the system has analytical solutions determined by the Kepler Hamiltonian H_{Kepler}. Integration along the delta functions allows one to obtain the explicit symplectic mapping which takes into account the effect of interaction. The mapping is approximately an order of magnitude faster than the traditional methods of integration.

Symplectic mappings can be refined to make them close to the original Hamiltonian system (Wisdom and Holman (1991)). It was accomplished by using the refined sum of delta functions $\Psi(t)$ instead of simple sum $\delta_T(t)$:

$$\Psi(t) = T \sum_{i=0}^{N-1} a_i \delta_T(t - \tau_i) ,$$

where N is the number of delta functions per mapping period T, a_i and τ_i are the corresponding amplitudes and phases ($0 < \tau_i < T$). They are found by imposing some constraints on the function $\Psi(t)$. The symplectic mappings obtained by this scheme can be treated also as a symplectic numerical integration algorithm. A similar scheme has been independently proposed by Kinoshita et al. (1991).

The method of delta functions and its shortcomings were discussed in Sect. 3.3. We recall that the mapping variables do not coincide with the variables of the original system. This is because that replacing a perturbation by the series of delta functions introduces artificial singularities at the periodic time instants where variables becomes undefined. This is origin of that symplectic integration with obtained mapping yields trajectories which exhibit spurious oscillation in energy and variables. This problem was later recognized by Wisdom et al. (1996) who introduced so-called *symplectic correctors* to relate the mapping variables with original variables.

This mapping method has been successfully applied to study many problems of a chaotic evolution of planetary systems which require the integration over several hundreds millions years. Among them there were the explanation of the Kirkwood gaps in the distribution of semimajor axes of the asteroids (Wisdom (1982, 1983)), confirmation that the evolution of the Solar system as a whole in chaotic with a time scale of exponential divergence of about 4 million years (Sussman and Wisdom (1988, 1992); Wisdom (1992) and others) (see Šidlichovský (1997); Hadjidemetriou (1998); Lissauer (1999) and references therein).

Mappings with mixed variable generating functions. Another mapping method based on mixed variable generating functions has been proposed by Hadjidemetriou (1991, 1993) (see also a review by Hadjidemetriou (1998)). It belongs to the class of methods discussed in Sect. 3.2. Below we briefly present the main idea of this approach.

Consider the resonant motion of a nearly integrable system with two degrees of freedom. By appropriately defining action-angle variable, $(\vartheta_1, \vartheta_2, I_1, I_2)$, the Hamiltonian of the system can be presented as

$$H = H_0(I_1, I_2) + \epsilon H_1(\vartheta_1, \vartheta_2, I_1, I_2) , \qquad (13.34)$$

where $H_0(I_1, I_2)$ is the integrable part, H_1 is the resonant perturbation, and ϵ is the small parameter. Suppose that ϑ_1 is the slowly varying angle near the resonance, and ϑ_2 is a fast angle. By a canonical change of variables $(\vartheta_1, \vartheta_2, I_1, I_2) \to (\psi_1, \psi_2, J_1, J_2)$ we eliminate the fast angle ψ_2 from the new Hamiltonian $\mathcal{H}$:

$$\mathcal{H} = \mathcal{H}_0(J_1, J_2) + \epsilon \mathcal{H}_1(\psi_1, J_1, J_2) + O(\epsilon^2) . \qquad (13.35)$$

This procedure is described by von Zeipel's perturbation method given in Sect. 2.1.4. If we succeed to eliminate the phase ψ_2 in the new averaged Hamiltonian (13.35) up to terms of order of ϵ^m then the action J_2 is an approximate constant of motion with accuracy of $O(\epsilon^{m+1})$.

Hadjidemetriou (1991) constructed the map $(\psi_k, J_k) \to (\psi_{k+1}, J_{k+1})$ in the symplectic form

$$J_{k+1} = J_k - \frac{\partial F(\psi_k, J_{k+1})}{\partial \psi_k} ,$$
$$\psi_{k+1} = \psi_k + \frac{\partial F(\psi_k, J_{k+1})}{\partial J_{k+1}} , \qquad (13.36)$$

defining the generating function $F(\psi_k, J_{k+1})$ as

$$F(\psi_k, J_{k+1}) = \psi_k J_{k+1} + T\mathcal{H}(\psi_k, J_{k+1}, J_2) . \qquad (13.37)$$

Here (ψ_k, J_k) are the values of $(\psi_1(t), J_1(t))$ taken at the discrete time moments $t = kT$, $(k = 0, \pm1, \pm2, \ldots)$, T is the period of the fast angle ψ_2. In practical applications one restricts itself with the two lowest order terms of the averaged perturbed Hamiltonian, $\mathcal{H} = \mathcal{H}_0 + \epsilon\mathcal{H}_1$.

According to Hadjidemetriou (1991) the fixed points of the averaged Hamiltonian system (13.35) should coincide with the periodic orbits of the original system (13.34). This condition is necessary to obtain a realistic model by averaging the original system. However, in some cases when the perturbation series in the averaging procedure does not converge this condition may be violated (Hadjidemetriou (1999)). In such a case the averaged Hamiltonian (13.35) and the corresponding mapping are not realistic models for the original system.

This mapping method and its different generalizations have been successfully applied to study the dynamical behavior of asteroid resonances by Ferraz-Mello (1997), Roig and Ferras-Mello (1999), Hadjidemetriou (1999) Sándor et al. (2002) (see also references therein).

Method of separatrix mapping. Some problems of chaotic dynamics of asteroids and the rotational motion of satellites can be reduced to the well-known Hamiltonian problem, the dynamics of pendulum perturbed by time-periodic perturbation (Wisdom et al. (1984); Celletti (1990)). It allowed Shevchenko and Scholl (1997); Shevchenko (1998, 1999); Shevchenko and Kouprianov (2002) to apply a separatrix mapping approach to study these problems. In particularly, it has been shown that the intermittency in the chaotic motion of asteroids due to the stickiness of orbits to the marginal resonances can be explained by the separatrix mapping which allows to predict the conditions for intermittency (Shevchenko and Scholl (1997); Shevchenko (1998)). The separatrix mapping has been also applied to study the planar resonant rotational motion of a non-symmetric satellite in an elliptic orbit (Shevchenko (1999); Shevchenko and Kouprianov (2002)). It correctly reproduces the phase portraits of motion obtained using other mapping methods and direct numerical integration of equations of the rotational motion of non-symmetric satellites. The separatrix map runs a hundred times faster than other mapping methods, and it allows one also to analytically precalculate positions of resonances and estimate the borders of chaotic motion.

Other applications of mapping methods in dynamical astronomy can be found in review papers by Šidlichovský (1997) and Hadjidemetriou (1998). Particularly, the so-called Kepler map has been introduced in several works by Petrosky (1986); Sagdeev and Zaslavsky (1987); Petrosky and Broucke (1988); Chirikov and Vecheslavov (1989) to reveal the chaotic nature of motion of comets in nearly parabolic orbits in the Solar system (see also Sect. 6.4).

A The Second Order Generating Function

Here we calculate the second order generating function $S_2(J, \theta, t)$ for the Hamiltonian function

$$H(I, \vartheta, t) = H_0(I) + \epsilon H_1(I, \vartheta, t) \,,$$

$$H_1(I, \vartheta, t) = \epsilon \sum_m H_m(I) \cos(m \cdot \vartheta + \chi_m) \,, \tag{A.1}$$

where $\vartheta = (\vartheta_1, \ldots, \vartheta_n, -\Omega t)$, $m = (m_1, \ldots, m_n, n)$, $m \cdot \vartheta = m \cdot \theta - n\Omega t$. We write the first order generating function $S_1(J, \vartheta, t)$ in the form

$$S_1(J, \vartheta, t) = - \int_{t_0}^{t} H_1(J, \vartheta(t'), t') dt' = \sum_m \frac{H_m(J)}{m \cdot \omega}$$

$$\times \left[\sin(m \cdot \vartheta + m \cdot \omega(t_0 - t) + \chi_m) - \sin(m \cdot \vartheta + \chi_m) \right] . \tag{A.2}$$

According to (2.11) and (2.12) the second order generating function $S_2(J, \theta, t)$ is

$$S_2(J, \vartheta, t) = - \int_{t_0}^{t} F_2(J, \vartheta(t'), t') dt' \,, \tag{A.3}$$

where

$$F_2(I, \vartheta, t) = \frac{1}{2} \frac{\partial^2 H_0}{\partial J_i \partial J_j} \frac{\partial S_1}{\partial \vartheta_i} \frac{\partial S_1}{\partial \vartheta_j} + \frac{\partial H_1}{\partial J_i} \frac{\partial S_1}{\partial \vartheta_i} \,. \tag{A.4}$$

In (A.4) summation over repeating indexes i, j $(i, j = 1, \ldots, n)$ is assumed. Putting (A.1) and (A.2) into (A.4) we obtain

$$F_2(J, \vartheta, t) = \frac{1}{2} \sum_{m, m'} m_i \frac{\partial^2 H_0^{(1)}}{\partial J_i J_j} m'_j \frac{H_m(J) H_{m'}(J)}{(m \cdot \omega) (m' \cdot \omega)}$$

$$\times \left[\cos(m \cdot \vartheta + m \cdot \omega (t_0 - t) + \chi_m) - \cos(m \cdot \vartheta + \chi_m) \right]$$

S.S. Abdullaev: *Construction of Mappings for Hamiltonian Systems and Their Applications*, Lect. Notes Phys. **691**, 317–320 (2006)
www.springerlink.com

$$\times \left[\cos(m' \cdot \vartheta + m' \cdot \omega \, (t_0 - t) + \chi_{m'}) - \cos(m' \cdot \vartheta + \chi_{m'}) \right]$$

$$+ \sum_{m,m'} m'_j \frac{\partial H_m}{\partial J_j} \frac{H_{m'}(J)}{m' \cdot \omega} \cos(m\vartheta + \chi_m)$$

$$\times \left[\cos(m' \cdot \vartheta + m' \cdot \omega \, (t_0 - t) + \chi_{m'}) - \cos(m' \cdot \vartheta + \chi_{m'}) \right] . \qquad (A.5)$$

We replace the products of the trigonometric functions in (A.5) by their sum and integrate each of them using the integral

$$F_{ms} = \int_{t_0}^{t} dt' \cos \left(m \cdot \vartheta(t') + s \cdot \omega \, (t_0 - t') + \chi \right)$$

$$= a(x_m, x_s) \sin(m \cdot \vartheta + \chi) + b(x_m, x_s) \cos(m \cdot \vartheta + \chi) , \qquad (A.6)$$

where $x_m = m \cdot \omega \, (t - t_0)$ and

$$a(x,y) = -(t - t_0) \, \frac{\cos x - \cos y}{x - y} , \qquad b(x,y) = (t - t_0) \, \frac{\sin x - \sin y}{x - y} .$$

Then combining the trigonometric functions we obtain the following expression for the generating function (A.3):

$$S_2(I, \vartheta, t) = -\frac{(t - t_0)^3}{4} \sum_{m,m'} m_i \frac{\partial^2 H_0}{\partial J_i \partial J_j} m'_j H_m(J) H_{m'}(J)$$

$$\times \left\{ A_{m,m'}^{(+)} \sin \phi_{mm'}^{(+)} + B_{m,m'}^{(+)} \cos \phi_{mm'}^{(+)} + A_{m,m'}^{(-)} \sin \phi_{mm'}^{(-)} + B_{m,m'}^{(-)} \cos \phi_{mm'}^{(-)} \right\}$$

$$- \frac{(t - t_0)^2}{2} \sum_{m,m'} m'_j \frac{\partial H_m(J)}{\partial J_j} H_{m'}(J)$$

$$\times \left\{ C_{m,m'}^{(+)} \sin \phi_{mm'}^{(+)} + D_{m,m'}^{(+)} \cos \phi_{mm'}^{(+)} + C_{m,m'}^{(-)} \sin \phi_{mm'}^{(-)} + D_{m,m'}^{(-)} \cos \phi_{mm'}^{(-)} \right\} , $$

$$\qquad (A.7)$$

where $\phi_{mm'}^{(\pm)} = (m \pm m') \cdot \vartheta + \chi_m \pm \chi_{m'}$, and the coefficients $A_{m,m'}^{(\pm)}$, $B_{m,m'}^{(\pm)}$, $C_{m,m'}^{(\pm)}$ and $D_{m,m'}^{(\pm)}$ are defined as

$$A_{m,m'}^{(+)} = U(x_m, x_{m'}) , \qquad\qquad A_{m,m'}^{(-)} = -U(x_m, -x_{m'}) ,$$

$$B_{m,m'}^{(+)} = V(x_m, x_{m'}) , \qquad\qquad B_{m,m'}^{(-)} = -V(x_m, -x_{m'}) , \qquad (A.8)$$

$$C_{m,m'}^{(+)} = W(x_m, x_{m'}) , \qquad\qquad D_{m,m'}^{(+)} = Y(x_m, x_{m'}) ,$$

$$C_{m,m'}^{(-)} = -W(x_m, -x_{m'}) , \qquad\qquad D_{m,m'}^{(-)} = -Y(x_m, -x_{m'}) , \qquad (A.9)$$

are expressed in terms of four functions $U(x,y)$, $V(x,y)$, $W(x,y)$ and $Y(x,y)$ of two variable x, y:

$$U(x,y) = \frac{1}{xy}\left[\sin(x+y) + \frac{\cos(x+y) - \cos y}{x} \right.$$
$$\left. + \frac{\cos(x+y) - \cos x}{y} + \frac{1 - \cos(x+y)}{x+y} \right] \tag{A.10}$$

$$V(x,y) = \frac{1}{xy}\left[\cos(x+y) - \frac{\sin(x+y) - \sin y}{x} \right.$$
$$\left. - \frac{\sin(x+y) - \sin x}{y} + \frac{\sin(x+y)}{x+y} \right] \tag{A.11}$$

$$W(x,y) = -\frac{1}{y}\left[\frac{\cos(x+y) - \cos y}{x} + \frac{1 - \cos(x+y)}{x+y} \right] \tag{A.12}$$

$$Y(x,y) = \frac{1}{y}\left[\frac{\sin(x+y) - \sin y}{x} - \frac{\sin(x+y)}{x+y} \right] \tag{A.13}$$

These functions have the following asymptotics at $y \to x$:

$$U(x,x) = \frac{1}{x^2}\left[\sin 2x + \frac{1 - 4\cos x + 3\cos 2x}{2x} \right] \tag{A.14}$$

$$V(x,x) = \frac{1}{x^2}\left[\cos 2x + \frac{4\sin x - 3\sin 2x}{2x} \right], \tag{A.15}$$

$$U(x,-x) = 0, \qquad V(x,-x) = \frac{2}{x^2}\left(1 - \frac{\sin x}{x}\right), \tag{A.16}$$

$$W(x,x) = \frac{1 - \cos x}{x^2}\cos x, \qquad Y(x,x) = -\frac{1 - \cos x}{x^2}\sin x, \tag{A.17}$$

$$W(x,-x) = \frac{1 - \cos x}{x^2}, \qquad Y(x,-x) = -\frac{1}{x}\left(1 - \frac{\sin x}{x}\right). \tag{A.18}$$

The functions $U(x,y)$, $V(x,y)$, $W(x,y)$ and $Y(x,y)$ of two variables x, y are localized in the finite region $|x| \le \pi, |y| \le \pi$. They decay at large values of x, y. The functions $U(x,y)$, $V(x,y)$, $Y(x,y)$ and $W(x,y)$ are shown in Figs. A.1, A.2.

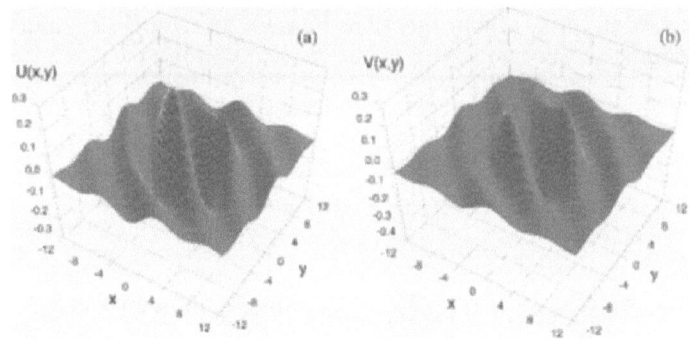

Fig. A.1. Functions: (a) $U(x, y)$; (b) $V(x, y)$

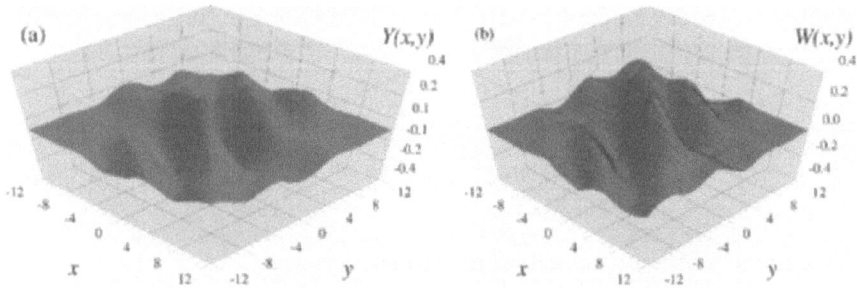

Fig. A.2. Functions: (a) $Y(x, y)$; (b) $W(x, y)$

If perturbation contains only one (m, n) mode then the function S_2 is reduced to

$$S_2(I, \vartheta, t) = -\frac{(t - t_0)^3}{4} \frac{\partial^2 H_0}{\partial J^2} m^2 H_m^2(J)$$

$$\times \left\{ A_{m,m}^{(+)} \sin 2(m \cdot \vartheta + \chi_m) + B_{m,m}^{(+)} \cos 2(m \cdot \vartheta + \chi_m) + B_{m,m}^{(-)} \right\}$$

$$-\frac{(t - t_0)^2}{2} \frac{\partial H_m(J)}{\partial J} m H_m(J)$$

$$\times \left\{ C_{m,m}^{(+)} \sin 2(m \cdot \vartheta + \chi_m) + D_{m,m}^{(+)} \cos 2(m \cdot \vartheta + \chi_m) + D_{m,m}^{(-)} \right\}.$$

$$(A.19)$$

B Asymptotic Estimations of the Integral $K(h)$ and $L(h)$ Near Separatrix

B.1 General Structure of Integrals

For simplicity we shall omit the subscript n of all quantities. Then we consider three types of the Melnikov type integrals defined by

$$R^+(h) = \int_0^{\pi/\omega(h)} V(h,\tau)e^{i\Omega\tau}d\tau,$$

$$R^-(h) = \int_{-\pi/\omega(h)}^0 V(h,\tau)e^{i\Omega\tau}d\tau,$$

$$R(h) = \int_{-\pi/\omega(h)}^{\pi/\omega(h)} V(h,\tau)e^{i\Omega\tau}d\tau = R^+(h) + R^-(h), \tag{B.1}$$

The integrals (B.1) are taken along the unperturbed orbits $(h, \vartheta(t))$ or $(q(t;h), p(t;h))$. We specify the time t in the following way: at $t = 0$ the orbit crosses the section Σ_c, and at $t = t_s = \pi/\omega(h)$ it crosses the section Σ_s. With such a definition the function $V(h,\tau)$ can be presented as a function of phase space coordinates:

$$V(h,\tau) \equiv H_1\left(q(\tau;h), p(\tau;h)\right) .$$

We study the asymptotics of $R(h)$, $R^+(h)$ and $R^-(h)$ near the separatrix, i.e., at $|h| \ll 1$. They can be presented as a sum of regular, $R^{(reg)}(h)$, and oscillatory, $R^{(osc)}(h)$, parts

$$R(h) = R^{(reg)}(h) + R^{(osc)}(h) ,$$

$$R^+(h) = R^{(reg)}(h) + R^{(osc+)}(h) ,$$

$$R^-(h) = R^{(reg)}(h) + R^{(osc-)}(h) , \tag{B.2}$$

where the regular parts

$$R^{(reg)}(h) = F(h, -0) - F(h, +0) ,$$

S.S. Abdullaev: *Construction of Mappings for Hamiltonian Systems and Their Applications*, Lect. Notes Phys. **691**, 321–333 (2006)
www.springerlink.com

$$R^{(reg)+}(h) = F(h, -0) \,,$$

$$R^{(reg)-}(h) = F(h, +0) \,, \tag{B.3}$$

and the oscillatory parts

$$R^{(osc)}(h) = F(h, \pi/\omega(h)) - F(h, -\pi/\omega(h)) \,,$$

$$R^{(osc)+}(h) = F(h, \pi/\omega(h)) \,,$$

$$R^{(osc)-}(h) = -F(h, -\pi/\omega(h)) \,, \tag{B.4}$$

are defined through the function $F(h, \tau)$,

$$\frac{dF(h, \tau)}{d\tau} = V(h, \tau)e^{i\Omega\tau} \,. \tag{B.5}$$

Further we suppose that the function $V(h, \tau)$ vanishes at the saddle points, (q_s, p_s), i.e., $V(0, \infty) = H_1(q_s, p_s) = 0$. Then we have $F(0, \infty) = 0$. In general cases, this condition can be satisfied by subtracting from the Hamiltonian a term $H_1(q_s, p_s, t)$ which does not affect the equations of motion.

Below we show that near the separatrix of a Hamiltonian system with hyperbolic fixed points the oscillatory parts of the integrals (B.1) have a generic asymptotic behavior. In the following sections we derive the asymptotical formulae for $R^{(osc)}(h)$, $R^{(osc)\pm}(h)$ in the limit $|h| \to 0$.

B.1.1 Unperturbed Orbits Near the Separatrix

Suppose $H_0(q, p)$ is unperturbed Hamiltonian and (q_s, p_s) is a saddle point where

$$\frac{\partial H_0}{\partial q}(q_s, p_s) = 0 \,, \qquad \frac{\partial H_0}{\partial p}(q_s, p_s) = 0 \,. \tag{B.6}$$

One can always choose that $H_0(q_s, p_s) = 0$. The unperturbed Hamiltonian $H_0(q, p)$ can be expanded in a series of powers of $(q - q_s)$, $(p - p_s)$ near the saddle point:

$$H_0(q, p) = \frac{1}{2}H_{qq}(q - q_s)^2 + H_{qp}(q - q_s)(p - p_s)$$
$$+ \frac{1}{2}H_{pp}(p - p_s)^2 + O[(q - q_s)^3, (p - p_s)^3] \,, \tag{B.7}$$

where H_{qq}, H_{qp}, and H_{pp} are second derivatives of H_0 with respect to (q, p) taken at the hyperbolic fixed point. By the linear transformation of variables:

$$\xi = (q - q_s)\cos\alpha + (p - p_s)\sin\alpha \,,$$

$$\eta = -(q - q_s)\sin\alpha + (p - p_s)\cos\alpha ,$$

$$q - q_s = \xi\cos\alpha - \eta\sin\alpha ,$$
$$p - p_s = \xi\sin\alpha + \eta\cos\alpha , \tag{B.8}$$

the Hamiltonian can be diagonalized:

$$h(\xi,\eta) \equiv H_0(q,p) = -\frac{\alpha^2}{2}\xi^2 + \frac{\beta^2}{2}\eta^2 + O(\xi^3,\eta^3) , \tag{B.9}$$

where $\alpha = \sqrt{|\lambda_1|}$, $\beta = \sqrt{\lambda_2}$, and $\lambda_1 < 0$ and $\lambda_2 > 0$:

$$(\lambda_1,\lambda_2) = \frac{H_{qq} + H_{pp}}{2} \pm \sqrt{\frac{(H_{qq} - H_{pp})^2}{4} + H_{qp}^2}$$

are the eigenvalues of the matrix

$$\begin{pmatrix} H_{qq} & H_{qp} \\ H_{qp} & H_{pp} \end{pmatrix} \equiv \begin{pmatrix} \frac{\partial^2 H_0}{\partial q^2} & \frac{\partial^2 H_0}{\partial q \partial p} \\ \frac{\partial^2 H_0}{\partial p \partial q} & \frac{\partial^2 H_0}{\partial p^2} \end{pmatrix}\Bigg|_{q=q_s,p=p_s} .$$

The angle α is determined by $\tan\alpha = 2H_{qp}/(H_{qq} - H_{pp})$.
The equations of motion in the coordinates (ξ,η) are

$$\frac{d\xi}{dt} = \frac{\partial h(\xi,\eta)}{\partial\eta} = \beta^2\eta , \qquad \frac{d\eta}{dt} = -\frac{\partial h(\xi,\eta)}{\partial\xi} = \alpha^2\xi . \tag{B.10}$$

Phase space of this system near the saddle point is shown in Fig. B.1. Its
solutions $(\xi(t;h),\eta(t;h))$, $0 < t < t_s$ which cross the section Σ_s at the time
moment $t \to t_s = \pi/\omega(h)$ along the branches I and III of the separatrix are

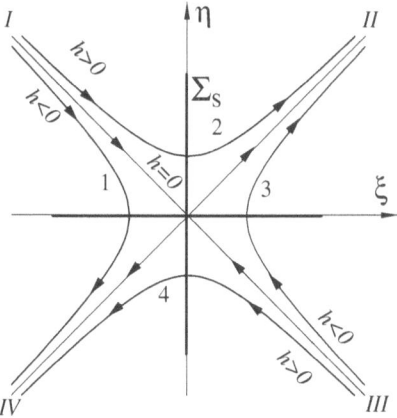

Fig. B.1. Phase curves of Hamiltonian system near the saddle point

$$\xi(t;h) = \mp \frac{\sqrt{2|h|}}{\alpha} \begin{cases} \cosh\left[\gamma\left(t_s - t\right)\right], & \text{for } h < 0, \\ \sinh\left[\gamma\left(t_s - t\right)\right], & \text{for } h > 0, \end{cases} \tag{B.11}$$

$$\eta(t;h) = \pm \frac{\sqrt{2|h|}}{\beta} \begin{cases} \sinh\left[\gamma\left(t_s - t\right)\right], & \text{for } h < 0, \\ \cosh\left[\gamma\left(t_s - t\right)\right], & \text{for } h > 0, \end{cases} \tag{B.12}$$

where $\gamma = \alpha\beta$. The upper and lower signs in (B.11), (B.12) correspond to the solution along the branches I and III, respectively. Similarly, solutions $(\xi(t;h), \eta(t;h))$, $0 > t > -t_s$ which cross the section Σ_s at the time moment $t \to -t_s = \pi/\omega(h)$ along the branches II and IV of the separatrix are given by

$$\xi(t;h) = \pm \frac{\sqrt{2|h|}}{\alpha} \begin{cases} \cosh\left[\gamma\left(t_s + t\right)\right], & \text{for } h < 0, \\ \sinh\left[\gamma\left(t_s + t\right)\right], & \text{for } h > 0, \end{cases} \tag{B.13}$$

$$\eta(t;h) = \pm \frac{\sqrt{2|h|}}{\beta} \begin{cases} \sinh\left[\gamma\left(t_s + t\right)\right], & \text{for } h < 0, \\ \cosh\left[\gamma\left(t_s + t\right)\right], & \text{for } h > 0. \end{cases} \tag{B.14}$$

These orbits are plotted in the upper half plane ($\eta > 0$) shown in Fig. B.1. Using the asymptotics

$$\frac{2\pi}{\omega(h)} = \frac{1}{\gamma} \ln \frac{A}{|h|},$$

we obtain the following asymptotical behavior of orbits at the separatrix, $h = 0$,

$$\xi(t;0) = \mp \frac{\sqrt{A/2}}{\alpha} e^{\mp\gamma t}, \qquad \eta(t;0) = \frac{\sqrt{A/2}}{\beta} e^{\mp\gamma t},$$

which asymptotically approach to the saddle point ($\xi = 0, \eta = 0$) in the limit $t \to \pm\infty$, respectively.

B.1.2 Perturbation Hamiltonian in Normal Coordinates ξ, η Near the Saddle Points

We expand the perturbation Hamiltonian $V(h,\tau) = H_1(q,p)$ near the saddle point, (q_s, p_s), in series of powers of $(q - q_s), (p - p_s)$:

$$H_1(q,p) = \sum_{j=1}^{\infty} V_j(q,p), \tag{B.15}$$

$$V_j(q,p) = \sum_{i=0}^{j} c_i^{(j)} (q - q_s)^i (p - p_s)^{j-i},$$

$$c_i^{(j)} = \left. \frac{\partial H_1}{\partial q^i \partial p^{j-i}} \right|_{q=q_s, p=p_s},$$

where the zero-order term $H_1(q_s, p_s)$ can be neglected in the evaluation of the integrals $R(h)$, since they do not affect the equations of motion. By a linear transformation of variables (B.8) the perturbation Hamiltonian (B.15) is rewritten in terms of variables (ξ, η):

$$H_1(q,p) = \sum_{j=1}^{\infty} H_j(\xi, \eta) \ ,$$

$$H_j(q,p) = \sum_{k=0}^{j} b_k^{(j)} \xi^k \eta^{j-k} \ , \qquad (B.16)$$

where the coefficients $b_k^{(j)}$ are linear combinations of $c_i^{(j)}$. If we retain only terms up to the second order in powers of ξ and η, (B.16) can be presented as

$$H_1(z, p_z) = a_\xi \xi + a_\eta \eta + b_{\xi\xi} \xi^2 + b_{\xi\eta} \xi\eta + b_{\eta\eta} \eta^2 + O(\delta^3) \ , \qquad (B.17)$$

where

$$a_\xi = a_q \cos\alpha + a_p \sin\alpha \ ,$$
$$a_\eta = -a_q \sin\alpha + a_p \cos\alpha \ ,$$

$$b_{\xi\xi} = b_{qq} \cos^2\alpha + \frac{1}{2} b_{zp} \sin 2\alpha + b_{pp} \sin^2\alpha \ ,$$
$$b_{\xi\eta} = -b_{qq} \sin 2\alpha + b_{qp} \cos 2\alpha + b_{pp} \sin 2\alpha \ ,$$
$$b_{\eta\eta} = b_{qq} \sin^2\alpha - \frac{1}{2} b_{qp} \sin 2\alpha + b_{pp} \cos^2\alpha \ .$$

Here $\delta \sim |q - q_s|, |p - p_s|$. The coefficients $a_q, \ldots, b_{pp}$ are given by

$$a_q = \left. \frac{\partial H_n(q,p)}{\partial q} \right|_{q=q_s, p=p_s} \ , \qquad a_p = \left. \frac{\partial H_n(q,p)}{\partial p_q} \right|_{q=q_s, p=p_s} \ ,$$

$$b_{qq} = \left. \frac{1}{2} \frac{\partial^2 H_n(q,p)}{\partial q^2} \right|_{q=q_s, p=p_s} \ , \qquad b_{qp} = \left. \frac{\partial^2 H_n(q,p)}{\partial q \partial p} \right|_{q=q_s, p=p_s} \ ,$$

$$b_{pp} = \left. \frac{1}{2} \frac{\partial^2 H_n(q,p)}{\partial p^2} \right|_{q=q_s, p=p_s} \ .$$

B.1.3 Integrals Over the Powers of Orbits $\xi(t,t), \eta(t,h)$ Near the Separatrix

The oscillatory parts of $R(h)$ (B.2) are given by the integrals $F(h,t)$ taken at the values $t = \pm t_s$, $t_s = \pi/\omega(h)$. According to (B.16) we have expressed them through the integrals over the powers of orbits $\xi(t,t), \eta(t,h)$

$$X^{(k,j-k)} = \int^{\pm t_s} \xi^k(t;h) \eta^{j-k}(t;h) \exp(i\Omega t) dt, \qquad (0 \le k \le j) \ . \quad (B.18)$$

Below we estimate the these integrals up to the second order ($j \leq 2$):

$$X^{(1,0)}(h, \pm t_s) = \int^{\pm t_s} \xi(t; h)e^{i\Omega t}dt ,$$

$$X^{(0,1)}(h, \pm t_s) = \int^{\pm t_s} \eta(t; h)e^{i\Omega t}dt ,$$

$$X^{(2,0)}(h, \pm t_s) = \int^{\pm t_s} [\xi(t; h)]^2 e^{i\Omega t}dt ,$$

$$X^{(1,1)}(h, \pm t_s) = \int^{\pm t_s} \xi(t; h)\eta(t; h)e^{i\Omega t}dt ,$$

$$X^{(0,2)}(h, \pm t_s) = \int^{\pm t_s} [\eta(t; h)]^2 e^{i\Omega t}dt . \tag{B.19}$$

It is easy to show that

$$\int^{\pm t_s} \sinh\left[\gamma(t_s \mp t)\right] e^{i\Omega t}dt = \pm\frac{\gamma}{\gamma^2 + \Omega^2}e^{\pm i\Omega t_s} ,$$

$$\int^{\pm t_s} \cosh\left[\gamma(t_s \mp t)\right] e^{i\Omega t}dt = \frac{i\Omega}{\gamma^2 + \Omega^2}e^{\pm i\Omega t_s} .$$

Using these integrals and the solutions for $\xi(t; h)$, $\eta(t; h)$ near the saddle point given by (B.11), (B.12) one obtains the following expressions for $X^{(1,0)}(h, t_s)$ and $X^{(0,1)}(h, t_s)$ along the branches I and III, respectively:

$$X^{(1,0)}(h, t_s) = \mp e^{i\Omega t_s}\frac{\sqrt{2|h|}}{\alpha(\gamma^2 + \Omega^2)} \begin{cases} i\Omega, & h < 0 \\ \gamma, & h > 0 , \end{cases}$$

$$X^{(0,1)}(h, t_s) = \pm e^{i\Omega t_s}\frac{\sqrt{2|h|}}{\beta(\gamma^2 + \Omega^2)} \begin{cases} \gamma, & h < 0 \\ i\Omega, & h > 0 . \end{cases} \tag{B.20}$$

Similarly, the corresponding functions along the the branches II and IV are given by

$$X^{(1,0)}(h, -t_s) = \pm e^{-i\Omega t_s}\frac{\sqrt{2|h|}}{\alpha(\gamma^2 + \Omega^2)} \begin{cases} i\Omega, & h < 0 \\ -\gamma, & h > 0 , \end{cases}$$

$$X^{(0,1)}(h, -t_s) = \pm e^{-i\Omega t_s}\frac{\sqrt{2|h|}}{\beta(\gamma^2 + \Omega^2)} \begin{cases} -\gamma, & h < 0 \\ i\Omega, & h > 0 , \end{cases} \tag{B.21}$$

respectively.

The second order integrals $X^{(2,0)}(h, \pm t_s)$, $X^{(1,1)}(h, \pm t_s)$, $X^{(0,2)}(h, \pm t_s)$ along the all four branches I–IV are given by

$$X^{(2,0)}(h, \pm t_s) = -i e^{i\Omega t_s} \frac{2|h|}{\alpha^2 \Omega} C_1(h) ,$$

$$X^{(1,1)}(h, \pm t_s) = e^{i\Omega t_s} \frac{2|h|}{4\gamma^2 + \Omega^2} ,$$

$$X^{(0,2)}(h, \pm t_s) = i e^{i\Omega t_s} \frac{2|h|}{\beta^2 \Omega} C_2(h) , \qquad (B.22)$$

where

$$C_1(h) = \begin{cases} \left(1 + \frac{\Omega^2}{4\gamma^2 + \Omega^2}\right), & h < 0 \\ \left(-1 + \frac{\Omega^2}{4\gamma^2 + \Omega^2}\right), & h > 0 , \end{cases} \qquad (B.23)$$

$$C_2(h) = \begin{cases} \left(-1 + \frac{\Omega^2}{4\gamma^2 + \Omega^2}\right), & h < 0 \\ \left(1 + \frac{\Omega^2}{4\gamma^2 + \Omega^2}\right), & h > 0 . \end{cases} \qquad (B.24)$$

B.1.4 Oscillatory Parts of $R(h)$

According to the relations (B.4) these quantities are expressed through the functions $F(h, \pm t_s)$ (B.5), which using the expansion (B.17) can be reduced to

$$F(h, t) = \int V(h, t) e^{i\Omega t} dt = a_\xi X^{(1,0)}(h, t) + a_\eta X^{(0,1)}(h, t)$$
$$+ b_{\xi\xi} X^{(2,0)}(h, t) + b_{\xi\eta} X^{(1,1)}(h, t) + b_{\eta\eta} X^{(0,2)}(h, t) + O(\delta^3) . \quad (B.25)$$

First we consider separately the integrals along each branches, I–IV, of the separatrix.

The First (I) and Third (III) Branches

Using the relations for the integrals $X^{(k,j-k)}$ given by (B.20), we obtain the following expressions for $R^{(osc)-}(h)$ and its the real and imaginary parts, $K^{(osc)-}(h)$ and $L^{(osc)-}(h)$:

$$R_1^{(osc)+}(h) = F(h, t_s) = \sqrt{|h|} e^{i\pi\Omega//\omega(h)}$$

$$\times \left(\pm \begin{cases} -i\Omega A_\xi + \gamma A_\eta, & \text{for } h < 0 \\ -\gamma A_\xi + i\Omega A_\eta, & \text{for } h > 0 \end{cases} + \sqrt{|h|} \, [C - iB(h)] \right) , \quad (B.26)$$

$$K_1^{(osc)+}(h) = \text{Re } R^{(osc)-}(h) = \pm\sqrt{|h|}$$

$$\times \left\{ \begin{array}{ll} \Omega A_\xi \sin[\pi\Omega/\omega(h)] + \gamma A_\eta \cos[\pi\Omega/\omega(h)], & \text{for } h < 0 \\ -\gamma A_\xi \cos[\pi\Omega/\omega(h)] - \Omega A_\eta \sin[\pi\Omega/\omega(h)], & \text{for } h > 0 \end{array} \right\}$$

$$+ |h| \left[C \cos\left(\frac{\pi\Omega}{\omega(h)}\right) + B(h)\sin\left(\frac{\pi\Omega}{\omega(h)}\right) \right], \tag{B.27}$$

$$L_1^{(osc)+}(h) = \text{Im } R^{(osc)-}(h)$$

$$= \pm\sqrt{|h|} \left\{ \begin{array}{ll} -\Omega A_\xi \cos[\pi\Omega/\omega(h)] + \gamma A_\eta \sin[\pi\Omega/\omega(h)], & \text{for } h < 0 , \\ -\gamma A_\xi \sin[\pi\Omega/\omega(h)] + \Omega A_\eta \cos[\pi\Omega/\omega(h)], & \text{for } h > 0 \end{array} \right\}$$

$$+ |h| \left[C \sin\left(\frac{\pi\Omega}{\omega(h)}\right) + B(h)\cos\left(\frac{\pi\Omega}{\omega(h)}\right) \right]. \tag{B.28}$$

The coefficients A_ξ, A_η, B_1, B_2, and C are defined by

$$A_\xi = \frac{\sqrt{2}a_\xi}{\alpha(\gamma^2 + \Omega^2)} , \qquad A_\eta = \frac{\sqrt{2}a_\eta}{\beta(\gamma^2 + \Omega^2)} ,$$

$$B(h) = \frac{2}{\Omega^2} \left(-b_{\eta\eta}\frac{C_1(h)}{\alpha^2} + b_{\xi\xi}\frac{C_2(h)}{\beta^2} \right) = \left\{ \begin{array}{ll} B_1, & \text{for } h < 0 , \\ B_2, & \text{for } h > 0 , \end{array} \right.$$

$$B_1 = \frac{2}{\Omega^2} \left[\frac{b_{\xi\xi}}{\alpha^2} - \frac{b_{\eta\eta}}{\beta^2} + \frac{\Omega^2}{4\gamma^2 + \Omega^2}\left(\frac{b_{\xi\xi}}{\alpha^2} + \frac{b_{\eta\eta}}{\beta^2}\right) \right] ,$$

$$B_2 = \frac{2}{\Omega^2} \left[-\frac{b_{\xi\xi}}{\alpha^2} + \frac{b_{\eta\eta}}{\beta^2} + \frac{\Omega^2}{4\gamma^2 + \Omega^2}\left(\frac{b_{\xi\xi}}{\alpha^2} + \frac{b_{\eta\eta}}{\beta^2}\right) \right] ,$$

$$C = \frac{2b_{\xi\eta}}{4\gamma^2 + \Omega^2} . \tag{B.29}$$

The Second (II) and Fourth (IV) Branches

Using (B.21) one obtains the following expressions for $R^{(osc)+}(h)$, $K^{(osc)+}(h)$ and $L^{(osc)+}(h)$:

$$R_1^{(osc)-}(h) = -F(h, -t_s) = -\sqrt{|h|}e^{-i\pi\Omega//\omega(h)}$$

$$\times \left(\pm\left\{ \begin{array}{ll} i\Omega A_\xi - \gamma A_\eta, & \text{for } h < 0 \\ -\gamma A_\xi + i\Omega A_\eta, & \text{for } h > 0 \end{array} \right\} + \sqrt{|h|}\left[C - iB(h)\right] \right) , \tag{B.30}$$

The corresponding integrals are

$$K_1^{(osc)-}(h) = \text{Re } R^{(osc)-}(h) = \mp\sqrt{|h|}$$

$$\times \left\{ \begin{array}{ll} \Omega A_\xi \sin[\pi\Omega/\omega(h)] - \gamma A_\eta \cos[\pi\Omega/\omega(h)], & \text{for } h < 0 \\ -\gamma A_\xi \cos[\pi\Omega/\omega(h)] + \Omega A_\eta \sin[\pi\Omega/\omega(h)], & \text{for } h > 0 \end{array} \right\}$$

$$+|h|\left[-C\cos\left(\frac{\pi\Omega}{\omega(h)}\right) + B(h)\sin\left(\frac{\pi\Omega}{\omega(h)}\right) \right], \tag{B.31}$$

$$L_1^{(osc)-}(h) = \text{Im } R^{(osc)+}(h) = \mp\sqrt{|h|}$$

$$\left\{ \begin{array}{ll} \Omega A_\xi \cos[\pi\Omega/\omega(h)] + \gamma A_\eta \sin[\pi\Omega/\omega(h)], & \text{for } h < 0 \,, \\ \gamma A_\xi \sin[\pi\Omega/\omega(h)] + \Omega A_\eta \cos[\pi\Omega/\omega(h)], & \text{for } h > 0 \,, \end{array} \right\}$$

$$+|h|\left[C\sin\left(\frac{\pi\Omega}{\omega(h)}\right) - B(h)\cos\left(\frac{\pi\Omega}{\omega(h)}\right) \right]. \tag{B.32}$$

The expressions for the integrals $R^{(osc)}(h)$ depend on the saddle–saddle connection and can be obtained through the integrals $R^{(osc)\pm}(h)$ given above. Below we consider the four type of homoclinic saddle–saddle connections when the system has a single hyperbolic fixed point: (*i*) the branch II is connected with the branch I of the separatrix; (*ii*) the branch IV is connected with the branch I; (*iii*) the branch II is connected with the branch III; (*iv*) the branch IV is connected with the branch III (see Fig. B.1).

The case (*i*). Adding the expressions for $R^{(osc)+}(h)$ and $R^{(osc)-}(h)$ taken with the upper signs we have

$$K^{(osc)}(h) = \text{Re } R^{(osc)}(h) = K^{(osc)+}(h) + K^{(osc)-}(h)$$

$$2\left\{ \begin{array}{ll} \sqrt{|h|}A_\eta\gamma\sin[\pi\Omega/\omega(h)] + |h|B_1\sin[\pi\Omega/\omega(h)], & \text{for } h < 0 \,, \\ -\sqrt{|h|}A_\eta\Omega\cos[\pi\Omega\omega(h)] + |h|B_2\sin[\pi\Omega/\omega(h)], & \text{for } h > 0 \,, \end{array} \right. \tag{B.33}$$

$$L^{(osc)}(h) = \text{Im } R^{(osc)}(h) = L^{(osc)+}(h) + L^{(osc)-}(h)$$

$$= 2\left\{ \begin{array}{ll} \sqrt{|h|}A_\xi\Omega\cos[\pi\Omega/\omega(h)] + C|h|\sin[\pi\Omega/\omega(h)], & \text{for } h < 0 \,, \\ \sqrt{|h|}A_\xi\gamma\sin[\pi\Omega/\omega(h)] + C|h|\sin[\pi\Omega/\omega(h)], & \text{for } h > 0 \,. \end{array} \right. \tag{B.34}$$

The case (*ii*). Adding $R^{(osc)+}(h)$ with the upper signs to $R^{(osc)-}(h)$ taken with the lower signs one obtains

$$K^{(osc)}(h) = 2 \begin{cases} \sqrt{|h|}A_\xi \Omega \sin[\pi\Omega/\omega(h)] + |h|B_1 \sin[\pi\Omega/\omega(h)], & \text{for } h < 0 , \\ -\sqrt{|h|}A_\xi \gamma \cos[\pi\Omega\omega(h)] + |h|B_2 \sin[\pi\Omega/\omega(h)], & \text{for } h > 0 , \end{cases}$$
$$(B.35)$$

$$L^{(osc)}(h) = 2 \begin{cases} \sqrt{|h|}A_\eta \gamma \sin[\pi\Omega/\omega(h)] + C|h| \sin[\pi\Omega/\omega(h)], & \text{for } h < 0 , \\ \sqrt{|h|}A_\eta \Omega \cos[\pi\Omega/\omega(h)] + C|h| \sin[\pi\Omega/\omega(h)], & \text{for } h > 0 . \end{cases}$$
$$(B.36)$$

In the **case** (iii) the formulae for $K^{(osc)}(h)$ and $L^{(osc)}(h)$ are given by (B.35), (B.36), respectively, taken with the opposite signs, and in the **case** (iv) $K^{(osc)}(h)$ and $L^{(osc)}(h)$ are obtained from (B.33), (B.34) by changing the sign.

B.2 Periodically–Driven Pendulum

In this section we obtain the asymptotical formulae for the oscillatory parts of the integrals $K_1(h)$ and $L_2(h)$ in the problem of the periodically–driven pendulum (see Sect. 6.2). According to the definitions of the functions $V_n(h, t)$ given by (6.33) we have

$$V_1(h, t - t_c) = (A + B)[1 + \cos x(\vartheta; h)] ,$$
$$V_2(h, t - t_c) = (A - B) \sin x(\vartheta; h) .$$

The hyperbolic fixed points are $(x_s = \pm\pi, p_s = 0)$. The first two terms in expansion in series of ξ, η are

$$V_1(h, t) = (A + B)\frac{1}{2}\,\xi^2(t; h) , \qquad V_2(h, t) = -(A - B)\xi(t; h) ,$$

Figure B.2 shows the unperturbed orbits along which integrations are taken. Then using the integrals (B.20), (B.22), we obtain

$$K_1^{(osc)\pm}(h) = \int^{\pm t_s} V_1(h, t) \cos(\Omega t) dt$$

$$= \frac{A + B}{2} \int^{\pm t_s} \xi^2(t; h) \cos(\Omega t) dt = \frac{A + B}{2}\mathrm{Re}X^{(2,0)}(h, \pm t_s)$$

$$= (A + B)\frac{|h|}{2\Omega\alpha^2} \begin{cases} \left(1 + \frac{\Omega^2}{4\gamma^2 + \Omega^2}\right) \sin[\pi\Omega/\omega(h)], & \text{for } h < 0 \\ -\left(1 - \frac{\Omega^2}{4\gamma^2 + \Omega^2}\right) \sin[\pi\Omega/\omega(h)], & \text{for } h > 0 , \end{cases} \quad (B.37)$$

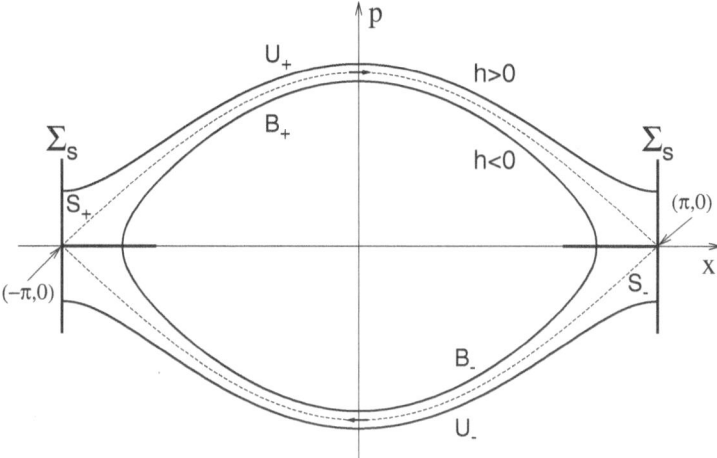

Fig. B.2. Phase curves of unperturbed pendulum near separatrix. Here U_+ and B_+ stand for the untrapped ($h > 0$) and trapped ($h < 0$) orbits near the separatrix S_+ connecting the saddle points $(-\pi, 0)$ and $(\pi, 0)$ in the upper half of phase space ($p > 0$). Similarly, U_- and B_- are untrapped ($h > 0$) and trapped ($h < 0$) orbits near the separatrix S_- connecting the saddle points $(\pi, 0)$ and $(-\pi, 0)$ in the lower half of phase space ($p < 0$)

$$L_2^{(osc)\pm}(h) = \int^{\pm t_s} V_2(h, t) \sin(\Omega t) dt$$

$$= -(A - B) \int^{\pm t_s} \xi(t; h) \sin(\Omega t) dt = -(A - B) \mathrm{Im} X^{(1,0)}(h, \pm t_s)$$

$$= \pm (A - B) \frac{\sqrt{2|h|}}{\alpha(\gamma^2 + \Omega^2)} \begin{cases} \Omega \cos[\pi \Omega/\omega(h)], & \text{for } h < 0 \\ \gamma \sin[\pi \Omega/\omega(h)], & \text{for } h > 0 \, . \end{cases} \quad \text{(B.38)}$$

The integrals $K_1^{(osc)}(h)$, $L_1^{(osc)}(h)$ taken along orbits near the separatrix connecting the saddle point $(-\pi, 0)$ with the other saddle point $(-\pi, 0)$ are obtained from the integrals (B.37), (B.38) using the relations $K_1^{(osc)}(h) = K_1^{(osc)-}(h) - K_1^{(osc)+}(h)$ and $L_2^{(osc)}(h) = L_2^{(osc)-}(h) - L_2^{(osc)+}(h)$:

$$K_1^{(osc)}(h) = (A + B) \frac{|h|}{\Omega \alpha^2} \begin{cases} \left(1 + \frac{\Omega^2}{4\gamma^2 + \Omega^2}\right) \sin[\pi \Omega/\omega(h)], & \text{for } h < 0 \\ -\left(1 - \frac{\Omega^2}{4\gamma^2 + \Omega^2}\right) \sin[\pi \Omega/\omega(h)], & \text{for } h > 0 \, , \end{cases} \quad \text{(B.39)}$$

$$L_2^{(osc)}(h) = \mp (A - B) \frac{2\sqrt{2|h|}}{\alpha(\gamma^2 + \Omega^2)} \begin{cases} \Omega \cos[\pi \Omega/\omega(h)], & \text{for } h < 0 \\ \gamma \sin[\pi \Omega/\omega(h)], & \text{for } h > 0 \, . \end{cases} \quad \text{(B.40)}$$

B.3 The Integral $K(h)$ in the Problem of Driven Morse Oscillator

The integrals in (6.72) can be written as

$$R(h) = 2a \int_0^{\pi} f(x)e^{i\eta x}dx, \qquad f(x) = \frac{\sin x}{1 - a\cos x}, \qquad (B.41)$$

where $R(h) = K(h) + iL(h)$. In (B.41) the following notations are introduced $a = \sqrt{1 - |h|}$, $\eta = \lambda/|h|^{1/2}$. Since the corresponding system does not have hyperbolic saddle points we cannot use the method used in the previous sections. Below we apply a standard asymptotical method to estimate these integrals.

The regular part of the integral (B.41) can be found by integrating it for integer values of $\eta = m$ corresponding to the primary resonances, $m\omega(h) = \Omega$. Then using the integral

$$\int_0^{\pi} \frac{\cos[(\eta \pm 1)x]}{1 - a\cos x}dx = \frac{\pi}{\sqrt{1 - a^2}}\left(\frac{1 - \sqrt{1 - a^2}}{a}\right)^{\eta \pm 1},$$

and continuing it to the noninteger values of η, we obtain

$$R^{(reg)}(h) = 2\pi \left(\frac{1 - \sqrt{1 - a^2}}{a}\right)^{\eta}. \qquad (B.42)$$

Then the integral (B.41) can be presented as a sum

$$R(h) = R^{(reg)}(h) + R^{(osc)}(h), \qquad (B.43)$$

where $R^{(osc)}(h)$ is the oscillatory part of the integral $R(h)$. We seek $R^{(osc)}(h)$ for large value of $\eta \gg 1$. Then it can be found by asymptotic expansion of the integral in a series of power of η^{-1}. We will find the asymptotic expansion by integration by part.

Integrating (B.41) by part N times, one can obtain

$$\int_0^{\pi} f(x)e^{i\eta x}dx = e^{i\eta\pi}\sum_{k=1}^{N}\frac{1}{(i\eta)^k}f^{(k-1)}(x)\bigg|_{x=\pi} + O(\eta^{-N-1}),$$

where $f^{(k)}(x)$ is the k-the derivative of the function $f(x)$. One can show that $f^{(2s)}(0) = f^{(2s)}(\pi) = 0$, $(s = 0, 1, 2, \ldots)$. For the odd $k = 2s + 1$ the derivatives $f^{(2k+1)}(0)$ have non-zero values. For the first two non-zero derivatives we obtain

$$a_1 \equiv f^{(1)}(0) = \frac{1}{1 - a}, \qquad a_2 \equiv f^{(3)}(0) = -\frac{1 + a + 2a^2}{(1 - a)^3},$$

$$b_1 \equiv f^{(1)}(\pi) = -\frac{1}{1 + a}, \qquad b_2 \equiv f^{(3)}(\pi) = \frac{1 - a - 2a^2}{(1 + a)^3}. \qquad (B.44)$$

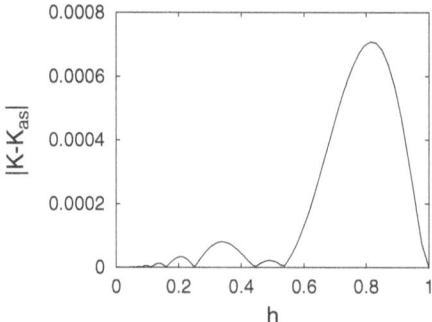

Fig. B.3. Deviation of the asymptotic formula (B.46) for $K(h)$ from the numerically integrated one. Parameter $\lambda = 2$

Therefore, the expansion of $K^{(osc)}(h)$ in a series of power of $1/\eta$ up to fourth terms is given by

$$K^{(osc)}(h) = \operatorname{Re} R^{(osc)}(h) = \frac{2a}{\eta^2}\left(b_1 - \frac{b_2}{\eta^2}\right)\sin(\pi\eta) + O\left(\eta^{-6}\right) . \quad (B.45)$$

Putting $a = \sqrt{1 - |h|}$ and $\eta = \lambda/|h|^{1/2}$, we have obtained the following asymptotic formula for $K(h)$:

$$K_{as}(h) = 2\pi \left(\frac{1 - |h|^{1/2}}{\sqrt{1 - |h|}}\right)^{\lambda/|h|^{1/2}}$$

$$+ \frac{2|h|\sqrt{1 - |h|}}{\lambda^2}\,\sin\left(\frac{\pi\lambda}{|h|^{1/2}}\right)\left(b_1 - b_3\frac{|h|}{\lambda^2}\right) + O\left(\frac{|h|^3}{\lambda^6}\right) . \quad (B.46)$$

At the limit $|h| \to 0$, we have

$$K_{as}(h) \to K_0 = 2\pi \exp(-\lambda) .$$

The asymptotics of the integral $K(h)$ given by (B.46) is plotted in Fig. 6.12 at the fixed value of the parameter $\lambda = 4$. The asymptotic formula (B.46) is in sufficiently good agreement with the values of $K(h)$ obtained by the numerical integration of the integral (B.41). The deviation of (B.46) from the numerical $K(h)$ is shown in Fig. B.3. The maximal deviation, $\max|K(h) - K_{as}(h)|$ is less than 8×10^{-4}.

C Proof of Rescaling Invariance of the Equations of Motion

C.1 The Case of Linear Approximation

For convenience we omit the parameter t_0 in (8.13), and consider the following linear equation with periodic coefficients:

$$\frac{dx}{dt} = -\gamma x + \epsilon[p_1(t) + c_{11}(t)x + c_{12}(t)y] ,$$

$$\frac{dy}{dt} = \gamma y + \epsilon[p_2(t) + c_{21}(t)x + c_{22}(t)y] . \qquad \text{(C.1)}$$

We will seek the solutions of (C.1) in the form

$$x(t) = A(t)\exp\left(-\gamma t + \epsilon \int_0^t c_{11}(t')dt'\right) ,$$

$$y(t) = B(t)\exp\left(\gamma t + \epsilon \int_0^t c_{22}(t')dt'\right) . \qquad \text{(C.2)}$$

For the unknown coefficients $A(t)$, $B(t)$, we obtain the following equations

$$\frac{dA}{dt} = \epsilon p_1(t)\exp\left(\gamma t - \epsilon \int_0^t c_{11}(t')dt'\right)$$

$$+\epsilon c_{12}(t)B(t)\exp\left(2\gamma t + \epsilon \int_0^t [c_{22}(t') - c_{11}(t')]\,dt'\right) ,$$

$$\frac{dB}{dt} = \epsilon p_2(t)\exp\left(-\gamma t - \epsilon \int_0^t c_{22}(t')dt'\right)$$

$$+\epsilon c_{21}(t)A(t)\exp\left(-2\gamma t + \epsilon \int_0^t [c_{11}(t') - c_{22}(t')]\,dt'\right) .$$

$$\text{(C.3)}$$

The solutions of (C.3) up to the first order of ϵ have the form

$$A(t) = A(0) + \epsilon B(0)\int_0^t e^{2\gamma t'}c_{12}(t')dt' + \epsilon \int_0^t e^{\gamma t'}p_1(t')dt' + O(\epsilon^2) ,$$

$$B(t) = B(0) + \epsilon A(0)\int_0^t e^{-2\gamma t'}c_{21}(t')dt' + \epsilon \int_0^t e^{-\gamma t'}p_2(t')dt' + O(\epsilon^2) . \text{ (C.4)}$$

S.S. Abdullaev: *Construction of Mappings for Hamiltonian Systems and Their Applications*, Lect. Notes Phys. **691**, 335–340 (2006)
www.springerlink.com

Now we replace the perturbation amplitude, ϵ, by its rescaled value $\epsilon_\lambda = \epsilon/\lambda$. Introducing the notations

$$A_\lambda(t) = A(t)\lambda_\epsilon^{-1/2}, \qquad B_\lambda(t) = B(t)\lambda_\epsilon^{-1/2}, \qquad (C.5)$$

$$t_{0\lambda} = \frac{1}{2\gamma}\ln\lambda ,$$

and replacing the integration variable $t' \to t' - t_{0\lambda}$, the solution, (C.4), for A can be transformed to

$$A_\lambda(t) = A_\lambda(0)$$
$$+\epsilon_\lambda B_\lambda(0)\int_{t_{0\lambda}}^{t+t_{0\lambda}} e^{2\gamma t'}c_{12}(t'-t_{0\lambda})dt'$$
$$+\epsilon_\lambda \int_{t_{0\lambda}}^{t+t_{0\lambda}} e^{\gamma t'}p_1(t'-t_{0\lambda})dt' + O(\epsilon^2) . \qquad (C.6)$$

Transform (C.6) by the following way:

$$A_\lambda(t) = A_\lambda(0) + \epsilon_\lambda B_\lambda(0)\int_0^t e^{2\gamma t'}c_{12}(t'-t_{0\lambda})dt'$$
$$+\epsilon_\lambda \int_0^t e^{\gamma t'}p_1(t'-t_{0\lambda})dt'$$
$$+\epsilon_\lambda B_\lambda(0)\left(\int_t^{t+t_{0\lambda}} - \int_0^{t_{0\lambda}}\right) e^{2\gamma t'}c_{12}(t'-t_{0\lambda})dt'$$
$$+\epsilon_\lambda \left(\int_t^{t+t_{0\lambda}} - \int_0^{t_{0\lambda}}\right) e^{\gamma t'}p_1(t'-t_{0\lambda})dt' . \qquad (C.7)$$

We note that

$$\int_t^{t+t_{0\lambda}} e^{n\gamma t'}f(t'-t_{0\lambda})dt' = e^{n\gamma t}\int_0^{t_{0\lambda}} e^{n\gamma t'}f(t'+t-t_{0\lambda})dt' ,$$
$$e^{n\gamma t} - 1 = n\gamma \int_0^t e^{n\gamma t'}dt', \qquad (C.8)$$

where $f(t) = c_{12}(t), n = 2$ or $f(t) = p_1(t), n = 1$.

We will consider Poincaré sections of the solutions $x(t), y(t)$, taken at the sections $t = Tm$, $(m = 0, 1, 2, ...)$. Then using the periodicity of the coefficients $c_{12}(t), p_1(t)$ and the relations (C.8), (C.7) can be reduced to

$$A_\lambda(t) = A_\lambda(0) + \epsilon_\lambda B_\lambda(0)\int_0^t e^{2\gamma t'}\left[c_{12}(t'-t_{0\lambda}) + \gamma c_{12}^0\right] dt'$$
$$+\epsilon_\lambda \int_0^t e^{\gamma t'}\left[p_1(t'-t_{0\lambda}) + \gamma p_1^0\right] dt' + O(\epsilon^2) , \qquad (C.9)$$

where the constant coefficients, c_{12}^0, and p_1^0, are

$$c_{12}^0 = 2 \int_0^{t_{0\lambda}} e^{2\gamma t} c_{12}(t - t_{0\lambda}) dt \ , \qquad p_1^0 = \int_0^{t_{0\lambda}} e^{\gamma t} p_1(t - t_{0\lambda}) dt \ .$$

In a similar way, we can obtain the solution for B_λ:

$$B_\lambda(t) = B_\lambda(0) + \epsilon_\lambda A_\lambda(0) \int_0^t e^{-2\gamma t'} \left[c_{21}(t' + t_{0\lambda}) + \gamma c_{21}^0 \right] dt'$$

$$+ \epsilon_\lambda \int_0^t e^{-\gamma t'} \left[p_2(t' + t_{0\lambda}) + \gamma p_2^0 \right] dt' + O(\epsilon^2) \ , \quad \text{(C.10)}$$

where

$$c_{21}^0 = 2 \int_{-t_{0\lambda}}^0 e^{-2\gamma t} c_{21}(t + t_{0\lambda}) dt \ , \qquad p_2^0 = \int_{-t_{0\lambda}}^0 e^{-\gamma t} p_2(t + t_{0\lambda}) dt \ .$$

The expressions, (C.9), (C.10), correspond to the solutions of the equations

$$\frac{d\xi_\lambda}{dt} = -\gamma \left[\xi_\lambda - \epsilon_\lambda(p_1^0 + c_{12}^0 \eta_\lambda) \right] + \epsilon_\lambda \left[p_1(t - t_{0\lambda}) \right.$$
$$\left. + c_{11}(t - t_{0\lambda})\xi_\lambda + c_{12}(t - t_{0\lambda})\eta_\lambda \right] \ ,$$

$$\frac{d\eta_\lambda}{dt} = \gamma \left[\eta_\lambda + \epsilon_\lambda(p_2^0 + c_{21}^0 \xi_\lambda) \right] + \epsilon_\lambda \left[p_2(t + t_{0\lambda}) \right.$$
$$\left. + c_{21}(t + t_{0\lambda})\xi_\lambda + c_{22}(t + t_{0\lambda})\eta_\lambda \right] \ , \quad \text{(C.11)}$$

where

$$\xi_\lambda(t) = A_\lambda(t) \exp \left(-\gamma t + \epsilon_\lambda \int_0^t c_{11} \left(t' - t_{0\lambda} \right) dt' \right) \ ,$$

$$\eta_\lambda(t) = B_\lambda(t) \exp \left(\gamma t + \epsilon_\lambda \int_0^t c_{22} \left(t' + t_{0\lambda} \right) dt' \right) \ . \quad \text{(C.12)}$$

By the linear transformation of variables

$$x_\lambda = \xi_\lambda - \epsilon_\lambda \left(p_1^0 + c_{12}^0 \eta_\lambda \right) \ ,$$

$$y_\lambda = \eta_\lambda + \epsilon_\lambda \left(p_2^0 + c_{21}^0 \xi_\lambda \right) \ . \quad \text{(C.13)}$$

Equation (C.11) with the accuracy up to the first order of ϵ_λ may be rewritten in the form

$$\frac{dx_\lambda}{dt} = -\gamma x_\lambda + \epsilon_\lambda \left[p_1(t - t_{0\lambda}) + c_{11}(t - t_{0\lambda})x_\lambda + c_{12}(t - t_{0\lambda})y_\lambda \right] + O(\epsilon_\lambda^2) \ ,$$

$$\frac{dy_\lambda}{dt} = \gamma y_\lambda + \epsilon_\lambda \left[p_1(t + t_{0\lambda}) + c_{11}(t + t_{0\lambda})x_\lambda + c_{12}(t + t_{0\lambda})y_\lambda \right] + O(\epsilon_\lambda^2) \ .$$
$$\text{(C.14)}$$

Therefore, the transformation $\epsilon \to \epsilon_\lambda = \lambda \epsilon$ of the perturbation amplitude, ϵ, preserves the form of the the equations if the periodic coefficients c_{ik}, and p_i

have the same arguments in both equations, (C.14). Since the coefficients are periodic functions of t with period T this condition is satisfied, for

$$2t_{0\lambda} = sT , \qquad (s = \pm1, \pm2, \ldots) .$$

Using the definition of $t_{0\lambda}$, (C.5), we obtain the possible values of the rescaling parameter λ:

$$\lambda(s) = \exp(\gamma T s) = \lambda^s, \qquad (s = \pm1, \pm2, \ldots) , \qquad (C.15)$$

where λ is the rescaling parameter given by (8.7). Then using the condition (8.10) for the coefficients $c_{ii}(t)$, and (C.5), (C.11), we have

$$\xi_\lambda(t = 2\pi m) = \exp(-\gamma t)A_\lambda(t) = \exp(-\gamma t)A(t)\lambda^{-1/2} = x(t)\lambda^{-1/2} ,$$

$$\eta_\lambda(t = 2\pi m) = \exp(+\gamma t)B_\lambda(t) = \exp(+\gamma t)B(t)\lambda^{-1/2} = y(t)\lambda^{-1/2} .$$
$$(C.16)$$

Finally, putting (C.15) into (C.13), we obtain the relation, (8.15), between the solutions of (C.1) and (C.14) at the sections $t = mT$, $(m = 0, 1, 2, \ldots)$.

C.2 The Case of Nonlinear Approximation

Consider the nonlinear differential equation, (8.8) with the perturbation functions $g_i(x, y, t)$, $(i = 1, 2)$ (8.16). The solution of this equation we will seek in the form

$$x(t) = A(t)\exp(-\gamma t) , \qquad y(t) = B(t)\exp(\gamma t) . \qquad (C.17)$$

For the unknown coefficients $A(t), B(t)$ one can obtain the following equations

$$\frac{dA}{dt} = \epsilon \sum_{k=0}^{n} c_{1k}(t)(A))^k(B)^{n-k}e^{(n-2k+1)\gamma t} ,$$

$$\frac{dB}{dt} = \epsilon \sum_{k=0}^{n} c_{2k}(t)(A)^k(B)^{n-k}e^{(n-2k-1)\gamma t} . \qquad (C.18)$$

The solution of (C.18) with the accuracy up to the first order of ϵ may be written as

$$A(t) = A(0) + \epsilon \sum_{k=0}^{n}(A(0))^k(B(0))^{n-k}\int_0^t e^{(n-2k+1)\gamma t'}c_{1k}(t')dt' ,$$

$$B(t) = B(0) + \epsilon \sum_{k=0}^{n}(A(0))^k(B(0))^{n-k}\int_0^t e^{(n-2k-1)\gamma t'}c_{2k}(t')dt' . \quad (C.19)$$

First consider positive values of $t > 0$. To evaluate the order of the integral in (C.19), we have specified the periodic coefficients as $c_{ik}(t) = \cos(t + \chi)$. Then, for $t = Tm$, one can obtain

$$\int_0^{2\pi m} e^{n\gamma t} \cos(t + \chi) dt = (\lambda^{nm} - 1)\frac{n\gamma \cos\chi + \sin\chi}{n^2\gamma^2 + 1} \sim \lambda^{nm} , \qquad (n \geq 2) ,$$

(C.20)

where $\lambda = \exp(\gamma T)$. Since $\gamma > 0$, the rescaling parameter $\lambda > 1$. Usually, the square of λ has an order of 10, i.e., λ^2, and may be considered as a large parameter. We can seek the asymptotic behavior of the solutions for $\lambda^2 \gg 1$, expanding (C.19) in series of inverse powers of λ^2. The main terms in (C.19) are

$$A(t) = A(0) + \epsilon(B(0))^n \lambda^{nm} C_{1n}[1 + O(\lambda^{-2m})] ,$$

$$B(t) = B(0) + \epsilon(B(0))^n \lambda^{(n-2)m} C_{2n}[1 + O(\lambda^{-2m})] ,$$

(C.21)

where C_{1n} and C_{2n} are constants of order $O(1)$. Therefore, for the for positive $t > 0$ the asymptotic expansion of the solutions, (C.19), for $\lambda^2 \gg 1$ may be written as

$$A(t) = A(0) + \epsilon(B(0))^n \int_0^t e^{(n+1)\gamma t'} c_{1n}(t')dt' \left[1 + O(\lambda^{-2})\right] ,$$

$$B(t) = B(0)\left[1 + \epsilon O(\lambda^{-2})\right] .$$

(C.22)

In a similar way, one can write the asymptotic expansion of the solution, (C.19), for the negative values of t $(t < 0)$:

$$A(t) = A(0)\left[1 + \epsilon O(\lambda^{-2})\right] ,$$

$$B(t) = B(0) + \epsilon(A(0))^n \int_0^t e^{-(n+1)\gamma t'} c_{2n}(t')dt' \left[1 + O(\lambda^{-2})\right] .$$

(C.23)

Now we will study the behavior of the solutions, (C.22), (C.23), in respect to the transformation, $\epsilon \to \epsilon_\lambda = \epsilon/\lambda$, of the perturbation parameter, ϵ. Using the notations in (C.5), and performing the transformations of (C.6) similar to those ones in the previous section, one can obtain the following expressions for the transformed solutions, $A_\lambda(t), B_\lambda(t)$, at the positive axis of $t > 0$:

$$A_\lambda(t) = A_\lambda(0) + \epsilon_\lambda(B_\lambda(0))^n \int_0^t e^{(n+1)\gamma t'} \left[c_{1n}(t' - t_{0\lambda}) + \gamma c_{1n}^0\right] dt'$$
$$\times \left[1 + O(\lambda^{-2})\right] ,$$

$$B_\lambda(t) = B_\lambda(0)\left[1 + \epsilon_\lambda O(\lambda^{-2})\right] ,$$

(C.24)

where

$$c_{1n}^0 = (n+1) \int_0^{t_{0\lambda}} e^{(n+1)\gamma t} c_{1n}(t - t_{0\lambda}) dt \ .$$

By the similar procedure one can transform the solution (C.23), and obtain the following expressions for $A_\lambda(t), B_\lambda(t)$ for the negative $t < 0$:

$$A_\lambda(t) = A_\lambda(0) \left[1 + \epsilon_\lambda O(\lambda^{-2})\right],$$

$$B_\lambda(t) = B_\lambda(0) + \epsilon_\lambda (A_\lambda(0))^n \int_0^t e^{-(n+1)\gamma t'} [c_{2n}(t' + t_{0\lambda}) + \gamma c_{2n}^0] dt'$$
$$\times \left[1 + O(\lambda^{-2})\right], \qquad (C.25)$$

with

$$c_{2n}^0 = (n+1) \int_{-t_{0\lambda}}^0 e^{-(n+1)\gamma t} c_{2n}(t + t_{0\lambda}) dt \ .$$

The transformed solutions, (C.24), (C.25), corresponds to the asymptotic solutions of the following nonlinear equations for the large parameter $\lambda^2 \gg 1$:

$$\frac{dx_\lambda}{dt} = -\gamma \left[x_\lambda + \epsilon_\lambda c_{1n}^0 (y_\lambda)^n\right] + \epsilon_\lambda \sum_{k=0}^n c_{1k}(t - t_{0\lambda})(x_\lambda)^k (y_\lambda)^{n-k} \ ,$$

$$\frac{dy_\lambda}{dt} = \gamma[y_\lambda + \epsilon_\lambda c_{1n}^0 (x_\lambda)^n] + \epsilon_\lambda \sum_{k=0}^n c_{2k}(t + t_{0\lambda})(x_\lambda)^k (y_\lambda)^{n-k} \ . \ (C.26)$$

The terms $\epsilon_\lambda c_{kn}^0 (a_\lambda^\pm)^n$, $(k = 1, 2)$ before the coefficient γ in (C.26) corresponds to the higher order corrections in the expansion of the unperturbed Hamiltonian, $H_0(x, y)$, in series of power of x, y near the hyperbolic fixed point, $(x = 0, y = 0)$. Since we neglected with these higher order corrections in $H_0(x, y)$, we can neglect them also in (C.26). Thanks to periodicity of the coefficients $c_{ik}^{(n)}(t)$, (8.17) and (C.26) have the same form for the value of the phase $t_{0\lambda} = \pi$. From the latter condition automatically follows the value of the rescaling parameter λ. The relation, (8.18), between the solutions $(x_a(t), y_a(t))$ and $(x_b(t), y_b(t))$ of the (8.17) and (C.26) follows from (C.5), which is valid for the $\lambda^2 \gg 1$.

D Relation Between ϑ and θ

The term under the integral (10.25) depending on the poloidal angle θ is

$$f(\varepsilon, \theta) = \frac{1}{(1 + \varepsilon \cos \theta)^2 (1 + \Lambda \varepsilon \cos \theta)} .$$

Expansion of $f(\varepsilon, \theta)$ in a series of powers of ε leads to

$$f(\varepsilon, \theta) = \sum_{m=0}^{\infty} a_m \varepsilon^m \cos^m \theta ,$$

$$a_m = (-1)^m \sum_{k=0}^{m} (m - k + 1) \Lambda^k . \tag{D.1}$$

Using the series (D.1) and integrating (10.25), one obtains the following relationship between ϑ and θ as a series of powers of ε:

$$\vartheta(\theta, \epsilon) = \theta + \sum_{m=1}^{M} \alpha_m \sin m\theta + O(\epsilon^{M+1}) , \tag{D.2}$$

where the coefficients α_m are also series in powers of ε. For $M = 4$ they are given by (10.30). By similar integration of (10.26) we obtain a series for the safety factor $q(r)$ (10.31).

Now we consider the inversion of the relation $\vartheta = \vartheta(\theta, \varepsilon)$ with respect to θ. Suppose that we have a relation in the form

$$\vartheta(\theta, \epsilon) = \theta + \sum_{m=1}^{\infty} \alpha_m \sin m\theta , \tag{D.3}$$

with given coefficients α_m. Note that $\alpha_m / \alpha_{m+1} \sim \varepsilon$. A similar expression exists for the inverse relation $\theta = \theta(\vartheta, \varepsilon)$:

$$\theta(\vartheta, \epsilon) = \vartheta + \sum_{m=1}^{\infty} \alpha_m^* \sin m\vartheta , \tag{D.4}$$

with the unknown coefficients α_m^*. From (D.3) and (D.4) follows the relation between the coefficients α_m^* and α_m:

S.S. Abdullaev: *Construction of Mappings for Hamiltonian Systems and Their Applications*,
Lect. Notes Phys. **691**, 341–343 (2006)
www.springerlink.com

$$\alpha_m^* = -\sum_{m'=1}^{\infty} A_{mm'} \alpha_{m'} \, , \qquad (D.5)$$

where

$$A_{mm'} = \frac{1}{2\pi} \int_0^{2\pi} e^{-i(m\vartheta - m'\theta(\vartheta))} d\vartheta = \frac{1}{2\pi} \int_0^{2\pi} e^{-i(m\vartheta(\theta) - m'\theta)} \frac{d\vartheta}{d\theta} d\theta \, .$$

We expand the integrand in a series of powers of ε. Since $\alpha_m \sim \varepsilon^m$, using (D.3) one can obtain

$$\exp(-im\vartheta)\frac{d\vartheta}{d\theta} = e^{-im\theta} \sum_{k=0}^{3} F_m^{(k)}(\theta) + O(\varepsilon^4) \, ,$$

where

$$F_m^{(0)} = 1, \qquad F_m^{(1)}(\theta) = \alpha_1(\cos\theta - im\sin\theta) \, ,$$

$$F_m^{(2)}(\theta) = -\frac{m^2}{4}\alpha_1^2 + (2\alpha_2 + \frac{m^2}{4}\alpha_1^2)\cos 2\theta - im(\frac{\alpha_1^2}{2} + \alpha_2)\sin 2\theta \, ,$$

$$F_m^{(3)}(\theta) = -im\left(\alpha_3 + \frac{3}{2}\alpha_1\alpha_2 + \frac{1}{24}m^2\alpha_1^3\right)\sin 3\theta$$

$$+ i\frac{m}{2}\left(\alpha_1\alpha_2 + \frac{1}{4}m^2\alpha_1^3\right)\sin\theta - \frac{m^2}{2}\left(\alpha_1\alpha_2 + \frac{1}{4}\alpha_1^3\right)\cos\theta$$

$$+ \left(3\alpha_3 + \frac{1}{2}m^2\alpha_1\alpha_2 + \frac{1}{8}m^2\alpha_1^3\right)\cos 3\theta \, .$$

Then the matrix $A_{mm'}$ can be presented in the form:

$$A_{mm'} = \frac{1}{2\pi} \int_0^{2\pi} e^{-i(m'\vartheta(\theta) - m\theta)} \frac{d\vartheta}{d\theta} d\theta = A_{mm'}^{(0)} + A_{mm'}^{(1)} + A_{mm'}^{(2)} + A_{mm'}^{(3)} + \cdots \, ,$$

where

$$A_{mm'}^{(0)} = \delta_{mm'}, \qquad A_{mm'}^{(j)} = \frac{1}{2\pi} \int_0^{2\pi} e^{-i(m\theta - m'\theta)} F_m^{(j)}(\theta) d\theta \, , \qquad j = 1, 2, 3, \ldots \, .$$

The terms of the first order of ε are

$$A_{11}^{(1)} = 0 \, , \qquad A_{21}^{(1)} = \alpha_1 \, , \qquad A_{12}^{(1)} = -\frac{\alpha_1}{2} \, , \qquad A_{32}^{(1)} = \frac{3\alpha_1}{2} \, ,$$

$$A_{23}^{(1)} = -\alpha_1 \, , \qquad A_{43}^{(1)} = 2\alpha_1 \, , \qquad A_{34}^{(1)} = -\frac{3}{2}\alpha_1 \, .$$

The second order terms are

$$A_{11}^{(2)} = -\frac{1}{4}\alpha_1^2 \, , \qquad A_{21}^{(2)} = 0; \qquad A_{31}^{(2)} = \frac{3}{8}\alpha_1^2 + \frac{3}{2}\alpha_2 \, ,$$

$$A_{12}^{(2)} = 0 \, , \qquad A_{22}^{(2)} = -\alpha_1^2 \, , \qquad A_{32}^{(2)} = 0 \, , \qquad A_{42}^{(2)} = \alpha_1^2 + 2\alpha_2 \, ,$$

$$A_{13}^{(2)} = \frac{3}{8}\alpha_1^2 - \frac{1}{2}\alpha_2 , \quad A_{23}^{(2)} = 0, \quad A_{33}^{(2)} = -\frac{9}{4}\alpha_1^2 , \quad A_{43}^{(2)} = 0 ,$$

$$A_{53}^{(2)} = \frac{15}{8}\alpha_1^2 + \frac{5}{2}\alpha_2 , \quad A_{23}^{(2)} = \alpha_1^2 - \alpha_2 .$$

Finally the third order terms are

$$A_{11}^{(3)} = 0 , \quad A_{21}^{(3)} = -\frac{1}{2}(\alpha_1\alpha_2 + \frac{1}{4}\alpha_1^3), \quad A_{31}^{(3)} = 0 ,$$

$$A_{41}^{(3)} = \frac{1}{12}\alpha_1^3 + \alpha_1\alpha_2 + 2\alpha_3 , \quad A_{12}^{(3)} = -\frac{1}{2}\alpha_1\alpha_2 + \frac{1}{4}\alpha_1^3, A_{22}^{(3)} = 0 ,$$

$$A_{32}^{(3)} = -\frac{5}{4}\alpha_1\alpha_2 - \frac{9}{16}\alpha_1^3, \quad A_{42}^{(3)} = A_{13}^{(3)} = 0 ,$$

$$A_{23}^{(3)} = -2\alpha_1\alpha_2 + \frac{13}{8}\alpha_1^3 , \quad A_{32}^{(3)} = -\frac{1}{2}\alpha_3 + \alpha_1\alpha_2 - \frac{1}{3}\alpha_1^3 .$$

Finally, using (D.5) and neglecting the terms of order higher than ε^5 we have obtained the following expressions for the expansion coefficients α_m^* in (D.4) in terms of α_k:

$$-\alpha_1^* = \alpha_1 + \alpha_1\alpha_2 - \frac{1}{4}\alpha_1^3 + O(\epsilon^5) ,$$

$$-\alpha_2^* = \alpha_2 - \frac{1}{2}\alpha_1^2 - \frac{3}{2}\alpha_1^2\alpha_2 + \frac{1}{4}\alpha_1^4 + \frac{3}{2}\alpha_1\alpha_3 + O(\epsilon^6) ,$$

$$-\alpha_3^* = \alpha_3 - \frac{3}{2}\alpha_1\alpha_2 + \frac{3}{8}\alpha_1^3 + O(\epsilon^5) ,$$

$$-\alpha_4^* = \alpha_4 - 2\alpha_1\alpha_3 + 2\alpha_1^2\alpha_2 + \alpha_2^2 - \frac{1}{4}\alpha_1^4 . + O(\epsilon^6) .$$

E Asymptotic Estimation of the Integral $S_{mm'}$ (10.49)

We write down the integral (10.49) as the Fourier integral

$$S_{mm'}(\psi) = \frac{1}{2\pi} \int_0^{2\pi} f(\vartheta) e^{im'\Phi(\vartheta)} d\vartheta \,, \tag{E.1}$$

where

$$\Phi(\vartheta) = \theta(\vartheta) - \frac{m}{m'}\vartheta \,, \qquad f(\vartheta) = 1 + \varepsilon \cos\theta(\vartheta)$$

are slowly varying functions of ϑ. Integrals of type (E.1) may be evaluated using the methods of asymptotic expansions in a series of inverse powers of $m' \gg 1$. However, as we will see below, the method of stationary phase cannot be directly applied to estimate the integral (E.1) for the values of m' being of interest because of the specific behavior of the phase function $\Phi(\vartheta)$.

According to the localization principle (see Fedoryuk (1989)) for $m' \gg 1$ the integral (E.1) is equal to sum of the contributions at the critical points for $S_{mm'}$. There are two critical points for the phase function $\Phi(\vartheta)$:

$$\vartheta_1 = 0 \,, \qquad \vartheta_2 = \pi \,.$$

As will be shown below, for $m' < m$ the main contribution to the integral comes from the first critical point, ϑ_1, and for $m' > m$ the second critical point contributes to the integral.

Consider first the case $m' > m$. One can expand the angle θ in terms of $(\vartheta - \pi)$ around the second critical point $\vartheta_2 = \pi$:

$$\theta(\vartheta) \approx \pi + \beta_1(\vartheta - \pi) + \frac{1}{6}\beta_3(\vartheta - \pi)^3 \,, \tag{E.2}$$

where β_1, β_3 are defined by (10.50). In (E.2) we have taken into account that $d^2\theta/d\vartheta^2\big|_{\theta=\pi} = 0$. Since $0 < \beta_1 < 1$ and $\beta_3 > 0$ the first derivative

$$\frac{d\Phi(\vartheta)}{d\vartheta} = (\beta_1 - m/m') + \frac{1}{2}\beta_3(\vartheta - \pi)^2 \,,$$

has two real zeros

S.S. Abdullaev: *Construction of Mappings for Hamiltonian Systems and Their Applications*,
Lect. Notes Phys. **691**, 345–347 (2006)
www.springerlink.com

$$\vartheta_{1,2} = \pi \pm \left(\frac{2|m'\beta_1 - m|}{m'\beta_3} \right)^{1/2} .$$

when $m' < m/\beta_1$, and complex roots when $m' > m/\beta_1$. For the values of m' which are sufficiently close to m/β_1 the singular points $\vartheta_{1,2}$ are close to each other, and the integral cannot be estimated by the ordinary method of stationary phase. In the case of degenerate stationary points one should apply the method described in Fedoryuk (1989).

Using the expansion (E.2), and introducing the integration variable $x = \vartheta - \pi$, the integral (E.1) may be written as

$$S_{mm'} = (-1)^{m+m'} \frac{1}{2\pi} \int_{-\pi}^{\pi} f(x + \pi) e^{i\lambda(\alpha x + x^3/3)} dx \qquad \text{(E.3)}$$

where

$$\alpha = \frac{m'\beta_1 - m}{m'\beta_3/2}, \qquad \lambda = m'\beta_3/2 .$$

For large values of λ and for the small α, the leading term of the asymptotic expansion of the integral (E.3) may be estimated by replacing $f(x + \pi)$ by $f(\pi)$ and expressing the integral by the Airy function $\mathrm{Ai}(z)$:

$$S_{mm'}(\psi) = f(\pi)(-1)^{m+m'} \frac{1}{2\pi} \int_{-\pi}^{\pi} e^{i\lambda(\alpha x + x^3/3)} dx$$

$$= (-1)^{m+m'} \left(\frac{2}{\beta_3 m'} \right)^{1/3} \mathrm{Ai}\left(\frac{\beta_1 m' - m}{(\beta_3 m'/2)^{1/3}} \right), \qquad \text{(E.4)}$$

where

$$\mathrm{Ai}(z) = \frac{1}{\pi} \int_0^{\infty} \cos(zt + t^3/3) dt$$

is the Airy function (see Abramowitz and Stegun (1965)).

The asymptotic formula (E.4) is valid for small values of α. Comparison with the exact numerical calculations of the integral (E.1) shows that (E.4) is a good approximation for $S_{mm'}$ in the interval

$$m' - m/\beta_1 > -c(m'\beta_3/2)^{1/3} . \qquad \text{(E.5)}$$

where $c \approx 3$.

A similar asymptotic estimation of $S_{mm'}$ can be obtained for the small values of m' satisfying the condition

$$m' - m/\gamma_1 < c(m'|\gamma_3|/2)^{1/3} . \qquad \text{(E.6)}$$

There is the following formula for these m':

$$S_{mm'}(\psi) = f(0)\left(\frac{2}{|\gamma_3|m'}\right)^{1/3} \mathrm{Ai}\left(-\frac{\gamma_1 m' - m}{(|\gamma_3|m'/2)^{1/3}}\right), \qquad (E.7)$$

where $\gamma_1 > 1$ and $\gamma_3 < 0$ are defined by (10.50).

Outside the intervals (E.5) and (E.6) the integrals $S_{mm'}(\psi)$ may be estimated by the method of stationary phase. These integrals are fast oscillating functions of m', and have an order of $(m')^{-1/2}$. We will not evaluate them here because of their small contribution.

The formulas (E.4) and (E.7) are the leading terms of an asymptotic expansion into a series of inverse powers of m'. The full asymptotic expansions may be found by the method described by Fedoryuk (1989).

F Sample Program for Implementing a Mapping Procedure

Below we present a program for the numerical implementation of the mapping for the generic Hamiltonian system of one-degree-of-freedom perturbed by time-periodic perturbation. The program is written in C-language.

We consider a Hamiltonian system

$$\frac{d\vartheta}{dt} = \frac{\partial H}{\partial I} = \omega(I) + \epsilon\frac{\partial H_1}{\partial I} \ , \qquad \frac{dI}{dt} = -\epsilon\frac{\partial H_1}{\partial I} \ , \tag{F.1}$$

given by the Hamiltonian function

$$H = H_0(I) + \epsilon H_1(I,\vartheta,t) \ ,$$

$$H_0(I) = \int \omega(I)dI \ ,$$

$$H_1(I,\vartheta,t) = \sum_{m=M_1}^{m=M_2} H_m(I)\cos(m\vartheta - \Omega t + \chi) \ , \tag{F.2}$$

where Fourier coefficients, $H_m(I)$, are real functions, and the phase χ is a constant. In (F.2) a finite number of terms numbering from M_1 to M_2, ($M_1 \leq M_2$), are taken into account.

Let $t_k = k\Delta T$, ($k = 0, \pm 1, \pm 2, \cdots$), be a time sequence with a step ΔT. Then the evolution of Hamiltonian system (F.1) along this time sequence, i.e.,

$$(\vartheta_{k+1}, I_{k+1}) = \hat{M}(\vartheta_k, I_k) \ , \tag{F.3}$$

where $(\vartheta_k, I_k) = (\vartheta(t_k), I(t_k))$, is given by the mappings (4.6), (4.7), (4.8). In the first order of perturbation theory the latter can be written as

$$J_k = I_k - F(J_k, \vartheta_k, t_k) \ , \qquad \psi_k = \vartheta_k + G(J_k, \vartheta_k, t_k) \ , \tag{F.4}$$

$$\psi_{k+1} = \psi_k + \omega(J_k)\Delta T \ , \tag{F.5}$$

$$I_{k+1} = J_k + F(J_k, \vartheta_{k+1}, t_{k+1}) \ , \qquad \vartheta_{k+1} = \psi_{k+1} - G(J_k, \vartheta_{k+1}, t_{k+1}) \ \tag{F.6}$$

S.S. Abdullaev: *Construction of Mappings for Hamiltonian Systems and Their Applications*,
Lect. Notes Phys. **691**, 349–357 (2006)
www.springerlink.com

where

$$F(J, \vartheta, t) = \epsilon \frac{\partial S(J, \vartheta, t)}{\partial \vartheta} \ ,$$
$$G(J, \vartheta, t) = \epsilon \frac{\partial S(J, \vartheta, t)}{\partial J} \ , \qquad (F.7)$$

are perturbation functions determined by the generating function

$$S(\vartheta, J, t) = - (t - t_k - \Delta T/2) \sum_{m=M_1}^{m=M_2} H_m(I)$$
$$\times \left[a(\alpha_m(J)) \sin(m\vartheta - \Omega t + \chi) + b(\alpha_m(J)) \cos(m\vartheta - \Omega t + \chi) \right] , \quad (F.8)$$

The functions $a(x)$, $b(x)$, and $\alpha_{mn}(J)$ are given by

$$a(x) = \frac{1 - \cos x}{x} \ , \qquad b(x) = \frac{\sin x}{x} \ ,$$
$$\alpha_m(J) = [m\omega(J) - \Omega] (t - t_k - \Delta T/2) \ . \qquad (F.9)$$

In (F.8), (F.9) the time parameter t_0 is taken in the middle of the interval $[t_k, t_{k+1}]$, i.e., $t_0 = (t_k + t_{k+1})/2 = t_k + \Delta T/2$.

The implicit mapping procedure presented by (F.4)–(F.6) is implemented by the procedure $MappingStep$(Y,Z,time).

```
MappingStep(LPLDOUBLE Y, LPLDOUBLE Z, LPLDOUBLE time)
{
  LDOUBLE F,G,F_J,G_T,f,f_J,f_t;
  LDOUBLE J,theta,U[2],V[2],t_0, eps_acc=1.0E-12;
  int L=0, N_iter=20;
      t_0=*time+MapPeriod/2.0;
/*---------------The first step -----------------------*/
  GeneratingFunction(&Y[0],&Y[1],&(*time),&t_0,&F,&G,&F_J,&G_T);
  J = Y[0]-F;
  theta = Y[1];
Iter1:
  GeneratingFunction(&J,&theta,&(*time),&t_0,&F,&G,&F_J,&G_T);
  f=J-Y[0]+F;
  f_J=1.0+F_J;
  V[0]=J-f/f_J;
  L++;
  if ((f_J LE 0) OR (L GT N_iter)) {sig1 = 0;  goto exit; }
  if (fabs(V[0] - J) GT eps_acc )
    {J = V[0]; goto Iter1; }
  V[1] = Y[1] + G;
/*--------------------The second step---------------------*/
          U[1] = V[1] + MapPeriod*omega(&V[0]);
          *time+= MapPeriod;
```

```
/*-------------------The third step----------------------*/
  L=0;
  GeneratingFunction(&V[0],&U[1],&(*time),&t_0,&F,&G,&F_J,&G_T);
  theta=U[1]-G;
Iter2:
  GeneratingFunction(&V[0],&theta,&(*time),&t_0,&F,&G,&F_J,&G_T);
  f =theta-U[1]+G;
  f_t=1.0+G_T;
  Z[1]=theta-f/f_t;
  L++;
  if ((f_t LE 0) OR (L GT N_iter)) {sig2 = 0;  goto exit; }
  if (fabs(Z[1]-theta) GT eps_acc)
    {theta=Z[1]; goto Iter2; }
  Z[0] = V[0] + F;
exit:  L = 0;
}
```

Two-dimensional variables $(Y[0],Y[1])$ and $(Z[0],Z[1])$ stand for the initial (I_k, ϑ_k) and the final $(I_{k+1}, \vartheta_{k+1})$ variables, respectively, and time stands for t_k. In the first step the map (F.4) is written as an algebraic equation,

$$f(J_k) = J_k - I_k + F(J_k, \vartheta_k, t_k) = 0 \;,$$

with respect to variable J_k, and it is solved using the Newton-Raphson method Press et al. (1992). The initial value of J_k in the iterative process is taken $J_k^{(0)} = I_k - F(I_k, \vartheta_k, t_k)$. The iteration is stopped in cases when the sufficient accuracy, eps_acc, is reached. It is interrupted if the derivative $f'(J_k)$ becomes negative, or number of iteration exceeds some number N_iter (it is taken equal to 20).

Similarly, in the third step the map (F.6) the variable ϑ_{k+1} is found as a root of the algebraic equation

$$f(\vartheta_{k+1}) = \vartheta_{k+1} - \psi_k + G(J_k, \vartheta_{k+1}, t_{k+1}) = 0 \;.$$

This procedure calls the subroutine $GeneratingFunction$ $(J, theta, time, t_0, F, G, F_J, G_T)$ in which the functions $F(J, \vartheta, t), G(J, \vartheta, t)$ (F.7) and their derivatives $\partial F/\partial J$, $\partial G/\partial \vartheta$ are calculated.

```
GeneratingFunction(J,theta,time,t_0,F,G,F_J,G_T)
LPLDOUBLE J,theta,time,t_0,F,G,F_J,G_T;
{
    int m;
    LDOUBLE S=0.0, S_J=0.0, S_th=0.0, S_thJ=0.0;
    LDOUBLE phi_m,sinm,cosm,eps_t, x,x_J;
    LDOUBLE mm;
      Spectrum_Of_Perturbation(&(*J));
      eps_t = - Epsilon * (*time-*t_0);
    for ( m = M_1; m LE M_2; m++)
```

```
        {
        mm = (LDOUBLE)m;
        x   = alpha(&(*J), m)*(*time-*t_0);
        x_J = alpha_J(&(*J),m)*(*time-*t_0);
        phi_m = fmod(mm* *theta - Omega* *time + phase, pi2);
        sinm = sin(phi_m);
        cosm = cos(phi_m);
/*      S    +=   Hm[m] * (a(&x)  * sinm + b(&x)  * cosm); */
        S_th +=mm*Hm[m] * (a(&x)  * cosm - b(&x)  * sinm);
        S_J  +=   Hm[m] * (a_J(&x)* sinm + b_J(&x)* cosm)*x_J
             +   Hm_J[m]*(a(&x)   * sinm + b(&x)  * cosm);
        S_thJ+= mm*Hm[m]* (a_J(&x)* cosm - b_J(&x)* sinm)*x_J
             + mm*Hm_J[m]*(a(&x)  * cosm - b(&x)  * sinm);
        }
        *F = eps_t * S_th;
        *G = eps_t * S_J;
        *F_J = eps_t * S_thJ;
        *G_T = *F_J;
}
```

This function procedure uses the predefined functions $a(x)$, $b(x)$, and $\alpha(x)$ (F.9) which are defined below. To avoid singularities at the extremely small value of x the functions $a(x)$ and $b(x)$ are expanded into series of powers of x in a small neighborhood of x.

```
LDOUBLE alpha(LPLDOUBLE J, int m)
    { return((LDOUBLE)m*omega(&(*J))- Omega); }
LDOUBLE alpha_J(LPLDOUBLE J, int m)
    { return((LDOUBLE)m*omega_J(&(*J))); }

LDOUBLE b(LPLDOUBLE x)
    { LDOUBLE y2;
      y2 = pow(*x,2.0);
      if ( fabs(*x) LT 0.01 )
        return(1.0-y2/6.0*(1-y2/20.0*(1-y2/42.0
                    *(1-y2/72.0*(1-y2/110.0)))));
      else  return(sin(*x)/ *x);
    }
LDOUBLE b_J(LPLDOUBLE x)
    { LDOUBLE y2;
      y2 = pow(*x,2.0);
      if (fabs(*x) LT 0.01)
        return(-*x/3.0*(1-y2/10.0*(1-y2/28
                    *(1-y2/54.0*(1-y2/88.0)))));
      else  return((cos(*x) - sin(*x)/ *x)/ *x);
    }
```

```
LDOUBLE a(LPLDOUBLE x)
  { LDOUBLE y2;
    y2 = pow(*x,2.0);
    if (fabs(*x) LT 0.01)
       return(0.5* *x*(1.0-y2/12.0*(1-y2/30.0
                *(1.0-y2/56.0*(1-y2/90.0)))));
    else return((1.0-cos(*x))/ *x);
  }
LDOUBLE a_J(LPLDOUBLE x)
  { LDOUBLE y2;
    y2 = pow(*x,2.0);
    if (fabs(*x) LT 0.01)
       return( 0.5 - y2/8.0 *(1-y2/18.0*(1-y2/40.0
                        *(1-y2/70.0))));
    else return((sin(*x)-(1.0-cos(*x))/ *x)/ *x);
  }
```

Below we give an example program for calculating Poincaré section of Hamiltonian system. Specifically, we consider the system with unperturbed frequency of motion $\omega(I) = 1/I$. The Fourier coefficients $H_m(I)$ of the perturbation Hamiltonian are constants, $H_m(I)$.

```
/* Mapping for plotting Poincare section */
#include <math.h>
#include <stdio.h>
#include "mapping.h"
/*-----------Global variables declaration---------*/
long tempOffset = 0;
long curPos = 0;

LDOUBLE Hm[50],Hm_J[50],Phase[50];
LDOUBLE Epsilon, Omega, time, phase;
LDOUBLE TimePeriod, MapPeriod, D_time;
LDOUBLE pi, pi2;
int M_1, M_2;
int sig1, sig2;

main(argc, argv )
int argc;
char *argv[];
  {
    FILE *startup, *outX;
    int i,j,Trax,NumberOfTrax,MaxSteps;
    int MapSteps;
    BOOL AutoStart = FALSE;
    LDOUBLE Y[4], Z[4];
```

```
    LDOUBLE J_i, J_f, theta_i, theta_f;
    char tStr[30];
    char XFileName[200];
    memset(tStr,0,30);
    memset(XFileName,0,200);

/*------------ loading initial values into -------------*/
    if (argc NE 2)
      { printf("usage: map _XXXX.ini\n\r");
        return 0; }
    else
        { int namelen = strlen(argv[1]);
          strncpy(tStr,&argv[1][1],namelen - 5);
          XFileName[0] = 'x';
          strcat(XFileName,tStr);
          strcat(XFileName,".dat"); }

  if ((startup = fopen(argv[1],"rt")) NE NULL)
   {float fl1 = 0, fl2 = 0, fl3 = 0;
    fscanf(startup,"%d %d\n\r",&NumberOfTrax,&MaxSteps);
    printf("NumberOfTrax= %d,MaxSteps= %d\n",
                            NumberOfTrax,MaxSteps);
    fscanf(startup,"%f %f %f\n\r",&fl1,&fl2,&fl3);
    Epsilon = fl1;
    Omega = fl2;
    phase = fl3;
    printf("Epsilon=%5.2f Omega= %5.2f phase= %f\n",
                            Epsilon,Omega,phase );
    fscanf( startup, "%d %d %d\n\r",&MapSteps, &M_1, &M_2);
    printf("Step= %d M_1=%d M_2=%d\n\r",MapSteps,M_1,M_2);
    fscanf( startup, "%f %f\n\r",&fl1, &fl2);
    theta_i = fl1;
    J_i = fl2;
    fscanf( startup, "%f %f\n\r",&fl1, &fl2);
    theta_f = fl1;
    J_f = fl2;
    printf("J_i= %f, J_f= %f\n",(float)J_i, (float)J_f );
    printf("th_i= %f, th_f= %f\n",(float)theta_i,
                            (float)theta_f );
    fclose( startup ); }

/*------------defining Global variables-----------------*/
        pi = 4.0 * atan( 1.0 );
        pi2 = 2.0 * pi;
```

```
        TimePeriod = pi2/Omega;
        MapPeriod = TimePeriod/(LDOUBLE)MapSteps;
/*----------------------------------------------------------------*/
        if ( NumberOfTrax EQ 1 ) Trax = 1;
        else Trax = 0;

    for ( i = Trax; i LE NumberOfTrax; i++ )
        { long plotNumber=0;
        sig1=1; sig2=1;
        time=0.0;
        Y[0]=J_i+(J_f-J_i)*(LDOUBLE)i/(LDOUBLE)NumberOfTrax;
        Y[1]=pi2*(theta_i+(theta_f-theta_i)*(LDOUBLE)i
                        /(LDOUBLE)NumberOfTrax);
        printf("J=%f theta=%f\n",Y[0],Y[1]);
DoIteration:
        for (j = 1; j LE MapSteps; j++)
            {
            MappingStep(Y,Z,&time);
            Z[1]=fmod(Z[1],pi2);
            if (Z[1] LT 0.0) Z[1]+=pi2;
            Y[0]=Z[0];
            Y[1]=Z[1];
            if ((sig1 EQ 0) OR (sig2 EQ 0) ) continue;
            }
        plotNumber++;
        if (outX=fopen(XFileName,"a+t") )
            { fprintf(outX,"%e %e %e\n",time,Z[1]/pi2,Z[0]); }
            fclose(outX);
        if ( plotNumber GT MaxSteps )
            { printf("End of trax i=%d\n", i);
            plotNumber=0;
            continue; }
        goto DoIteration;
    }
    printf( "end of computations\n" );
    printf("\n");
}
```

The frequency of motion $\omega(I)$ and its derivative $d\omega(I)/dI$ are defined by the functions $omega(I)$ and $omega_J(I)$. The procedure $SpectrumOfPerturbation(J)$ defines the perturbation spectrum $H(I)$.

```
LDOUBLE omega(LPLDOUBLE J)
    { return( 1.0/ *J ); }

LDOUBLE omega_J(LPLDOUBLE J)
```

```
{ return(-1.0/pow(*J,2.0)); }

Spectrum_Of_Perturbation(LPLDOUBLE J)
    { int m;
      for (m = M_1; m LE M_2; m++ )
          { Hm[m] = 1.0;
            Hm_J[m] = 0.0; }
}
```

The above program uses a heading file "mapping.h", where the definitions of some operations and functions are given:

```
#define FAR _far
#define FALSE 0
#define TRUE  1

#define  AND &&
#define  OR    ||
#define  EQ    ==
#define  NE    !=
#define  LE    <=
#define  LT    <
#define  GE    >=
#define  GT    >
#define  ENDIF
#define  VER   -100
#define  HOR   -101
#define  ROUND_UP  -102
#define  min(a,b) (((a)<(b)) ? (a) : (b))
#define  max(a,b) (((a)>(b)) ? (a) : (b))
typedef struct tgRECT  { float left;
                         float top;
                         float right;
                         float bottom;} FRECT;
typedef  FRECT *      LPFRECT;
typedef struct tgPOINT  { float x; float y;} FPOINT;
typedef FPOINT *      LPFPOINT;
typedef double        LDOUBLE;
typedef int           BOOL;
typedef LDOUBLE *  LPLDOUBLE;
typedef LDOUBLE **  LP2LDOUBLE;

/*---------- Definition of functions ----------------*/
LDOUBLE omega(LPLDOUBLE x);
LDOUBLE omega_J(LPLDOUBLE x);
LDOUBLE alpha(LPLDOUBLE,int);
```

```
LDOUBLE alpha_J(LPLDOUBLE,int);
LDOUBLE a(LPLDOUBLE);
LDOUBLE a_J(LPLDOUBLE);
LDOUBLE b(LPLDOUBLE);
LDOUBLE b_J(LPLDOUBLE);
```

Finally we give an example of a file with initial data of parameters.

```
0010 1000
0.0010000 1.00000 0.0000000
2 1 10
0.100000 1.050000
0.200000 9.050000
```

References

Main textbooks, monographs and reviews

Andronov, A.A., Vitt, A.A. and Khaikin, S.E. (1981) *Theory of oscillations*, (Moscow: Nauka)

Arnold, V.I. (1963a) "Proof of a theorem of A.N. Kolmogorov's on the invariance of quasi-periodic motions under small perturbations of the Hamiltonian", *Usp. Mat. Nauk*, **18** (5), 13–40 (Russian); English trans.: *Russian Math. Surveys*, **18**, 9–36

Arnold, V.I. (1963b) "Small denominators and the problems of stability of motion in classical and celestial mechanics", *Usp. Mat. Nauk*, **18** (6), 90–192 (Russian); English trans.: *Russian Math. Surveys*, **18** (6), 85–191.

Arnold V.I. (1989) *Mathematical Methods of Classical Mechanics* (Berlin: Springer)

Arnold, V.I., Kozlov, V.V., and Neishtadt, A.I. (1988) *Mathematical Aspects of Classical and Celestial Mechanics*, in "Encyclopedia of Mathematical Sciences. Dynamical Systems", v. 3 (Berlin: Springer) xxi+285.

Balescu, R. (1988) *Transport processes in plasmas*, (North–Holland, Amsterdam).

Birkhoff, G.D. (1927) *Dynamical Systems*, AMS Coll. Publications vol. 9 (reprinted 1966).

Bogolyubov, N.N. and Mitropol'skij, Yu.A. (1958) *Asymptotic methods in the theory of nonlinear oscillations*, 2-nd Ed. (Moscow: Nauka) [English transl.: Gordon and Breach Science Publ., New York, 1961]

Chirikov, B.V. (1979) "A universal instability of many-dimensional oscillator systems", *Physics Reports*, **52**, 265–379.

Goldstein, H. *Classical Mechanics*, 2-nd Ed. (Addison–Wesley, Reading, MA, 1980).

Corben, H.C., and Stehle, Ph. (1974) *Classical Mechanics* (New York: Robert E. Krieger Publishing Co.) Ch. 13.

Guckenheimer, J. and Holmes, P. *Nonlinear oscillations, dynamical systems, and bifurcations of vector fields*, Series Appl. Mathematical Sciences, v. 42 (Springer: New York, 1983).

Hamilton, W.R. (1828) "Theory of systems of rays", *Transactions of the Royal Irish Academy*, **15**, 69–174. (see http://www.emis.de/classics/Hamilton/).

Hamilton, W.R. (1834) "On general method in dynamics", *Philosophical Transactions of the Royal Society*, part II, 247–308; (see also *Mathematical papers*, Part 1 and Part 2, (Cambridge University Press: Cambridge, 1940) and http://www.emis.de/classics/Hamilton/).

Hilborn, R.C. (2000) *Chaos and Nonlinear dynamics. An Introduction for Scientist and Engineers*, 2-nd Edition (Oxford University Press: New York).

Jensen, R.V. (1992) *Chaos*, in "Encyclopedia of Physical Sciences and Technology", (Academic Press Inc.,) v. 3.

Kolmogorov, A.N. (1954) "On conservation of conditionally periodic motions under small perturbations of the Hamiltonian", *Dokl. Akad. Nauk USSR*, **98**, 527–530.

Kolmogorov, A.N. (1957) "General theory of dynamical systems and classical mechanics", *Proc. Int. Congress of Math., 1954, Amsterdam*, **1**, 315–333.

Kolmogorov, A.N. (1958) "A new metric invariant of transient dynamical systems and automorphisms in Lebesque spaces", *Dokl. Akad. Nauk USSR*, **119**, 861–864.

Lanczos, C. (1962) *The Variational Principles of Mechanics* (Toronto: University of Toronto Press) xxv+367.

Lichtenberg, A.J. and Lieberman, M.A. (1992) *Regular and Stochastic Motion* 2-nd Ed. (Springer, New York).

MacKay, R.S. and Meiss, J.D. (eds) (1987) *Hamiltonian Dynamical Systems – A Reprint Selection* (Bristol: Adam Hilger).

Meiss, J.D. (1992) "Symplectic maps, variational principles, and transport", *Reviews of Modern Physics*, **64**, 795–848.

Melnikov, V.K. (1963) "On the stability of the center for time-periodic perturbations", *Trans. Moscow Math. Soc.*, **12**, 1–56.

Moser, J. (1962) "On invariant curves of area–preserving mappings of an annulus", *Nachr. Akad. Wiss., Göttingen, Math.-Phys. Kl.*, No. 1, 1–20.

Moser, J. (1973) *Stable and Random Motion in Dynamical Systems* (Princeton Univ. Press, Princeton, 1973), Annals of Mathematics Studies, No. 77.

Nayfeh, A. (1973) *Perturbation Methods* (New York: John Wiley) xii+418.

Ott, E. (1993) *Chaos in Dynamical Systems* (Cambridge University Press: Cambridge). x+385.

Poincaré, H. (1892/1893/1899) *Les méthodes de la mécanique céleste* vols. 1–3 (Gauthier-Villars: Paris), [Engl. trans.: *New Methods of Celestial Mechanics*, vols. 1–3 (AIP, New York, 1992).

Sagdeev, R.Z., Usikov, D.A., and Zaslavsky, G.M. (1988) *Nonlinear Physics. From the Pendulum to Turbulence and Chaos* (Harwood Academics: Chur, Switzeland).

Siegel, C.L. and Moser, J.K. (1995) *Lectures on Celestial Mechanics*, 2nd edition (Springer Verlag: Berlin) 290 pp.

Zaslavsky, G.M. (1985) *Stochasticity of dynamical systems*, (Harwood Academics: Chur, Switzeland).

General references

Abdullaev, S.S. (1991) "Classical chaos and nonlinear dynamics of rays in inhomogeneous media", *Chaos*, **1**, 212–219

Abdullaev, S.S. (1993B) *Chaos and dynamics of rays in waveguide media*, (Gordon and Breach Science Publ.: Yverdon) xvi+312.

Abdullaev, S.S. (1994a) "Classical chaotic dynamics of sound rays in 3-D inhomogeneous waveguide media", *Chaos*, **4**, 63–73.

Abdullaev, S.S. (1994b) "Two-dimensional model of kicked oscillator: Motion with intermittency", *Chaos*, **4**, 569–581.

Abdullaev, S.S. (1997) "Renormalization invariance of a Hamiltonian system near the saddle point", *Physics Letters* A, **234**, 281–290

Abdullaev, S.S. (1999) "A new integration method of Hamiltonian systems by symplectic maps", *J. Phys. A: Math. & Gen.*, **32**, 2745–2766.

Abdullaev, S.S. (2000) "Structure of motion near saddle points and chaotic transport in Hamiltonian systems", *Phys. Rev. E*, **62**, 3508–3528.

Abdullaev, S.S. (2002) "The Hamilton–Jacobi method and Hamiltonian maps", *J. Phys. A: Math. & Gen.*, **35**, 2811–2832.

Abdullaev, S.S. (2004a) "On mapping models of field lines in a stochastic magnetic field", *Nuclear Fusion*, **44**, S12-S27.

Abdullaev, S.S. (2004b) "Canonical maps near separatrix in Hamiltonian systems", *Phys. Rev. E*, **70**, 046202.

Abdullaev, S.S. (2005) "Asymptotical forms of canonical mappings near separatrix in Hamiltonian systems", *Phys. Rev. E*, **72**, 046202.

Abdullaev, S.S. and Finken, K.H. (1998) "Widening the magnetic footprints in a poloidal divertor tokamak: a proposal", *Nuclear Fusion*, **38**, 531–544.

Abdullaev, S.S. and Spatschek, K.H. (1999) "Rescaling invariance and anomalous transport in a stochastic layer", *Phys. Rev. E*, **60**, R6287–R6290.

Abdullaev, S.S. and Zaslavsky, G.M. (1988) "Fractals and ray dynamics in a longitudinally inhomogeneous media" *Akust. Zh.* **34**, 578–582. [*Sov. Phys. Acoust*, **34**, 334–336]

Abdullaev, S.S. and Zaslavsky G.M. (1991) "Classical nonlinear dynamics and chaos of rays in problems of wave propagation in inhomogeneous media". *Uspekhi Fiz. Nauk*, **161** (8), 1–43 [*Sov. Phys. Uspekhi*, **34**, 645–664]

Abdullaev, S.S. and Zaslavsky, G.M. (1994) "New properties of the ergodic layer", *Bull. Amer. Phys. Soc.*, **39**, 1659.

Abdullaev, S.S. and Zaslavsky, G.M. (1995b) "Self-similarity of stochastic magnetic field lines near the X-point", *Phys. Plasmas* **2**, 4533–4540.

Abdullaev, S.S. and Zaslavsky, G.M. (1996) "Application of the separatrix map to study perturbed field lines near the separatrix", *Phys. Plasmas* **3**, 516–528.

Abdullaev, S.S., Finken, K.H., and Spatschek, K.H. (1999) "Asymptotical and mapping methods in study of ergodic divertor magnetic field in a toroidal system", *Phys. Plasmas*, **6**, 153–174.

Abdullaev, S.S., Finken, K.H., Kaleck, A., and Spatschek, K.H. (1998) "Twist mapping for the dynamics of magnetic field lines in a tokamak ergodic divertor", *Phys. Plasmas*, **5**, 196–210.

Abdullaev, S.S., Eich, Th., and Finken, K.H. (2001) "Fractal structure of the magnetic field in the laminar zone of the Dynamic Ergodic Divertor of the Torus Experiment for Technology-Oriented Research (TEXTOR-94)", *Phys. Plasmas*, **8**, 2739–2749.

Abdullaev, S.S., Finken, K.H., Jakubowski, M.W., Kasilov, S.V., Kobayashi, M., Reiser, D., Reiter, D., Runov, A.M., and Wolf, R. (2003) "Overview of magnetic structure induced by the TEXTOR-DED and the related transport", *Nuclear Fusion*, **43**, 299–313.

Abramowitz, M. and Stegun, I. Eds. (1965) *Handbook on Mathematical Functions* (Dover Publ., New York).

Ahn, T. and Kim, S. (1994) "Separatrix map analysis of chaotic transport in planar periodic vortical flows *Phys. Rev. E* **49**, 2900–2911.

Ahn, T., Kim, S., and Kim, G. (1996) "Analysis of the separatrix map in Hamiltonian systems", *Physica*, D **89**, 315–328.

Ahn, T. and Kim, S. (1997) "Chaotic transport and anomalous diffusion in a planar flow with a chain of vortices", *Int. J. Bifurcation and Chaos* **7**, 1025–1033.

Albert, J.M. and Boozer, A.H. (1988) "Discrete mappings and resonant ripple transport in a tokamak", *Physics of Fluids*, **31**, 1811–1812.

Ali, H., Punjabi, A., Ali, H., Boozer, A.H., and Evans, T. (2004) "The low MN map for single–null divertor tokamaks", *Phys. Plasmas* **11**, 1908–1919.

Aref, H. (1991) "Stochastic particle motion in laminar flows", *Phys. Fluids* A **3**, 1009–1016.

Arnold, V.I. (1964) "Instability of dynamical systems with several degrees of freedom", *Sov. Math. Dokl.*, **5**, 581–585.

Artuso, R. (1999) "Correlation time and return time statistics", *Physica*, **D 131**, 66–77.

Aubry, S. (1978) "The new concept of transitions by breaking of analyticity in crystallographic model" in *Soliton and Condensed Matter physics*, ed. A. Bishop and T. Schneider (Springer-Verlag: New York), pp. 264–277.

Bak, P. (1986) "The Devil's Staircase", *Physics Today*, **39** (12), 38.

Bak, P. and Bruinsma, R. (1982) "One-Dimensional Ising Model and the Complete Devil's Staircase", *Phys. Rev. Lett.*, **49**, 249–251.

Balescu, R. (1997) "Continuous time random walk model for standard map dynamics", *Phys. Rev. E*, **55**, 2465–2474.

Balescu, R. (1998) "Hamiltonian non-twist map for magnetic field lines with locally reversed shear in a toroidal geometry", *Phys. Rev. E* **58**, 3781–3792.

Balescu, R. (2000a) "Kinetic theory of an area-preserving maps: Application to the standard map in the diffusive regime", *J. Statist. Phys.*, **98**, 1169–1234.

Balescu, R. (2000b) "Kinetic theory of the standard map in the localized weak-stochasticity diffusive regime", *J. Plasma Phys.*, **64**, 379–396.

Balescu, R., Vlad M., and Spineanu (1998) "Tokamap: A Hamiltonian twist map for magnetic field lines in a toroidal geometry", *Phys. Rev. E* **58**, 951–964.

Bazzani, A., Todesco, E., Turchetti, G., and Servizi, G. (1994) "A normal form approach to the theory of nonlinear betatronic motion", CERN Report No. 94–02, 225 pp.

Bazzani, A., Malavasi, M., and Siboni, M. (1989) "Poincaré map and anomalous transport in a magnetically confined plasma", *Il Nuovo Cimento*, **103**, 659–668.

Bazzani, A., Di Sebastiano, A., and Turchetti, G. (1998) "Diffusion of magnetic field lines a confined RFP plasma", *Il Nuovo Cimento*, D **20**, 1795–1818.

Bénisti, D. and Escande, D.F. (1997) "Origin of diffusion in Hamiltonian dynamics" *Phys. Plasmas*, **4**, 1576–1581.

Bénisti D. and Escande D.F. (1998) "Nonstandard diffusion properties of the standard map" *Phys. Rev. Lett.*, **80**, 4871–4874

Benkadda, S., Sen, A., and Shklyar, D.R. (1996) "Chaotic dynamics of charged particles in the field of two monochromatic waves in a magnetized plasma", *Chaos*, **6**, 451–460.

Benkadda, S., Kassibrakis, S., White, R.B., and Zaslavsky, G.M. (1997a) "Self-similarity and transport in the standard map", *Phys. Rev. E* **55**, 4909–4917.

Berg, S.J., Warnock, R.L., Ruth, R.D., and Forest, É. (1994) "Construction of symplectic maps for nonlinear motion of particles in accelerators", *Phys. Rev. E* **49**, 722–739

Beringer, R.P., Meyers, S.D., and Swinney, H.L. (1991) "Chaos and mixing in a geostrophic flow", *Phys. Fluids* A **3**, 1243.

Bertozzi, A.L. (1988) "Heteroclinic orbits and chaotic dynamics in planar fluid flows", *SIAM J. Math. Anal.* **19**, 1271–1294.

Bleher, S., Ott, E., and Grebogi, C. (1989) "Routes to chaotic scattering", *Phys. Rev. Lett.*, **63**, 919–922.

Bleher, S., Grebogi, C., and Ott, E. (1990) "Bifurcation to chaotic scattering", *Physica*, **D 46**, 87–121.

Bogolyubov, N.N. (1945) *On some statistical problems in mathematical physics* (Kiev: The Academy of Sciences of Ukraine).

Boozer, A.H. (1983) "Evaluation of the structure of ergodic zone", *Phys. Fluids*, **26**, 1288–1291.

Boozer, A.H. (1992) *Plasma confinement*, in "Encyclopedia of Physical Science and Technology" (Academic Press, New York), vol. 13.

Boozer, A.H. (2004) "Physics of magnetically confined plasmas", *Reviews of Modern Physics*, **76**, 1071–1141.

Boozer, A.H. and Rechester, A.B. (1978) "Effect of magnetic perturbations on divertor scrape-off width", *Phys. Fluids*, **21**, 682–689.

Born, M. and Wolf, E. (1986) *Principled of optics: electromagnetic theory of propagation, interference and diffraction of light*, 6-th Ed., Reprinted with corrections (Pergamon Press: Oxford).

Bouchaud, J.-Ph. and Georges, A. (1990) "Anomalous diffusion in disordered media: statistical mechanisms, models and physical applications", *Physics Reports*, **195**, 127–293.

Boud, P.T. and McMillan, S.L.W. (1993) "Chaotic scattering in the gravitational three-body problem", *Chaos*, **3**, 507–523.

Brahic, A. (1971) "Numerical study of a simple dynamical dynamical system. 1. Associated plane area–preserving mapping *Atsron. Astrophys.*, **12**, 98.

Brekhovskih, L.M. and and Lysanov, Y.P. (2003) *Fundamentals of Ocean Acoustics*, 3-rd Ed. (Amer. Inst. Phys.).

Brown M.G., Tappert F.D., and Goñi G. (1991a) "An investigation of sound ray dynamics in the ocean volume using an area-preserving mapping", *Wave Motion*, **14**, 93–99.

Brown M.G., Tappert F.D., Goñi G., and Smith K.B. (1991b) "Chaos in underwater acoustics", in "*Ocean Variability and Acoustic Propagation*", edited by J. Potter and A. Warn-Varnas (Kluwer Academic, Dordrecht, 1991), 139–160.

Brown, M.G., Colosi, J.A., Tomsovic, S., Virovlyansky, A.L., Wolfson, M.A., and Zaslavsky, G.M. (2003) "Ray dynamics in long-range deep ocean sound propagation", *J. Acous. Soc. Amer.*, **113**, 2533–2547.

Bullet, S. (1986) "Invariant circles for the piecewise linear standard map", *Commun. Math. Phys.*, **107**, 241–262.

Caldas, I.L., Pereira, J.M., Ullmann, K., and Viana, R.L. (1996) "Magnetic field line mappings for a tokamak with ergodic limiters", *Chaos, Solitons and Fractals*, **7**, 991–1010.

Cartwright, J.H.E. and Piro, O. (1992) "The dynamics of Runge–Kutta methods", *Int. J. Bifurcations and Chaos* **2**, 427–449.

Cary, J.R. (1981) "Lie transform perturbation theory for Hamiltonian systems", *Phys. Reports*, **79**, 129–159.

Cary, J.R. (1982) "Vacuum magnetic fields with dense flux surfaces", *Phys. Rev. Lett.*, **49**, 276–279.

Cary, J.R. (1984) "Construction of three-dimensional vacuum magnetic fields with dense nested flux surfaces", *Phys. Fluids*, **27**, 119–128.

Cary, J.R. and Littlejohn, R.G. (1983) "Noncanonical Hamiltonian mechanics and its applications to magnetic field line flow", *Annals of Physics* (N.Y.) **151**, 1–34.

Casati, G., Chirikov, B.V., Izraelev, F.M., and Ford, J. (1979) "Stochastic behavior of a quantum pendulum under a periodic perturbation", *Lect. Notes Phys.*, **93**, 334–352.

Casati, G., Cuarneri, I., and Shepelyansky, D.L. (1987) "Exponential photonic localization for the hydrogen atom in a monochromatic field", *Phys. Rev.*, A **36**, 3501–3504.

Casati, G., Cuarneri, I., and Shepelyansky, D.L. (1988) "Hydrogen atom in monochromatic field: chaos and dynamical photonic localization", *IEEE: J. Quany. Electr.*, **24**, 1420–1444.

Casati, G., Cuarneri, I., and Shepelyansky, D.L. (1990) "Classical chaos, quantum localization and fluctuations: a unified view", *Physica*, A **163**, 205–214.

Celletti, A. (1990) "Analysis of resonances in the spin–orbit problem in celestial mechanics: The synchronous resonance. Part I", *ZAMP*, **41**, 174–204.

Chirikov, B.V. (1969) USSR Academy of Science, Siberian Branch, Report No. 267, CERN translation No. 71-40, 1969 (unpublished).

Chirikov, B.V. (1983) "Chaotic dynamics in Hamiltonian systems with divided phase space", *Lect. Notes Physics*, **179**, 29–46.

Chirikov, B.V. (1977) *Nonlinear Resonance* (Novosibirsk State (Gos.) Univ., Novosibirsk).

Chirikov, B.V. (1991) "Patterns in chaos", *Chaos, Solitions & Fractals*, **1**, 79–103.

Chirikov, B.V. and Shepelyansky, D.L. (1984) "Correlation properties of dynamical chaos in Hamiltonian systems", *Physica*, D **13**, 395–400.

Chirikov, B.V. and Shepelyansky, D.L. (1999) "Asymptotic statistics of Poincaré recurrences in Hamiltonian systems with divided phase space", *Phys. Rev. Lett.*, **82**, 528–531.

Chirikov, B.V. and Vecheslavov, V.V. (1989) "Chaotic dynamics of comet Halley", *Astron. Astrophys.*, **221**, 146–154.

Chirikov, B.V. and Vecheslavov, V.V. (2002) "Fractal diffusion in smooth systems with virtual invariant curves", *J. Exper. Theor. Phys.*, **95**, 560–571.

Chirikov, B.V., Keil, E., and Sessler, A. (1971) "Stochasticity in many-dimensional nonlinear oscillating systems", *J. Stat. Phys.*, **3**, 307–321.

Chugurin, V.V., Erukhimov, L.M., and Ryndyk, E.Y. (1994) "Ray focusing and diffusion in the regular two-periodic system", *Waves in random media*, **4**, 21–28.

Dana, I. (1996) "Resonances and diffusion in periodic Hamiltonian maps", *Phys. Rev. Lett.*, **62**, 233–236.

D'Angelo, F. and Paccagnella, R. (1996) "The stochastic diffusion process in reversed-field pinch", *Phys. Plasmas*, **3**, 2353–2364.

D'Angelo, F. and Paccagnella, R. (1999) "Stochastic diffusivity and heat transport in the presence of a radial dependence of the perturbed magnetic field in the reversed field pinch", *Plasma Phys. Control. Fusion*, **41**, 941–954.

Da Silva, E.C., Caldas, I.L., and Viana, R.L. (2001) "Field line diffusion and loss in a tokamak with an ergodic magnetic limiter", *Phys. Plasmas*, **8**, 2855–2865.

Da Silva, E.C., Caldas, I.L., and Viana, R.L. (2001) "The structure of chaotic magnetic field lines in a tokamak with the external non-symmetric magnetic perturbations", *IEEE Trans. on Plasma Science*, **29**, 617–631.

da Silva, E.C., Caldas, I.L., and Viana, R.L. (2002) "Bifurcations and onset of chaos on the ergodic magnetic limiter mapping", *Chaos, Solitons and Fractals*, **14**, 403–423.

da Silva, E.C., Caldas, I.L., Viana, R.L., and Sanjuán, M.A.F. (2002) "Escape patterns, magnetic footprints, and homoclinic tangles due to ergodic magnetic limiter", *Phys. Plasmas*, **9**, 4917–4928.

de Rover, M., Lopes Cardozo, N.J., and Montvai, A. (1996) "Motion of relativistic particles in axially symmetric and perturbed magnetic fields in a tokamak", *Phys. Plasmas*, **3**, 4478–4488.

Davis, M.J. and Wyatt, R.E. (1982) "Surface-of-surface analysis in the classical theory of multiphoton absorption", *Chem. Phys. Lett.*, **86**, 235–241.

del-Castillo-Negrete, D. and Morrison, P.J. (1993a) "Chaotic transport by Rossby waves in shear flow", *Phys. Fluids A* **5**, 948–965.

del-Castillo-Negrete, D. (1998) "Asymmetric transport and non-Gaussian statistics of passive scalars in vortices in shear", *Phys. Fluids* **10**, 576–594.

del-Castillo-Negrete, D. (2000b) "Chaotic transport in zonal flows in analogous geophysical and plasma systems, *Phys. Plasmas*, **7**, 1702–1711.

del-Castillo-Negrete, D. and Morrison, P.J. (1993b) in *Chaotic Dynamics and Transport in Fluids and Plasmas*, ed. I. Prigogine (Americal Institute of Physics: New York)

del-Castillo-Negrete, D., Greene, J.M., and Morrison, P.J. (1996) "Area preserving non-twist maps: periodic orbits and transition to chaos", *Physica*, D **91**, 1–23.

del-Castillo-Negrete, D., Greene, J.M., and Morrison, P.J. (1997) "Renormalization and transition to chaos in area preserving non-twist maps", *Physica*, D **100**, 311–329.

Deprit, A. (1969) "Canonical transformations depending on a small parameter", *Celestial Mech.* **1**, 12–30.

Dewar, R.L. (1976) "Renormalized canonical perturbation theory for stochastic propagators", *J. Phys. A: Math. Gen.*, **9**, 2043–2057.

Douady, R. (1982) "Une démonstration directe de l'équivalence des théorémes de tores invariants pour difféomorphismes et champs des vecteurs", *C.R. Acad. Sci., Paris, Sér.* I, **295**, 201–204.

Dragt, A.J. (1979) "A method of transfer maps for linear and nonlinear beam elements", *IEEE Transactions on Nuclear Science*, **NS-26** No.3, 3601.

Dragt, A.J. (1996) "Summary of the working group on maps", *Particle accelerators*, **55**, [499–530]/253–281.

Dragt, A.J. (2000) *Lie methods for Nonlinear Dynamics with Application to Accelerator Physics*, (Center for Theoretical Physics, Department of Physics, University of Maryland: College Park)

Dragt, A.J. and Abel, D.T. (1996) "Symplectic maps and computation of orbits in particle accelerators", "*Integration algorithms for classical mechanics*", Eds. J.E. Marsden, G.W. Patrick and W.F. Shedwick (Field Institute Communications: Rhode Island, v. 10), pp. 59–85.

Dragt, A.J. and Finn, J.M. (1976) "Lie series and invariant functions for analytic symplectic maps" *J. Math. Phys.*, **17**, 2215–2227.

Easton, R.W., Meiss, J.D., and Carver, S. (1993) "Exit times and transport for symplectic twist maps", *Chaos*, **3**, 153–165.

Eberhard, M. (1999) "Transition from time-continuous systems to discrete mappings", *Europhysics Conference Abstracts*, **23J**, 781–784.

Eckhard, B. (1988) "Irregular scattering", *Physica*, **D 33**, 89–98.

Edwards, D.A. and Syphers, M.J. (1993) *An Introduction to the Physics of High Energy Accelerators*, (Wiley & Sons: New York). xii+ 292 pp.

Eich, Th., Reiser, D., and Finken, K.H. (2000) "Two dimensional modeling approach to transport properties of the TEXTOR-DED laminar zone", *Nucl. Fusion*, **40**, 1757–1772.

Eilenberger, G. and Schmidt, K. (1992) "Poincaré maps of Duffing-type oscillators and their reduction to circle maps: I. Analytical results", *J. Phys. A: Math. Gen.*, **25**, 6335–6356.

Elskens, Y. and Escande, D.F. (2002) "Proof of quasilinear equations in the chaotic regime of the weak warm beam instability", *Phys. Lett.*, A **302**, 110–118.

Elskens, Y. and Escande, D.F. (2002) *Macroscopic Dynamics of Plasmas and Chaos*, (IOP Publ. : Bristol) 300 pp.

Engelhardt, W. and Feneberg, W. (1978) "Influence of an ergodic magnetic limiter on the impurity content in a tokamak", *J. Nucl. Mater.* **76 & 77**, 518–520.

Escande, D.F. (1985) "Stochasticity in classical Hamiltonian systems: universal aspects", *Physics Reports*, **121**, 165–261.

Escande, D.F. "Hamiltonian chaos and adiabaticity "in *Proceedings of the International Workshop on Plasma Theory and Nonlinear and Turbulent Processes in Physics, Kiev, 1987*, edited by V.G. Bar'yakhtar, V.M. Chernousenko, N.S. Erokhin, A.G. Sitenko, and A.V. Zakharov (World Scientific, Singapore, 1988) pp. 398–430.

Evans, T.E., Moyer, R.A., and Monat, P. (2002) "Modeling of stochastic magnetic flux loss from the edge of poloidally diverted tokamak", *Phys. Plasmas*, **9**, 4957–4967.

Evans, T.E., Moyer, R.A., and Thomas, P.R. et al. (2004) "Suppression of large edge-localized modes in high-confinement DIII-D plasmas with a stochastic magnetic boundary", *Phys. Rev. Lett.*, **92**, 235003.

Fedoryuk, M.V. (1989) *Asymptotic Methods in Analysis*, in "Encyclopaedia of Mathematical Sciences", v. 13, (Springer, Berlin), pp. 83–192.

Feneberg, W., and Wolf, G.H. (1981) "A helical magnetic limiter for boundary-layer control in large tokamaks", *Nucl. Fusion* **21**, 669–676.

Feng, K. and Wang, D.-L. (1994) *Contemporary Mathematics* **163** eds. Z.-C. Shi and C.-C. Yang.

Feng, Y., Sardei, F., and Grigull, P. et al. (2002) "Transport in island divertors: physics, 3D modelling and comparison to first experiments on W7-AS", *Plasma Phys. Control. Fusion*, **44**, 611–625.

Fermi, E. (1949) "On the origin of the cosmic radiation", *Phys. Rev.* **75**, 1169–1174.

Féron S. and Ghendrih Ph. (1997) "Boundary temperature profile and core energy confinement in Ergodic Divertor configuration", *Europhysics Conference Abstracts*, **21 A**, part I, 185–188.

Ferraz-Mello, S. (1997) "A symplectic mapping approach to the study of the stochasticity of asteroidal resonances", *Celest. Mech. & Dynam. Astron.*, **65**, 421–437.

Filonenko, N.N., Sagdeev, R.Z., and Zaslavsky, G.M. (1967) "Destruction of magnetic surfaces in tokamaks by magnetic field irregularities: Part II", *Nuclear Fusion*, **7**, 253–266.

Filonenko, N.N. and Zaslavsky, G.M. (1968) "Stochastic instability of trapped particles and conditions of applicability of the quasi-linear approximation", *Zh. Eksp. Teor. Fiz.* **54**, 1590–1602 [*Sov. Phys. JETP*, **27** (5), 851–857].

Finken, K.H. (ed.) (1997) Special issue "The Dynamic Ergodic Divertor", *Fusion Eng. Design*, **37**, 335–448.

Finken, K.H., Abdullaev, S.S., Kaleck, A., and Wolf, G.H. (1999) "Operational space of the Dynamic Ergodic Divertor for TEXTOR-94", *Nuclear Fusion*, **39**, 637–661.

Finn, J.M. (1975) "The destruction of magnetic surfaces in tokamaks by current perturbations", *Nuclear Fusion*, **15**, 845–854.

Fischer, O. and Cooper, W.A. (1998) "Mapping of a stochastic magnetic field in toroidal systems", *Plasma Phys. Reports*, **24**, 727–731.

Freis, R.P., Hartman, C.W., Hanzeh, F.M., and Lichtenberg, A.J. (1973) "Magnetic island formation and destruction in a Levitron", *Nuclear Fusion*, **13**, 533–548.

Garrido, L.M. (1968) "General interaction picture from action principle for mechanics" *J. Math. Phys.* **10**, 1045.

Geisel, T., Zacherl, A., and Radons, G. (1987) "Generic 1/f noise in chaotic Hamiltonian dynamics", *Phys. Rev. Lett.*, **59**, 2503–2506.

Gentle, K.W. (1981) "The Texas experimental tokamak (TEXT) facility", *Nucl. Technol. Fusion*, **1**, 479.

Ghendrih, Ph., Grossman, A., and Capes, H. (1996) "Theoretical and experimental investigations of stochastic boundaries in tokamaks", *Plasma Phys. Control. Fusion*, **38**, 1653–1724.

Giovannozzi, M., Scandale, W., and Todesco, E. (1998) "Dynamic aperture in the presence of tune modulation", *Phys. Rev. E*, **57**, 3432–3443.

Goggin, M.E. and Milonni, P.W. (1988) "Driven Morse oscillator: classical chaos, quantum theory, and photodissociation", *Phys. Rev. A*, **37**, 796–806.

Gontis, V. and Kaulakys, B. (1987) "Stochastic dynamics of hydrogenic atoms in the microwave field: modelling by maps and quantum description", *J. Phys.*, **B 20**, 5051–5064.

Greene J.M. (1984) "Renormalization and the breakup of magnetic surfaces" *Statistical Physics and Chaos in Fusion Plasmas* ed. C.W. Horton, Jr., and L.E. Reichl (John Wiley & Sons: New York) pp. 3–20.

Gurevich, A.V. and Tsedilina, E.E. (1985) *Long distance propagation of HF radio waves*, (Springer-Verlag: Berlin).

Hadjidemetriou, J.D. (1991) "Mapping models for Hamiltonian systems with application to resonant asteroid motion", In A.E. Roy (ed.), *Predictability, Stability, and Chaos in N-Body Dynamical systems*, (Plenum Press: New York) pp. 157–175.

Hadjidemetriou, J.D. (1999) "Asteroid motion near the 3:1 resonance", *Celest. Mech. & Dyn. Astron.*, **56**, 563–599.

Hadjidemetriou, J.D. (1998) "Symplectic Maps and their use in Celestial Mechanics", in C. Froeschle and D. Benest (eds.) *Analysis and Modelling of Discrete Dynamical Systems*, (Gordon and Breach Publ) ch. 9, pp. 249–282.

Hadjidemetriou, J.D. (1999) "A symplectic mapping model as a tool to understand the dynamics of 2/1 resonant asteroid motion", *Celest. Mech. & Dyn. Astron.*, **73**, 63–76.

Hamzeh, F.M. (1974) "Magnetic surface destruction in toroidal systems", *Nuclear Fusion*, **14**, 523–536.

Hatori, T., Kamimura, T., and Ichikawa, Y.H. (1985) "Turbulent diffusion for the radial twist map", *Physica*, D **14**, 193–202.

Hénon, M. (1969) "Numerical study of quadratic area preserving mappins", *Q. Appl. Math.*, **27**, 291–312.

Hénon, M. and Heiles, C. (1964) "The applicability of the third integral of motion: some numerical experiments", *Astrophys. J.*, **69**, 73–79.

Hénon, M. and Wisdom, J. (1983) "The Benettin-Strelcyn oval billiard revisited", *Physica*, **D 8**, 157–169.

Hinton, F.L., and Hazeltine, R.D. (1976) "Theory of plasma transport in toroidal confinement systems", *Rev. Mod. Phys.*, **48**, 239–308.

Holman, M. and Wisdom, J. (1993) "Dynamical stability in the outer solar system and the delivery of short period comets", *Astron. J.*, **105**, 1987–1999.

Hori, G. (1966) "Theory of general perturbations with unspecified canonical variables", *Publ. Astron. Soc. Japan* **18**, 287–296.

Horton, W. (1990) "Nonlinear drift waves and transport in magnetized plasma", *Physics Reports*, **192**, 1–177.

Horton, W., Park, H.-B., Kwon, J.-M., Strozzi, D., Morrison, P.J., and Choi, D.-I. (1998) "Drift wave test particle transport in reversed shear profile", *Phys. Plasmas*, **5**, 3910–3917.

Howard, J.E. and Hohs, S.M. (1984) "Stochasticity and reconnection in Hamiltonian systems", *Phys. Rev. A* **29**, 418–421.

Ichikawa, Y.H., Kamimura, T., and Hirose, K. (1987) "Stochastic diffusion in the standard map", *Physica D* **29**, 247–255.

Jakubowski, M., Abdullaev, S.S., Finken, K.H., and TEXTOR Team, (2004) "Modelling of the magnetic field structures and first measurements of heat fluxes for TEXTOR-DED operation", *Nuclear Fusion*, **44**, S1-S11.

Jensen, R.V., Leopold, J.G., and Richards, D. (1988) "High-frequency microwave ionisation of excited hydrogen atoms", *J. Phys.*, **B 21**, L527–L531.

Jensen, R.V., Susskind, M.M., and Sanders, M.M. (1991) "Chaotic ionization of highly excited hydrogen atom: comparison of classical and quantum theory with experiment", *Phys. Reports*, **201**, 1–56.

Kadomtsev, B.B. (1988) *Collective phenomena in plasmas*, (Nauka, Moscow) (in Russian).

Kaleck, A., Hassler, M., and Evans, T. (1997) "Ergodization of the magnetic field at the plasma edge by the dynamic ergodic divertor", *Fusion Eng. Design*, **37**, 353–378.

Karney, C.F.F. (1983) "Long-time correlations in the stochastic regime", *Physica*, D **8**, 360–380.

Karney, C.F.F., Rechester, A.B., and White, R.B. (1982) "Effect of noise in the standard mapping", *Physica*, D **4**, 425–438.

Kasilov, S.V., Moiseenko, V.E., and Heyn, M.F. (1997) "Solution of the drift kinetic equation in the regime of weak collisions by stochastic mapping techniques", *Phys. Plasmas*, **4**, 2422–2435.

Kaulakys, B. and Vilutis, G. (1999) "Kepler map", *Physica Scripta*, **59**, 251–256.

Kawamura, T., Abe, Y., and Tazima, T. (1982) "Formation of magnetic islands and ergodic magnetic layers in wall-lapping plasma as a non-divertor concept for a reactor – relevant tokamak", *J. Nuclear Mater.*, **111–112**, 268–273.

Kerst, D.W. (1962) "The influence of errors on the plasma-confining magnetic fields", *Journal Nuclear Energy*, **4C**, 253–262.

Kinoshita, H., Yoshida, H., and Nakai, H. (1991) "Symplectic integrators and their application to dynamical astronomy, *Celest. Mech. Dyn. Astron.*, **50**, 59–71.

Klafter, J., Shlesinger, M., and Zumofen, G. (1996) "Beyond Brownian motion", *Physics Today*, **49**, 33–39.

Kuznetsov, L. and Zaslavsky, G.M. (1997) "Hidden renormalization group for the near-separatrix Hamiltonian dynamics, *Physics Reports*, **288**, 457–485.

Kuznetsov, L. and Zaslavsky, G.M. (1998) "Regular and chaotic advection in the flow field of a three-vortex system", *Phys. Rev.*, E **58**, 7330–7349.

Kuznetsov, L. and Zaslavsky, G.M. (2002) "Scaling invariance of the homoclinic tangle", *Phys. Rev.*, E **66** No. 046212.

Kwon, J.-M., Horton, W., Zhu, P., Morrison, P.J., Park, H.-B., and Choi, D.-I. (2000) "Global drift wave map test particle simulations", *Phys. Plasmas*, **7**, 1169–1180.

LaHaye, R.J. (1991) "Calculations of the effects of field errors on diverted magnetic-field lines in the DIII-D tokamak", *Nuclear Fusion*, **31**, 1550–1555.

Lai, Y.-C., Ding, M., and Grebogi, C. (1993) "Controlling Hamiltonian chaos", *Phys. Rev.*, E **47**, 86–92.

Latka, M. and West, B.J. (1995) "Structure of the stochastic layer of a perturbed resonant triad" *Phys. Rev.* E **52**, 3252–3255.

Lichtenberg, A.J. and Wood, B. (1989) "Diffusion through a stochastic web" *Phys. Reviews A*, **39**, 2153–2159.

Lichtenberg, A.J., Lieberman, M.A., and Cohen, R.H. (1980) "Fermi acceleration revisited " *Physica*, **D 1**, 291–305.

Lieberman, M.A. and Lichtenberg, A.J. (1972) "Stochastic and adiabatic behavior of particles accelerated by periodic forces", *Phys. Rev*, A **5**, 1852–1866.

Lissauer, J. (1999) "Chaotic motion in the Solar system", *Reviews of Modern Phys.*, **71**, 835–845.

Litaudon, X. (1998) "Profile control for steady-state operation", *Plasma Phys. Control. Fusion* **40**, A251–A268.

Luo, A.C.J. (2002) "Resonant layers in a parametrically excited pendulum", *Int. J. Bifurcation & Chaos*, **12**, 409–419.

Luo, A.C.J. and Han, R.P.S. (2000) "The dynamics of stochastic and resonant layers in a periodically driven pendulum" *Chaos, Solitons & Fractals* **11**, 2349–2359.

Luo, A.C.J. and Han, R.P.S. (2001) "The resonance theory for stochastic layers in nonlinear dynamic systems", *Chaos, Solitons & Fractals* **12**, 2493–2508.

MacKay, R.S. (1983) "A renormalization approach to invariant circles in area-preserving maps", *Physica*, D **9**, 289–300.

MacKay, R.S. (1993) *Renormalization in are-preserving maps* (corrected and annotated), (World Sci. Publ.: Singapore), xix+304.

Mandelbrot, B. (1982) *The Fractal Geometry of Nature*, (W.H. Freeman: New York).

Marcuse, D. (1982) *Light Transmission Optics*, 2-nd Ed. (Krieger: Malabar, Fl.).

370   References

Martin, T.J. and Taylor, J.B. (1984) "Ergodic behavior in a magnetic limiter", *Plasma Phys. Contr. Fusion*, **26**, 321–334.

Matsuda, S. and Yoshikawa, M. (1975) "Magnetic island formation due to error field in the JFT-2 tokamak", *Japanese J. Applied Phys.*, **14**, 87.

McCool, S.C., Wootton, A.J., Kotschenreuther, M., Aydemir, A.Y., Bravenec, R.V., DeGrassie, J.S., Evans, T.E., Hickok, R.L., Richards, B., Rowan, W.L., and Schoch, P.M. (1990) "Particle transport studies with the applied resonant fields on TEXT", *Nuclear Fusion*, **30**, 167–173

McLachlan, R. and Atela, P. (1992) "The accuracy of symplectic integrators", *Nonlinearity*, **5**, 541–562.

Meiss, J.D., Cary, J.R., Grebogi, C., Crawford, J.D., Kaufman, A.N., and Abarbanel, H.D.I. (1983) "Correlations of periodic, area-preserving maps", *Physica*, D **6**, 375–384.

Mendonça, J.T. (1991) "Diffusion of magnetic field lines in a toroidal geometry", *Phys. Fluids B* **3**, 87–94.

Metzler, R. and Klafter, J. (2000) "The random walk's guide to anomalous diffusion", *Phys. Rep.*, **339**, 1–77.

Misguich, J.H. (2001) "Dynamics of chaotic magnetic field lines: Intermittency and noble internal barrier in the tokamap", *Phys. Plasmas*, **8**, 2132–2138.

Misguich, J.H., Reuss, J.D., Consrtantinescu, D., Steinbrecher, G., Vlad, M., Spineanu, F., Weyssow, B., and Balescu, R. (2002) "Noble internal transport barriers and radial subdiffusion of toroidal magnetic lines", (CEA/Cadarache). Preprint EUR-CEA-FC-1724.

Misguich, J.H., Reuss, J.D., Consrtantinescu, D., Steinbrecher, G., Vlad, M., Spineanu, F., Weyssow, B., and Balescu, R. (2002) "Noble Cantor sets acting as partial internal transport barriers in fusion plasmas", *Plasma Physics and Controlled Fusion*, **44**, L29–L35.

Miskane, F., Dezairi, D., Saifaoui, D., Imzi, H., Imrane, H., and Benharraf, M. (2001) "Contribution of electrostatic and magnetic turbulence to anomalous transport in tokamak", *Eur. Phys. J.: Appl. Phys.*, **13**, 205–223.

Montvai, A., and Düchs, D.F. (1993) in *Proc. Physics Computing'92. Prague, 1992* (World Scientific: Singapore) p. 417.

Morozov, A.I. and Solov'ev, L.S. (1966) "The structure of magnetic fields", in *"Reviews of Plasma Physics*, vol. 2 Ed. M.A. Leontovich, (Consultants Bureau: New York). pp. 1–101.

Morrison, P.J. (2000) "Magnetic field lines, Hamiltonian dynamics, and nontwist systems", *Phys. Plasmas*, **7**, 2279–2289.

Murray, N.W. (1991) "Critical function for the standard map", *Physica*, D **52**, 220–245.

Murray, N.W. Lieberman, M.A., and Lichtenberg, A.J. (1985) "Corrections to quasilinear diffusion in area-preserving maps", *Phys. Rev. A* **32**, 2413–2424.

Nauenberg, M. (1990) "Canonical Kepler map", *Europhys. Lett.*, **13**, 611–616.

Nguyen, F., Ghendrih, Ph., and Samain, A. (1995) "Calculation of Magnetic Field Topology of Ergodized Zone in Real Tokamak Geometry. Application to the Tokamak TORE-SUPRA through the MASTOC Code", (CEA, Cadarache)DFRC/CAD Preprint EUR-CEA-FC-1539.

Ottino, J.M. (1989) *The kinematics of Mixing: Stretching, Chaos, and Transport* (Cambridge University Press: Cambridge). xiii+357.

Pakoński and Zakrzewski, J. (2001) "Kepler map for H atom driven by microwave with arbitrary polarization", *Acta Physica Polonica* B **32**, 2801–2812.

Paladin, G. and Vulpiani, A. (1987) "Anomalous scaling laws in multifractal objects", *Physics Reports* **156**, 147–225.

Pedlosky, J. (1982) *Geophysical Fluid Dynamics* (Springer-Verlag, New York). xiv+706.

Percival, I.C. (1979) *Nonlinear dynamics and beam-beam interaction– 1979*, ed. M. Month and J.C. Herrera, AIP Conference Proceedings No. 57 (American Institute of Physics, New York, 1979). p. 302.

Petrosky, T.Y. (1986) "Chaos and cometary clouds in the Solar system", *Physics Lett.* **117**, 328–332.

Petrosky, T.Y. and Broucke, R. (1988) "Area-preserving mappings and deterministic chaos for nearly parabolic motion", *Celest. Mech.* **42**, 53–79.

Petzold, L.R., Jay, L.O., and Yen, J. (1997) "Numerical solutions of highly oscillatory ordinary differential equations", *Acta Numerica*, (Cambridge Univ. Press, Cambridge). pp. 437–483.

Pommois, P., Veltri, P., and Zimbardo, G. (1999) "Anomalous and Gaussian transport regimes in anisotropic three-dimensional magnetic turbulence", *Phys. Rev* E, **59**, 2244–2252.

Pomphrey, N. and Reiman, A. (1992) "Effect of nonaxisymmetric perturbations on the structure of a tokamak poloidal divertor", *Phys. Fluids* B **4**, 938–948.

Portela, J.S.E., Viana, R.L., and Caldas, I.L. (2003) "Chaotic magnetic field lines in tokamaks with ergodic limiter", *Physica* A, **317**, 411–431.

Press, W.H., Teukolsky, S.A., Vetterling, W.T., and Flannery, B.P. (1992) *Numerical Recipes in C. The Art of Scientific Computing* 2nd edn (Cambridge: Cambridge University Press).

Punjabi, A., Verma, A., and Boozer, A.H. (1992) "Stochastic broadening of the separatrix of tokamak divertor", *Phys. Rev. Lett.* **69**, 3322–3325.

Punjabi, A., Verma, A., and Boozer, A.H. (1994) "Tokamak divertor map", *J. Plasma Phys.* **52**, 91–111.

Punjabi, A., Verma, A., and Boozer, A.H. (1996) "The simple map for a single-null divertor tokamap", *J. Plasma Phys.*, **56**, 569–603.

Punjabi, A., Ali, H., and Boozer, A.H. (1997) "Symmetric simple map for a single-null divertor tokamak", *Phys. Plasmas* **4**, 337–345.

Punjabi, A., Ali, H., and Boozer, A.H. (2003) "Effects of dipole perturbation on the stochastic layer and magnetic footprint in single-null divertor tokamaks", *Phys. Plasmas* **10**, 3992–4003.

Rax, J.M. and White, R.B. (1992) "Effective diffusion and nonlocal heat transport in a stochastic magnetic field", *Phys. Rev. Lett.*, **68**, 1523–1526.

Regianni, N. and Sakanaka, P.H. (1994) "The effect of the magnetic limiter current on the peripheral chaotic region of a tokamak", *Plasma Phys. Control. Fusion*, **36**, 513–522.

Reiman, A. (1996) "Singular surfaces in the open field line region of a divertor tokamak", *Phys. Plasmas*, **3**, 906–913.

Rechester, A.B. and White, R.B. (1980) "Calculation of turbulent diffusion for the Chirikov-Taylor model", *Phys. Rev. Lett.*, **44**, 1586–1589.

Rechester, A.B., Rosenbluth, M.N., and White, R.B. (1979) "Calculation of the Kolmogorov entropy for motion along a stochastic magnetic field", *Phys. Rev. Lett.*, **42**, 1247–1250.

Rechester, A.B., Rosenbluth, M.N., and White, R.B. (1981), "Fourier-space paths applied to the calculation of diffusion for the Chirikov -Taylor model", *Phys. Rev. A*, **23**, 2664–2672.

Roig, F. and Ferras-Mello, S. (1999) "A symplectic mapping approach of the dynamics of the Hecuba", *Planetary and Space Science*, **47**, 653–664.

Rom-Kedar, V. (1994) "Homoclinic tangles – classification and applications", *Nonlinearity*, **7**, 441–473.

Rom-Kedar, V. (1995) "Secondary homoclinic bifurcation theorems", *Chaos*, **5**, 385–401.

Rosenbluth, M.N., R.Z. Sagdeev, Taylor J.B. and Zaslavsky G.M. (1966) "Destruction of magnetic surfaces by magnetic field irregularities", *Nuclear Fusion*, **6**, 297–300.

Rokhlin V.A. (1961), "Exact endomorphisms of a Lebesgue space" (Russian), *Izv. Akad. Nauk Mat.* **25**, 499–530.

Runov, A.M., Reiter, D., Kasilov, S.V., Heyn, M.F., and Kernbicher, W. (2001) "Monte Carlo study of heat conductivity in stochastic boundaries: Application to the TEXTOR ergodic divertor", *Phys. Plasmas*, **8**, 916–930.

Ruth, R.D. (1987) "Single-particle dynamics in circular accelerators", *Physics of Particle Accelerators*, edited by M. Month and M. Dienes, AIP Conf. Proc. No. 153 (AIP: New York) Pt.1

Sagdeev, R.Z. and Zaslavsky, G.M. (1987) "Stochasticity in the Kepler problem and a model of possible dynamics of comets in the Oort cloud", *Il Nuovo Cimento*, **97**, 119–130.

Sagdeev, R.Z., and Galeev, A.A. (1969) *Nonlinear Plasma Theory*, ed. T.M. O'Neil, & D.L. Book (New York: Benjamin).

Samain, A., Grosman, A., Feneberg, W. (1982) "Plasma motion and purification in an ergodic divertor", *J. Nucl. Mater.* **111 & 112**, 408–412.

Sanders, M.M. and Jensen, R.V. (1996) "Classical theory of the chaotic ionization of highly excited hydrogen atom", *Amer. Journ. of Physics*, **64**, 21–31.

Sándor, Z., Érdi, B., and Murray, C.D. (2002) "Symplectic mappings of co-orbital motion in the restricted problem of three bodies", *Celest. Mech. & Dyn. Astron.*, **84**, 355–368.

Sanz-Serna, J.M. (1991) "Symplectic integrators for Hamiltonian problems: an overview". *Acta Numerica*, 243–286.

Sanz-Serna, J.M. and Calvo, M.P. (1993) "Symplectic numerical methods for Hamiltonian problems", *Int. J. Modern Physics* C, **4**, 385–392.

Sanz-Serna, J.M. and Calvo, M.P. (1994) "Numerical Hamiltonian problems", *Applied mathematics and mathematical computation*, v. 7 (Chamman and Hall).

Schmitd, K. and Eilenberger, G. (1998) "Poincaré maps of Duffing-type oscillators and their reduction to circle maps: II. Methods and numerical results", *J. Phys. A: Math. Gen.*, **31**, 3903–3927.

Schwagerl, M. and Krug, J. (1991) "Subdiffusive transport in stochastic webs", *Physica*, D **52**, 143–156.

Shevchenko, I.I. (1998) "Marginal resonances and intermittent behavior in the motion in the vicinity of a separatrix", *Phys. Scripta*, **57**, 185–191.

Shevchenko, I.I. (1999) "The separatrix algorithmic map: Application to the spin-orbit motion", *Celes. Mech. Dyn. Astronomy* **73**, 259–268.

Shevchenko, I.I. (2000) "Geometry of a chaotic layer" *J. Exp. Theor. Phys.* **91**, 615–625.

Shevchenko, I.I. and Scholl, H. (1997) "Intermittent trajectories in the 3/1 Jovian resonance", *Celes. Mech. Dyn. Astronomy* **68**, 163–175.

Shevchenko, I.I. and Kouprianov, V.V. (2002) "On the chaotic rotation of planetary satellites: The Lyapunov spectra and the maximum Lyapunov exponents" *Astronomy & Astrophysics*, **394**, 663–674.

Shlesinger, M.F., Zaslavsky, G.M., and Klafter, J. (1993) "Strange kinetics", *Nature*, (London) **363**, 31–37.

Šidlichovský, M. (1997) "Mapping and dynamical systems", *Celes. Mech. & Dynam. Astron.*, **65**, 69–84.

Smirnov, I.P., Virovlyanski, A.L., and Zaslavsky, G.M. (2001) "Theory and applications ray chaos to underwater acoustics", *Phys. Rev.*, E **64**, 036221, 1–20.

Soskin, S.M. (1994) "Nonlinear resonance for the oscillator with a nonmonotonic dependence of eigenfrequency on energy", *Phys. Rev.*, E **50**, R44-R46.

Soskin, S.M., Mannella, R., and McClintock, P.V.E. (2003) "Zero-dispersion phenomena in oscillatory systems", *Phys. Rep.*, **373**, 247–408.

Stanley, H.E. and Meakin P. (1988) "Multifractal phenomena in physics and chemistry", *Nature*, **335**, 405–409.

Sussman, G.J. and Wisdom, J. (1988) "Numerical evidence that the motion of Pluto is chaotic". *Science*, **241**, 22 July 1988, 433–437.

Sussman, G.J. and Wisdom, J. (1992) "Chaotic evolution of the solar system", *Science*, **257**, 3 July, 56–62.

Tabet, R., Saifaoui, D., Dezairi, D., and Raouak, A. (1998) "Contribution to the study of the non-Gaussian dynamics of stochastic magnetic field lines in a toroidal geometry", *Eur. Phys. J.: Appl. Phys.*, **4**, 329–336.

Tabet, R., Imrane, H., Saifaoui, D., Dezairi, D., and Miskane, F. (2000) "Stochastic magnetic field lines diffusion in a toroidal configuration (Tokamak)", *Eur. Phys. J.: Appl. Phys.*, **12**, 145–153.

Takens, F. (1971) "A C^1– counterexample to Moser's twist theorem", *Indag. Math.*, **33**, 379–386.

Tappert F.D., Brown M.G., and Goñi G. (1991c) "Weak chaos in an area-preserving mapping for sound ray propagation", *Phys. Lett. A* **153**, 181–185.

Tappert, F.D. and Tang, X. (1996) "Ray chaos and eigenrays", *J. Acous. Soc. Am*, **99**, 185–194.

Tel, T. and Ott, E. (1993) "Chaotic Scattering: An Introduction", *Chaos*, **3**, 417–426.

Todesco, E. (1999) "Overview of single-particle nonlinear dynamics" in: *Nonlinear and collective effects in particle accelerators*, edited by C. Pellegrini and M. Cornacchia (AIP, New York) 157–72.

Tomita Y., Seki S., and Momota H. (1977) "Destruction of magnetic surfaces in a divertor region attributed to a discrete structure of magnetic coils", *J. Phys. Soc. Japan*, **42**, 687–693.

Tomita Y., Momota H., and Itatani R. (1978) "Destruction of magnetic surfaces near a separatrix of stellarator attributed to perturbations of magnetic fields", *J. Phys. Soc. Japan*, **44** (2), 637–642.

Tongue, B.H. (1987) "On obtaining global nonlinear system characteristics through interpolated cell mapping", *Physica* D, **28**, 401–408.

Touma, J. and Wisdom, J. (1993) "The chaotic obliquity of Mars", *Science*, **259**, 26 Feb., 1294–1297.

Touma, J. and Wisdom, J. (1994a) "Lie-Poisson integrators for the rigid body dynamics in the solar system", *Astron. J.*, **107**, 1189–1202.

Touma, J. and Wisdom, J. (1994b) "Evolution of the Earth- Moon system", *Astron. J.*, **108**, 1943–1961.

Touma, J. and Wisdom, J. (1998) "Resonances in the early evolution of the Earth-Moon system", *Astron. J.*, **115**, 1653–1663.

Treshev, D. (1998) "Width of stochastic layers in near-integrable two-dimensional symplectic maps", *Physica* D **116**, 21–43.

Treschev, D. (2002) "Multidimensional symplectic separatrix maps", *J. Nonlinear Sciences*, **12**, 27–58.

Treschev, D. (2004) "Evolution of slow variables in a priori unstable Hamiltonian systems", *Nonlinearity*, **17**, 1803–1841.

Ullmann, K. and Caldas, I.L. (2000) "A symplectic mapping for the ergodic magnetic limiter and its dynamical analysis", *Chaos, Solitons and Fractals*, **11**, 2129–2140.

Vatrunin, V.E., Dubinov, A.E., Sadovoi, S.A., and Selemir, V.D. (1997) "Chaos of electronagnetic ray trajectories in a strongly corrugated waveguide partially filled with isotropic plasma", *Plasma Phys. Reports*, **23**, 413–418.

Vecheslavov, V.V. (1996) "Formation of a chaotic layer of nonlinear resonance by a two-frequency perturbation" *JETP Lett.* **63** (12) 1047–1053.

Vecheslavov, V.V. (1999) "Chaotic layer of nonlinear resonance driven by quasi-periodic perturbation", *Physica* D **131**, 55–67.

Vecheslavov, V.V. (2000) "Dynamics of sawtooth mapping: 1. New numerical results", Preprint No. 2000–27 (Budker Inst. of Nuclear Physics, Novosibirsk, 2000) (in Russian). E-print. nlin. CD/0005048.

Vecheslavov, V.V. (2001) "Suppression of dynamic chaos in Hamiltonian systems", *J. Exp. Theor. Phys.*, **92** (4) 744–751.

Vecheslavov, V.V. (2002) "The chaotic layer of a nonlinear resonance under low-frequency perturbation", *Tech. Phys.*, **47** (2) 160–167.

Vecheslavov, V.V. and Chirikov, B.V. (2001) "Separatrix conservation mechanism for nonlinear resonance in strong chaos", *J. Exp. Theor. Phys.*, **93** (3) 649–656.

Vecheslavov, V.V. and Chirikov, B.V. (2002) "Diffusion in smooth Hamiltonian systems", *J. Exp. Theor. Phys.*, **95** (1) 154–165.

Viana, R.L. (2000) "Chaotic magnetic field lines in a tokamak with resonant helical windings", *Chaos, Solitons and Fractals*, **11**, 765–778.

Viana, R.L. and Caldas, I.L. (1992) "Peripheral stochasticity in tokamak with an ergodic magnetic limiter", *Z. Naturforsch.* **47a**, 941–944.

Viana, R.L. and Vasconcelos, D.B. (1997) "Field-line stochasticity in tokamak with an ergodic magnetic limiter", *Dynamics Stability Systems*, **12**, 75.

Wan, W. and Cary, J.R. (1998) "Increasing the dynamic aperture of accelerator lattices", *Phys. Rev. Lett.*, **81**, 3655–3658.

Wan, W. and Cary, J.R. (2001) "Method for enlarging the dynamic aperture of accelerator lattices", *Phys. Rev. STAC*, **4**, 084001, 1–10.

Warnock, R.L. and Berg, S.J. (1996) "Fast symplectic mapping, quasi-invariants, and long-term stability in the LHC", *Particle Accelerators*, **54**, 213–222.

Warnock, R.L. and Berg, S.J. (1997) "Fast symplectic mapping and long–term stability near broad resonances", in AIP Conference Proceedings, 395 (Amer. Inst. Phys.: Woodbury, N.Y.) 423–445.

Warnock, R.L. and Ellison, J.A. (1999) "From symplectic integrator to Poincaré map: spline expansion of a map generator in Cartesian coordinates", *Applied Numerical Mathematics*, **29**, 89–98.

Weeks, E.R. and Swinney, H.L. (1999) "Anomalous diffusion resulting from strongly asymmetric random walks", *Phys. Rev.* E **57** (5), 4915–4920.

Weeks, E., Urbach, J.S., and Swinney, H. (1996) "Anomalous diffusion in asymmetric random walks with a quasi-geostrophic flow example", *Physica* D **97**, 291–310.

Weiss, J.B. (1991) "Transport and mixing in traveling waves", *Phys. Fluids* A **3** (5), 1379–1384.

Weiss, J.B. and McWilliams, J.C. (1991) "Nonergodicity of point vortices", *Physics of Fluids* A, 3 (5) pp. 835–844.

Weiss, J.B. (1994) "Hamiltonian maps and transport in structured fluids", *Physica* D **76**, 230–238.

Weiss, J.B. and Knobloch, E. (1989) "Mass transport and mixing by modulated traveling waves", *Phys. Rev.* A **40** (5), 2579–2589.

Wesson, J. (2004) *Tokamaks*, 3-rd Edition, Oxford Engineering Science Series: **48**, (Clarendon Press: – Oxford).

Weyssow, B. and Misguich, J.H. (1999) "Hamiltonian map for guiding centres in a perturbed toroidal magnetic geometry", Proc. of the 26-th EPS Conf. on Controlled Fusion and Plasma Physics, Maastricht, June 14–18. *Europhysics Conference Abstracts*, **23J**, 793–796.

White, R.B. (1998) "Chaos in trapped particle orbits", *Phys. Rev.* E **58**, 1774–1779.

Wiggins, S. (1990a) *Introduction to Applied Nonlinear Dynamical Systems and Chaos* (Springer, New York) Ch. 1.

Wiggins, S. (1990b) *Chaotic transport in dynamical systems* (Springer, New York).

Wisdom, J. (1982) "The origin of the Kirkwood gaps: a mapping for asteroid motion near the 3/1 commensurability", *Astron. J*, **87** (3), 577–593.

Wisdom, J. (1983) "Chaotic behavior and the origin of the 3/1 Kirkwood gap", *Icarus* 56, 51–57.

Wisdom, J. (1992) "Long term evolution of the solar system", *"Chaos, Resonances and Collective Dynamical Phenomena in the Solar System"* Ed. S. Ferraz-Mello, IAU, pp. 17–24.

Wisdom, J. and Holman, M. (1991) "Symplectic maps for the *n*-body problem". *Astron. J*, **102**(4), 1528–1538.

Wisdom, J. and Holman, M. (1992) "Symplectic maps for the *n*-body problem: stability analysis", *Astron. J.*, **104** (5), 2022–2029.

Wisdom, J., Peale, S.J. and Mignard, F. (1984) "The chaotic rotation of Hyperion", *Icarus"*, **58**, 137–152.

Wisdom, J., Holman, M., and Touma, J. (1996) "Symplectic Correctors", *"Integration algorithms for classical mechanics"*, Eds. J.E. Marsden, G.W. Patrick and W.F. Shedwick (Field Institute Communications: Rhode Island, v. 10), 217–244.

Wobig, H. (1987) "Magnetic surfaces and localized perturbations in the Weldelstein VII-A stellarator", *Z. Naturforsch.*, **42a**, 1054–1066.

Wobig, H. and Fowler, R.H. (1988) "The effect of magnetic surface destruction on test particle diffusion in the Weldelstein VII-A stellarator", *Plasma Phys. Contr. Fusion*, **30**, 721–741.

Wobig, H. and Pfirsch, D. (2001) "On guiding centre orbits of particles in toroidal systems", *Plasma Phys. Control. Fusion*, **43** (5) 695–716.

Wojtkowski, M. (1981) "A model problem with the coexistence of stochastic and integrable behavior", *Commun. Math. Phys.*, **80**, 453–464.

Wolf, R. (2003) "Internal transport barriers in tokamak plasmas", *Plasma Phys. Control. Fusion*, **45** R1–R91.

Wootton, A.J., McCool, S.C., and Zheng S.B. (1991) "Test particle calculations for the Texas Experimantal Tokamak with resonant magnetic fields", *Fusion Technology*, **19** (5), 473–491.

Yamagishi, T. (1995) "Evaluation of the stochastic layer near separatrix in toroidal divertor", *Fusion Technology*, **27**, 505–508.

Zaslavsky, G.M. (2002) "Chaos, fractional kinetics, and anomalous transport", *Physics Reports*, **371**, 461–580.

Zaslavsky, G.M. (2005) *Hamiltonian Chaos and Fractional Dynamics*, (Oxford University Press: Oxford).

Zaslavsky, G.M. and Abdullaev, S.S. (1995) "Scaling properties and anomalous transport of particles inside the stochastic layer" *Phys. Rev. E*, **51** (5), 3901–3910.

Zaslavsky, G.M. and Chirikov, B.V. (1965), "Fermi acceleration mechanism in 1-dimensional case", *Sov. Phys. Dokl.*, **9**, 989.

Zaslavsky, G.M. and Chirikov, B.V. (1971) "Stochastic instability of non-linear oscillations", *Usp. Fiz. Nauk*, **105** (3), 3–40 [*Sov. Phys.: Uspekhi*, **14** (5), [549–568].

Zaslavsky, G.M., Sagdeev, R.Z., Usikov, D.A., and Chernikov, A.A. (1991) *Weak chaos and quasi-regular patterns* (Cambridge University Press: Cambridge).

Zosimov, V.V. and Lyamshev, L.M. (1995) "Fractals in wave processes", *Physics Uspekhi*, **38**, 347–384.

Index

Action-angle variables 8
  many degrees of freedom 9
  one degree of freedom 9
Aphelion 132
Arnold diffusion 149
Aspect ratio 224
Averaging principle 21, 22

Backward map 282
Betatron oscillations 310

Chaos 145
  global 149
Chirikov criteria 149
Clebsch representation 221
Correlation function 198
Correlation time 198

Devil's staircase 306
Diffusion coefficient 200
  field lines 265
  global 265
  local 265
  quasilinear 200, 266
Diffusion process 201
  normal 201
  subdiffusive 201
  superdiffusive 201
Dynamical aperture 308
Dynamical entropy 199

Eikonal equation 301
Elliptic fixed points 12
Ergodic divertor 255
Ergodic motion 197
Ergodic zone 255

Fermat's principle 299

Fermi acceleration mapping 49
Forward map 282
Fundamental problem of dynamics 22

Generating function 5
  Lie 26
  mixed variable 26

Hamilton
  equations 1
  Hamilton's function 1
Hamilton–Jacobi equation 6
  complete integral 6
Hamiltonian 1
Heteroclinic orbit 83
High field side 257
Homoclinic orbit 83
Hyperbolic fixed points 12

Integrability 7
Integral invariants
  Poincaré-Cartan 3
Intermittence 166
Invariant curve 141
Invariants of motion 2

Jacobi's theorem 6

KAM theory 23, 139
Kepler map 132, 135, 138, 316
Kolmogorov entropy 199
Kolmogorov's technique 36

Laminar zone 267
Levý flight 213
Liouville theorem 7
Low field side 257
Lyapunov exponent 147

Magnetic field
  equilibrium   220
Magnetic footprints   270, 296
Magnetic stochasticity   230
Magnetic surfaces   220
Major radius   222
Map
  Full turn transfer   310
  Melnikov type integrals   94–96
  asymptotics   321
Minor radius   224
Mixing   198

Nonlinear resonance   142
  width   143
Nonsmooth map   168
Nontwist map   243
Nontwist standard map   158
Nontwist systems   151
  condition   151

Optical length   307

Paraxial approximation   301
Pendulum   12
Perihelion   132
Perturbation theory   21
Perturbation series   23
  Lindstedt method   23
Poincaré map   39
Poincaré recurrence   203
Poincaré–Birkhoff theorem   142
Poloidal angle
  intrinsic   220
Poloidal divertor
  double-null   275
  single-null   275
Poloidal field   224
Poloidal flux   220
Primary resonant approximation   98

Quasilinear theory   200

Rescaling invariance   175
  in parameter space   182
  proof   179
  universal   178
Rescaling parameter   118, 175, 178
Residence time   202
Revtokamap   243

Safety factor   221
Separatrix   13, 83
  magnetic   275, 278
Separatrix map   83, 85, 86
  algorithmic   112
Shafranov shift   226
Shearless curve   151
Smooth perturbations   150, 173
Smoothness parameter   150
Splitting of separatrices   144
Stable and unstable manifolds   144
Standard magnetic field   224
Standard map   47, 69, 241
  symmetric   69
Stellarator   219
Stochastic layer   83, 84
Stochastic web   165
Stroboscopic map   39
Symmetric map   56
Symplectic map   3
Symplectic correctors   47
Symplectic integration   17
Synchrotron oscillations   310

Taylor map   311
Tokamak   219
  Poloidal divertor   275
Tokamak divertor maps   277
Tokamap   242
  symmetric   247
Toroidal angle   220
Toroidal field   224
Toroidal flux   220
Torus   7
  non-resonant   139
  resonant   140
Tracking   310
Twist condition   43, 140
Twist map
  perturbed   43, 55
  radial   46
  symmetric   67

Volume-preserving map   3

Weak chaos   165
Whisker map   83, 85, 86
Winding number   221

X-point   275

Lecture Notes in Physics

For information about earlier volumes
please contact your bookseller or Springer
LNP Online archive: springerlink.com

Vol.644: B. Grammaticos, Y. Kosmann-Schwarzbach, T. Tamizhmani (Eds.) Discrete Integrable Systems

Vol.645: U. Schollwöck, J. Richter, D. J. J. Farnell, R. F. Bishop (Eds.), Quantum Magnetism

Vol.646: N. Bretón, J. L. Cervantes-Cota, M. Salgado (Eds.), The Early Universe and Observational Cosmology

Vol.647: D. Blaschke, M. A. Ivanov, T. Mannel (Eds.), Heavy Quark Physics

Vol.648: S. G. Karshenboim, E. Peik (Eds.), Astrophysics, Clocks and Fundamental Constants

Vol.649: M. Paris, J. Rehacek (Eds.), Quantum State Estimation

Vol.650: E. Ben-Naim, H. Frauenfelder, Z. Toroczkai (Eds.), Complex Networks

Vol.651: J. S. Al-Khalili, E. Roeckl (Eds.), The Euroschool Lectures of Physics with Exotic Beams, Vol.I

Vol.652: J. Arias, M. Lozano (Eds.), Exotic Nuclear Physics

Vol.653: E. Papantonoupoulos (Ed.), The Physics of the Early Universe

Vol.654: G. Cassinelli, A. Levrero, E. de Vito, P. J. Lahti (Eds.), Theory and Appplication to the Galileo Group

Vol.655: M. Shillor, M. Sofonea, J. J. Telega, Models and Analysis of Quasistatic Contact

Vol.656: K. Scherer, H. Fichtner, B. Heber, U. Mall (Eds.), Space Weather

Vol.657: J. Gemmer, M. Michel, G. Mahler (Eds.), Quantum Thermodynamics

Vol.658: K. Busch, A. Powell, C. Röthig, G. Schön, J. Weissmüller (Eds.), Functional Nanostructures

Vol.659: E. Bick, F. D. Steffen (Eds.), Topology and Geometry in Physics

Vol.660: A. N. Gorban, I. V. Karlin, Invariant Manifolds for Physical and Chemical Kinetics

Vol.661: N. Akhmediev, A. Ankiewicz (Eds.) Dissipative Solitons

Vol.662: U. Carow-Watamura, Y. Maeda, S. Watamura (Eds.), Quantum Field Theory and Noncommutative Geometry

Vol.663: A. Kalloniatis, D. Leinweber, A. Williams (Eds.), Lattice Hadron Physics

Vol.664: R. Wielebinski, R. Beck (Eds.), Cosmic Magnetic Fields

Vol.665: V. Martinez (Ed.), Data Analysis in Cosmology

Vol.666: D. Britz, Digital Simulation in Electrochemistry

Vol.670: A. Dinklage, G. Marx, T. Klinger, L. Schweikhard (Eds.), Plasma Physics

Vol.671: J.-R. Chazottes, B. Fernandez (Eds.), Dynamics of Coupled Map Lattices and of Related Spatially Extended Systems

Vol.672: R. Kh. Zeytounian, Topics in Hyposonic Flow Theory

Vol.673: C. Bona, C. Palenzula-Luque, Elements of Numerical Relativity

Vol.674: A. G. Hunt, Percolation Theory for Flow in Porous Media

Vol.675: M. Kröger, Models for Polymeric and Anisotropic Liquids

Vol.676: I. Galanakis, P. H. Dederichs (Eds.), Half-metallic Alloys

Vol.677: A. Loiseau, P. Launois-Bernede, P. Petit, S. Roche, J.-P. Salvetat (Eds.), Understanding Carbon Nanotubes

Vol.678: M. Donath, W. Nolting (Eds.), Local-Moment Ferromagnets

Vol.679: A. Das, B. K. Chakrabarti (Eds.), Quantum Annealing and Related Optimization Methods

Vol.680: G. Cuniberti, G. Fagas, K. Richter (Eds.), Introducing Molecular Electronics

Vol.681: A. Llor, Statistical Hydrodynamic Models for Developed Mixing Instability Flows

Vol.682: J. Souchay (Ed.), Dynamics of Extended Celestial Bodies and Rings

Vol.683: R. Dvorak, F. Freistetter, J. Kurths (Eds.), Chaos and Stability in Planetary Systems

Vol.684: J. Dolinsek, M. Vilfan, S. Zumer (Eds.), Novel NMR and EPR Techniques

Vol.685: C. Klein, O. Richter, Ernst Equation and Riemann Surfaces

Vol.686: A. D. Yaghjian, Relativistic Dynamics of a Charged Sphere

Vol.687: J. W. LaBelle, R. A. Treumann (Eds.), Geospace Electromagnetic Waves and Radiation

Vol.688: M. C. Miguel, J. M. Rubi (Eds.), Jamming, Yielding, and Irreversible Deformation in Condensed Matter

Vol.689: W. Pötz, J. Fabian, U. Hohenester (Eds.), Quantum Coherence

Vol.690: J. Asch, A. Joye (Eds.), Mathematical Physics of Quantum Mechanics

Vol.691: S. S. Abdullaev, Construction of Mappings for Hamiltonian Systems and Their Applications